ALLE ZEIT WACH 1842

Dedicated to our wives Eva and Dale

K. Weber · W. Burgdorfer (Eds.) · Coeditor: G. Schierz

Aspects of
Lyme Borreliosis

With 42 Tables and 138 Illustrations

Springer-Verlag
Berlin Heidelberg New York London Paris Tokyo
Hong Kong Barcelona Budapest

Dr.med.Klaus Weber
Dermatologist in Private Practice
Rosenstr.6
8000 München 2
Germany

Willy Burgdorfer, Ph.D., M.D. (h.c.)
Laboratory of Vectors and Pathogens
Rocky Mountain Laboratories
Hamilton, MT 59840
USA

ISBN-13:978-3-642-77616-8 e-ISBN-13: 978-3-642-77614-4
DOI: 10.1007/978-3-642-77614-4

Softcover reprint of the hardcover 1st edition 1993

Typesetting: Fotosatz Reiter, Berchtesgaden
27/3145-543210 - Printed on acid-free paper

Preface

The story of Lyme borreliosis started in Europe at the end of the last century and is still continuing today. The aim of this book is to present current knowledge of Lyme borreliosis primarily as it occurs in Europe. Despite providing such a regional view, observations made in the United States of America and elsewhere are also considered. Thus "Aspects of Lyme Borreliosis" may serve within these limits as a presentation of the disease in general.

The various stages of preparation in completing this book included sending invitations to authors from different departments and even different countries as far as this was practicable, and arranging manuscript evaluation by one or two anonymous reviewers and at least one of the editors. Dr.W. Burgdorfer was mainly responsible for editing the nonclinical part, and Dr.K. Weber for editing the clinical part. The clinical chapters, when feasible follow a specific outline.

The term "Lyme borreliosis" has only recently been introduced. European authors describing aspects of this disease find themselves, therefore, to a certain extent, in a nomenclatural dilemma (see Chap. 7). The majority of the authors of this book elected to use the designation "Lyme borreliosis" rather than "erythema migrans borreliosis" or both terms concomitantly.

We are deeply indebted to our fellow authors for their contributions. Our gratitude is also extended to the reviewers and to Mrs. U. Perschau-Jendreizig und Mrs. K. Baade in the editorial office.

April 1992 *Klaus Weber* *Willy Burgdorfer*

List of Contributors

Aberer, Elisabeth, II. Universitäts-Hautklinik Wien, Alserstraße 4, 1090 Wien, Austria
Aeschlimann, A., University of Neuchâtel, Zoological Institute, Chantemerle 22, 2007 Neuchâtel, Switzerland
Åsbrink, Eva, Department of Dermatology, Karolinska Institute at Södersjukhuset, 11883 Stockholm, Sweden
Ballmer, P.E., Department of Internal Medicine, University of Bern, Inselspital, 3010 Bern, Switzerland
Büchner, S.A., Dermatologische Universitätsklinik, Kantonsspital Basel, Petersgraben 4, 4031 Basel, Switzerland
Burgdorfer, W., Department of Health and Human Services, Public Health Service, National Institutes of Health, National Institute of Allergy and Infectious Diseases, Laboratory of Vectors and Pathogens, Hamilton, MT 59840, USA
Christen, H.-J., Kinderklinik und Poliklinik, Universität Göttingen, Robert-Koch-Straße 40, W-3400 Göttingen, Germany
Detmar, U., Dermatologist in Private Practice, Hauptstraße 12, 8939 Bad Wörishofen, Germany
Duray, P.H., Department of Pathology, Brigham and Women's Hospital, 75 Francis Street, Boston, Mass. 02115, USA
Gern, Lise, University of Neuchâtel, Zoological Institute, Chantemerle 22, 2007 Neuchâtel, Switzerland
Hanefeld, F., Kinderklinik und Poliklinik, Universität Göttingen, Robert-Koch-Straße 40, W-3400 Göttingen, Germany
Hassler, D., General Practioner in Private Practice, Untere Hofstatt 3, 7527 Kraichtal, Germany
Hayes, S.F., Department of Health and Human Services, Public Health Service, National Institutes of Health, National Institute of Allergy and Infectious Diseases, Laboratory of Vectors and Pathogens, Hamilton, MT 59840, USA
Herzer, P., Medizinische Poliklinik der Universität München, Pettenkoferstraße 8a, 8000 München 2, Germany
Hovmark, A., Department of Dermatology, Karolinska Institute at Södersjukhuset, 11883 Stockholm, Sweden
Kaudewitz, P., Dermatologische Klinik und Poliklinik, Universität München, Frauenlobstraße 9-11, 8000 München 2, Germany
Kohlhepp, W., Neurologische Klinik, Universität Würzburg, Josef-Schneider-Straße 11, 8700 Würzburg, Germany
de Koning, J., Laboratorium voor de Volksgezondheid in Friesland, Postbus 21020, 8900 JA Leeuwarden, The Netherlands
Krampitz, H.E., Abteilung Infektions- und Tropenmedizin, Ludwig-Maximilians-Universität München, Germany (deceased)

Kristoferitsch, W., Neurologische Abteilung, SMZ-Ost-Krankenhaus, Langobardenstraße 122, 1220 Wien, Austria
van der Linde, M.R., Department of Cardiology, Thoraxcenter, University Hospital of Groningen, Oostersingel 59, 9713 EZ Groningen, The Netherlands
Maciejewski, W., Städtisches Krankenhaus München-Schwabing, Kölner Platz, 8000 München 440, Germany
Marget, W., Kinderklinik, Universität München, Editor-in-Chief, Infection, Münchner Str. 1, 8000 München 70, Germany
Martin, R., Neurologische Klinik, Universität Tübingen, Hoppe-Seyler-Straße 3, 7400 Tübingen, Germany
Mayer-Berger, W., Klinische Abteilung IV, Klinik Höhenried für Herz- und Kreislaufkrankheiten, 8139 Bernried, Germany
Meier, C., St. Gallische Höhenklinik Walenstadtberg, 8881 Knoblisbühl, Switzerland
Mertens, H.G., Neurologische Klinik, Universität Würzburg, Josef-Schneider-Straße 11, 8700 Würzburg, Germany
Neubert, U., Dermatologische Klinik und Poliklinik, Universität München, Frauenlobstraße 9-11, 8000 München 2, Germany
Pfister, H.-W., Neurologische Klinik der Universität München, Klinikum Großhadern, Marchioninistraße 15, 8000 München 70, Germany
Preac-Mursic, Vera, Max von Pettenkofer-Institut für Hygiene und Medizinische Mikrobiologie der Ludwig-Maximilians-Universität München, Pettenkoferstraße 9a, 8000 München 2, Germany
Reimers, C.D., Friedrich Baur-Institut bei der Medizinischen Klinik Innenstadt, Ludwig Maximilians-Universität München, Ziemssenstraße 1A, 8000 München 2, Germany
Rufli, T., Dermatologische Universitätsklinik, Kantonsspital Basel, Petersgraben 4, 4031 Basel, Switzerland
Satz, N., Medizinische Klinik, Kantonsspital Winterthur, Winterthur, Switzerland
Schierz G., Max von Pettenkofer-Institut für Hygiene und Medizinische Mikrobiologie der Ludwig-Maximilians-Universität München, Pettenkoferstraße 9a, 8000 München 2, Germany
Schmutzhard, E., Klinik für Neurologie, Universität Innsbruck, Anichstraße 35, 6020 Innsbruck, Austria
Schönherr, U., Augenklinik mit Poliklinik der Universität Erlangen-Nürnberg, Schwabachanlage 6, 8520 Erlangen, Germany
Sköldenberg, Birgit, Department of Infectious Diseases, Karolinska Institute, Danderyd Hospital, 18288 Danderyd, Sweden
Stanek, G., Hygiene-Institut der Universität Wien, Kinderspitalgasse 15, 1095 Wien, Austria
Strle, F., University Medical Center Ljubljana, Department of Infectious Diseases, Japljeva 2, 61000 Ljubljana, Slovenia
Weber, K., Dermatologist in Private Practice, Rosenstraße 6, 8000 München 2, Germany
Wilske, Bettina, Max von Pettenkofer-Institut für Hygiene und Medizinische Mikrobiologie der Ludwig-Maximilians-Universität München, Pettenkoferstraße 9a, 8000 München 2, Germany

Contents

I History

1 History of Lyme Borreliosis in Europe

K. WEBER and H.-W. PFISTER

1.1 Introduction

Knowledge of Lyme borreliosis in Europe has roughly developed in three phases. Between 1883 and 1945, basic descriptions of dermatological and neurological manifestations were presented. The beneficial effect of penicillin and other antibiotics noted in the late 1940s and early 1950s gave a new impetus to the study of the cause and nature of these manifestations, especially the dermatological ones; this phase lasted until 1975. Following the first description of Lyme disease in 1977 and the discovery of the causative organism in the United States in 1982, a third and this time broader look into the disease and much of its surroundings ensued in Europe.

1.2 First Phase (1883-1945)

Knowledge of the various dermatological and neurological manifestations developed mostly independently from each other during the first phase.

1.2.1 Acrodermatitis Chronica Atrophicans

In 1883, Alfred Buchwald from Breslau, Germany, described "diffuse idiopathic skin atrophy" of 16 years duration in a 36-year-old man; some histologic features were presented. We now regard this report as the earliest description of acrodermatitis chronica atrophicans (ACA), a late manifestation of Lyme borreliosis. Until 1902, more than a dozen case reports appeared in the literature (cited in Herxheimer and Hartmann 1902). Buchwald and others reported on the atrophic phase of ACA, while a variety of authors described the first inflammatory phase of ACA shortly afterwards. In 1902, Karl Herxheimer and Kuno Hartmann combined these observations, added a further 12 patients of their own, and discussed at some length that acrodermatitis chronica atrophicans, a term introduced by them, develops from an early inflammatory phase into a late atrophic one. By about 1906, most of the important clinical signs associated with ACA had been desbribed, i.e., hair loss (Buchwald 1883), macular atrophy (Pellizari 1884), which was later called anetoderma by Jadassohn, sclerodermatous changes (Touton 1886; Rusch 1906), ulnar bands (Herxheimer and Hartmann 1902), and fibrous nodules (Herxheimer 1905, cited in Herxheimer and Schmidt 1910). Pospelow described atrophic ACA skin as resembling "crumpled cigarette paper" (1886).

In a monograph on atrophic conditions of the skin, Finger and Oppenheim (1910) reported on 202 patients with atrophia cutis, 134 of whom can be regarded as having had ACA. Most of these patients were collected from the literature. The authors summarized the findings on age, sex ratio and occupation, mentioned the rare involvement of the face, sole of the foot, palms, trunk, mucous membranes and the entire skin and discussed the scleroderma-like changes, macular atrophy and the histologic picture. Joint and bone involvement was mentioned by several authors. Jessner (1921) and later Ehrmann and Falckenstein (1925) observed that a few patients had arthralgia prior to developing ACA, but they did not discuss a connection between the two phenomena. Jessner and Loewenstamm (1924) mentioned that 9 out of their 66 patients had changes in the joints, particularly thickening. They believed to have observed atrophy of the bones in the affected areas in 10 out of the 17 patients examined. Mild periosteal thickening and atrophy of the bone of the proximal phalanges of the fingers in one man were noted by Piorkowski (1926). Hövelborn (1931) presented the first detailed description of the arthropathy seen in ACA; he believed that, analogous to the development of the ACA skin process, the synovia of the affected joints first became inflamed, and later widened and atrophic. A case of ACA with involvement of the subcutaneous layers and a dermatochalasis-like appearance was reported by Löwenfeld (1932).

The characteristic histopathological changes found in ACA were already described at the end of the last and the beginning of this century. Buchwald (1883) described atrophy of cutaneous structures. Huber (1900) noted plasma cells and diminishing of elastic fibers. Krzysztalowicz distinguished between three histologic stages: an early inflammatory stage, a late inflammatory stage with atrophy, and a final atrophic stage without inflammation (Krzysztalowicz 1901, cited in Finger and Oppenheim 1910). Herxheimer and Hartmann (1902) separated only an early inflammatory from a late atrophic stage. A particularly interesting study was published by Ehrmann and Falkenstein (1925); these authors were impressed by the analogy of the histologic picture of both syphilis and ACA, especially with regard to the great numbers of plasma cells and the spread of the inflammatory infiltrate along the perivenous lymphatics; furthermore, they envisaged local spread of the presumed "virus" along blood vessels and lymphatics, and they believed in occasional dissemination via the blood stream (Ehrmann and Falkenstein 1925). Ehrmann had already stressed the early lymphangitis found in some of his patients (Ehrmann 1905 and 1909, cited in Finger and Oppenheim 1910).

Another interesting aspect of the history of ACA was its occurence in the United States. Apparently seen primarily in European immigrants, ACA was first described in the United States by Bronson and Elliot in 1895 (cited in Sweitzer and Laymon 1935). Sweitzer and Laymon (1935) reported that one of their five patients had involvement of the entire skin and several mucous membranes, arthropathy of many finger and toe joints, and atrophy of the bones of the hands and wrist. Three out of the 45 patients seen by Montgomery and Sullivan (1945) were said to have generalized erythroderma, and seven had arthralgia or arthritis before the ACA started. The authors doubted that the arthropathy, atrophy of the bones, and cardiac abnormalities found in their patients were related to ACA. A few of the American patients were said to be natives (Sweitzer and Laymon 1935; Montgomery and Sullivan 1945).

1.2.2 Erythema Migrans

A case of "erythema migrans" in a woman which developed after a tick bite was presented by Arvid Afzelius at a dermatology meeting in Stockholm in October 1909 (Afzelius 1910). In 1910, Wilhelm Balban from Wien, Austria, described in some detail three patients with an expanding bright-red annular erythema with a bluish-red center; his first case can be regarded as an erythema migrans of short duration (lasting about 10 days). Balban noted small papules in the center of the two adjacent annular erythemas of his patient, and he inferred from this that the patient might have been exposed to a vector. A few years later, Lipschütz (1913) described a case of "erythema chronicum migrans", a slowly expanding erythema lasting for at least 7 months. Commenting on Lipschütz's case presentation, G. Riehl (1913) stated that he had seen several patients with these sometimes huge, slowly expanding lesions in previous years, that these lesions clear centrally, that most of them occur on the lower legs, and that, once the borders of this type of lesion have merged after expanding around an extremity, they tend to spread distally and proximally in a ring-like manner. Nonspecific histologic features were described by Lipschütz (1913). Kerl (1915) and Strandberg (1920) described central lymphocytoma-like lesions and a surrounding erythema migrans in individual children. Strandberg noted regional lymphadenopathy. Afzelius (1921) stressed the importance of tick bites, and also stated that the erythema migrans in his six patients had a self-limited course usually lasting a few months, although in one instance it lasted for at least 15 months. He described the erythema starting from a small plaque and commented that, when it spread, the red ring became paler. Furthermore, he stated that when the lesions lasted for a longer period of time, they would then become irregular in shape, and certain portions would disappear before others. Lipschütz (1923) observed a patient with two erythema migrans lesions. Stadelmann (1934) described "myalgia, fatigue and severe arthralgia" in a 34-year-old man and mentioned concomitant myocarditis in a 52-year-old man. A few patients with erythema migrans and subsequent lymphadenosis benigna cutis were observed by Bäfverstedt (1943; cases 18, 19, 21, and possibly 11).

1.2.3 Borrelial Lymphocytoma

The term borrelial lymphocytoma has been recently introduced, and the condition is not always identical with lymphadenosis benigna cutis or lymphocytoma in general (Weber et al. 1984a, 1985; see Chap. 9). However, the solitary nodules and plaques found in some patients which were formerly described as lymphocytomas or solitary lesions of lymphadenosis benigna cutis, can be regarded as having been due to *Borrelia burgdorferi*. Patients with multiple lesions usually have to be excluded here, with the exception of those associated with ACA.

The first description of a solitary lymphocytoma was published by Jean Louis Burckhardt from Basel, Switzerland, in 1911. He described an erythematous plaque, 6 x 2 cm in diameter, present for several weeks on the upper arm of a 60-year-old woman. Histologically, lymphoid tissue with germinal centers like those seen in lymph nodes and other pertinent features were described and interpreted as probably being due to a chronic inflammation (Burckhardt 1911). The term lymphocytoma was first used by Biberstein (1923), who referred to a nomenclatural proposal made by M. Kaufmann-Wolf. Mulzer and Keining (1929) first observed multiple (follicular) lesions in association with ACA. A variety of case reports appeared in the literature until 1943, when Bäfverstedt published his monograph.

Bo Bäfverstedt from Stockholm, Sweden, selected a group of pseudolymphomas of the skin, both solitary and multiple, with benign course and characteristic histologic picture. Using essentially these criteria, he was able to present 41 of his own patients in detail and to discuss additional 101 patients from the literature. Two-thirds of the solitary lesions were found to be confined to he head, while the areas of predilection of solitary lesions were the ear lobe, the nipple, the scrotum, and the vulva. A relationship to an erythema migrans was found in four patients of his series, and multiple lesions were seen in three of his patiens with ACA, in one of whom these lesions apparently preceded and occurred concomitantly with the ACA. This patient also lost 30 kg in weight and had scleroderma of the lower arms and hands, leading to sclerodactyly and osseous demineralization (Bäfverstedt 1943). It remains uncertain how many of Bäfverstedt's patients (1943) had lesions of borrelial origin, and whether the lymphocytoma described by Burckhardt (1911) was a borrelial lymphocytoma.

1.2.4 Neurological Disorders

In 1922, the French authors Garin and Bujadoux reported on a patient who developed a marked erythema at the site of a previous tick bite. This was followed by radiculoneuritis associated with severe radicular pain and meningitic symptoms. However, the erythema was not classified as erythema migrans. Sven Hellerström from Stockholm, Sweden (1930), reported on a patient who developed meningoencephalitis 3 months after the onset of an erythema migrans; the erythema migrans spread over the entire body of his patient within 4 months. Hellerström was the first to stress the relationship between neurological symptoms, tick bite and erythema migrans.

In 1941, Alfred Bannwarth from München, Germany, reported on a syndrome which he called "chronic lymphocytic meningitis with the clinical syndrome of neuralgia or neuritis". Bannwarth carefully described the histories of 15 patients; most of the patients were men. He described three different groups: the main symptoms were in the first group of seven patients intense radicular pain; in the second group of six patients, four of whom were between the ages of 8 and 17 years, facial palsy; and in the third group of two patients "chronic lymphocytic meningitis with cerebral symptoms" such as severe headache and vomiting. The common feature in all patients was lymphocytic pleocytosis in the cerebrospinal fluid (CSF). Bannwarth reported on the results of 76 lumbar punctures performed in these 15 patients up to 6 years after the acute stage of the disease. The highest cell count was 1090 cells /µl (patient 6). The CSF usually contained about 90% lymphocytes. The cerebral meninges were also affected, as demonstrated by examination of the cisternal CSF in five patients. In one patient (patient 3), symmetrical internal hydrocephalus was detected by means of air encephalography.

Bannwarth found it to be typical that the severe radicular pain occurring in the extremities and on the trunk was experienced mainly at night and that it was of migrating character. Unilateral facial palsy was observed in two patients, and bilateral facial palsy in four patients. Transient palatal paresis and slight bilateral papilledema were each found once. Occasionally, fever, chills, cough, tachycardia, and arthralgia were observed. The symptoms usually resolved within several weeks or months. None of the patients of the groups one and two presented with typical symptoms of acute meningitis such as headache, stiff neck, or vomiting. An additional interesting finding was that months to years after recovery of the meningoradiculoneuritis, recurrences with intense multilocated pain occurred without a concomitant increase in cell count (patients 1,2,5,6). Despite complaints of intense radicular pain similar to the initial ones, the CSF was always normal or showed only very slight changes. Bannwarth thought that the tendency for a chronic course, the weather-dependent fluctuation of the pain intensity, the migrating character of the pain, and the benign course of the disease were further evidence for a rheumatic disease, which he postulated for other reasons.

One patient from the third group with headache, vomiting and lymphocytic meningitis had papilledema producing changes of 1–2 diopters. The other patient had slight right-sided abducent paresis. One patient who had recurrent headache and vomiting 2 years after the initial disease showed inflammatory CSF changes (26 cells/µl) at the time of clinical recurrence. Unfortunately, Bannwarth missed the causal relation between the chronic meningitis and tick bite and erythema migrans, even though one of his patients seemed to have had this skin eruption (patient 7).

In 1944, Bannwarth reported on 11 more patients with lymphocytic meningitis. Huismann (1944) also favored a rheumatic-allergic cause of the disease. Gelbjerg-Hansen (1945) reported on lymphocytic meningitis following erythema migrans, and thought of a toxic-allergic etiology or viral transmission by a tick bite.

1.2.5 Concluding Remarks

By 1945, knowledge of the three "classical" skin manifestations and the early neurological disorders had accumulated mainly independently of each other, although a relationship was occasionally noticed. Erythema migrans and solitary lymphocytoma were widely regarded as harmless, self-limited disorders. ACA was held to be a chronic inflammatory skin condition, which only occasionally gave rise to the development of malignancy; the association of ACA with joint and bone changes was disputed by serious investigators. Despite a careful description of "chronic lymphocytic meningitis", Bannwarth failed to elucidate the relationship to tick bite and erythema migrans. However, such a relationship was observed by a few other investigators.

1.3 Second Phase (1946-1976)

1.3.1 Acrodermatitis Chronica Atrophicans

As in the first phase, initially the most progress was made with ACA in the second phase, too. In 1946, following a suggestion from Adlercreutz, Svartz treated two ACA patients with penicillin because of an elevated erythrocyte sedimentation rate (ESR). Niels Thyresson (1949) from Stockholm reported on 57 patients followed up for up to 2 years after penicillin therapy; 7 were cured, 13 improved considerably, 15 improved quite well, and 5 showed only slight improvement. Furthermore, he also described the disappearance of the fibroid (fibrous) nodules and infiltrations in all patients in whom these lesions were present, and he noted a beneficial effect in some patients with sensory disturbances; the elevated ESR partly or totally normalized. Several other investigators confirmed these results in the following years using penicillin (Brunner 1951; Götz and Ludwig 1951; Pirilä 1951) and to a lesser extent tetracyclines (Gougerot et al. 1951; Ludwig 1955) and chloramphenicol (Götz 1952). The results regarding the effect of streptomycin were somewhat contradictory (Hauser 1955; Ludwig 1955). Götz and Ludwig (1951) observed a patient with a generalized ACA whose temperature rose shortly after initiation of treatment with penicillin up to 38.8°C, although she was not allergic to penicillin. This fever of "unknown origin" can now in retrospect be interpreted as a Jarisch-Herxheimer reaction.

The success of the antibiotic therapy greatly stimulated the search for the causative organism of ACA (see Chap.2) and led to a renewed interest in clinical work, too. Hauser (1952) found inflammatory changes in the bone marrow of 25 patients with ACA (increase in the number of eosinophils, plasma cells, lymphoid cells, and mast cells) and an elevation in gamma globulins in serum protein electrophoresis. A woman reported on by Gans and Landes (1952) had "arthritis" in the left hip and both knees 6 months prior to developing ACA. Examination of the patient revealed lymphadenopathy, psychic changes, an intraventricular conduction abnormality, myositis, arthropathy of most fingers associated with sclerosis and osseous demineralization, impairment of movement in both shoulders and elbows, and leukocytosis with 13.100 leukocytes/ mm^3.

Based on the evaluation of 234 patients with ACA, Walter Hauser (1955) from Würzburg, Germany, described more clinical details. He presented a patient with a preceding erythema migrans and another with a concomitant erythema migrans, and noted that ACA occurred mainly in areas of the geographical distribution of Ixodes ricinus. Furthermore, he discussed the relationship to lymphocytoma based on previous experiences as well as his own. All this led him to the concept of ACA being a general disease. He found an arthrosis deformans in only 3% of 101 patients with ACA examined for bone and joint abnormalities and doubted the specificity of the arthropathy described previously.

In 1954 and 1955, Hans Götz published a report on the first successful transmission of ACA to himself and three other physicians of the Department of Dermatology, University of München, Germany. Götz and his colleagues experienced hyperesthesia, arthralgias in the joint, and sensitivity of the periosteum of the bone adjacent to the induced ACA. They saw an erythema migrans-like appearance of the ACA border and improvement of all these changes after antibiotic therapy. Götz assumed that the primary lesion induced by the transplantation was due to local accumulation of the causative agent and that the lymphogenous and hematogenous spread occurred in a similar way to that found in syphilis (Götz 1954/1955).

In 1959, Donnermann and Heite evaluated 840 ACA patients reported in the literature. They found the age of the patients at onset of the disease (among 621 patients) to follow an even Gauss-type distribution curve with a mean between 40 and 50 years (range 6–69). Fibroid nodules started to develop approximately 5 years after the bluish-red discoloration or the dermatosclerosis ($p<0.05$). These nodules were located predominantly on the arms, while dermatosclerosis and anetoderma were mainly found on the legs. Of the 840 patients, 70% were women. A qualitative increase in $ß^2$-macroglubin was found by Herrmann (1961) and Brehm (1963). Brehm described a patient with a positive rheumatoid factor test. Kraus et al. (1963) claimed to have detected cryoglobulins in 19 out of 26 untreated patients with ACA. Danda (1963) published a paper on the distribution of the ACA according to a world-wide evaluation; ACA was mainly found to occur in Central, Northern, and Eastern Europe, and rarely in other areas of Europe and the United States (see Chap.29).

In 1966, Hanns Christian Hopf presented convincing evidence based on previous case reports and his own findings in 92 patients that a peripheral neuropathy of the distal type is part of the clinical picture in a considerable number of patients with ACA; 26 of 31 patients in whom a lumbar puncture was performed had a normal CSF. Hopf presented in detail the course of 27 patients with ACA followed for several years. Of these, seven had a history or concomitant arthritis or arthralgia, five had signs and symptoms of cardiac disease, and three had cerebral involvement; a correlation of these changes to the ACA was not, however, suggested. A woman with ACA (patient 19) had a history of a facial palsy followed by headache, several cerebral symptoms, fatigue, and dizziness, while a man with ACA (patient 25) had arthritis, weight loss of 13 kg, intermittent tachycardia, dyspnea, dizziness, lymphadenopathy, splenomegaly, arthropathy of fingers and toes, and probably encephalitis. This patient experienced prompt improvement of his illness after three courses of parenteral penicillin treatment (Schulz, cited in Hopf 1966). In addition, Hopf reported on the beneficial effect of penicillin on the peripheral neuropathy, but he also observed that sometimes the peripheral neuropathy redeveloped, despite disappearance of the skin lesions.

Hopf and Stroux (1968) reported on clustering of ACA based on an epidemiological study of 314 patients in Lower Franconia, Germany, and discussed a possible relationship to tick activity in these areas. Hardmeier (1968) found the histopathology of the fibroid nodules to resemble other reversible fibrotic diseases, but to be different from the changes seen in rheumatoid nodules. Another woman with ACA, which resembled to some extent that described in the patients reported by Gans and Landes (1952) and Hopf (1966; patient 25), had morphea, followed by joint involvement with deformities of the feet and hands, arthralgias in the knees, wrists, and shoulders, depression, electroencephalogram (EEG) changes, myositis, elevation of ESR, creatinin and creatin, and positive antinuclear antibodies; most abnormalities promptly improved after antibiotic therapy (Müller 1969). Gebhart et al. (1971), in agreement with earlier light-microscopical findings reported by A. Wiedmann (1955, cited in Gebhart et al. 1971), were unable to find neural structures in an anesthetic subepidermal area of the skin of a patient with ACA suffering from neuropathy and bone changes. They discussed the possibility that the bone and joint abnormalities described in ACA patients might be neurogenic in origin in analogy to leprosy. Goor and Ott (1971) described a patient with ACA, polyradiculoneuritis and increased CSF protein; one of their patients had a positive fluorescent treponemal antibody (FTA) test, but otherwise negative syphilis serology. Kaiser (1972) reported on relapse of the peripheral neuropathy in ACA patients treated with penicillin.

1.3.2 Erythema Migrans

Bruder (1950) observed a man with an erythema migrans lasting almost 2 years, the longest duration reported so far; a similarly long duration was also noted by Sonck some years later (1965). Based on the seemingly positive spirochetal staining results which Lennhoff claimed to have obtained in a number of skin diseases, Hollström from Stockholm started to treat various dermatological disorders with penicillin in the late 1940s. In 1951, Hollström was able to report on the successful treatment of some patients with erythema migrans, including a patient with meningitis, using comparably low doses of parenteral penicillin. The median resolution of the erythema migrans in six penicillin recipients was 2 (1 to 8) weeks, whereas the resolution in nine bismuth or arsenic salt recipients was usually longer. After the successful transmission of ACA by Götz and colleagues, Binder, Doepfmer, and Hornstein from Würzburg were able to transmit erythema migrans to themselves and to each other (1955). Hollström (1958) reported that the erythema migrans disappeared in 89% of his 65 patients within 2 weeks of starting penicillin therapy. He observed a recurrence in two patients treated with low doses of penicillin and described the first case of reinfection 5 years after primary infection. He also stated that most of his patients were women, and that the median age was approximately 43 years. Sedlacek (1960) confirmed many of these findings on the basis of an examination of 89 patients. He found the incidence of the erythema migrans particularly high during spring and fall, roughly following the two-peak activity of *Ixodes ricinus* in Europe, and claimed to have observed post-treatment cyanosis, strongly resembling or even representing ACA in 11 of his patients; the latter finding has not been confirmed by other investigators so far.

Among 164 patients collected by Hauser (1965), in 42% the erythema migrans was on the legs, in 28% on the trunk, in 13% on the arms, and in 7% on the face; the patients had up to six lesions. Hauser also provided more details on the geographical distribution of erythema migrans in Europe and elsewhere. Most of 197 patients were found to be between 30 and 60 years old (range 1.5–61 years), and 58% of 114 patients were women. In his handbook article (1965), Hauser grouped ACA, erythema migrans, and lymphadenosis benigna cutis (lymphocytoma) together and discussed the close relationship between them.

Up to six erythema migrans lesions were also noted by Sonck (1965). The nymph of *Ixodes ricinus* was suggested by Thöne (1968) as an often unnoticed vector of the disease. Sandbank and Feuerman (1972) reported on ultrastructural changes in two patients with multiple erythema migrans-like lesions; the most striking finding was a pronounced phagocytosis by macrophages. Discrete changes in the differential blood count and lack of leukocytosis were observed in 27 patients by Weber and Braun-Falco (1974).

1.3.3 Borrelial Lymphocytoma

For definitions of borrelia lymphocytoma see Sect. 1.2.3 and Chap.9. In 1950, Bianchi reported on the beneficial effect of penicillin in patients with lymphocytoma. A few years later, the first case of an association between a lymphocytoma (consisting of a large plaque following a tick bite) and a meningoradiculoneuritis was described (Paschoud 1954). Höfer (1956), discussing 143 recent patients collected from the literature, confirmed by and large the findings of Bäfverstedt.

Paschoud (1957/1958) from Lausanne, Switzerland, was successful in transmitting borrelial lymphocytoma from human to human. The incubation period was ususally 6 weeks and a typical erythema migrans developed in one instance from a transplantation site. The borrelial lymphocytomas observed by Paschoud were mostly of the solitary type, but consisted of several small nodules in one patient. Paschoud emphasized the close relationship between erythema migrans and lymphocytoma. Two lymphocytoma lesions were followed by ACA in a patient observed by Bäfverstedt over a number of years (1960). An uneven global distribution of the lymphocytoma with clustering in Europe was discussed by Hauser (1965). Mach and Wilgram (1966) stressed the importance of a large pale reticulum cell characteristically containing Fleming bodies; they distinguished five histologic subtypes in their series of 115 cases. The lymphocytoma was found to contain both B and T lymphocytes (Braun-Falco and Burg 1975). However, it must again be emphasized that some of the patients presented by Bianchi, Höfer and the last five authors may not have been of borrelial origin, so that confirmation is needed that their findings were indeed borrelial lymphocytoma.

1.3.4 Neurological Disorders

Additional patients with erythema migrans and meningitis were reported by Sälde (1946), Zellweger (1946), Gsell (1947), Dalsgaard-Nielsen and Kierkegaard (1948), Schirduan (1950) and Hellerström (1951). In 1949, Georges Schaltenbrand from Würzburg reported on eight patients with "subacute lymphocytic meningitis"; the highest cell count was 1675 cells/µl. One patient had transient arthritis (patient 4), while one of five additional patients with "chronic aseptic meningitis" of unknown origin, a 36-year-old woman, died and was found at autopsy to have had pronounced meningovasculitis and a lesion in the right thalamus due to vasculitis (representing a possible case of chronic central nervous system involvement). Two reports were published on the clinical recovery of a patient with an erythema migrans associated with meningitis (Hollström 1951; Bjornstad and Mossige 1955).

Putkonen et al. (1962) described 31 patients with erythema migrans; none of their patients had meningeal symptoms or clinical evidence for central nervous system involvement. Lumbar puncture, which was performed in 13 patients with erythema migrans lasting a median of 4 months (range 1–8 months) after the onset of the eruption showed normal CSF values in all instances.

Schaltenbrand (1962) described three patients with painful radiculomeningitis after a tick bite. He found that the pain as well as the neurological deficit usually occurred in the same topographical region as the previous tick bite or erythema. Schaltenbrand suggested that the etiologic agent might cause an inflammatory cutaneous reaction 8–14 days after a tick bite. He further speculated the agent would migrate along the peripheral nerves and lymphatics to the central nervous system and would cause a circumscribed inflammation of meninges and radices. Schaltenbrand reported on the prompt improvement of the radicular pain after therapy with corticosteroids and tetracycline in individual patients. In 1965, Bammer and Schenk described four patients with tick bite, erythema migrans, and subsequent neurological symptoms including central nervous system involvement; signs of pyramidal tract lesions (patients 1 and 2) and a focus on EEG (patient 2) were observed. Keuth and Mennicken (1966) reported on three children with peripheral facial palsy due to "subacute-chronic lymphocytic meningitis Bannwarth"; the authors found seven additional case reports in the literature involving children. In 1967, Schaltenbrand reported on 50 patients who developed neurological symptoms after arthropod bites, usually tick (82%) or horse fly bites (10%). Schaltenbrand found the time interval between a tick bite and the onset of the neurological disease to be between a few weeks and 8 months. The most common findings were neuritis (78%), myelitis (32%), and cranial nerve palsies (32%). However, Schaltenbrand included patients with normal CSF findings (29%). This group of patients presumably had neurological disorders of various origins including viral tick-borne encephalitis.

Erbslöh and Kohlmeyer (1968) reported on 12 patients, all of whom had a history of a tick or insect bite and an erythema migrans followed by meningoradiculoneuritic symptoms and lymphocytic meningitis. They pointed out that encephalitic symptoms may occur. They found patients with "symptomatic psychosis" (patient 1), an affective disorder (patient 6), and paranoid hallucinations for 3 days (patient 8). In addition, stiff neck, headache, and somnolence were observed.

Bammer and Schenk (1965), Meyer-Rienecker and Hitzschke (1969), and Hörstrup and Ackermann (1973) mentioned a relatively high percentage of plasma cells in the CSF. The time sequence of the disease was carefully described by Rancurel (1969), Arnould et al. (1970), and Hörstrup and Ackermann (1973). Hörstrup and Ackermann stressed the importance of severe pain as a hallmark of the disease in their 47 patients. The pain developed a mean of 3 weeks before the neurological deficit occurred and lasted a mean of 6 weeks; they found that the erythema migrans lasted only for about 1 week in most of their patients. Wolf (1970; 32 patients) and Wenig (1975; 65 patients) reported on neuritis of the cranial nerves II–XII. Two of the patients described by Wolf (1970) attempted suicide because of excruciating pain. Rohmer et al. (1974) performed electromyo- and neurographic studies in 36 patients with meningoradiculoneuritis and found marked signs of denervation, but normal nerve conduction velocities and distal latencies; the authors interpreted their results as radiculitis without neuritis. In 1974, Weber reported on another patient with meningitis which developed despite previous oral penicillin therapy for erythema migrans. The meningitis was promptly cured with high doses of intravenous penicillin; a bacterial origin was postulated and borreliae, rickettsiae, and *Pasteurella (Francisella) tularensis* were discussed as possible causative agents.

1.3.5 Concluding Remarks

By 1976, three dermatological manifestations had been linked by their relation to the tick *Ixodes ricinus*, geographical clustering in certain areas of Europe, and close clinical relations to each other. Their infectious nature had been proven by successful transmission experiments. Furthermore, a relation to certain cases of meningoradiculoneuritis and meningitis had been acknowledged. Several authors had considered an infectious disease (Götz 1954; Binder et al. 1955; Paschoud 1957/1958; Hauser 1965; Sonck 1965; Erbslöh und Kohlmeyer 1968; Weber 1974; Weber and Braun-Falco 1974). To emphasize the close relationship between dermatological and neurological manifestations, a nomenclatural proposal was made (Weber 1974).

1.4 Third Phase (1977 to the Present)

In 1977, Allen C. Steere and coworkers from Yale Universitiy reported on their observations in North American patients with Lyme arthritis, later called Lyme disease. They noted joint, heart and neurological abnormalities besides the erythema migrans (Steere et al. 1977). A series of publications followed describing in more detail the clinical features, laboratory abnormalities, and staging of Lyme disease in the United States (Steere et al. 1986; Steere 1989).

The initial impression gained by many observers was that joint and heart manifestations were much less frequently seen in Europe than in the United States. This view had to be at least partly revised when more and more European patients with arthritis and carditis became known. Furthermore, a variety of additional clinical observations have been made in Europe in recent years. Greatly instrumental for this developement was the availability of specific serological tests after *B. burgdorferi* had been discovered by the Swiss-American scientist Willy Burgdorfer and coworkers at the Rocky Mountain Laboratories, Hamilton, USA, in 1982.

1.4.1 European Work from 1977 Until the Discovery of *B. burgdorferi*

A European patient with arthritis analogous to Lyme arthritis was described by Ackermann et al. in 1980. Krüger et al. (1981) noted oligoclonal banding in the CSF electrophoresis in 17 of 18 patients with lymphocytic meningoradiculoneuritis. Weber (1981) postulated that persistence of the causative organism was important for the pathogenesis of the various manifestations of the disease and recommended appropriate antibiotic therapy. Felgenhauer (1982) demonstrated intrathecal immunoglobulin M (IgM) antibody production in patients with lymphocytic meningoradiculoneuritis. A prospective (long-term) study described various signs and symptoms in 71%, arthralgia and arthritis in 32%, neurological symptoms in 23%, and signs and symptoms suggestive of cardiac involvement in 10% of 31 patients with "erythema migrans disease", a term used on the analogy of the previous designation erythema chronicum migrans meningitis (Weber et al. 1983). Meanwhile, Schmoeckel et al. (1977) had reported on the ultrastructure of patients with lymphadenosis benigna cutis, and Burg and Braun-Falco (1983) presented a summary of current knowledge on cutaneous pseudolymphomas, to which lymphadenosis benigna cutis and borrelial lymphocytoma belong. Kristoferitsch et al. (1983) described more details on the meningoradiculoneuritis: several patients had persistent pain and dysesthesia for more than 1 year, some had elevated serum IgM and complement C3 levels, and several patients had persistently high

CSF IgM levels. In 1983, Sköldenberg et al. published results describing the beneficial effect of intravenous penicillin G in 21 patients with aseptic meningitis; later, these patients were proven to suffer from neurological manifestations of Lyme borreliosis (Stiernstedt et al. 1984).

1.4.2 European Work After the Discovery of *B. burgdorferi*

Using *Borrelia duttonii* as an antigen, Rudolf Ackermann (1983) from Köln, Germany, was able to find elevated antiborrelial antibody titers in six patients with erythema migrans and eight patients with meningoradiculoneuritis. Ryberg et al. (1983) using *B. burgdorferi* as an antigen found increased antibody titers in six out of nine patients with meningoradiculoneuritis. Frithz and Lagerholm (1983) claimed to have detected spirochetes by silver staining in the three classical skin manifestations, but the specificity of these findings remains questionable. In November 1983, Weber et al. (1983/1984) announced at the German ADF meeting in Kiel and at the First International Symposium at Yale University that all nine patients with ACA, all four patients with spirochetal lymphocytoma, and some of 42 patients with erythema migrans had elevated antibody titers against *B. burgdorferi*. These serological findings were independently confirmed by several European groups in the following year (Ackermann et al. 1984 a, b; Asbrink et al. 1984a, b; Neubert and Krampitz 1984; Weber et al. 1984a, b; Wilske et al. 1984). It was demonstrated that only a portion of patients with erythema migrans showed elevated antibody titers against *B. burgdorferi*; a rise and fall of IgM and IgG antibody titers in patients with erythema migrans and meningoradiculoneuritis was demonstrated (Ackermann et al. 1984a; Wilske et al. 1984). Patients with ACA were shown to have sometimes strongly elevated IgG antibody titers, which partly declined after antibiotic therapy (Weber et al. 1983/1984, 1984a, b; Ackermann et al. 1984b; Wilske et al. 1984), and a lack of increased IgM antibody titers (Weber et al. 1983/1984, 1984a, b; Wilske et al. 1984). Borrelial lymphocytoma patients were found to have elevated antibody titers of the IgG and/or IgM type (Weber et al. 1983/84, 1984a; Wilske et al. 1984). In addition, *B. burgdorferi* was isolated from the blood of a patient with erythema migrans (Ackermann et al. 1984d), from the CSF of a patient with meningoradiculoneuritis in the laboratory of Vera Preac-Mursic (Pfister et al. 1984; Preac-Mursic et al. 1984a, b), and from the skin of a patient with ACA (Asbrink et al. 1984b); a spirochete could be demonstrated by silver staining in a patient with erythema migrans (Neubert 1984).

These findings were in agreement with analogous reports from the United States. The sometimes high IgG antibody titers in all patients tested and the isolation of *B. burgdorferi* suggested that ACA is a chronic, late manifestation of Lyme borreliosis (Weber et al. 1983/1984, 1984a, b; Asbrink et al. 1984b).

Cardiac abnormalities were described in detail in some patients with erythema migrans (Corneau et al. 1984; Houwerzyl et al. 1984). A patient with ACA was found to have a history of erythema migrans followed by oligoarthritis (Weber et al. 1984b). A quite detailed description was presented of a Jarisch-Herxheimer reaction that occurred in four patients with erythema migrans treated with oral penicillin or minocycline (Weber 1984). Sural nerve biopsies of seven patients with lymphocytic meningoradiculoneuritis showed lymphocytes and plasma cells in epineurial and endoneurial vessels and loss of myelinated axons (Vallat et al. 1984). The specificity of the oligoclonal CSF band could be demonstrated (Stiernstedt et al. 1984).

In the following years, these early observations were extended and confirmed by the same and other European investigators. More details based on a couple of larger studies on erythema migrans (Asbrink and Olsson 1985; Weber and Neubert 1986) and early neurologic involvement (Pfister et al. 1986a; Stiernstedt et al. 1986) were presented. More and more patients were found to have arthritis and arthralgia (Herzer et al. 1986; Weber and Neubert 1986). Four patients with lymphocytoma were described in detail and the term borrelial lymphocytoma was introduced (Weber et al. 1985). *B. burgdorferi* was isolated from the borrelial lymphocytoma (Hovmark et al 1986a). A case report describing myositis in addition to meningoradiculoneuritis appeared (Schmutzhard et al. 1986). Reports on chronic central nervous system involvement were published (Kohlhepp and Mertens 1984/1985; Ackermann et al. 1985; Klenk et al. 1985). Knowledge of ACA was extended by Asbrink (1985), Asbrink et al. (1986) and Hovmark et al. (1986b), describing more patients with a history of erythema migrans, joint and nervous system involvement, and concomitant arthropathy.

Spirochetes could be demonstrated by silver staining in tissues of patients with Lyme borreliosis (De Koning and Hoogkamp-Korstanje 1986; De Koning et al. 1987).

Methods were developed to detect locally synthesized specific antibodies in the CSF (Rehse-Küpper and Ackermann 1986; Wilske et al. 1986b). *B. burgdorferi* was isolated from the skin of patients with erythema migrans (Asbrink and Hovmark 1985; Stanek et al. 1985a; Preac-Mursic et al. 1986), from the blood of a patient with erythema migrans (Neubert et al. 1986), from a patient with borrelial lymphocytoma (Hovmark et al. 1986a), and from the skin of patients with ACA (Asbrink and Hovmark 1985; Preac-Mursic et al. 1985, 1986). Reinfection was described in a small series of three patients (Weber et al. 1985b, 1986a, b) and in patients with erythema migrans following meningoradiculoneuritis (Neubert et al. 1986; Pfister et al. 1986b). Of 242 forest workers, 16.5% were found to have elevated specific antibody titers, although many of them seemed to lack clinical manifestations (Wilske et al. 1985b). Evidence for the presence of a cellular immune response in erythema migrans and ACA was presented (Tjerulund et al. 1986; Büchner and Rufli 1987).

Preac-Mursic et al. (1986, 1987) tested a variety of antibiotics against a considerable number of *B. burgdorferi* strains; in addition, gerbils were used to evaluate the most important antibiotics in vivo. Differences between European and American *B. burgdorferi* strains were noted (Ackermann et al. 1984a; Stanek et al. 1985b; Wilske et al. 1985a, 1986a; Hovind-Hongen et al. 1986). Wilske et al. (1986a, 1988) reported that *B. burgdorferi* isolates of erythema migrans and ACA belonged to a certain serotype, whereas CSF isolates belonged to a different serotype and were heterogeneous. A SCID mouse model showed that low-passage *B. burgdorferi* isolates retained their virulence in vivo and that the experimentally induced arthritis and carditis developed independently of lymphocyte function (Schaible et al. 1989). Postic et al. (1990) suggested that besides *B. burgdorferi* senso stricto, another closely related, but different borrelial species could exist. An improvement in serological tests has recently been reported by several groups of investigators (Chap. 21).

Sural nerve biopsies of patients with ACA-associated neuropathy revealed similar histopathological findings as in lymphocytic meningoradiculoneuritis (Kristoferitsch et al. 1988). An ocular involvement was observed in a woman (Bialasiewicz et al. 1988). A considerable number of children with facial palsy turned out to suffer from Lyme borreliosis (Olsson et al. 1988; Christen et al. 1990). Extrapyramidal, parkinsonism-like signs and symptoms were noted in five patients with encephalitis (Kohlhepp et al. 1989). Between 1987 and 1990, several case reports appeared demonstrating a stroke-like picture due to cerebral vasculitis (Hänny and Hänselmann 1987; Midgard and Hofstad 1987; Uldry et al. 1987; Weder et al. 1987 and others, cited in May and Jabbari 1990), (see Chap. 15). A recent evaluation has shown that cardiac involvement is more common in Europe than previously thought (see Chap. 10). A case of congenital infection with demonstration of *B. burgdorferi* in noninflamed brain and liver was observed (Weber et al. 1988b). Therapeutic trials in patients with erythema migrans demonstrated that there was no statistically significant difference between oral penicillin, tetracycline, and amoxicillin-clavulanic acid (Weber et al. 1986a, 1988a); in addition, ceftriaxone was found to be superior to oral penicillin in a subgroup of patients with erythema migrans (Weber et al. 1990). Recently, Pfister et al. (1989, 1991) obtained evidence that ceftriaxone and cefotaxime, but not intravenously administered penicillin G, were present in a sufficiently high concentration in the CSF if calculated on the basis of a previous in vitro study (Preac-Mursic et al. 1987). Cefotaxime was found to be superior to intravenous penicillin G in patients with late Lyme borreliosis (Hassler et al. 1990).

Much more information has accumulated in various fields in the past few years, but a detailed discussion of these observations would be beyond the scope of this chapter and will be presented elsewhere in this book. In conclusion, knowledge of the clinical features, treatment, and laboratory diagnosis of Lyme borreliosis in Europe has greatly expanded in recent years. It has become apparent that differences between the European and American forms of Lyme borreliosis seem to be more quantitative in nature.

References

Ackermann R (1983) Erythema chronicum migrans und durch Zecken übertragene Meningopolyneuritis (Garin-Bujadoux-Bannwarth): Borrelien-Infektionen? Dtsch Med Wochenschr 108:577-580

Ackermann R, Runne U, Klenk W, Dienst C (1980) Erythema chronicum migrans mit Arthritis. Dtsch Med Wochenschr 105:1779-1781

Ackermann R, Kabatzki J, Boisten HP, Steere AC, Grodzicki RL, Hartung S, Runne U (1984a) Spirochäten-Ätiologie der Erythema-chronicum-migrans-Krankheit. Dtsch Med Wochenschr 109:92-97

Ackermann R, Boisten HP, Kabatzki J, Runne U, Krüger K, Herrmann WP (1984b) Serumantikörper gegen *Ixodes-ricinus*-Spirochäte bei Acrodermatitis chronica atrophicans (Herxheimer). Dtsch Med Wochenschr 109:6-10

Ackermann R, Gollmer E, Rehse-Küpper B (1985) Progressive Borrelien-Enzephalomyelitis. Dtsch Med Wochenschr 110:1039-1042

Afzelius A (1910) Verhandlungen der Dermatologischen Gesellschaft zu Stockholm, 28 Oct 1909. Arch Dermatol Syph 101:404

Afzelius A (1921) Erythema chronicum migrans. Acta Derm Venereol (Stockh) 2:120-125

Arnould G, Cuny G, Weber M, Peniu F, André JM, Cagniart MJ (1970) Méningopolyradiculonévrite à tiques à propos d'une observation. Ann Med Nancy 9:629

Asbrink E (1985) Erythema chronicum migrans Afzelius and acrodermatitis chronica atrophicans. Acta Derm Venereol (Stockh) (Suppl) 118:1-63

Asbrink E, Hovmark A (1985) Successful cultivation of spirochetes from skin lesions of patients with erythema chronicum migrans Afzelius and acrodermatitis chronica atrophicans. Acta Pathol Microbiol Immunol Scand (B) 93:161-163

Asbrink E, Olsson I (1985) Clinical manifestations of erythema chronicum migrans Afzelius in 161 patients. Acta Derm Venereol (Stockh) 65:43-52

Asbrink E, Hederstedt B, Hovmark A (1984a) The spirochetal etiology of erythema chronicum migrans Afzelius. Acta Derm Venereol (Stockh) 64:291-295

Asbrink E, Hovmark A, Hederstedt B (1984b) The spirochetal etiology of acrodermatitis chronica atrophicans Herxheimer. Acta Derm Venereol (Stockh) 64:506-512

Asbrink E, Hovmark A, Olsson I (1986) Clinical manifestations of acrodermatitis chronica atrophicans in 50 Swedish patients. Zentralbl Bakteriol Hyg (A) 263:253-261

Bäfverstedt B (1943) Über Lymphadenosis benigna cutis. Acta Derm Venereol (Stockh) (Suppl) 24:1-202

Bäfverstedt B (1960) Lymphadenosis benigna cutis (LABC), its nature, course, and prognosis. Acta Derm Venereol (Stockh) 40:10-18
Balban W (1910) Erythema annulare entstanden durch Insektenstiche. Arch Dermatol Syph 105:423-430
Bammer H, Schenk K (1965) Meningo-Myelo-Radikulitis nach Zeckenbiß mit Erythem. Dtsch Z Nervenheilkd 187:25-34
Bannwarth A (1941) Chronische lymphocytäre Meningitis, entzündliche Polyneuritis und "Rheumatismus". Arch Psychiatr Nervenkr 113:284-376
Bannwarth A (1944) Zur Klinik und Pathogenese der "chronischen lymphocytären Meningitis". Arch Psychiatr Nervenkr 117:161–185
Bialasiewicz AA, Ruprecht KW, Naumann GOH, Blenk H (1988) Bilateral diffuse choroiditis and exsudative retinal detachments with evidence of Lyme disease (letter). Am J Ophthalmol 105:419-420
Bianchi GE (1950) Die Penicillinbehandlung der Lymphocytome. Dermatologica 100:270-273
Biberstein H (1923) Lymphozytome. Zentralbl Hautkr 6:70-71
Binder E, Doepfmer R, Hornstein O (1955) Experimentelle Übertragung des Erythema chronicum migrans von Mensch zu Mensch. Hautarzt 6:494-496
Bjornstadt RT, Mossige K (1955) Erythema chronicum migrans med meningopolyradiculitis. Tidsskr Norske Laegerfor 75:264-265
Braun-Falco O, Burg G (1975) Lymphoretikuläre Proliferation in der Haut. Hautarzt 26:124-132
Brehm G (1963) Symptomatische Makro- und Kryoglobulinämie bei Akrodermatitis chronica atrophicans. Hautarzt 14:75-79
Bruder K (1950) Zur Kenntnis des Erythema chronicum migrans (Lipschütz). Dermatol Wochenschr 121:337-344
Brunner N (1951) Zur Penicillinbehandlung bei Sklerodermie und Akrodermatitis atrophicans Herxheimer. Hautarzt 2:545-547
Büchner SA, Rufli T (1987) Erythema chronicum migrans: evidence for cellular immune reaction in the skin lesion. Dermatologica 174:144-149
Buchwald A (1883) Ein Fall von diffuser idiopathischer Haut-Atrophie. Arch Dermatol Syph 10:553-556
Burckhardt JL (1911) Zur Frage der Follikel- und Keimzentrenbildung in der Haut. Frankf Z Pathol 6:352-359
Burg G, Braun-Falco O (1983) Cutaneous lymphomas, pseudolymphomas, and related disorders. Springer, Berlin, Heidelberg, New York, pp 415-441
Burgdorfer W, Barbour AG, Hayes SF, Benach JL, Grunwaldt E, Davis JP (1982) Lyme disease – a tick-borne spirochetosis? Science 216:1317-1319
Christen HJ, Bartlau N, Hanefeld F, Eiffert H, Thomssen R (1990) Peripheral facial palsy in childhood – Lyme borreliosis to be suspected unless proven otherwise. Acta Paediatr Scand 79:1219-1224
Corneau C, Bardet M, Baudoin P, Daumas PL, Oblet B, Poirot G, Valois M (1984) Bloc auriculo-ventriculaire aigu, syncopal, au cours de la maladie de Lyme. Presse Med 13:888
Dahlsgaard-Nielsen T, Kierkegaard A (1948) Allergic meningitis and chronic erythema migrans Afzelii after bite by *Ixodes reduvius*. Acta Allerg 1:388-393
Danda J (1963) Die Weltfrequenz der Akrodermatitis chronica atrophicans. Hautarzt 14:337-340
DeKoning J, Hoogkamp-Korstanje JAA (1986) Diagnosis of Lyme disease by demonstration of spirochetes in tissue biopsies. Zentralbl Bakteriol Hyg (A) 263:179-188
DeKoning J, Bosma RB, Hoogkamp-Korstanje JAA (1987) Demonstration of spirochetes in patients with Lyme disease with a modified silver stain. J Med Microbiol 23:261-267
Donnermann C, Heite HJ (1959) Beitrag zur Symptomatologie der Akrodermatitis chronica atrophicans (Pick-Herxheimer). Arch Klin Exp Dermatol 208:516-527
Ehrmann S, Falkenstein F (1925) Über Dermatitis atrophicans und ihre pseudo-sklerodermatischen Formen. Arch Dermatol Syph 149:142-175
Erbslöh F, Kohlmeyer K (1968) Über polytope Erkrankungen des peripheren Nervensystems bei lymphocytärer Meningitis. Fortschr Neurol Psychiatr 36:321-342
Felgenhauer K (1982) Differentiation of the humoral immune response in inflammatory diseases of the CNS. J Neurol 228:223-227

Finger E, Oppenheim M (1910) Die Hautatrophien. Deuticke, Wien
Frithz A, Lagerholm B (1983) Acrodermatitis chronica atrophicans, erythema chronicum migrans and lymphadenosis beninga cuis – spirochetal diseases? Acta Derm Venereol (Stockh) 63:432-436
Gans O, Landes E (1952) Akrodermatitis atrophicans arthropatica. Hautarzt 3:151-155
Garin C, Bujadoux (1922) Paralysie par les tiques. J Med Lyon 71:765-767
Gebhart W, Lindemayr W, Partsch H (1971) Ulceromutilierende Neuropathie bei Acrodermatitis chronica atrophicans. Z Hautkr 46:171-178
Gelbjerg-Hansen G (1945) Erythema chronicum migrans Afzelii and meningitis after a tick bite. Acta Derm Venereol (Stockh) 25:458-463
Goor W, Ott F (1971) Zirkulationsstörung – Acrodermatitis chronica atrophicans Pick-Herxheimer? Schweiz Med Wochenschr 101:1334-1348
Götz H (1952) Die Acrodermatitis chronica atrophicans Herxheimer als neue Indikation für die Behandlung mit Chloromycetin. Hautarzt 3:310-312
Götz H (1954/1955) Die Acrodermatitis chronica atrophicans Herxheimer als Infektionskrankheit. Hautarzt 5:491-504, 6:249-252
Götz H, Ludwig E (1951) Die Behandlung der Akrodermatitis chronica atrophicans Herxheimer mit Penicillin. Hautarzt 2:6-14
Gougerot H, Meyer JJ, Flechner (1951) Acrodermatite de Pick-Herxheimer, au stade de début, traitée par l'auréomycine. Bull Soc Franc Derm Syph 58:386-387
Gsell O (1947) Beiträge zur Meningitis serosa und Radikulitis. Schweiz Arch Neurol Psychiatr 59:135-144
Hardmeier T (1968) Zur Histopathologie der fibroiden Knoten bei Akrodermatitis chronica atrophicans. Arch Klin Exp Dermatol 232:373-383
Hassler D, Zöller L, Haude M, Hufnagel HD, Heinrich F, Sonntag HG (1990) Cefotaxime versus penicillin in the late stage of Lyme disease – prospective, randomized therapeutic study. Infection 18:16-20
Hauser W (1952) Sternalmarkbefunde und ihre Beziehungen zur Blutsenkungsgeschwindigkeit bei Acrodermatitis chronica atrophicans. Arch Dermatol Syph 195:164-170
Hauser W (1955) Zur Kenntnis der Akrodermatitis chronica atrophicans. Arch Dermatol Syph 199:350-393
Hauser W (1965) Wahrscheinliche Infektionskrankheiten der Haut. (Handbuch Haut- u. Geschlechtskrankheiten, vol IV, part 1A), Springer, Berlin Heidelberg New York, pp 556-629
Hellerström S (1930) Erythema chronicum migrans Afzelii. Acta Derm Venereol (Stockh) 11:315-321
Herrmann WP (1961) Immunoelektrophoretische Untersuchungen an Hautkranken. Arch Klin Exp Dermatol 212:452-459
Herxheimer K, Hartmann K (1902) Über Acrodermatitis chronica atrophicans. Arch Dermatol Syph 61:57-76, 255-300
Herxheimer K, Schmidt W (1910) Über "strangförmige Neubildungen bei Acrodermatitis chronica atrophicans". Arch Dermatol Syph 105:145-168
Herzer P, Wilske B, Preac-Mursic V, Schierz G, Schattenkirchner M, Zöllner N (1986) Lyme arthritits: clinical features, serological and radiographic findings of cases in Germany. Klin Wochenschr 64:206-215
Höfer W (1956) Lymphadenosis benigna cutis. Arch Klin Exp Dermatol 203:23-40
Hollström E (1951) Successful treatment of erythema migrans Afzelius. Acta Derm Venereol (Stockh) 31:235-243
Hollström E (1958) Penicillin treatment of erythema chronicum migrans Afzelius. Acta Derm Venereol (Stockh) 38:285-289
Hopf HC (1966) Acrodermatitis chronica atrophicans (Herxheimer) und Nervensystem. (Monographien aus dem Gesamtgebiete der Neurologie und Psychiatrie, vol 114) Springer, Berlin Heidelberg New York
Hopf HC, Stroux B (1968) Die geographische Verteilung der Akrodermatitis chronica atrophicans (Herxheimer) in der Umgebung von Würzburg. Z Hautkr 43:41-48
Hörstrup P, Ackermann R (1973) Durch Zecken übertragene Meningopolyneuritis (Garin-Bujadoux, Bannwarth). Fortschr Neurol Psychiatr 41:583-606

Hövelborn C (1931) Gelenkveränderungen bei Acrodermatitis chronica atrophicans. Arch Dermatol Syph 164:349-356

Houwerzyl J, Root JJ, Hoogkamp-Korstanje JAA (1984) A case of Lyme disease with cardiac involvement in the Netherlands (letter). Infection 12:358

Hovind-Hougen K, Asbrink E, Stiernstedt G, Steere AC, Hovmark A (1986) Ultrastructural differences among spirochetes isolated from patients with Lyme disease and related disorders, and from *Ixodes ricinus*. Zentralbl Bakteriol Hyg (A) 263:103-111

Hovmark A, Asbrink E, Olsson I (1986a) The spirochetal etiology of lymphadenosis benigna cutis solitaria. Acta Derm Venereol (Stockh) 66:479-484

Hovmark A, Asbrink E, Olsson I (1986b) Joint and bone involvement in Swedish patients with *Ixodes ricinus* borne *Borrelia* infection. Zentralbl Bakteriol Hyg (A) 263:275-284

Huber A (1900) Über Atrophica cutis idiopathia im Gegensatz zur senilen Atrophie der Haut. Arch Dermatol Syph 52:71-90

Huismann G (1944) Über chronische lymphocytäre Meningitis mit Facialisbeteiligung. Arch Psychiatr Nervenkr 117:443-448

Jessner M (1921) Zur Kenntnis der Acrodermatitis chronica atrophicans. Arch Dermatol Syph 134:478-487

Jessner M, Loewenstamm A (1924) Bericht über 66 Fälle von Acrodermatitis chronica atrophicans. Dermatol Wochenschr 79:1169-1177

Kaiser M (1972) Neurologische Komplikationen bei Akrodermatitis chronica atrophicans (Herxheimer) und ihre Beeinflussung durch die Penicillintherapie. Dissertation, University of Göttingen

Kerl W (1915) Erythema chronicum migrans. Arch Dermatol Syph 119:301

Ketron LW (1913) Dermatitis atrophicans: with reports of a case showing a fibroid formation. Urol Cutan Rec Techn (Suppl) 1:286-294

Keuth U, Mennicken U (1966) Zur Kenntnis der "peripheren" Facialisparese mit subakut-chronisch lymphozytärer Meningitis (Bannwarth). Z Kinderheilkd 97:45-56

Klenk W, Heitmann R, Ackermann R (1985) Rezidivierende Erythema-chronicum-migrans-Krankheit des Nervensystems: Querschnittmyelitis als Rückfall einer Meningopolyneuritis Garin-Bujadoux-Bannwarth. Aktuel Neurol 12:20-23

Kohlhepp W, Mertens HG (1984/1985) Chronische Meningomyeloradikulitis durch *Ixodes-ricinus*-Spirochäten – eine der Metalues analoge Spirochätose? Springer, Berlin Heidelberg New York, pp 747-749 (Verhandlungen der Deutschen Gesellschaft für Neurologie, vol 3)

Kohlhepp W, Kuhn W, Krüger H (1989) Extrapyramidal features in central Lyme borreliosis. Eur Neurol 29:150-155

Kraus Z, Mateja F, Mazak J (1963) Myelogram and cryoglobulins in acrodermatitis chronica atrophicans. Acta Derm Venereol (Stockh) 43:163-175

Kristoferitsch W, Spiel G, Wessely P (1983) Zur Meningopolyneuritis (Garin-Bujadoux, Bannwarth). Nervenarzt 54:640-646

Kristoferitsch W, Sluga E, Graf M, Partsch H, Neumann R, Stanek G, Budka H (1988) Neuropathy associated with acrodermatitis chronica atrophicans. Ann NY Acad Sci 539:35-45

Krüger H, Englert D, Pflughaupt KW (1981) Demonstration of oligoclonal immunoglobulin G in Guillan-Barré syndrome and lymphocytic meningoradiculitis by isoelectric focusing. J Neurol 226:15-24

Lipschütz B (1913) Über eine seltene Erythemform (Erythema chronicum migrans). Arch Dermatol Syph 118:349-356

Lipschütz B (1923) Weiterer Beitrag zur Kenntnis des "Erythema chronicum migrans". Arch Dermatol Syph 143:365-374

Löwenfeld W (1932) Acrodermatitis und Dermatochalasis. Zentralbl Hautkr 41:562

Ludwig E (1955) Über die Wirksamkeit der Antibiotica Streptomycin und Tetracyclin bei der Behandlung der Acrodermatitis atrophicans Herxheimer. Arch Klin Exp Dermatol 201:495-506

Mach KW, Wilgram GF (1966) Characteristic histopathology of cutaneous lymphoplasia (lymphocytoma). Arch Dermatol 94:26-32

May EF, Jabbari B (1990) Stroke in neuroborreliosis. Stroke 21:1232-1235

Meyer-Rienecker HJ, Hietzschke B (1978) Lymphocytic meningoradiculitis. Elsevier, Amsterdam, pp 571-586 (Handbook of clin neurology, vol 34)
Montgomery H, Sullivan RR (1945) Acrodermatitis atrophicans chronica. Arch Dermatol 51:32-47
Müller H (1969) Akrodermatitis atrophicans arthropathica mit Pseudosklerodermie, Sklerodermia circumscripta und muskulärer Beteiligung. Z Hautkr 44:1-12
Mulzer P, Keining E (1929) Über miliare Lymphozytome der Haut. Dermatol Wochenschr 88:293-301
Neubert U (1984) Zur Ätiologie von Erytherma-migrans-Krankheit und Lyme-Erkrankung. Hautarzt 35:563-570
Neubert U, Krampitz HE (1984) Spirochaetal antibodies in patients with erythema chronicum migrans, meningoradiculitis Bannwarth and acrodermatitis chronica atrophicans (abstract). J Invest Dermatol 82:544-545
Neubert U, Krampitz HE, Engl H (1986) Microbiological findings in erythema chronicum migrans, acrodermatitis chronica atrophicans und lymphadenosis cutis benigna. Zentralbl Bakteriol Hyg (A) 263:237-252
Olsson J, Engervall K, Asbrink E, Carlsson-Nordlander B, Hovmark A (1988) Tick-borne borreliosis and facial palsy. Acta Otolaryngol (Stockh) 105:100-107
Paschoud JM (1954) Lymphocytom nach Zeckenbiss. Dermatologica 108:435-437
Paschoud JM (1957/1958) Die Lymphadenosis benigna cutis als übertragbare Infektionskrankheit. Hautarzt 8:197-211, 9:153-165, 263-269, 311-315
Pellizari C (1884) Atrophierendes Erythema urticatum. Arch Dermatol Syph 16:504-506
Pfister HW, Einhäupl KM, Preac-Mursic V, Wilske B, Schierz G (1984) The spirochetal etiology of lymphocytic meningoradiculitis of Bannwarth (Bannwarth's syndrome). J Neurol 231:141-144
Pfister HW, Einhäupl KM, Wilske B, Preac-Mursic V (1986a) Bannwarth's syndrome and the enlarged neurological spectrum of arthropod-borne borreliosis. Zentralbl Bakteriol Hyg (A) 263.343-347
Pfister HW, Neubert U, Wilske B, Preac-Mursic V, Einhäupl KM, Borasio GD (1986b) Reinfection with *Borrelia burgdorferi* (letter). Lancet II:984-985
Pfister HW, Preac-Mursic V, Wilske B, Einhäupl KM (1989) Cefotaxime vs penicillin G für acute neurologic manifestations in Lyme borreliosis. Arch Neurol 46:1190-1194
Pfister HW, Preac-Mursic V, Wilske B, Einhäupl KM, Schielke E, Sörgel F (1991) Randomized comparison of ceftriaxone and cefotaxime in Lyme borreliosis. J Infect Dis 163:311-318
Piorkowski (1926) Zwei Fälle von Acrodermatitis chronica atrophicans. Zentralbl Haut 19:362
Pirilä V (1951) The penicillin treatment of acrodermatitis atrophicans chronica. Acta Derm Venereol (Stockh) 31:576-591
Pospelow A (1886) Cas d'une atrophie idiopathique de la peau. Annales de Dermatol 7:505-510
Postic D, Edlinger, C, Richard C et al. (1990) Two genomic species in *Borrelia burgdorferi*. Res Microbiol 141:465-475
Preac-Mursic V, Schierz G, Pfister HW, Einhäupl KM, Wilske B, Weber K (1984a) Isolierung einer Spirochäte aus Liquor cerebrospinalis. Münch Med Wochenschr 126:275-276
Preac-Mursic V, Wilske B, Schierz G, Pfister HW, Einhäupl KM (1984b) Repeated isolation of spirochetes from the cerebrospinal fluid of a patient with meningoradiculitis Bannwarth. Eur J Clin Microbiol 3:564-565
Preac-Mursic V, Wilske B, Herzer P, Schierz G, Bauer M (1985) Acrodermatitis chronica atrophicans – eine Borreliose! Hautarzt 36:691-693
Preac-Mursic V, Wilske B, Schierz G (1986) European *Borrelia burgdorferi* isolated from humans and ticks: culture conditions and antibiotic susceptibility. Zentralbl Bakteriol Hyg (A) 263:112-118
Preac-Mursic V, Wilske B, Schierz G, Holmburger M, Süss E (1987) In vitro and in vivo susceptibility of *Borrelia burgdorferi*. Eur J Clin Microbiol 6:424-426
Putkonen T, Mustakallio KK, Salminen A (1962) Erythema chronicum migrans with meningitis: a rare coincidence of two tick-borne diseases? Dermatologica 125:184-188
Rehse-Küpper B, Ackermann R (1986) Demonstration of locally sythesized Borrelia antibodies in cerebrospinal fluid. Zentralbl Bakteriol Hyg (A) 263:407-411

Rancurel G (1969) Polyradiculonévrités à tiques. Concours Med 91:8876
Riehl G (1913) Arch Dermatol Syph 115:401
Rohmer F, Collard M, Jesel M, Warter JM, Coquillat G, Class JJ (1974) Les meningoradiculites: donnees cliniques, electromyographiques et etiologiques a propos de 36 observations. Limites nosologiques. Rev Neurol 130:415-431
Rusch P (1906) Über idiopathische Hautatrophie und Skerlodermie. Dermatol Z 13:749-774
Ryberg B, Nilsson B, Burgdorfer W, Barbour AG (1983) Antibodies to Lyme-diesease spirochaete in European lymphocytic meningoradiculitis (Bannwarth's syndrome) (letter). Lancet II:519
Sälde H (1946) cited in Hellerström S (1951) Erythema chronicum migrans Afzelius with meningitis. Acta Derm Venereol (Stockh) 31:227-234
Sandbank M, Feuermann EJ (1972) Ultrastructural observation of Rickettsia-like bodies in erythema chronicum migrans. J Cutan Pathol 6:253-264
Schaible UE, Kramer MD, Musteanu C, Zimmer G, Mossmann H, Simon MM (1989) The severe combined immunodeficiency (scid) mouse: a laboratory model for the analysis of Lyme arthritis and carditis. J Exp Med 170:1427-1432
Schaltenbrand G (1949) Chronische aseptische Meningitis. Nervenarzt 20:433-442
Schaltenbrand G (1962) Radikulomyelomeningitis nach Zeckenbiß. Münch Med Wochenschr 104:829-834
Schaltenbrand G (1967) Durch Artropoden übertragene Erkrankung der Haut und des Nervensystems. Verhandl Dtsch Ges Inn Med 72:975-1006
Schirduan M (1950) Verdacht auf Toxoplasmose bei Meningitis nach Erythema chronicum migrans bullosum. Arch Dermatol Syph 192:256-260
Schmoeckel C, Burg G, Wolff HH, Braun-Falco O (1977) The ultrastructure of lymphadenosis benigna cutis (pseudolymphoma cutis). Arch Dermatol Res 258:161-167
Schmutzhard E, Willeit J, Gerstenbrand F (1986) Meningopolyneuritis Bannwarth with focal nodular myositis. Klin Wochenschr 64:1204-1208
Sedlacek V (1960) Erythema chronicum migrans – poznamky ke klinice, etiopatogenezi a zarazeni. Cs Dermatol 35:386-399
Sköldenberg B, Stiernstedt G, Garde A, Kolmodin G, Carlström A, Nord CE (1983) Chronic meningitis caused by a penicillin-sensitive microorganism? Lancet II:75-78
Sonck CE (1965) Erythema chronicum migrans with multiple lesions. Acta Derm Venereol (Stockh) 45:34-36
Stadelmann R (1934) Ein Beitrag zum Krankheitsbild des Erythema chronicum migrans Lipschütz. Dissertation, University of Marburg
Stanek G, Wewalka G, Groh V, Neumann R (1985a) Isolation of spirochetes from the skin of patients with erythema chronicum migrans in Austria. Zentralbl Bakteriol Hyg (A) 260:88-90
Stanek G, Wewalka G, Groh V, Neumann R, Kristoferitsch W (1985b) Differences between Lyme disease and European arthropod-borne borrelia infections (letter). Lancet I:401
Steere AC (1989) Lyme disease. N Engl J Med 321:586-596
Steere AC, Malawista SE, Hardin JA, Ruddy S, Askenase PW, Andiman WA (1977) Erythema chronicum migrans and Lyme arthritis: the enlarging clinical spectrum. Ann Intern Med 86:685-698
Steere AC, Bartenhagen NH, Craft JE et al. (1986) Clinical manifestations of Lyme disease. Zentralbl Bakteriol Hyg (A) 263:201-205
Stiernstedt GT, Sköldenberg B, Vandvik B, Hederstedt B, Garde A, Kolmodin G, Jörbeck H, Svenungsson B (1984) Chronic meningitis and Lyme disease in Sweden. Yale J Biol Med 57:491-497
Stiernstedt G, Sköldenberg B, Garde A, Kolmodin G, Jörbeck, H, Svenungsson B, Carlström A (1986) Clinical manifestations of *Borrelia* infections of the nervous system. Zentralbl Bakteriol Hyg (A) 263:289-296
Strandberg J (1920) Regarding an unusual form of migratory erythema caused by tick bites. Acta Derm Venereol (Stockh) 1:422-427
Svartz N (1946) Penicillinbehandling vid dermatitis atrophicans Herxheimer. Nord Med 32:2783
Sweitzer SE, Laymon CW (1935) Acrodermatitis chronica atrophicans. Arch Dermatol Syph 31:196-211

Thöne AW (1968) Ixodes ricinus und erythema chronicum migrans (Afzelius). Dermatologica 136:57-60

Thyresson N (1949) The penicillin treatment of acrodermatitis chronica atrophicans (Herxheimer). Acta Derm Venereol (Stockh) 29:572-621

Tjernlund U, Scheynius A, Asbrink E, Hovmark A (1986) Expression of HLA-DQ antigens on keratinocytes in *Borrelia* spirochete-induced skin lesions. Scand J Immunol 23:383-388

Touton K (1886) Fall von erworbener idiopathischer Hautatrophie. Dtsch Med Wochenschr 12:6

Vallat JM, Leboutet MJ, Loubet A, Dumas P, Hugon J, Corvisier N (1984) Tick bite neuropathy: an analysis of nerve biopsies from seven cases. Neurology 34 (Suppl 1): 180-181

Weber K (1974) Erythema-chronicum-migrans-Meningitis – eine bakterielle Infektionskrankheit? Munch Med Wochenschr 116:1993-1998

Weber K (1981) Treatment of Lyme disease. Ann Intern Med 94:137 (letter)

Weber K (1984) Jarisch-Herxheimer-Reaktion bei Erythema-migrans-Krankheit. Hautarzt 35: 588-590

Weber K, Braun-Falco O (1974) Blutbildveränderungen bei Erythema chronicum migrans. Hautarzt 25:611-613

Weber K, Neubert U (1986) Clinical features of early erythema migrans disease and related disorders. Zentralbl Bakteriol Hyg (A) 263:209-228

Weber K, Puzik A, Becker T (1983) Erythema-migrans-Krankheit: Beitrag zur Klinik und Beziehung zur Lyme-Krankheit. Dtsch Med Wochenschr 108:1182-1190

Weber K, Schierz G, Wilske B, Preac-Mursic V, Burgdorfer W, Barbour AG (1983/1984) Antibodies against *Ixodes dammini* and *Ixodes ricinus* spirochetes in tick-borne disorders. 11th ADF meeting, Kiel, 12 Nov 1983 (abstract). Arch Dermatol Res 276:260

Weber K, Schierz G, Wilske B, Preac-Mursic V (1984a) European erythema migrans disease and related disorders. Yale J Biol Med 57:463–471

Weber K, Schierz G, Wilske B, Preac-Mursic V (1984b) Zur Klinik und Ätiologie der Acrodermatitis chronica atrophicans. Hautarzt 35:571-577

Weber K, Schierz G, Wilske B, Preac-Mursic V (1985) Das Lymphozytom – eine Borreliose? Z Hautkr 60:1585-1598

Weber K, Neubert U, Thurmayr R (1986a) Antibiotic therapy in early erythema migrans disease and related disorders. Zentralbl Bakteriol Hyg (A) 263:377-388

Weber K, Schierz G, Wilske B, Neubert U, Krampitz HE, Barbour AG, Burgdorfer W (1986b) Reinfection in erythema migrans disease. Infection 14:32-35; see also Zentralbl Hautkr (1985) 150:628-629

Weber K, Preac-Mursic V, Neubert U et al. (1988a) Antibiotic therapy of early European Lyme borreliosis and acrodermatitis chronica atrophicans. Ann NY Acad Sci 539:324-345

Weber K, Bratzke HJ, Neubert U, Wilske B, Duray PH (1988b) *Borrelia burgdorferi* in a newborn despite oral penicillin for Lyme borreliosis during pregnancy. Pediatr Infect Dis 7:286-289

Weber K, Preac-Mursic V, Wilske B, Thurmayr R, Neubert U, Scherwitz C (1990) A randomized trial of ceftriaxone versus oral penicillin for the treatment of early European Lyme borreliosis. Infection 18:91-96

Wenig C (1975) Chronische lymphocytäre Meningitis mit polytoper Beteiligung des Nervensystems (nach Zeckenbiß?). Psychiatr. Neurol Med Psychol 27:592-600

Wilske B, Schierz G, Preac-Mursic V, Weber K, Pfister HW, Einhäupl KM (1984) Serological diagnosis of erythema migrans disease and related disorders. Infection 12:331-337

Wilske B, Preac-Mursic V, Schierz G (1985a) Antigenic heterogeneity of European *Borrelia burgdorferi* strains isolated from patients and ticks (letter). Lancet I:1099

Wilske B, Münchhoff P, Schierz G, Preac-Mursic V, Roggendorf M, Zoulek G (1985b) Zur Epidemiologie der *Borrelia burgdorferi* Infektion bei Waldarbeitern aus Oberbayern. Munch Med Wochenschr 127:171-172

Wilske B, Preac-Mursic V, Schierz G, von Busch K (1986a) Immunochemical and immunological analysis of European *Borrelia burgdorferi* strains. Zentralbl Bakteriol Hyg (A) 263:92-102

Wilske B, Schierz G, Preac-Mursic V, von Busch K, Kühbeck R, Pfister HW, Einhäupl K (1986b) Intrathecal production of specific anbibodies against *Borrelia burgdorferi* in patients with lymphocytic meningoradiculitis (Bannwarth's syndrome). J Infect Dis 153:304-314

Wilske B, Preac-Mursic V, Schierz G, Gueye W, Herzer P, Weber K (1988a) Immunochemische Analyse der Immunantwort bei Spätmanifestationen der Lyme Borreliose. Zentralbl Bakteriol Hyg (A) 267:549-558
Wilske B, Preac-Mursic V, Schierz G, Kühbeck R, Barbour AG, Kramer M (1988b) Antigenic variability of *Borrelia burgdorferi*. Ann NY Acad Sci 539:126-143
Wolf G (1970) Über die chronische lymphozytäre Meningitis unter dem Bilde der Polyneuritis (Bannwarth). Fortschr Neurol Psychiatr 38:221-234
Zellweger H (1946) Über die chronische allergische Meningitis. Helv Paediatr Acta 1:417-426

2 The Historical Road to the Discovery of *Borrelia burgdorferi*

W. BURGDORFER

This chapter is a historical account of events that eventually led to the discovery of the spirochetal etiology of the disease complex now known as Lyme spirochetosis or Lyme borreliosis. It is written to credit those whose names are rarely mentioned in today's literature, despite the fact that they significantly contributed to one of the most exciting detective stories in microbiology. The earliest reference to a disease manifestation of Lyme borreliosis dates back to 1883 when the German physician Dr. Alfred Buchwald described a cutaneous syndrome as *diffuse idiopathic skin atrophy*. In 1902, Herxheimer and Hartmann introduced the name, acrodermatitis chronica atrophicans (ACA). This condition occurs commonly in Europe but rarely in the United States.

In October 1909, the Swedish physician Dr. Arvid Afzelius, demonstrated to the Swedish Dermatological Society in Stockholm a patient with a ringlike lesion measuring 0.5–2.0 cm that slowly migrated by expansion of its peripheral borders and was accompanied by central clearing (Afzelius, 1910). This lesion was thought to have developed as a result of a tick bite (*Ixodes ricinus*) and was referred to as erythema migrans. Afzelius did not publish his observation until 1921 when he reported on a patient seen in 1908 and on four other patients he had seen since. In 1913, Dr. Lipschütz, from the Dermatology Department of the Hospital Wieden in Vienna, Austria, described, in a 29-year-old woman, a similar lesion that lasted for 7 months. Although Lipschütz did not acknowledge Afzelius, he named the lesion erythema chronicum migrans (ECM).

By the beginning of the 1920s, these and other physicians had seen several cases of ECM. Since the etiology was still unknown, Afzelius, whose cases were all associated with tick bite, speculated that the infection could have been produced either by a virus transmitted by the ticks or by a toxic substance contaminating the ticks. Lipschütz (1923) who gives no reference to tick bites, speculated that ECM might be due to a pathogen. A toxin appeared unlikely because of the affliction's long duration. "Perhaps," he stated "we are dealing with a skin infection caused through the bite of a tick. Therefore, attention should be directed towards microscopic/bacteriologic investigations of the intestinal tract and of the salivary gland secretions of the tick." No one followed his suggestion.

The first evidence of nervous system involvement associated with erythema migrans was reported in 1922 by the French neurologists Garin and Bujadoux. They reported a typical, painful meningoradiculitis after a bite by the ixodid tick *I. hexagonus* – a bite that resulted in a reddish skin lesion. It should be noted, however, that Garin and Bujadoux diagnosed their case as *tick paralysis*, and that it was not until 1930 that Hellerström, who was the first to describe a case of ECM with meningitis, recognized it as ECM with nervous system involvement. Of interest is also the conclusion by Garin and Bujadoux: "If this illness is due to a virus (probably a spirochete), as is suspected, our observation indicates that this virus should be investigated in the nervous system and not in the blood, etc".

In 1936, Dr. Askani from the Dermatology Clinic of the University of Heidelberg, Germany, discussed two additional cases of ECM and pointed out that about three times as many cases were attributed to tick bite than to other causes. He reviewed the significance of ticks as vectors of human and animal pathogens and concluded that ECM was tick-transmitted and caused by an agent from the tick's salivary glands, although it remained to be determined wheter this agent was a toxin or a living pathogen. The same conclusions were reached by Gelbjerg-Hansen, who in 1945 reported three additional ECM cases; one of these patients also had lymphocytic meningitis.

The cause of ECM remained in the dark until 1948, when Dr. Carl Lennhoff from the Dermatological Clinic at the Karolinska Institute in Stockholm presented a report entitled *Spirochetes in Aetiologically Obscure Diseases*. In this publication, Lennhoff claimed to have developed a staining technique that permitted him to demonstrate elements representing the morphological aspects of spirochetes in 27 diseases including psoriasis, zoster, varicella, rubella, lymphadenosis benigna cutis, various erythemas and lymphogranulomas, endocarditis, leukemia, and erythema migrans. The staining method used by Lennhoff was based on Ehrlich's theory that certain antibiotics specifically interact with a causative agent, and that this interaction could be made visible with certain chemicals producing deposits on pathogens such as spirochetes or plasmodia. Lennhoff used mercury chloride interaction with spirochetes – made visible by ammonium sulfide – resulting in a deposit of mercury sulfide. Lennhoff concentrated his efforts on demonstrating spirochetes as the cause of psoriasis, and found, in 72 of 80 German patients and in 1 of 52 Swedish patients, typical spirochetes, which in 95% of all cases could also be demonstrated by dark-field microscopy.

Lennhoff's paper caused excitement among dermatologists, because never before had it been so easy to demonstrate spirochetes – especially not in so many distinct clinical entities with unknown etiology. One of several physicians who questioned Lennhoff's findings was Dr. Schuppli from the Dermatological Clinic of the State Hospital in Basel, Switzerland. He went to Lennhoff's laboratory in Stockholm to acquaint himself with the staining method and to verify the spirochetal etiology of psoriasis. He was successful in staining spirochetes in sera, blood, and cultures, although fine structures of the organisms were never detectable because of heavy mercury sulfide precipitations. He also found the technique to be nonspecific, i.e., other microorganisms and fungi stained as well.

Of 50 psoriasis cases, one showed a single spirochete under dark-field illumination. In thick drops, on the other hand, spirochetal elements similar to those illustrated by Lennhoff could be seen in 12 of 50 cases. Whether these were spirochetes or spirochete-like fibers could not be determined. The same was true for spirochete-like elements in histological sections. Schuppli also examined dark-field preparations from glands of lymphatic leukemia patients and in three of eight found spirochetes of typical morphology. Organisms could also be found in tissue imprints, although they could not be differentiated from tissue fibers. Schuppli's conclusion (Schuppli 1950) was that Lennhoff's staining method might be well-suited for the demonstration of spirochetes in cultures, etc, but was not specific, since tissue fibers, mycelium of fungi, and other microbial agents stained as well. Needless to say, Lennhoff's staining method did not find wide approval and has not been found to be a useful technique in demonstrating spirochetes in erythema migrans lesions.

In 1949, Dr. Hellerström from the Karolinska Institute in Stockholm, presented a talk entitled *ECM Afzelius with Meningitis* at the 43rd annual meeting of the Southern Medical Association in Cincinnati, Ohio, USA (Hellerström 1950). He not only reviewed a number of cases in which, after tick bite, both erythema and subsequent meningial symptoms occurred – the first such case was described by him in 1930, –

but also reported on the successful administration of penicillin in the treatment of ECM by his colleague, Dr. Hollström and presented his thoughts on the still unsolved etiology of ECM.

"Personally ... " he wrote "... I feel inclined towards interpreting erythema chronicum migrans Afzelius with or without meningitis as being the manifestation of an infective agent with allergizing properties, which is transmitted by ticks, and possibly other insects ... Circular erythemas ... subsiding in the center and advancing peripherally with obviously local phenomena of allergization (and immunization?) as well as meningoencephalitic processes occur not infrequently in association with diseases due to spirochetes or other infections where the causal organism is present in *loco morbi* both in the skin and nervous system ... (It) seems reasonable to raise the question of whether the ticks are carriers of spirochetes with allergizing (and immunizing) properties." In the summary, he stated, "Our investigations tend to demonstrate that a spirochete is the cause of the disease. Definite evidence is still lacking, but Lennhoff's findings of spirochetes in histologic sections prepared from lesions of erythema migrans are remarkable" (Hellerström 1950).

The first indisputable evidence that ECM is caused by an infectious agent susceptible to penicillin was provided by Binder and coworkers a few years later in 1955. They successfully transplanted pieces of skin taken from the peripheral zone of a lesion of an ECM patient into three volunteers who developed typical erythema migrans lesions within 1-3 weeks – in fact, a subpassage to a fourth person and among the volunteers was equally successful. Treatment with penicillin led in each case to the prompt disappearance of the lesions within a few days.

In 1962, Schaltenbrand described the case histories of four patients who, following tick bites, developed redness and swelling of their skin around the bite regions and subsequently a nervous system disorder he referred to as radiculomyelomeningitis. In one patient, there was a bilateral Bell's palsy, and in another a quadraplegia with respiratory and sensitivity disorders and paralysis of the pyramidal tract. Tick-borne encephalitis was suspected as the causative virus of illness. After studying additional cases, Schaltenbrand in 1967 showed that the nervous system disorders associated with ECM were not identical to those of the Central European tick-borne encephalomyelitis, and he assumed that ECM meningitis was caused by an as yet unidentified viral agent, possibly related to, if not identical with a virus (later determined to be related to the virus of tick-borne encephalitis) that had been recovered by his coworker W.K. Müller from *I. ricinus*.

In 1974, Dr. Klaus Weber presented a clinical report of a female patient with ECM meningitis that could be successfully treated with antibiotics (penicillin and chloramphenicol). Based on his and on two additional similar reports in the literature, Weber suggested that ECM meningitis was a bacterial and not a viral infectious disease. In discussing the etiology, he ruled out spirochetes of the genus *Borrelia* because "(these spirochetes) are transmitted by argasid and not ixodid ticks (*I. ricinus*), and in addition, spirochetes could not be demonstrated in the skin" (Hard 1966). His remarks were taken from a review article by Felsenfeld (1965) in which it is stated: "Ticks propagating human relapsing fever belong to the Argasidae never to the Ixodidae." Weber also referred to the negative findings by Hard (1966), who evaluated Lennhoff's claim (see p. 22) and allowed larval, nymphal, and adult *I. ricinus* to feed on the erythematous ring of an ECM lesion. Subsequent dark-field examination of tick tissues revealed no spirochetes and equally negative were transmission experiments with nymphal ticks that had been allowed to feed as larvae on the skin of a patient with ECM.

A rickettsial etiology had been suggested as early as 1962, when based on Giroud's slide agglutination test, Dègos et al. reported six of seven ECM patients with antibodies ($\geq$ 1:160) against the epidemic typhus agent *Rickettsia prowazekii*, the murine typhus rickettsia *R. mooseri*, and/or boutonneuse fever agent *R. conorii*. Although these findings could not be confirmed by other investigators, the rickettsial etiology of ECM was kept alive by the electron microscopic demonstration of rickettsia-like microorganisms in the macrophages of two ECM patients (Sandbank and Feuermann 1979). This induced Weber (1981) to have the sera of 13 ECM patients evaluated against 14 rickettsial antigens by two different laboratories, namely the Institute of Virology in Bratislava, Czechoslovakia, where the complement fixation (CF) and microagglutination tests were used, and the Rocky Mountain Laboratories in Hamilton, Montana, where, in addition to the CF test, the microimmunofluorescence (MIF) test was performed. With the exception of a few nondiagnostic low titers against the rickettsialpox agent *R. akari* and against *Coxiella burnetii*, the cause of Q fever, all tests were negative.

In the United States, the first case of ECM was not observed until 1969, when it occurred in a physician who had been bitten by a tick while hunting grouse in north central Wisconsin. The attending physician, Dr. Rudolf Scrimenti, familiar with the ECM lesions described in Europe, accurately diagnosed the case and treated it effectively with penicillin (Scrimenti 1970). A cluster of four additional cases of ECM were recorded in midsummer 1975 in southeastern Connecticut (Mast and Burrows 1976).

According to Mrs. Polly Murray who had lived in the community of Lyme, Connecticut, USA, since 1959, diseases of unknown origin had occurred there since the mid-1960s. She and Mrs. Judith Mensch of Old Lyme, Connecticut did not accept the medical diagnosis of rheumatoid arthritis, and they made known to the state health department their suspicions that something unusual was occurring. It was not until 1975, however, that a research team headed by Dr. Allen Steere of Yale's Rheumatology Department began its retrospective investigations (Steere et al. 1977). Of the 51 residents, 39 children and 12 adults, who had developed attacks of asymmetric swelling and pain in large joints, particularly the knee, 13 had expanding lesions thought to fit the description of European ECM. However, since arthritis, with few exceptions, had not been reported to be associated with ECM, Steere and his associates believed that the patients had been suffering from a previously unrecognized clinical entity and named it Lyme arthritis at first, and later Lyme disease (Steere and Malawista 1979).

Because arthritis is known to occur in certain viral infections such as rubella, hepatitis B, mumps and influenza and often is found in mosquito-borne group A arboviral infections, the search for a causative agent concentrated on the serologic evaluation of acute and convalescent sera of patients for antibodies against adenoviruses, coxsackieviruses, cytomegaloviruses, hepatitis B, herpes simplex, influenza, lymphocytic choriomeningitis, mumps and rubella viruses, and 216 arboviruses, 38 of which were tick-borne. All tests were essentially negative, as were serological tests against four rickettsial agents, i.e. *R. akari, R. rickettsii, R. mooseri* and *C. burnetii* (Steere et al. 1977). Also negative for viruses were 62 pools of ticks collected from vegetation in communities east and west of the Connecticut River, where the incidence of Lyme disease was high and low, respectively. In that study, 622 *I. dammini* – first considered to be *I. scapularis*, but later described as a new species – were evaluated (Wallis et al. 1978).

Also unsolved for almost 100 years, remained the etiology of ACA, which had been considered a distinct clinical entity although many patients remembered having been bitten by a tick (*I. ricinus*), and some reported skin lesions similar to erythema migrans (Hauser 1955; Kristoferitsch 1989). ACA develops months to years after initial infection as a primary chronic disease, or after the initial and secondary stages of Lyme borreliosis. Regarding the cause of ACA, there have been two major hypotheses: (a) the disease is caused by an infectious agent, and (b) it may be caused as a result of hormonal-vegetative defects initiated by traumatic and hypothermal factors.

An important observation was made in 1942 by R.H. Kahle, at the Dermatological Clinic of the Martin Luther University in Halle, Wittenberg, Germany. In his thesis, directed towards evaluation of the Pallida reaction, Kahle noted that six of seven patients with ACA reacted positively against *Treponema pallidum*. Dr. Grüneberg from the same institute confirmed this 10 years later and speculated that the reactions must have been due to a group-specific spirochete as the causative agent of ACA (Grüneberg 1952). In the meantime, by 1946, penicillin had been found to be very effective in the treatment of ACA, and all research efforts were now directed towards the identification of a "living" agent.

In 1954 and 1955, Dr. Hans Götz, from the Dermatological Clinic at the University of Munich, Germany, was the first to show that ACA is indeed an infectious disease. He successfully transplanted tissues of affected skin from an ACA patient into four volunteers (including himself) and recorded a clinical picture similar to that of ACA, i.e., chronic inflammation that lasted for as long as 312 days and disappeared only after treatment with penicillin. Götz speculated that a large virus may be the cause of the disease but, 2 years later, he found no evidence of a viral etiology and concluded that the cause of ACA continued to be an open question (Götz and Nasemann 1956). Successful transfer of spirochetes from the blood of ACA patients into mice was claimed by Lohel (1955), who, like Kahle and Grüneberg, used the Pallida reaction to determine the production of antibodies against spirochetes. Meinicke (1957) repeated Lohel's experiments 2 years later and, with the exception of a few low titers, was unable to confirm the serological responses in mice; he questioned Lohel's claim of ACA being a spirochetosis.

Thus, at the beginning of the 1980s, the causative agent(s) of ECM and ACA in Europe and of Lyme disease in the United States remained elusive, primarily because no one followed Lipschütz's 1923 recommendation that attention should be directed toward microscopic/bacteriologic investigations of the intestinal tract and of the salivary gland secretions of the tick.

That is exactly what I did in 1981 (Burgdorfer et al. 1982), when Dr. Jorge Benach and I evaluated ticks from Long Island as vectors of the spotted fever agent, *R. rickettsii*. Because we were not successful in detecting this agent in the well established tick vector, the American dog tick *Dermacentor variabilis*, we speculated that other ticks such as *I. dammini*, which occurs abundantly on Long Island and especially on offshore islands, might play an up to that time unestablished role in the ecology of spotted fever. Consequently, several hundred *I. dammini* were collected by flagging and were examined for rickettsiae by the hemolymph test; none were infected with rickettsiae. In October 1981, Benach provided additional collections of *I. dammini* from Shelter Island, New York, where Lyme disease was known to be endemic. Although none of 44 males and females had rickettsiae, two females contained microfilariae similar to *Dipetalonema (Wehrdickmannia) rugosicauda*, a deer microfilaria associated with *I. ricinus* in Europe (Aeschlimann et al. 1979). It was during the search of additional microfilariae in the tissues of these two ticks that I encountered in smears of midgut diverticula the poorly stained, individual or clumped, rather long, irregularly coiled spirochetes. Of 124 additional ticks examined similarly, 75 (60%) contained spirochetes that were limited in distribution to the midgut and, by dark-field microscopy, revealed a sluggish and slow movement. The triturates of four infected ticks were then given to Dr. Alan Barbour, a molecular biologist and physician at the Rocky Mountain Laboratories, who inoculated suspensions into a modified Kelly medium. After 5 days of incubation at 35°C, all cultures contained spirochetes that could be regularly subcultured.

The antigenic relatedness of the *I. dammini* spirochete to the etiologic agent of Lyme disease was first established by indirect immunofluorescence with the sera of several Lyme disease patients seen by Dr. Edgar Grunwaldt, a practicing physician on Shelter Island, New York. Spirochetes in midgut smears of infected ticks served as antigens.

The presence of similar if not identical spirochetes in *I. ricinus*, the incriminated vector of ECM in Europe, was first established with smears of ticks that had been collected during a tick/rickettsial survey on the Swiss Plateau in 1978 (Burgdorfer et al. 1979). The slides had been stored as reference material and had to be stained by Giemsa for reexamination. The smears of 13 (17%) of 135 ticks contained spirochetes that morphologically appeared similar to those discovered in *I. dammini*. In the spring of 1982, these findings could be confirmed with live ticks collected near Neuchâtel, Switzerland. Of 390 adult ticks, 112 (29%) proved infected (Burgdorfer et al. 1983).

Convinced that ECM in Europe and Lyme disease in the United States are expressions of one and the same etiologic agent, we directed our attention toward the west coast of the United States where the first case of ECM had been contracted in Sonoma County, California in 1975, and where the blacklegged western tick *I. pacificus* had been incriminated as the possible vector. A tick survey in southeastern Oregon and in north/central California yielded 2402 ticks of which 39 contained spirochetes similar if not identical to those discovered in *I. dammini* and *I. ricinus* (Burgdorfer et al. 1985).

At the time that we made our serendipitous discovery, other investigators in the United States as well as abroad were looking for the causative agent. Thus, at the Virology Department of the University Neurological Clinic in Cologne, Germany, Dr. Rudolf Ackermann speculated that successful treatment of ECM and related disorders with penicillin and tetracycline, as well as the prolonged occasionally recurring clinical manifestations could be due to a bacterial agent, possibly a relapsing fever-like spirochete. He evaluated by indirect immunofluorescence the sera of 14 patients with ECM or meningopolyneuritis against the African relapsing fever spirochete, *Borrelia duttonii*, and found significant immunoglobulin M (IgM) and IgG antibody titers in sera and of IgG antibodies in the spinal fluids. In addition, he was able to demonstrate by indirect immunofluorescence spirochetes in *I. ricinus* collected in June 1982 in two disease foci of West Germany. Although Ackermann's findings were not reported until 1983, he was the first to confirm our discovery in the European tick vector *I. ricinus*, of the spirochete now known als *B. burgdorferi* (Johnson et al. 1984).

Prompt isolation and identification of the Lyme disease spirochete would not have been possible without availability of Kelly's medium, as modified by Stoenner and coworkers (1982). In 1971, Richard Kelly described the successful primary growth and serial subculture of the tick-borne relapsing fever spirochete *B. hermsii* in an artificial medium that consisted of tryptone, peptone, bovine albumin, N-acetyl glucosamine, pyruvate, and rabbit serum in highly buffered medium thickened with gelatin. For their study on the biology of *B. hermsii*, Stoenner and coworkers (1982) "fortified" Kelly's medium, by adding a yeast extract and concentrated tissue culture mixture (CMRC 1066) containing amino acids, vitamins, nucleotides, and other growth factors. It was this medium that provided the initial isolation of the Lyme disease spirochete from *I. dammini*. Additional modifications were subsequently proposed and led to the description of the Barbour-Stoenner-Kelly II (BSKII) medium (Barbour 1984) now commonly used for the isolation of *B. burgdorferi* from ticks, animal hosts, and patients.

References

Ackermann R (1983) Erythema chronicum migrans und durch Zecken übertragene Meningopolyneuritis (Garin-Bujadoux-Bannwarth): Borrelien Infektionen? Dtsch Med Wochenschr 108:577-580

Aeschlimann A, Burgdorfer W, Matile H, Pèter O, Wyler R (1979) Aspects nouveaux du rôle de vecteur joué par *Ixodes ricinus* L en Suisse. Acta Trop (Basel) 36:181-191

Afzelius A (1910) Verhandlungen der Dermatologischen Gesellschaft zu Stockholm, 28 Oct 1909. Arch Dermatol Syph 101:404

Afzelius A (1921) Erythema chronicum migrans. Acta Derm Venereol 2:120-125

Askani H (1936) Zur Ätiologie des Erythema chronicum migrans. Dermatol Wochenschr 102:125-131

Barbour AG (1984) Isolation and cultivation of Lyme disease spirochetes. Yale J Biol Med 57:521-525

Binder E, Doepfmer R, Hornstein O (1955) Experimentelle Übertragung des Erythema chronicum migrans von Mensch zu Mensch. Hautarzt 6:494-496

Buchwald A (1883) Ein Fall von diffuser idiopathischer Haut Atrophie. Arch Dermatol Syph 10:553-556

Burgdorfer W, Aeschlimann A, Pèter O, Hayes SF, Philip RN (1979) *Ixodes ricinus*: vector of a hitherto undescribed spotted fever group agent in Switzerland. Acta Trop (Basel) 36:357-367

Burgdorfer W, Barbour AG, Hayes SF, Benach JL, Grunwaldt E, Davies JP (1982) Lyme disease – A tick-borne spirochetosis? Science 216:1317-1319

Burgdorfer W, Barbour AG, Hayes SF, Pèter O, Aeschlimann A (1983) Erythema chronicum migrans – a tickborne spirochetosis. Acta Trop (Basel) 40:79-83

Burgdorfer W, Lane RS, Barbour AG, Gresbrink RA, Anderson JR (1985) The western black-legged tick, *Ixodes pacificus:* a vector of *Borrelia burgdorferi.* Am J Trop Med Hyg 34:925-930

Dègos R, Touraine R, Arouete J (1962) L'erythema chronicum migrans. Syph 89:247-260

Felsenfeld O (1965) *Borreliae,* human relapsing fever, and parasite-vector-host relationships. Bacteriol Rev 29:46-74

Garin Ch, Bujadoux (1922) Paralysie par les tiques. J Med Lyon 71:765-767

Gelbjerg-Hansen G (1945) Erythema chronicum migrans Afzelii and meningitis after a tick bite. Acta Derm Venereol (Stockh) 25:458-463

Götz H (1954) Die Acrodermatitis chronica atrophicans Herxheimer als Infektionskrankheit. Hautarzt 5:491-504

Götz H (1955) Die Acrodermatitis chronica atrophicans Herxheimer als Infektionskrankheit – Ergänzung zur I. Mitteilung. Hautarzt 6:249-253

Götz H, Nasemann T (1956) Zur Frage der Virusätiologie der Akrodermatitis chronica atrophicans Herxheimer. Hautarzt 7:349-353

Grüneberg T (1952) Zur Frage der Ätiologie der Acrodermatitis chronica atrophicans. Dermatol Wochenschr 126:1041-1046

Hard S (1966) Erythema chronicum migrans (Afzelli) associated with mosquito bite. Acta Derm Venereol (Stockh) 46:473-476

Hauser W (1955) Zur Kenntnis der Akrodermatitis chronica atrophicans. Arch Dermatol Syph 199:350-393

Hellerström S (1930) Erythema chronicum migrans Afzelii. Acta Derm Venereol (Stockh) 11:315-321

Hellerström S (1950) Erythema chronicum migrans Afzelius with meningitis. South Med J 43:330-334

Johnson RC, Schmid GP, Hyde FW, Steigerwalt AG, Brenner DJ (1984) *Borrelia burgdorferi* sp.nov: etiologic agent of Lyme disease. Int J Syst Bacteriol 34:496-497

Kahle RH (1942) Pallida-Reaktion bei peripheren Durchblutungsstörungen der Haut, insbesondere bei Acrodermatitis atrophicans. Dissertation, University of Halle

Kelly R (1971) Cultivation of *Borrelia hermsii.* Science 173:443-444

Kristoferitsch W (1989) Neuropathien bei Lyme-Borreliose. Springer, Vienna, New York

Lennhoff C (1948) Spriochetes in aetiologically obscure diseases. Acta Derm Venereol (Stockh) 28:295-324

Lipschütz B (1913) Über eine seltene Erythemform (Erythema chronicum migrans). Arch Dermatol Syph 118:349-356
Lipschütz B (1923) Weiterer Beitrag zur Kenntnis der "Erythema chronicum migrans." Arch Dermatol Syph 143:365-374
Lohel H (1955) Tierexperimentelle Untersuchungen zur Ätiologie der Acrodermatitis chronica atrophicans Herxheimer. Klin Wochenschr 33:185-186
Mast WE, Burrows WM (1976) Erythema chronicum migrans in the Unites States. JAMA 236:859-860
Meinicke K (1957) Zur Frage der Spirochätenätiologie der Akrodermatitis chronica atrophicans Herxheimer. Hautarzt 8:128-131
Sandbank M, Feuerman EF (1979) Ultrastructural observation of rickettsia-like bodies in erythema chronicum migrans. J Cutan Pathol 6:253-264
Schaltenbrand G (1962) Radikulomyelomeningitits nach Zeckenbiss. Munch Med Wochenschr 104:829-834
Schaltenbrand G (1967) Durch Arthropoden übertragene Erkrankung der Haut und des Nervensystems. Verh Dtsch Ges Inn Med 72:975-1006
Schuppli R (1950) Zur Frage des Nachweises von Spirochäten bei ätiologisch unabgeklärten Krankheiten. Dermatologica 101:193-212
Scrimenti RJ (1970) Erythema chronicum migrans. Arch Dermatol 102:104-105
Steere AC, Malawista SE (1979) Cases of Lyme disease in the United States: locations correlated with distribution of *Ixodes dammini*. Ann Intern Med 91:730-733
Steere AC, Malawista SE, Snydman DR, Shope RE, Andiman WA, Ross MR, Steele RM (1977) Lyme arthritis. An epidemic of oligoarticular arthritis in children and adults in three Connecticut communities. Arthritis Rheum 20:7-17
Stoenner HG, Dodd T, Larsen C (1982) Antigenic variation of *Borrelia hermsii*. J Exp Med 156:1297-1311
Wallis RC, Brown SE, Kloter KO, Main AJ, Jr (1978) Erythema chronicum migrans and Lyme arthritis: field study of ticks. Am J Epidemiol 108:322-327
Weber K (1974) Erythema chronicum migrans Meningitis- eine bakterielle Infektionskrankheit? Munch Med Wochenschr 116:1993-1998
Weber K (1981) Serological study with rickettsial antigens on erythema chronicum migrans. Dermatologica 163:460-467

II Characteristics of *Borrelia burgdorferi*

3 Ultrastructure of *Borrelia burgdorferi*

S. F. Hayes and W. Burgdorfer

3.1 Introduction

The Lyme borreliosis spirochete *Borrelia burgdorferi* represents the first ixodid tick-associated spirochete pathogenic to humans. When first discovered in and isolated from the midgut tissues of *Ixodes dammini* from Shelter Island, New York (Burgdorfer et al. 1982) and shortly thereafter from *I. ricinus* collected from the Seewald Forest in Switzerland (Burgdorfer et al. 1983; Barbour et al. 1983), this spirochete was described as a treponeme-like organism, irregularly coiled, 10–30 µm in length with a diameter of 0.18–0.25 µm and with as many as eight endoflagella. Early electron microscopy studies showed spirochetes with 6-8 endoflagella; more recent examinations of cultured spirochetes, however, have revealed spirochetes possessing as many as 8–14 (Ackermann et al. 1984; Anderson et al. 1989; Hulinská et al. 1989). Immunochemical and molecular investigations have identified this microorganism as a hitherto undescribed member of the genus *Borrelia* (Johnson et al. 1984a, b; Schmid et al. 1984; Barbour et al. 1986). Although ultrastructural analysis of *B. burgdorferi* has been the subject of several investigations in the United States and also in Europe, descriptions vary. Differences in morphological criteria, such as end shape and the number of endoflagella, among isolates of diverse origins have led to the speculation that additional spirochetal species may be involved in the etiology of Lyme borreliosis (Hovind-Hougen et al. 1987; Karlsson et al. 1990). The present chapter summarizes our observations on the ultrastructure of *B. burgdorferi*, based upon investigations of multiple tick and human isolates from the United States as well as from Europe.

3.2 Material and Methods

3.2.1 Isolates of *B. burgdorferi*

The isolates of *B. burgdorferi* studied included (a) type strains B31 and IRS, recovered initially from *I. dammini* from Shelter Island, New York in 1981 and from *I. ricinius* from Seewald, Switzerland in 1982, respectively (Burgdorfer et al. 1982, 1983; Barbour et al. 1983), (b) the Sh-2-82 isolate from *I. dammini* from Shelter Island, New York, 1982 and (c) six isolates from Lyme borreliosis patients: two from cerebrolspinal fluid, two from erythema migrans, and two from acrodermatitis chronica atrophicans lesions. All human isolates were provided by courtesy of Dr. Vera Preac-Mursic of the Pettenkofer Institute, Munich, Germany. They were fixed and processed for transmission electron microscopy (TEM) as described by Hayes and Burgdorfer (1979). For negative staining, we initially followed the methods of Hovind-Hougen and Birch-Andersen (1971). They subjected some samples to treatment with either Teepol or 1% deoxycholate or both in distilled water, or AL-1 enzyme from *myxo-*

bacter, to disrupt them. Treatment times varied from 30 s to 4 min. Other samples were subjected to fixation with low concentrations of glutaraldehyde prior to staining. The samples were then stained with 1% ammonium molybdate, at pH 7. The same procedures were later followed by Hovind-Hougen (1976) in her monograph proposing taxonomic criteria for differentiating borreliae from treponemes.

Pertinent to the present study of *B. burgdorferi* are the following modifications of the methods outlined by Hovind-Hougen:

1. We did not use Teepol or any enzymes for disruption of spirochetes.
2. For negative staining, the spirochetes were centrifuged from Barbour-Stoenner-Kelly (BSK)-II media at 3000 rpm using a Tomy MTX-150 table model, refrigerated centrifuge equipped with a swinging bucket rotor, for 20 min at 20°C to keep the S-layer intact. Whenever the S-layer was not desired, greater speeds and shorter times were used.
3. To avoid mechanical removal of the S-layer, resuspension was gently done in Tyrode's, Tyrode's/Na cacodylate/$CaCl_2$, or in stabilizing buffer as described by Robinson et al. (1987). Tyrode's buffer represents a modified Ringer's solution and has been commonly used in studies involving blood cells that are highly susceptible to osmotic and physiologic influences (Cutts 1970). This buffer is prepared by adding the salts in the order given on a gram per liter basis: NaCl 8.0g, KCI 0.2g, $CaCl_2$ 0.2g, $MgCl_2$ 0.1g, $NaH_2HPO_4.H_2O$ 0.05g, $NaHCO_3$ 1.0g, glucose 1.0g and DH_2O to 1 l volume. Sodium cacodylate/$CaCl_2$/sucrose was made as a 0.1 *M* Na cacodylate solution containing 20 m*M* $CaCl_2$ and sucrose to 0.1 M in DH_2O.

Thin-section TEM was according to methods previously published (Wagner 1976; Hayes and Burgdorfer 1979). Cryosectioning was according to Hayes et al. (1990), and freeze fracture was performed as outlined by Swanson and Zeligs (1974) and Todd et al. (1984). Scanning electron microscopy (SEM) methods were as previously noted (Hitchcock et al. 1985).

Spirochetal membrane blebs were removed from supernates according to protocols previously described by Dorward and Judd (1988) and Dorward and Garon (1990). Endoflagella bundles were released from spirochetes by treatment of borreliae adsorbed to a parlodion film with 0.00001% sodium dodecyl sulfate (SDS) for 30 seconds, followed by washing with distilled H_2O and then by staining with 0.5% NH_4Mo for 30 seconds.

3.3 Results

3.3.1 General Description

The Lyme borreliosis spirochete *B. burgdorferi* measures from 10 to 30 µm in length while maintaining a width of 0.2–0.25 µm. Its end-shape profile varies from blunt to pointed. The general morphology of *B. burgdorferi* is illustrated in Fig.1a. It possesses a surface or S-layer that shows no substructural pattern, a trilaminar limiting outer membrane surrounding the periplasmic space that contains varying numbers of endoflagella, and an innermost compartment, the protoplasmic cylinder. These morphological components are described in the following:

3.3.2 Surface or S-Layer

The S-layer is in immediate apposition to the outer leaflet of the outer membrane, and does not appear to have any substructural pattern (Fig.1). It is easily lost in the pres-

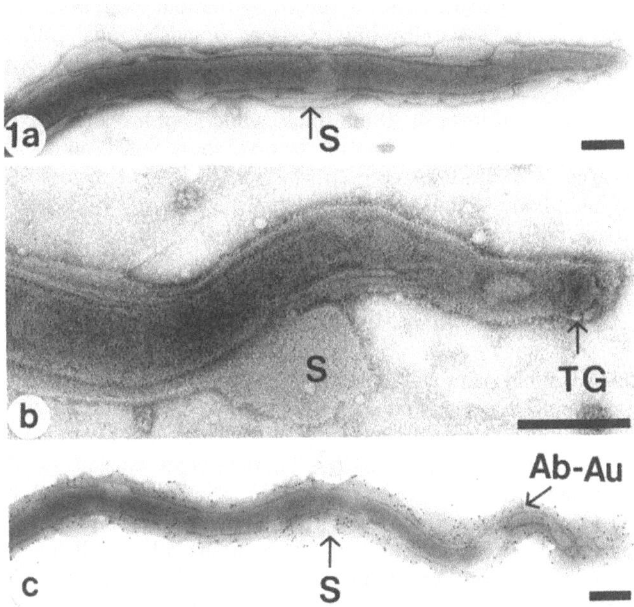

Fig. 1a-c. Negatively stained profiles of *B. burgdorferi*, (Sh-2-82, and B31). **a** Sh-2-82 gently resuspended in Tyrodes's buffer, adsorbed to grids, interacted with a non-specific antibody gold complex (*Ab-Au*), washed with stabilizing buffer, and stained with 0.5% ammonium molybdate.
b Prototype B31 spirochete washed one time with phosphate buffer solution (PBS) $MgCl_2$, adsorbed and stained with 0.5% phosphotungstic acid, TG terminal granule.
c Sh-2-82 spirochete from Tyrode's as in *a* but reacted with a monovalent polyclonal rabbit antisera complexed with 10 nm Au against the 83-kDa protein described by Dorward et al. (1990). In all planels, S denotes the surface or S-layer. Bars = 0.25 µm.

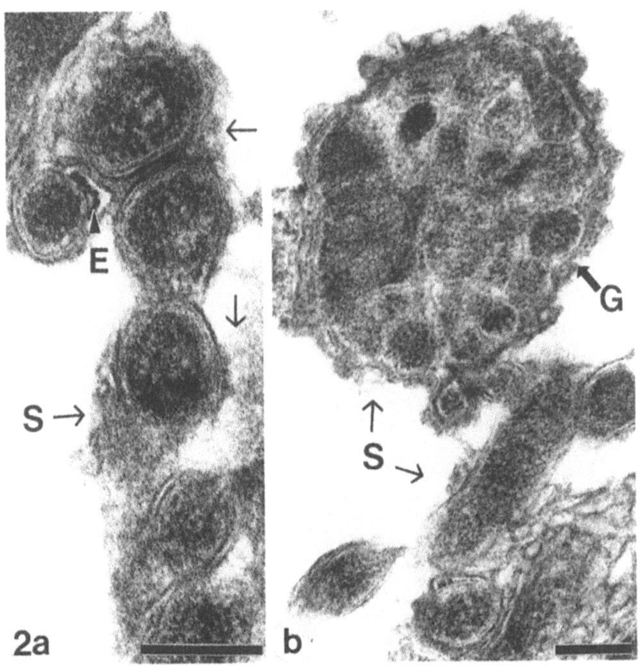

Fig. 2a,b. *B. burgdorferi* from isolates B31 **a** and IRS **b** reacted with ruthenium red included in the glutaraldehyde fixative at 0.1 mg/ml. Electron density at S denoted by arrows, show interaction with the S-layer; *E* endoflagella; *G* gemma (broad arrow) with internalized gemma granules. Bars = 0.25 µm.

ence of aqueous solutions. As shown by Barbour et al. (1983) and Barbour and Hayes (1986), it can be stabilized by exposure to anti-borrelial antibody or by fixation with formaldehyde and then labeled with specific gold-antibody complexes to various surface antigens (Fig. 1c). After negative staining, the S-layer appears as a homogeneous electron lucent layer 4–30 nm thick, while samples prestained with ruthenium red or alcian blue dyes at the time of fixation, show this layer as an electron dense fuzzy material 2–10 nm thick (Fig.2). Most of the S-layer material remains in contact with the main body of the cell if gently resuspended in Tyrode's or Na cacodylate/$CaCl_2$/-sucrose buffer or a combination of these (Fig. 1a,c).

3.3.3 Trilaminar Outer Membrane

The trilaminar outer membrane of *B. burgdorferi* is highly flexible, similar to the limiting cytoplasmic membrane of eukaryotic cells. Thin-section TEM shows this membrane to possess an electron dense outer leaflet and a less electron dense inner leaflet (Fig. 3). After negative staining, this membrane appears as an electron lucent single contoured membrane (refer to Fig. 1b). The spirochetal end shapes produced by the course of this membrane vary from blunt to pointed (Fig. 1). These variations were consistently observed in all isolates, including B31 and IRS which were cloned several times by limiting dilution and then examined at different periods of growth. Similar results were obtained when negative staining profiles were compared against profiles obtained by TEM of spirochetes in tick tissues or against profiles obtained by SEM (Hayes et al. 1988). This finding differs from that reported by Hovind-Hougen et al. (1987) who described sharply pointed ends in relapsing fever borreliae. It also should

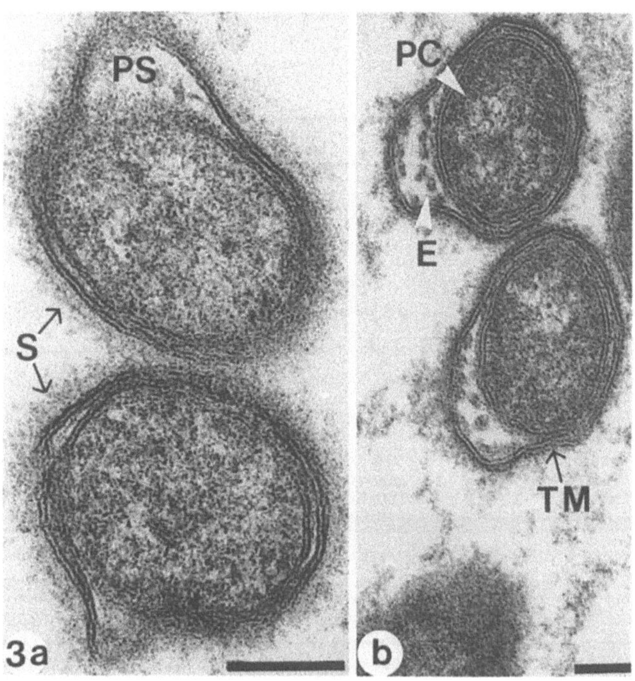

Fig. 3a, b. Transmission electron microscopy cross-section profiles of *B. burgdorferi* (isolate Sh-2-82) utilizing in **a** glutaraldehyde-paraformaldehyde fixative plus ruthenium red in a Na cacodylate/sucrose buffer. In **b** the fixation was according to Hayes and Burgdorfer (1979), and employed the use of tri-nitro-cresol according to Karnovsky (see above citation). Evident in *a* is the S-layer, which is not as demonstrable in *b*, while *b* gives better profiles of the endoflagella E, the trilaminar outer membrane profile (*TM*), and the protoplasmic cylinder (PC). (PS) represents the periplasmic space occupied by the endoflagella. Bars = 0.1 µm be mentioned that recent *B. burgdorferi* isolates from Sweden, one from a tick and one from a human, have double-contoured outer membranes (Stiernstedt et al. 1985; Karlsson et al. 1990).

3.3.4 Periplasmic Space

Between the outer membrane and the periplasmic cylinder is the periplasmic space that contains the endoflagella – a prominent feature of all spirochetes (Figs. 3b, 4a, 4c and 6b). These give rise to the organism's serpentine profile and to its screw-like motility. Each endoflagellum consists of a basal disk, neck, hook, and filament portion (fig. 4b). The endoflagella of *B. burgdorferi*, like those of other borreliae resemble the external flagella of Gram-negative bacteria. At the point of their insertion into the periplasmic membrane, the basal disks at the apical end of the endoflagella become an integral part of the periplasmic cylinder (Fig. 4c). This disk connects distally with a small diameter neck that then fuses with the hook which has a honeycombed appearance and whose distal portion attaches to the endoflagellum's mains shaft. Figure 4c illustrates the orientation of the endoflagella at the terminus of the spirochete. With the S-layer present and *B. burgdorferi* well preserved, assessment of the number of endoflagella

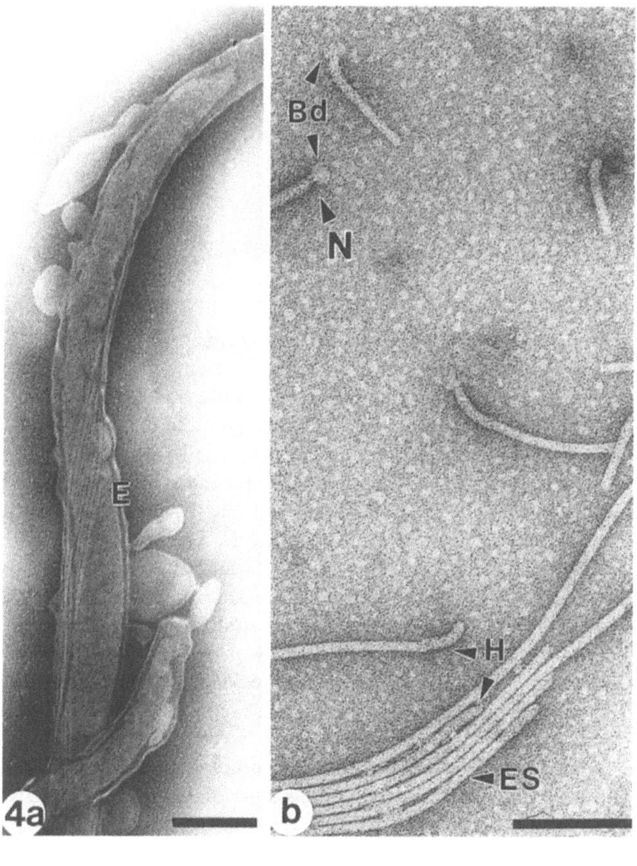

Fig. 4. a, b. The serpentine profile of the spirochete B31 is demonstrated along with resolution of the endoflagellar band (E). Surface blebbing is evident in these phosphate buffer solution (PBS)/$MgCL_2$ washed cells. **b** The described endoflagella. These were purified according to Barbour et al. (1986), and stained with 1.0% uranyl acetate (aqueous pH 3.9); *Bd*, basal disk; *N*, neck, *H*, hook with honeycombed appearance; *ES*, endoflagella shaft.
Bars = 0.25 μm (a), 0.1 μm (b).

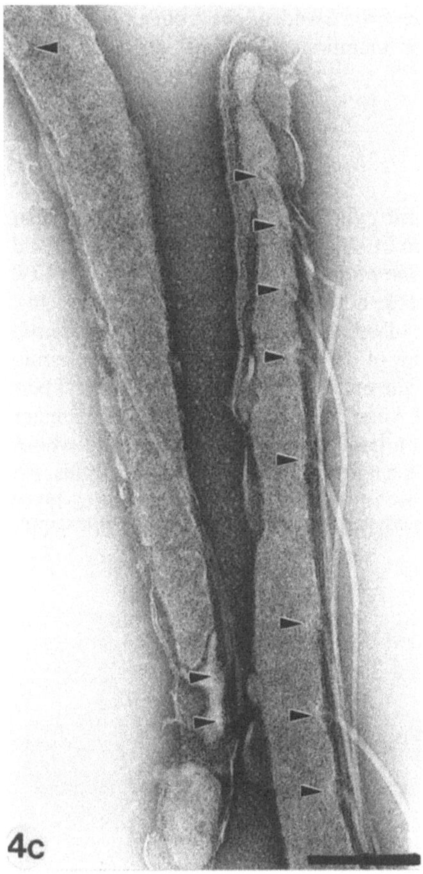

Fig. 4c The termini and the exposed insertion points of two vortexed and washed *B. burgdorferi* (B31) organisms. The left terminus possesses 3 and the right 8 insertion points (*arrow heads*). Bar = 0.25 µm.

and of the insertion points is difficult (Figs. 1, 4a). However, detection of the insertion points and determination of endoflagella numbers are made easier when the S-layer is removed, and the spirochetes become more translucent because of plasmolysis or disruption due to washing, vortexing, and/or rapid centrifugation (Fig. 4a,c). Our findings show endoflagella counts to range from 3 to 18 (Figs. 4c, 5), depending upon the physiological status of the spirochete, and far more importantly upon the region of the spirochete evaluated. This is illustrated in Fig. 5, which shows a released bundle of endoflagella varying from 4 to 18 in number. The release of this bundle was achieved as noted in Sect. 3.2. Asterisks in Fig. 5 denote the spacing of insertion points, with an increase in endoflagella number towards the center of the bundle (from bottom to top), and a decrease towards the opposite end. From this profile it is difficult to determine whether this bundle of endoflagella was derived from a dividing organism. Although the prominent basal disks usually associated with the insertion end of each endoflagellum were removed by the SDS treatment, one remaining basal disk is marked by a circle.

Endoflagella when stained with certain negative stains were found to vary in size and morphological appearance because of varying amounts of intercalation into and surrounding each endoflagellum (Fig. 6a). Size variance was most apparent with the

Fig. 5a-c. The location of insertion points, illustrating an endoflagella bundle of isolate Sh-2-82 that was released by sodium dodecyl sulfate (SDS) treatment. The *large arrows* at the lower left of each part (**a**-**c**) indicate the direction of traverse (*bottom* to *top*) for the bundle. *Asterisks* denote endoflagellar ends with their attendant hook. *Small directional arrows* show which direction the hook opens. The *circle* denotes an endoflagellum with its basal disk still attached. *Broad arrows* denote two possible sites where division may occur. Bars = 0.1 μm.

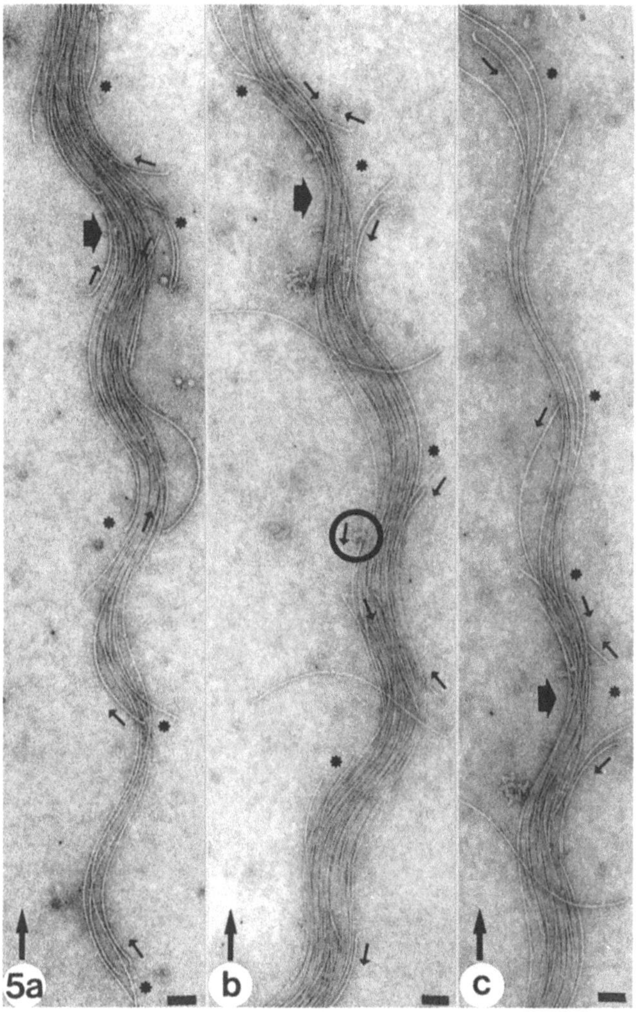

use of phosphotungstic acid, less so with ammonium molybdate or methylamine tungstate, and little with uranyl acetate (aqueous). To avoid these effects, a 0.5% rather than 2% stain concentration was used. Staining with negative stains occasionally revealed the presence of a fibrillar substructure of endoflagella (arrows on Fig. 6a). Application of modified TEM fixation and staining procedures for thin sectioning (Hayes and Burgdorfer 1979) and modifications to cryotomy methods (Hayes et al. 1990) have shown endoflagella to be hollow (Fig. 6b; see also Fig. 3b). They are unsheathed, as demonstrated initially by Hovind-Hougen (1974) in her study of the relapsing fever spirochetes *B. merionisi* and *B. recurrentis* and further described by Johnson (1977), and Holt (1978). Cytoplasmic fibrils characteristic of treponemes were never detected.

Recently, Hayes et al. (1990) described the presence of a granular element located

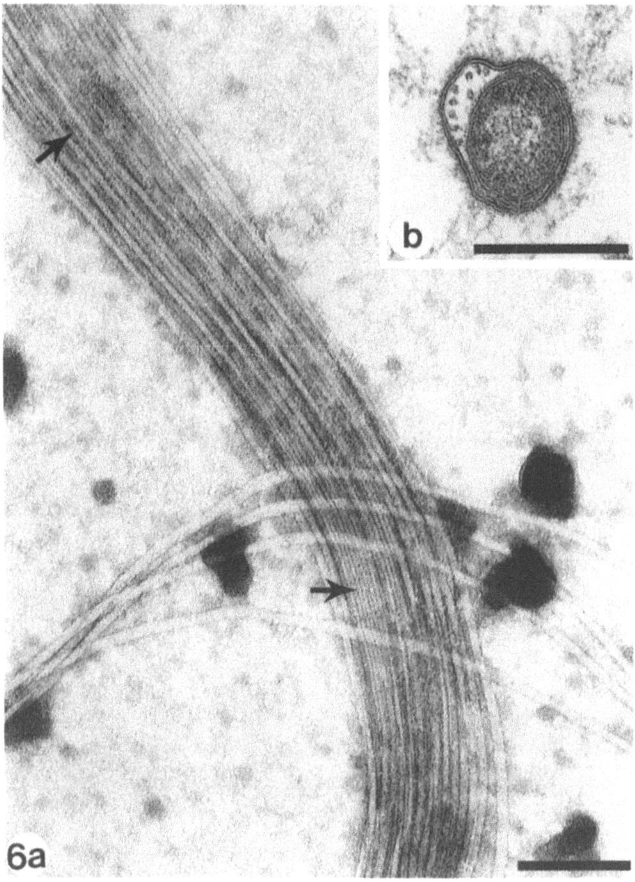

Fig. 6a, b. Ammonium molybdate (2%) negatively stained bundle of endoflagella showing variability in staining. The degree of intercalation of the stain into and surrounding each endoflagellum gives them the appearance of having variable diameters. *Arrows* point to areas where much narrower diameter fibrils (substructural elements) can be seen. **b** The hollow profile of endoflagella by thin sectioning. Bars = 0.1 μm (a) and 0.25 μm (b).

at the terminus of *B. burgdorferi* (Fig. 7). This granule, negatively stained in whole or cryosectioned organisms, appeared electron lucent (Fig. 1b), while in thin-section profiles where metal staining had occured, it appeared electron dense (Fig. 7). The presence of this granule beneath the outer membrane contributes to the rounded or truncated appearance of the end shape.

3.3.5 Protoplasmic Cylinder

The innermost compartment of *B. burgdorferi* consists of the protoplasmic cylinder. Limited by a trilaminar membrane, this structure contains ground substance, nucleoplasm, ribosomes, mesosomal components, and occasional crystalline elements as illustrated in Fig. 8. Eversion of cylinder components into a sac-like aneurysmic protrusion of the outer and protoplasmic membranes leads to the formation of gemmae.

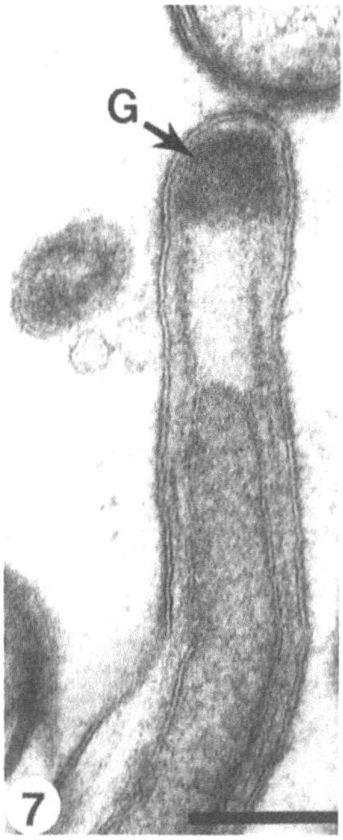

Fig. 7. Shows the appearance of the terminal granule and the rounded end shape of the spirochetal terminus (compare with 1b); *G* terminal granule. Bar = 0.25 μm

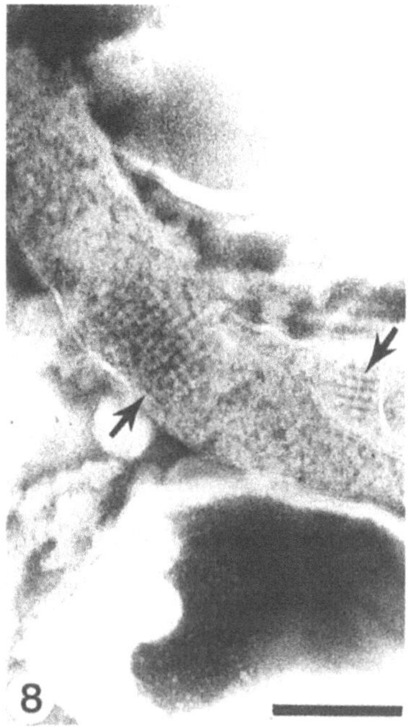

Fig. 8. Negatively stained with 0.5% phosphotungstic acid, this organism, reisolated from the bladder of a *Peromyscus leucopus* experimentally infected with Sh-2-82, shows the presence of crystalline lattice structures within the periplasmic cylinder of the spirochete (*arrows*). Bar = 0.25 μm

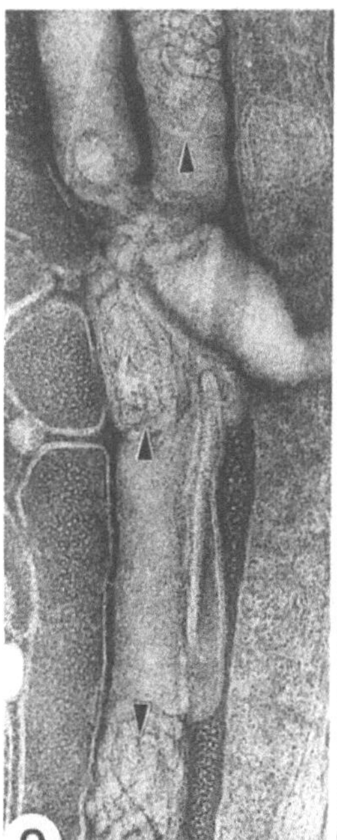

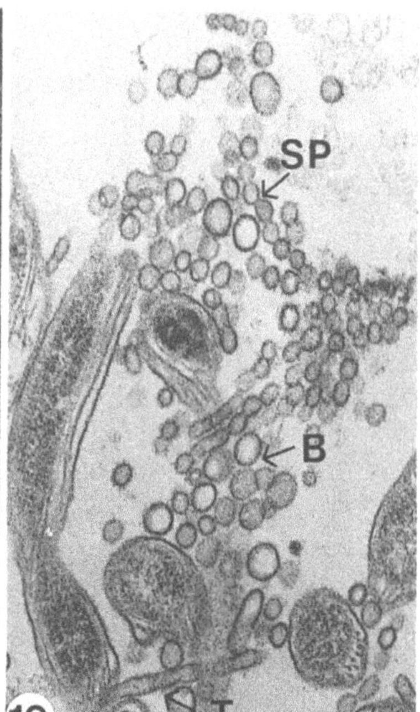

Fig. 9. *B. burgdorferi* (isolate Sh-2-82) after exposure to diluted sodium dodecyl sulfate (SDS) for 4 s and negative staining with 0.5% ammonium molybdate. The coiled element of the mesosomal complex (arrows) appears similar in morphology to the tubular bleb structures noted in the transmission electron microscopy profile of Fig. 10. Bar = 0.25 μm.

Fig. 10. Transmission electron microscopy profile of *B. burgdorferi* (B31 passage 2) showing variation in bleb morphology (*SP*, string-of-pearls; *B*, vesicle or bleb; *T*, tubule). Bar = 0.25 μm.

3.3.6 Blebs and Gemmae

Blebs originating from the trilaminar outer membrane and mesosomes are of varying morphological forms (Fig. 9). Burgdorfer and Hayes (1989) described them to resemble tubules, "string-of-pearls-like" structures, and vesicles. They have been found associated with spirochetes in ticks, as well as in BSK-II cultures. Blebs may arise at the terminus or the lateral margins, and from mesosomes within the periplasmic cylinder (Fig. 10). The production of tubular and string-of pearls-like structures in copious amounts early in the cultivation of each new isolate appears to coincide with the clumping phenomenon described by Barbour (1984).

The other consistently associated outer membrane feature is the large ball-like gemmae (Fig. 11), most often located at the terminus of the spirochete. Much larger than

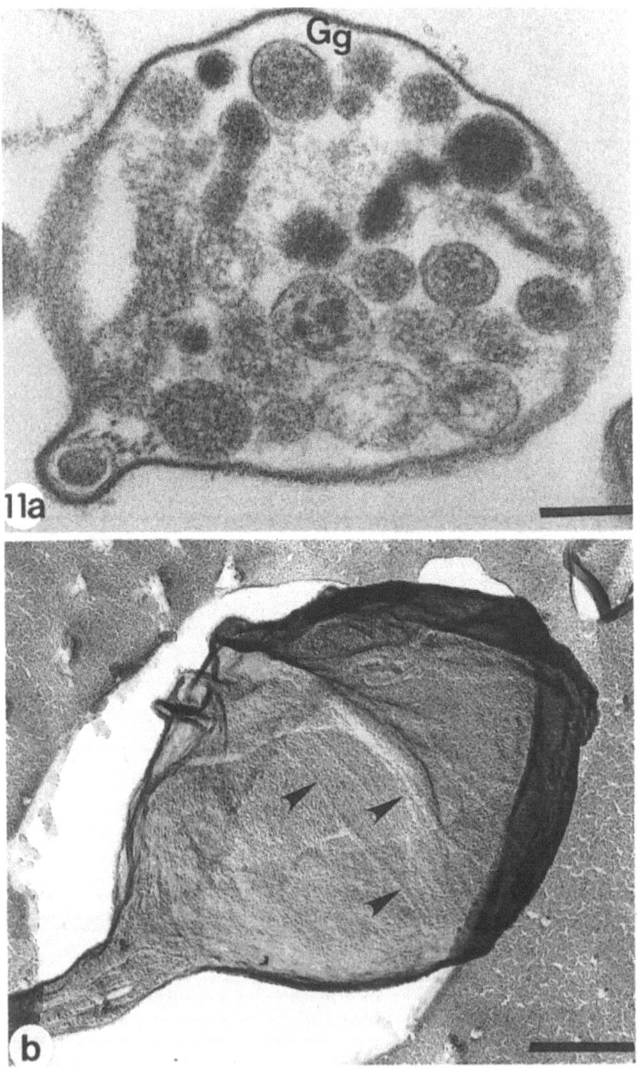

Fig. 11. a Thin section profile of *B. burgdorferi*, (B31 passage 2), gemma-containing granules (*Gg*). **b** The gemma by freeze fracture. *Arrows* note the endoflagellar elements on the exposed and etched surface. Bars = 0.25 μm

blebs, gemmae measure up to 2 µm in diameter. These structures were initially examined with the electron microscope by Hampp (1948, 1950) and subsequently by others, including Pillot et al. (1964a, b), Barbour and Hayes (1986), Burgdorfer and Hayes (1989), and Hayes et al. (1990). They are particularly noticeable in SEM micrographs and in TEM thin-section profiles (Barbour and Hayes 1986; Burgdorfer and Hayes 1989). Gemmae become more numerous in acidic older cultures and can be induced by rapid decreases of pH (Pillot et al. 1964a,b; Pillot and Ryter 1965). Within gemmae are granules that appear to form as the result of an active condensation of nucleoplasmic and cytosolic constituents followed by a membrane enrobement. Coiled and serpentine structures have also occasionally been observed within gemmae and appear to represent remnants of the protoplasmic cylinder as well as endoflagella. However, gemmae are not the same structures as the encysted forms that arise from encapsulation of the entire protoplasmic cylinder and endoflagella within a coating of stabilized or opsonized S-layer and outer membrane (Burgdorfer and Hayes 1989).

3.4 Discussion

The morphological appearance of spirochetes by electron microscopy, particularly the end shape, was considered a taxonomic criterion as early as 1945, when Lofgren and Soule reported *Spirochaeta novyi*, now considered *Borrelia recurrentis*, to have sharply pointed ends in contrast to the syphilitic spirochete *Sp. pallida*, now *Treponema pallidum* (Lofgren and Soule 1945). However, Babudieri and Bocciarelli (1948), investigating *Sp. noveyj* and *Sp. obermeieri* detected rounded and pointed forms, and demonstrated that end shape was dependent upon the rate of division. Pillot et al. (1964) and Pillot and Ryter thereafter in 1965 found little difference in the end shape of treponemes and borreliae. In the same year, (1965), Felsenfeld suggested that "variations in morphology (refering to borreliae) often appear in transfer to a heterologous animal." However, Hovind-Hougen (1976) claimed that negatively stained treponemes and borreliae could be distinguished from each other with reasonable certainty, according to end shape. Her report, however, showed treponemes to have a spectrum of shapes ranging from blunt to pointed, while *B. merionesi* and *B. recurrentis* were consistently pointed. No other species of borreliae were investigated. Holt (1978), however, was cautious in interpreting such electron microscopy findings, stating, "One must be cautious in interpreting details of fine structure because attention must be paid not only to the physiologic conditions of the cells being examined but also to physical and chemical treatments, including the stains used, their pH, and the methods of staining ..."

Our past and present investigations of *B. burgdorferi* tend to support Holt's statement, for we were unable to find certain features such as end shape or number of endoflagella characteristic for this spirochete. Spirochetes of all isolates including type strains B31 and IRS, which were cloned repeatedly by limiting dilution before examination, revealed endshape profiles varying from blunt to pointed. Thus, we cannot support Hovind-Hougen's claim that the presence of blunt and pointed spirochetes in one and the same culture suggests a mixture of two separate species of spirochetes (Hovind-Hougen 1984; Hovind-Hougen et al. 1987). Although we do not deny the possibility that as yet undefined species of spirochetes may be involved in the etiology of Lyme borreliosis – their identification and differentiation from *B. burgdorferi* must include molecular and genetic characterization rather than rely exclusively on ultrastructural morphology.

Structural analysis of the S-layer, whether possessing a regularly patterned substructure or not, has been suggested for use in differentiating borreliae from treponemes (Hovind-Hougen 1976). However, since treponemes within the subgroups of *Treponema* have both patterned and unpatterned S-layers, this criterion does not lend itself for taxonomic purposes to distinguish these spirochetes from borreliae, which exhibit only an unpatterned S-layer.

Similarly, the number(s) of endoflagella is also of questionable taxonomic value in differentiating borreliae from treponemes. Prior to the discovery of *B. burgdorferi* (Burgdorfer et al. 1982), it was postulated that relapsing fever borreliae had 15-30 endoflagella whereas treponemes had only 4-8. Our studies with *B. burgdorferi* initially suggested 4-8 endoflagella, based on early cross-section profiles, and later 8-14, based upon negative staining. Recent investigations (see Fig. 5 above) suggest perhaps even more.

Endoflagellar insertion patterns in the periplasmic cylinder membrane have also added little to the process of differentiation. The subgroups of the genus *Treponema* show variable patterns that correlate with their pathogenicity. The insertion point pattern of *B. burgdorferi* appears identical to that of pathogenic treponemes (Hovind-Hougen 1976), i.e., in a helically rotating row parallel to the long axis of the spirochete. This may imply some relationship between insertion point pattern and pathogenicity.

In contrast, there are ultrastructural characterstics that are useful in the differentiation of these spirochetes. Sheathedness of the endoflagella and the presence of cytoplasmic fibrils appear characteristic for all treponemes, except *T. genitallium*, which lack cytoplasmic tubules (Hovind-Hougen 1976). Borrelia lack these identifiers.

Few studies of the periplasmic space have been attempted. The function of the terminal granule recently detected by multiple electron microscopic techniques remains unclear. It may play a role in the penetration of tissues, provided the granule contains certain enzymes such as phospholipase or laminase for action upon cell membranes or the basal lamina.

Although the functions of spirochete membranous vesicles (blebs, gemmae) are not yet defined, the fact that intact plasmid DNA is packaged within some of them has led to the hypothesis that these structures may play a role in the protection and transfer of genetic markers (Garon et al. 1989; Dorward and Garon 1990). Previously, Schwan et al. (1988) reported a loss of virulence associated with plasmid loss in culture. However, it remains to be determined whether these elements play a role in the pathogenesis of spirochetal agents.

Gemmae were initially studied with electron microscopy by Hampp (Hampp et al. 1948; Hampp 1950), but it was DeLamater et al. (1950, 1951), who advanced the hypothesis that the granules within represented intermediate growth forms. This was discounted by Pillot et al. (1964), who reported that gemmae and their granules could regularly be induced in treponemes and borreliae by changes in pH, and postulated that they represented products of aging. Gemmae and their granules are again the subjects of intensive investigations.

Finally, immunoelectron microscopy applied to *B. burgdorferi* reveals that all spirochetal components contain numerous molecules that may play important roles in the interaction between *B. burgdorferi* and the host's immune system (Barbour et al. 1984; Barbour et al. 1986; Dorward et al. 1991; Simpson et al. 1991). For more information see Chapters 4 and 21.

3.5 Summary

The Lyme borreliosis spirochete *B. burgdorferi*, measuring 10-30 µm in length and 0.2-0.25 µm in width, reveals an ultrastructure similar to that of other spirochetes in the genus Borrelia. Its end-shape profiles vary from blunt to pointed. It possesses an unpatterned surface or S-layer, a trilaminar limiting outer membrane surrounding the periplasmic space which contains 3-18 hollow endoflagella, a granule at the terminus, and a protoplasmic cylinder. *B. burgdorferi*, like other borreliae, have unsheated endoflagella and lack cytoplasmic tubules, characteristics that differentiate it from treponemes. Brief reference is made to other ultrastructural elements such as membrane blebs, gemmae, gemmae granules, and cysts.

Acknowledgements. We thank Betty J. Kester for excellent secretarial help in the preparation of this manuscript and Dr. Claude F. Garon, Chief of the Laboratory of Vectors and Pathogens, and Dr. Patty Rosa, Laboratory of Microbial Structure and Function, for their critical review of our manuscript.

References

Ackermann R, Kabatzki J, Boisten HP, Steere AC, Grodzicki RL, Hartung S, Runne U (1984) *Ixodes ricinus* spirochete and European erythema chronicum migrans disease. Yale J Biol Med 57:573-580

Anderson JF, Magnarelli LA, LeFebvre RB, Andreadis TG, McAninch JB, Perng GC, Johnson RC (1989) Antigenically variable Borrelia burgdorferi isolated from cottontail rabbits and *Ixodes dentatus* in rural and urban areas. J Clin Microbiol 27:13-20

Babudieri B, Bocciarelli D (1948) Electron-microscope studies on relapsing fever spirochaetes. J Hyg 46:438-440

Barbour AG (1984) Isolation and cultivation of Lyme disease spirochetes. Yale J Biol Med 57:521-525

Barbour AG, Hayes SF (1986) Biology of *Borrelia* species. Microbiol Rev 50:381-400

Barbour AG, Burgdorfer W, Hyes SF, Péter O, Aeschlimann A (1983) Isolation of a cultivable spirochete from *Ixodes ricinus* ticks of Switzerland. Curr Microbiol 8:123-126

Barbour AG, Tessier SL, Todd WJ (1983) Lyme disease spirochetes and ixodid tick spirochetes share a common surface antigenic determinant defined by a monoclonal antibody. Infect Immun 41:795-804

Barbour AG, Tessier SL, Hayes SF (1984) Variation in a major surface protein of Lyme disease spirochetes. Infect Immun 45:94-100

Barbour AG, Hayes SF, Heiland RA, Schrumpf ME, Tessier SL (1986) A *Borrelia*-specific monoclonal antibody binds to a flagellar epitope. Infect Immuns 52:549-554

Burgdorfer W, Hayes SF (1989) Vector-spirochete relationships in louse-borne and tick-borne borrelioses with emphasis on Lyme disease. In: Harris KF (ed) Advances in disease vector Research Vol. 6. Springer, Berlin Heidelberg New York, pp 127-150

Burgdorfer W, Barbour AG, Hayes SF, Benach JL, Grundwaldt E, Davis JP (1982) Lyme disease - a tick-borne spirochetosis? Science 216:1317-1319

Burgdorfer W, Barbour AG, Hayes SF, Peter O, Aeschlimann A (1983) Erythema chronicum migrans - a tick borne spirochetosis. Acta Trop 40:79-83

Cutts JH (1970) Cell separation methods in hematology. Academic New York, p. 172

DeLamater ED, Newcomer VD, Haanes M, Wiggall RH (1950) Studies on the life cycle of spirochetes. I. The use of phase contrast microscopy. Am J Syph Gonorrhea Vener Dis 34:122-125

DeLamater ED, Newcomer VD, Haanes M, Wiggall RH (1951) Studies on the life cycle of spirochetes. VIII. Summary and comparision of observations on various organisms. J Invest Dermatol 16:231-256

Dorward DW, Garon CF (1990) DNA is packaged within membrane-derived vesicles of Gram-negative but not Gram-positive bacteria. Appl Environ Microbiol 56:1960-1962

Dorward DW, Judd RC (1988) The Isolation and partial characterization of naturally-evolved outer membrane blebs of *Neisseria gonorrhoeae*. In: Poolman JT, Zanen HC, Meyer TF, Heckels JE, Makela PRH, Smith H, Beuvery EC (eds) Gonococci and meningococci. Kluwer, Dordrecht, pp 349-356

Dorward DW, Schwan TG, Garon CF (1991) Immune capture and detection of *Borrelia burgdorferi* antigens in urine, blood, or tissues from infected ticks, mice, dogs, and humans. J Clin Microbiol 29:1162-1170

Felsenfeld O (1965) Borreliae, human relapsing fever, and parasite-vector-host relationships. Bacteriol Rev 29:46-74

Garon CF, Dorward DW, Corwin MD (1989) Structural features of *Borrelia burgdorferi* - the Lyme disease spirochete: silver staining for nucleic acids. Scanning Microsc Suppl 3:109-115

Hampp EG (1950) Morphologic characteristic of the smaller oral treponemes and *Borrelia vincenti* as revealed by stained smear, darkfield and electron microscopc techniques. J Am Dent Assoc 40:1-11

Hampp EG, Scott DB, Wyckoff RWG (1948) Spirochetal morphology and electron microscope. J Bacteriol 56:755-769

Hayes SF, Burgdorfer W (1979) Ultrastructure of *Rickettsia rhipicephali*, a new member of the spotted fever group rickettsiae in tissues of the host vector *Rhipicephalus sanguineus*. J Bacteriol 137:605-613

Hayes SF, Burgdorfer W, Barbour AG (1988) Electron microscope characterization of cloned and uncloned strains of *Borrelia burgdorferi*. Ann NY Acad Sci 539:383-385

Hayes SF, Corwin MD, Schwan TG, Dorward DW, Burgdorfer W (1990) New ultrastructural findings about *Borrelia burgdorferi* by cryofixation, negative staining, thin sectioning and scanning techniques. In: Peachey LD, Williams DB (eds) Proceedings of the XIIth International Congress on Electron Microscopy. San Francisco Press, San Francisco, pp 582-583

Hitchcock PJ, Brown TM, Corwin D, Hayes SF, Olszewski A, Todd WJ (1985) Morphology of three strains of contagious equine metritis organism. Infect Immun 48:94-108

Holt SC (1978) Anatomy and chemistry of spirochetes. Microbiol Rev 42:114-160

Hovind-Hougen K (1974) Electron microscopy of *Borrelia merionesi* and *Borrelia recurrentis*. Acta Pathol Microbiol Immunol Scand (B) 82:799-809

Hovind-Hougen K (1976) Determination by means of electron microscopy of morphologic criteria of value for clarification of some spirochetes, in particular, treponemes. Acta Pathol Microbiol Immunol Scand (B) 225:1-41

Hovind-Hougen K (1984) Ultrastructure of spirochetes isolated from *Ixodes ricinus* and *Ixodes dammini*. Yale J Biol Med 57:543-548

Hovind-Hougen K, Birch-Andersen A (1971) Electron microscopy of endoflagella and microtubules in *Treponema* Reiter. Acta Pathol Microbiol Immunol Scand (B) 79:37-50

Hovind-Hougen K, Asbrink E, Stiernstedt G, Steere AC, Hovmark A (1987) Ultrastructural differences among spirochetes isolated from patients with Lyme disease and related disorders, and from *Ixodes ricinus*. In: Stanek G, Flamm H, Barbour A, Burgdorfer W (eds) Lyme Borreliosis. Gustac Fischer, Stuttgart, pp 103-111

Hulínská D, Jirouš J, Valešová M, Herzogová J (1989) Ultrastructure of *Borrelia burgdorferi* in tissues of patients with Lyme disease. J Basic Microbiol 2:73-83

Johnson RC (1977) The spirochetes. Ann Rev Microbiol 31:89-106

Johnson RC, Hyde FW, Rumpel CM (1984a) Taxonomy of the Lyme disease spirochetes. Yale J Biol Med 57:529-537

Johnson RC, Schmid GP, Hyde FW, Steigerwalt AG, Brenner DJ (1984b) *Borrelia burgdorferi* sp.nov.: etiologic agent of Lyme disease. Int J Syst Bacteriol 34:496-497

Karlsson M, Hovind-Hougen K, Svenungsson B, Stiernstedt G (1990) Cultivation and characterization of spirochetes from cerebrospinal fluid of patients with Lyme borreliosis. J Clin Microbiol 28:473-479

Lofgren R, Soule MJ (1945) The structure of *Spirochaeta novyi* as revealed by the electron microscope. J Bacteriol 50:679-690

Pillot J, Ryter A (1965) Structure des Spirochetes. I. Étude des genres *Treponema, Borrelia*, et *Leptospira* au microscope electronique. Ann Inst Pasteur 108:791-804

Pillot J, Dupouey P, Ryter A (1964a) La signification des formes atypiques et la notion de cycle évolutif chez les spirochètes (Première partie). Ann Inst Pasteur 107:484-502

Pillot J, Dupouey P, Ryter A (1964b) La signification des formes atypiques et la notion de cycle évolutif chez les spirochètes (Deuxième partie). Ann Inst Pasteur 107:663-677

Robinson Jr EN, McGee ZA, Clemens CM (1987) Probing the surface of bacteria: use of gold sphere immunological probes. Microb Pathog 2:159-169

Schmid GP, Steigerwalt AG, Johnson SE, Barbour AG, Steere AC, Robinson IM, Brenner DJ (1984) DNA characterization of the spirochete that causes Lyme disease. J Clin Microbiol 20:155-158

Schwan TG, Burgdorfer W, Garon CF (1988) Changes in infectivity and plasmid profile of the Lyme disease spirochete, *Borrelia burgdorferi*, as a result of in vitro cultivation. Infect Immun 56:1831-1836

Simpson WJ, Schrumpf ME, Hayes SF, Schwan TG (1991) Molecular and immunological analysis of a polymorphic periplasmic protein of *Borrelia burgdorferi*. J Clin Microbiol 29:1940-1948

Stiernstedt G, Barbour AG, Hovind Hougen K, Hansson LO, Svenungsson B, Jörbeck H, Sköldenberg B (1985) *Borrelia* strains isolated from *Ixodes ricinus* in Sweden - studies on their ultrastructure, and reactivities with antibodies of patients with spirochetal meningitis. In: Stiernstedt G (ed) Tick-borne *Borrelia* infection in Sweden, vol II. Kongl Carolinska Medico Chirurgiska Institute, Stockholm, pp 1-13

Swanson J, Zeligs B (1974) Studies on gonococcus infection. VI. Electron microscopic study on in vitro phagocytosis of gonococci by human leukocytes. Infect Immun 10:645-656

Todd WJ, Wray GP, Hitchcock PJ (1984) Arrangement of pili in colonies of *Neisseria gonorrhoeae*. J Bacteriol 159:312-320

Wagner RC (1976) The effect of tannic acid on electron images of capillary endothelial cell membranes. J Ultrastruct Res 57:132-139

4 Biology of *Borrelia burgdorferi*

V. Preac-Mursic and B. Wilske

4.1 Introduction

Isolation and in vitro cultivation of *Borrelia burgdorferi* were first accomplished from *Ixodes dammini* ticks by Burgdorfer et al. (1982) and from patients by Benach et al. (1983) and Steere et al. (1983) in the Unites States. In Europe, identical microorganisms were detected and isolated from *I. ricinus* by Burgdorfer et al. (1983) and Barbour et al. (1983a) respectively. Isolates from patients were subsequently accomplished by Ackermann et al. (1984), Asbrink (1985), Asbrink et al. (1984) and Preac-Mursic et al. (1984). An important contribution had earlier been made by Kelly, who, in 1971 described a new medium permitting the cultivation of *B. hermsii* and other borreliae. Later on, Kelly's medium was repeatedly modified especially for the cultivation of *B. burgdorferi*. The ability to grow *B. burgdorferi* in vitro has greatly facilitated studies concerning its morphological and biological characteristics.

4.2 Taxonomy

The causative agent of Lyme borreliosis was classified according to its morphological and biological characteristics as belonging to the order *Spirochaetales*. The spirochetes isolated from ticks and patients constitute a group of phenotypically similar gram-negative bacteria and are defined as belonging to the genus *Borrelia*. Up to now, no definite schemes of biotyping or serotyping have been developed. However, preliminary results show that different types of *B. burgdorferi* can be recognized (Wilske et al. 1991).

4.3 Morphology

B. burgdorferi is a gram-negative helical-shaped bacterium with periplasmic flagella. The cells are 8 to 22 µm long, 0.25-0.30 µm wide and composed of 3-10 loose coils. *B. burgdorferi* as well as other spirochetes can be detected as living organisms by dark-field or phase-contrast microscopy. Borreliae can also be visualized by light microscopy after staining with Gram, Giemsa, or carbolfuchsin (Fig. 1 see color plate I after page 118). Active corkscrew-like motility and definite lateral oscillation have been observed in cells obtained from fresh preparations of infected blood and in actively growing cultures (Preac-Mursic et al. 1986, 1990; Stanek et al. 1986). Optimal conditions for motility are generally the same as for growth, but we have observed a faster motion at a temperature of 30°C than at 35°C and above (Preac-Mursic et al. 1986).

B. burgdorferi does not develop resistant endospores. Similar to other spirochetes, it forms spheroids under unfavorable conditions. We previously observed minute granu-

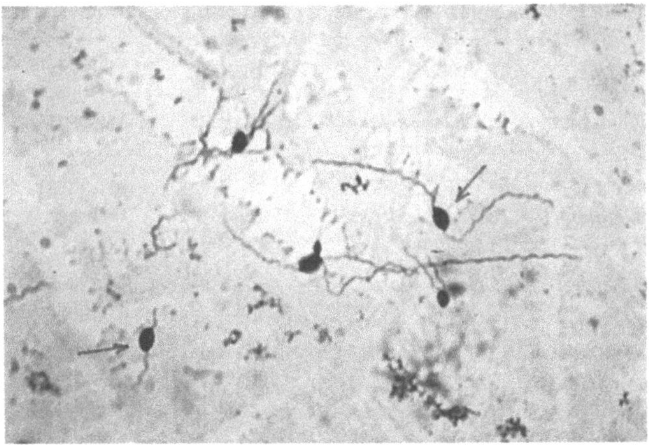

Fig. 2 Human strain of *Borrelia-burgdorferi* – blebs-like structures in aging culture, Fuchsin, x 1600

lar forms and other forms with terminal and/or central bleb-like structures (Preac-Mursic et al. 1986). These develop spontaneously in aging cultures, but their formation can also be produced artificially in young cultures using various chemical additives such as antibiotics (Preac-Mursic et al. 1987) (Fig. 2). These structures vary according to the type of medium and the functional state of the cells. The viability and nature of these structures has not yet been determined. Several possibilities such as accumulation of "food reserves", cysts, degeneration products, or a complex of deoxyribonucleic acid (DNA) scattered throughout the cytoplasma of the cell could be conceivable explanations. Initial examination by Burgdorfer et al. (personal communication) seemed to indicate DNA complexes. The authors identified two types of vesicles: blebs (200 nm in size), which develop as an eversion of the spirochete's outer membrane, and gemmae (2.0 µm in size) with eversion of the other membrane and cytoplasmic cylinder. Using chloroplatinum (II) chloride, a DNA-specific staining technique, DNA-containing vesicles (blebs) have been demonstrated on the surface of *B. burgdorferi* cells (Garon et al. 1990). The spheroids of treponemes were studied by Pillot (1965) and were found not to be evolutionary or vegetative forms, but rather to be produced by hyperhydration of the interstices between the outer membrane and the cytoplasmic body. Stoenner (1974) described the appearance of cyst-like structures at the end of the body of *B. hermsii*. These structures do not constitute a means of differentiation as may be the endospores of the genus *Clostridium*. *B. burgdorferi* cannot be differentiated morphologically from other borreliae when staining methods in light or dark-field microscopy are used.

4.4 Growth Conditions

4.4.1 Growth Conditions In Vitro

B. burgdorferi can be grown in vitro as can some other tick-borne *Borrelia* species (see Table 1). For the in vitro cultivation of borrelial organisms, various modifications of Kelly medium are used, e.g., Barbour-Stoenner-Kelly (BSK II) medium according to Barbour et al. (1983a) or modified Kelly medium according to Preac-Mursic (MKP) (Preac-Mursic et al. 1986). The culture medium is a complex liquid medium similar to

Table 1. In vitro cultivable species of the genus Borrelia (from Holt and Krieg 1984)

Species	Vector	Host	Distribution	Disease
B. burgdorferi	*Ixodes dammini* *Ixodes ricinus* *Ixodes persulcatus*	Rodents, man, birds	Worldwide	Lyme borreliosis
B. duttonii	*Ornithodoros moubata*	Man	Africa	Tick-borne relapsing fever
B. hermsii	*Ornithodoros hermsi*	Rodents, man	Western USA, Canada	Tick-borne relapsing fever
B. hispanica	*Ornithodoros erraticus*	Rodents, man	Spain, Portugal, Marocco, Tunisia, Algeria	Tick-borne relapsing fever
B. parkeri	*Ornithodoros parkeri*	Rodents, man	Western USA	Tick-borne relapsing fever
B. recurrentis	*Pediculus humanus*	Man	South Africa, Europe, Africa, Asia	Louse-borne relapsing fever
B. turicatae	*Ornithodoros turicata*	Rodents, man	USA, Mexico	Tick-borne relapsing fever

that used for tissue cultures containing a mixture of amino acids, vitamins, inorganic salts (CMRL-1066, Gibco), *N*-acethylglucosamine, serum albumin, and rabbit serum. *B. burgdorferi* are anaerobic bacteria as indicated by their preference to grow on the bottom of the medium tube and aerotolerance testing. The optimal temperature for multiplication and fermentative processes is 30°-34°C, with peak multiplication at 4 days. Low temperatures (4°C) are better tolerated than high temperatures (37°-42°C). The generation time, which is influenced by available nutrients and conditions of cultivation, is 7-20 h, and culture-adapted strains reach a cell concentration of 10^{6-8}/ml.

Cell concentration is determined by total counts or by viable counts, using a counting chamber (Petroff-Hausser) and dark-field microscopy. Estimation of cell density by photometry often leads to incorrect values; many strains, especially in initial cultures, grow in aggregates and/or plaits, making counting very difficult (Preac-Mursic et al. 1986, 1987). Colony counting, the method usually used for other bacteria, is difficult because growth of *B. burgdorferi* as isolated, surface macroscopic colonies on solid medium has not been achieved. However, Kurtti et al. (1987) cultivated *B. burgdorferi* on solid BSK medium with 1.5% agarose and obtained discrete colonies. The plates were incubated at 34°C for 2-5 weeks in a candle jar and examined with a dissecting microscope. We compared six different solid media including the medium of Kurtti et al. (1987) for the growth *B. burgdorferi,* quantitatively and qualitatively. The highest recoveries, as well as the best colony formation, though differing greatly in size, were observed on PMR agar (Preac-Mursic et al. 1991) after 2 weeks when plates were incubated in Gas Pack (Becton Dickinson, Heidelberg, Germany) at 33°C. The most characteristic feature was growth in the form of individual colonies on the agar surface. In general, colonies had two distinct forms of morphology with a diameter

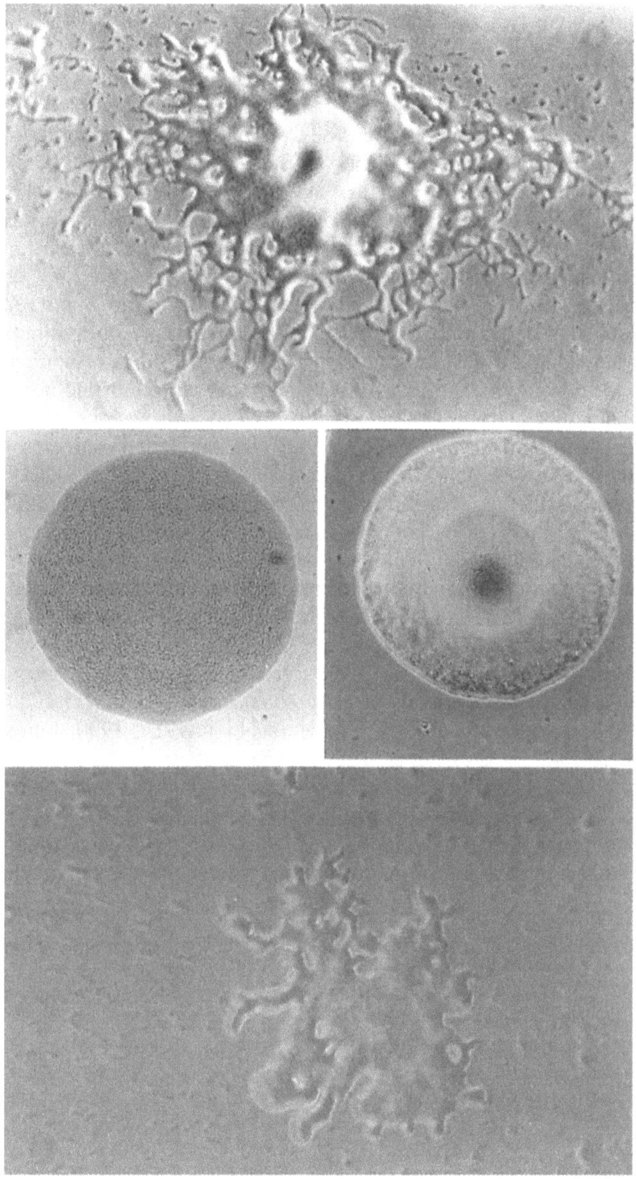

Fig. 3 Colony morphologies of Borrelia burgdorferi on PMR-agar after 2 weeks of anaerobic incubation at 33°C

from 0.3 to 3 mm: one form was gray and shallow with a convex dark center and smooth or rough edges, similar to the "fried egg" appearance of mycoplasma colonies, and the other was gray, smooth, or "rough diffuse", and without a dark center (Fig. 3). Isolates from different patients showed distinction in colony form. We noted also morphological variations among colonies obtained from isolates of single patients (Preac-Mursic et al. 1991). The growth of *B. burgdorferi* on solid media has proven beneficial in obtaining pure cultures of *B. burgdorferi* from cloned material.

4.4.2 Growth Conditions In Vivo

To study the biological parameters of *B. burgdorferi* in vivo several animal models have been established. Experiments with white rabbits by Burgdorfer (1984), Syrian hamsters by Johnson et al. (1988), rabbits by Kornblatt et al. (1984), dogs by Burgess (1986), splenectomized gerbils by Stanek et al. (1986), severe combined immunodeficient (SCID) mice by Schaible et al. (1989), rats by Barthold et al. (1988), hispid cotton rats by Burgdorfer and Gage (1987), and gerbils by Preac-Mursic et al. (1987, 1990) have shown that all these animals can be infected by intraperitoneal, intradermal, or intravenous inoculation of 10^{4-7}/ml *B. burgdorferi* or by exposing them to infected ticks. The reaction of the animals to a *B. burgdorferi* infection varies. Hamsters and immunologically intact mice show no symptoms of apparent illness, whereas infant gerbils are retarded in growth and development. Infant rats and SCID mice develop the largest number of clinical symptoms including arthritis. The subcutaneous inoculation of rabbits is followed by skin lesions similar in appearance to erythema migrans. Absence of obvious clinical signs in infected animals does not prove the nonpathogenicity of inoculated strains. Barthold et al. (1988) have found microscopic evidence of arthritis in rats without clinical symptoms and suggest that mild cases of the disease may not be recognized clinically. The in vivo studies show further that experimentally infected animals develop spirochetemias. After a first peak (3 days after infection) the number of borreliae in blood decreases and rises again a few days later. After subsequent peaks with alternating high and low concentrations, the bacteria disappear gradually (Burgdorfer 1984, 1991; Kornblatt et al. 1984; Stanek et al. 1986; Preac-Mursic et al. 1987, 1990; Johnson et al. 1988). *B. burgdorferi* has been recovered from several organs of infected animals (Burgdorfer 1984, 1991; Kornblatt et al. 1984; Stanek et al. 1986; Preac-Mursic et al. 1987, 1990; Johnson et al. 1988; Schwan et al. 1988; Schaible et al. 1989; Simpson et al. 1991). Our studies have shown that in infected gerbils the persistence of borreliae may last for months in several organs such as liver, spleen, kidney, heart, skin and joints (Table 2).

Intraperitoneal inoculation of *B. burgdorferi* induced significant histopathological changes in most of the major organ systems and the surrounding adipose and fibrous

Table 2. Persistence of *Borrelia burgdorferi* in 27 experimentally infected gerbils

Specimen	*Borrelia burgdorferi* isolation / Time post infection									
	0	1 day	3 days	7 days	10 days	14 days	1 month	3 months	6 months	1 year
Liver		+	+	+	+	+	+	+	–	–
Spleen		+	+	+	+	+	+	+	+	–
Kidney		–	+	+	+	+	+	+	–	–
Heart		+	+	+	+	+	+	+	–	–
Brain		–	–	–	–	–	–	–	–	–
Joint		–	–	–	–	+	+	+	+	+
Skin[a]		+	+	+	+	+	+	+	+	–
Blood		++	+++	+	++	(+)	–	–	–	–
Antibody-titer range										
from	<2	<2	/	32	/	512	1024	1024	512	32
to						2048	4096	2048	1024	256
Leukocyte	8500	13500	/	10600	/	10200	/	9500	7900	/

[a]After intradermal inoculation 1 x 10^6 cells/ml

Table 3. Comparison of histopathologic findings (organs involved in the inflammatory process) in the experimentally infected gerbils with the infected human tissue in patients with Lyme disease reported in the literature.

Examined organs	Experimentally infected gerbils	Patients with Lyme disease
Central nervous system	+	+
Sciatic and other peripheral nerves	+	+
Striate muscle	+	+
Eyes and orbital contents	+	+
Tongue	–	–
Heart	+	+
Vessels	+	+
Lungs	+	–
Lymph nodes	+	+
Liver	+	+
Spleen	+	+
Stomach	+	–
Intestine	+	–
Pancreas	–	–
Kidneys	+	–
Adrenal glands	–	–
Urinary bladder	+	–
Testis	+	+
Skin	+	+
Knee joint	+	+

For references see Chapt. 6

connective tissues (Preac-Mursic et al. 1987, 1990). No significant differences were noted between the histopathological findings in the animals infected with cells from subcultures no. 5 and no. 25 (Preac-Mursic et al. 1990). These findings seem to contrast with the results obtained by Johnson et al. (1984c) and Schwan et al. (1988a), who observed loss of infectivity after a lower number of passages. Possibly, these differences result from growth conditions, animal susceptibility, and strains. The persistence of *B. burgdorferi* and the large number of organs involved with slight to severe signs of inflammation in the gerbil model appear comparable to the persistence and multiorgan involvement seen in human Lyme borreliosis (Table 3). Thus gerbils can serve as suitable experimental animals to study the pathogenesis of Lyme borreliosis and the extent of organ damage caused by *B. burgdorferi*.

4.5 Physiology and Chemical Analysis

The ability to culture *B. burgdorferi* has permitted studies on the chemical composition, nutrition, and physiology of this pathogenic organism. Chemical analysis of the outer envelope of *B. burgdorferi* revealed a composition of 45.90% protein, 50.75% lipid and 3.33% carbohydrate (Coleman et al. 1986). Peptidoglycan was isolated as a component of the cell wall (Beck and Habicht 1988).

4.5.1 Biochemical Characteristics

Only a few studies concerning the biochemical characteristics of borreliae have been reported. *B. burgdorferi*, as investigated so far, apparently has a fermentative metabolism similar to *B. hermsii* and *B. parkeri*. The production of lactic acid as the main end product of fermentation is a characteristic of the genus *Borrelia*. In these characteristics they differ from spirochetes of the genus *Treponema* and *Leptospira* (Kelly 1976; Livermore et al. 1978). In studies to determine the biochemical activity of 30 B. *burgdorferi* isolates, five other borreliae, and five *Treponema* strains using the commercially available enzyme system APIZYM (Api Systems S.A.,France) we found a few tests to be attractive possibilities for rapid presumptive identification to the genus level, namely, leucine aminopeptidase, B-galactosidase and α-glucosidase. Unlike Fumarola et al. (1986) we did not find this system to provide a convincing means for detecting interspecies differences among borreliae (Preac-Mursic, unpublished).

4.5.2 DNA Analysis

Analysis of *B. burgdorferi* isolates by their genetic properties is still in an early stage. However, it has been shown that the chromosome of *B. burgdorferi* is linear and with 1000 kb very small in size (Baril et al. 1989; Ferdows and Barbour 1989). The presence of extrachromosomal DNA in the form of circular and linear plasmids has been detected (Barbour and Garon 1987; Barbour 1988; Hyde and Johnson 1988). The linear plasmids account for most of the extrachromosomal DNA and a 49-kb plasmid carries the genes encoding the two major outer surface proteins OspA and OspB (Barbour and Garon 1987; Barbour 1988). The function of circular plasmids is unknown. The results of DNA characterization have shown that *B. burgdorferi* is a separate species within the genus Borrelia (Hyde and Johnson 1984, 1988; Johnson et al. 1984a, b; Schmid et al. 1984). Using DNA/DNA hybridization and μRNA gene-restriction patterns of 12 strains, Postic et al. (1990) observed two genomic DNA hybridization groups: *B. burgdorferi* sensu stricto and a second genomic species referred to as

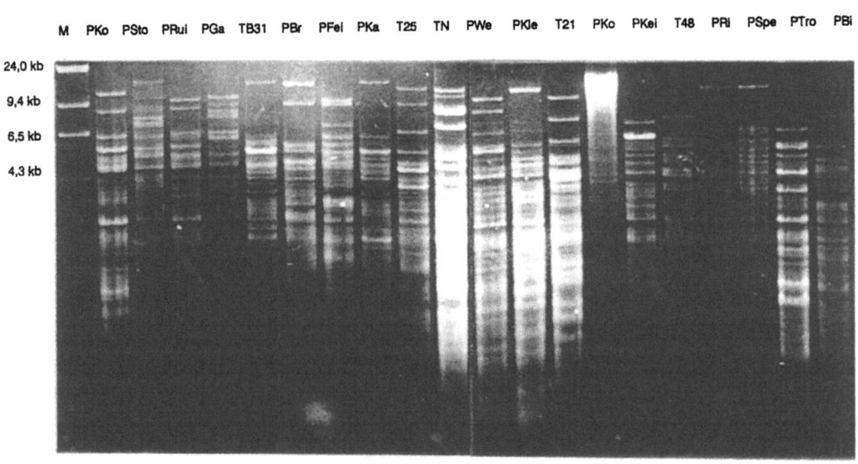

Fig. 4 DNA-profiles of 15 human (*P*) and 5 tick (*T*) isolates of *Borrelia burgdorferi* (HindIII-digested; TB31, American-type Strain; Pko, European-type strain)

Borrelia group 20047. Preliminary data suggested that strains differ in their DNA pattern. Results of DNA analysis of 20 isolates are given in Fig. 4. Comparison of their restriction enzyme Hind III, EcoRI, Hinc II, and Hpa I fragment patterns in agarose gel electrophoresis showed the *B. burgdorferi* strains to be somewhat different (Preac-Mursic, unpublished). In contrast, restriction endonuclease analysis in conjunction with Southern blot hybridization (with OspA, 30 bases probe; OspA 0.7 kb EcoRI-HincII fragment probe; p41, 1 kb BamHI-PstI fragment probe) was used more effectively to study the relatedness of various *B. burgdorferi* strains to the type strain B31 and to determine a genetically similar organism (Wilske et al. 1991) (Fig. 5).
Analysis of extrachromosomal plasmid DNA shows that there is considerable heterogeneity among strains in their plasmid profile, only two of the 13 strains from the United States and Europe exmined by Barbour (1988) had the same plasmid profile. Plasmid agarose gel analysis yielded plasmids ranging in size from 15 to 60 kb; the

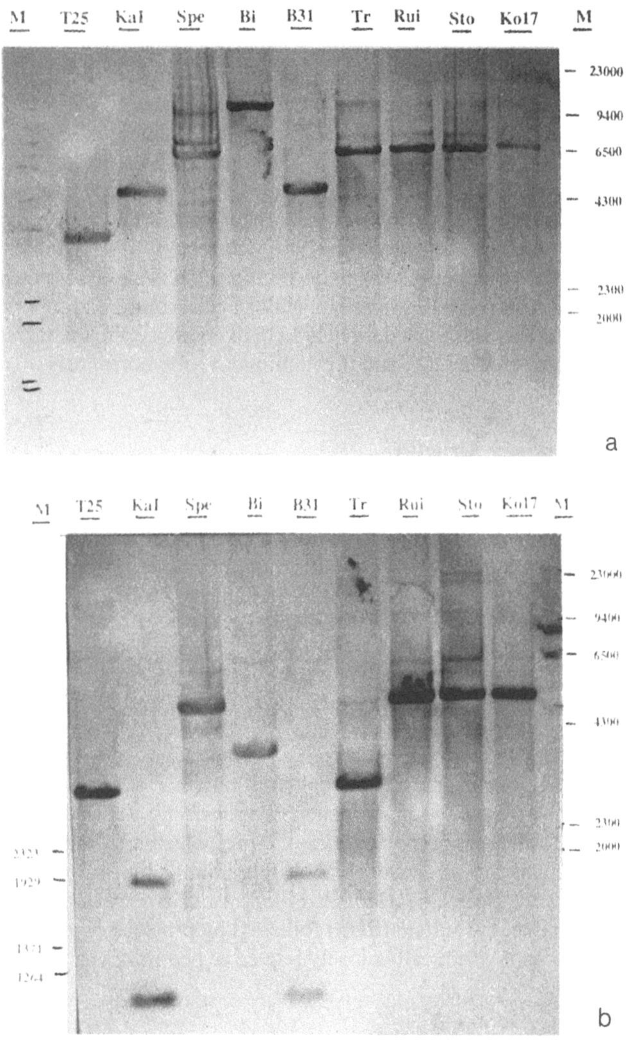

Fig. 5a, b. Southern blots of DNA probes. **a** Hybridization of *B. burgdorferi* BglII-digested DNA with p41 (1 kb BamHI-PstI fragment) probe. **b** Hybridization of *B. burgdorferi* HpaI-digested DNA with p41 (1 kb BamHI-PstI fragment) probe genome comprised several circular and linear plasmids; the number of plasmid bands ranged from three to seven; the majority of the strains had five plasmids (Barbour 1988; Baril et al. 1989).

The clinical significance and the exact role of the particular plasmids for biological and ecological differences have not yet been defined. It is known that the virulence factors of other bacteria are often plasmid mediated (Elwell and Shipley 1980). A detailed characterization of plasmids could, therefore, be useful in detecting virulence factors. Barbour (1988) has suggested plasmid profile analysis for typing of strains. However, some loss of plasmids in strains has been observed after only a few in vitro passages of *B. burgdorferi*. Plasmid loss or rearrangement during continued in vitro cultivation may play a role in the antigenic changes of OspB proteins (Schwan and Burgdorfer 1987) and in a decrease of virulence, as has been observed in hamsters and mice (Johnson et al. 1984; Schwan et al. 1988a). Possible instability of extrachromosomal DNA must be taken into consideration in determining virulence factors or for typing strains.

Use of DNA probes for species-specific genes such as the OspA-, OpsB-, pC-, and p 100-encoding genes and species-specific plasmid DNA may offer highly sensitive diagnostic tools. Recently the polymerase chain reaction (PCR) technique has successfully been used to amplify specific DNA sequences of *B. burgdorferi*. The PCR technique has been used to detect *B. burgdorferi* in its vector (Schwan and Barbour 1988; Rosa and Schwan 1989; Persing et al. 1990) and in specimens from patients (Goodman et al. 1991). Dot blot hybridization as well as PCR were used for characterization of isolates (Schwan and Barbour 1988; Rosa et al. 1991; Wilske et al. 1991). The sensitivity of this technique seems to be high. For final assessment of its diagnostic value for Lyme borreliosis, more data about the structure and function especially of plasmid DNA are needed. Such data may help to answer basic questions concerning pathogenesis, diagnosis, therapy, and prevention of Lyme borreliosis.

4.5.3 Antigen Structure

Molecular analysis of the chemical compounds of *B. burgdorferi* has revealed unique characteristics but also compounds rather closely related to other bacteria. Up to now most studies have concentrated on the characterization of proteins of *B. burgdorferi*, whereas only a few investigators have analyzed the nonproteinaceous compounds of the organism.

4.5.3.1 Nonproteinaceous Compounds

4.5.3.1.1 Peptidoglycan

Peptidoglycan is an essential cell-wall polymer of most bacteria and has been isolated from many species of spirochetes. Spirochetal peptidoglycan is associated with the cytoplasmic membrane covering the protoplasma cylinder. Due to its rigidity, it probably retains the screw-shaped configuration of the spirochete. Peptidoglycan has been isolated by Beck and Habicht (1988) from *B. burgdorferi* by treatment with warm sodium dodecyl sulfate (SDS) followed by protease digestion. The *B. burgdorferi* peptidoglycan is chemically closely related to peptidoglycan from other spirochetes.

4.5.3.1.2 Lipopolysaccharide-like Substances

Lipopolysaccharide (LPS) is a constitutive part of the outer cell wall of gram-negative bacteria. A LPS-like substance has been isolated by Habicht et al. (1986) using the petroleum-ether chloroform liquid phenol method, which amounts to 4% of the dry weight of *B. burgdorferi* and can be demonstrated as a diffuse band in high-percentage SDS-polyacrylamide gel in the 3-4 kDA molecular weight range. However, the presence of LPS in the borrelial cell wall is discussed controversially. Acording to Takayama et al. (1987), a classical LPS is not present in borreliae. The sugar- and lipid-containing compounds, therefore, should be addressed as LPS-like substances which, however, are not well characterized.

4.5.3.2 Proteinaceous Compounds

B. burgdorferi proteins have been intensively investigated. SDS lysates from whole cells reveal a broad pattern of more than 30 polypeptides in SDS-polyacrylamide gel electrophoresis (SDS-PAGE) (Barbour and Hayes 1986). Major protein bands are located at the 60 kDa, 40 kDa, 30 kDa and 20 kDa regions. As shown in Fig. 6, major protein bands of the 60 kDa and 40 kDa region have constant molecular weights. The 60 kDa and p 41 kDa major proteins are present within a wide range of *B. burgdorferi* isolates independent of geographical or biological sources (Wilske et al. 1988). The 60 kDa polypeptide p60 is the common antigen of *B. burgdorferi* responsible for broad crossreactions to other bacteria, and the gene encoding p60 has been cloned in *Escherichia coli*. It is supposed to be a cytoplasmatic protein belonging to the heat shock protein family (Hansen et al. 1988). However, it was shown by Cluss and Boothly (1990) that production of p60 cannot be induced by heat shock.

The 41 kDa polypeptide p41, the major protein of the 40 kDa region, represents the flagellin of *B. burgdorferi*, as has been shown by Barbour et al. (1986) by immunelectronmicroscopy with a gold-labeled monoclonal antibody. In contrast to treponemal flagella, which are composed of several distinct polypeptides, the *B. burgdorferi* flagellum possesses only one major polypeptide of about 41 kDa in size (Barbour et al. 1986). The flagella polypeptides of *B. burgdorferi* have common epitopes (recognized

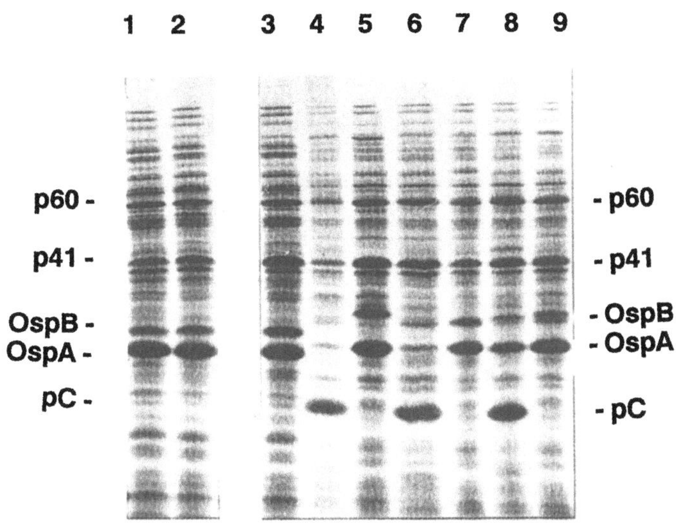

Fig. 6 SDS-PAGE analysis of *Borrelia burgdorferi* strains, coomassie blue. Lanes 1 und 2 are strains from North America, and Lanes 3–9 are European strains.

by a monoclonal antibody) which, however, are shared by other members of the genus *Borrelia* (Barbour et al. 1986). The gene encoding the flagella protein has been cloned in *E. coli* (Gassmann et al. 1989). The N-terminal amino acid sequence of the *B. burgdorferi* flagellum exhibitis a moderate homology to respective sequences encoding *Treponema pallidum* or *T. phagedenis* flagella. On the other hand, B. burgdorferi flagellin may exhibit serotype-specific epitopes. Flagella polypeptides of most European skin strains were strongly reactive with a certain monoclonal antibody, whereas European isolates from cerebrospinal fluid (CSF) or ticks and North American isolates were mostly negative or only weakly reactive (Wilske et al. 1990).

4.5.3.2.1 Outer Membrane Proteins

The outer membrane of *B. burgdorferi* contains at least two abundant proteins with molecular weights of about 30-36 kDa, designated OspA (lower molecular weight) and OspB (higher molecular weight) (Barbour et al. 1983b, 1984). Both proteins are embedded in the fluid outer membrane of *B. burgdorferi* and are encoded by a linear plasmid (Barbour and Garon 1987).

OspA and OspB are antigenically variable proteins. Whereas the first analyses of a limited panel of North American strains have revealed homogeneity of their OspA components, the same panel was variable in its reactivity with monoclonal antibodies to the OspB (Barbour et al. 1983b, 1984). Later on it was shown that especially European isolates were antigenically variable in their OspA components (Wilske et al. 1985, 1986; Barbour and Schrumpf 1986). A major OspB is often absent or only weakly expressed in European isolates. Recently, Anderson et al. (1989) have isolated strains from wild animals which are more heterogeneous than the previously described North American isolates. In contrast, a major OspA component is present in almost all *B. burgdorferi* strains. Therefore, a serotyping system based on the OspA protein has been proposed (Wilske et al. 1988). Using sets of monoclonal and polyclonal antibodies, seven different serotypes were identified among 25 isolates. Later, the OspA serotyping system was based solely on monoclonal antibody testing; Table 4 gives the distribution of OspA serotypes within a panel of 57 *B. burgdorferi* isolates (Wilske et al. 1991).

Table 4. Distribution of B. burgdorferi OspA serotypes

	United States	Europe			
OspA type		Tick	Skin	Cerebrospinal fluid	Total
1	7[a]	2	1	1	11
2	–	2	18	2	22
3	–	1	–	1	2
4	–	–	1	2	3
5	–	–	–	2	2
6	–	6	1	–	7
7	–	1	–	–	1
X[b]	3[c]	2	–	–	5
0[d]	–	1	1	2	4
total	10	15	22	10	57

[a] Four isolates from ticks, three human isolates from skin, CSF, and blood, respectievely.
[b] Other than types 1-7
[c] Isolates from ticks.
[d] OspA-negative isolates.

4.5.3.2.2 Other Proteins

Many European strains exhibit a major protein in the 20 kDa range when analyzed by SDA-PAGE. This protein was designated pC by Wilske et al. (1986). In contrast to OspA and OspB, which are only weakly immunogenic in patients, the pC is a major protein of the early immune response (see also Chap. 21). The pC component is recognized by a monoclonal antibody nonreactive with other proteins of *B. burgdorferi* and also nonreactive with proteins of relapsing fever borreliae (Wilske et al. 1991). Another monoclonal antibody has been only reactive with the pC of a few strains (Wilske et al. 1991) indicating that the pC is antigenically heterogeneous.

Nearly all *B. burgdorferi* strains exhibit a 80-100 kDa range protein (p 100) which is recognized by a monoclonal antibody (L 100 1D4). An important finding was the nonreactivity of this monoclonal antibody with relapsing fever borreliae and treponemes. This antibody, therefore, may recognize a species-specific common epitope of *B. burgdorferi* (Wilske et al. 1990).

4.5.3.2.3 Antigenic Variability

The well-known antigenic variability of relapsing fever borreliae derived from a single clone raises the question whether a comparable antigenic variability is also present in *B. burgdorferi*. Preliminary reports with uncloned cultures have shown that *B. burgdorferi* spirochetes may change their protein profile during subculturing. Quantitative changes between OspA and OspB on one side and pC on the other side or the loss of OspA and OspB have been observed in subcultured European isolates (Wilske et al. 1986, 1988). Schwan and Burgdorfer (1987) and Schwan et al. (1988) have described loss or antigenic changes of OspB during subculturing. The most convincing data for an antigenic variability of the progeny of a single *B. burgdorferi* cell have been provided by Bundoc and Barbour (1989). These authors showed that daughter clones change the phenotype of the OspB protein compared with the parental clones. This means that the progeny of a single cell may differ from the parental cell. However, the authors believe that the mutation rate in culture is very low. It would be very interesting to determine antigenic variation rates in animal models where because of the host's immune response antigen variation may occur at a higher rate. The immunological characterization of *B. burgdorferi* strains from different geographical and biological sources and the examination for antigenic variability is very important in respect to taxonomy, epidemiology, and serodiagnosis, including antigen detection, pathogenesis of the disease, and development of a vaccine.

References

Ackermann R, Kabatzki J, Boisten HP, Steere AC, Grodzicki RL, Hartung S, Runne U (1984) Spirochäten-Ätiologie der Erythema-chronicum-migrans-Krankheit. Dtsch Med Wochenschr 109:92-97

Anderson JF, Magnarelli LA, LeFebvre RB, Andreadis TG, McAnich JB, Perng GC, Johnson RC (1989) Antigenically variable *Borrelia burgdorferi* isolated from cottontail rabbits and *Ixodes dentatus* in rural and urban areas. J Clin Micobiol. 27:13-20

Asbrink E (1985) Erythema chronicum migrans Afzelius and acrodermatitis chronica atrophicans. Early and late manifestations of *Ixodes ricinus*-borne *Borrelia* spirochetes. Akademisk Avhandling, Stockholm

Asbrink E, Hederstedt B, Hovmark A (1984) The spirochetal etiology of erythema chronicum migrans Afzelius. Acta Derm Venereol (Stockh) 64:291-295

Barbour AG (1988) Plasmid analysis of *Borrelia burgdorferi*, the Lyme borreliosis agent. J Clin Microbiol 26:475-478

Barbour AG, Garon CF (1987) Linear plasmids of the bacterium *Borrelia burgdorferi* have covalently closed ends. Science 237:409-411
Barbour AG, Hayes SF (1986) Biology of *Borrelia* species. Microbiol Ref 4 (50) 381-400
Barbour AG, Schrumpf ME (1986) Polymorphisms of major surface proteins of *Borrelia burgdorferi*. Zentralbl Bakteriol Hyg (A) 263:83-91
Barbour AG, Burgdorfer W, Hayes SF, Péter O, Aeschlimann A (1983a) Isolation of a cultivable spirochete from *Ixodes ricinus* ticks of Switzerland. Curr Microbiol 8:123-126
Barbour AG, Tessier SL, Todd WJ (1983b) Lyme disease spirochetes and ixodid tick spirochetes share a common surface antigenic determinant defined by a monoclonal antibody. Infect Immun 41:795-804
Barbour AG, Tessier SL, Hayes SF (1984) Variation in a major surface protein of Lyme disease spirochetes. Infect Immun 45:94-100
Barbour AG, Hayes SF, Heiland RA, Schrumpf ME, Tessier SL (1986) A *Borrelia*-specific monoclonal antibody binds to a flagellar epitope. Infect Immun 52:549-554
Baril C, Richaud C, Baranton G, Saint Girons J (1989) Linear chromosome of *Borrelia burgdorferi*. Res Microbiol 140:507-516
Barthold SW, Moody KD, Terwilliger JA, Duray PH, Jacoby RO, Steere AC (1988) Experimental Lyme arthritis of rats infected with *Borrelia burgdorferi*. J Infect Dis 157:842-846
Beck G, Habicht GS (1988) Isolation and biological activity of *Borrelia burgdorferi* peptidoglycan. Ann NY Acad Sci 539:365-366
Benach JL, Bosler EM, Hanrahan JP, Coleman JL, Habicht GS, Bost TF, Cameron DJ, Ziegler JL, Barbour AG, Burgdorfer W, Edelman R, Kaslow RA (1983) Spirochetes isolated from the blood of two patients with Lyme disease. N Engl J Med 308:740-742
Bundoc VG, Barbour AG (1989) Clonal polymorphisms of outer membrane protein OspB of *Borrelia burgdorferi*. Infect Immun 57:2733-2741
Burgdorfer W (1984) The New Zealand white rabbit. An experimental host for infecting ticks with Lyme disease spirochetes. Yale J Biol Med 57:609-612
Burgdorfer W (1991) Lyme borreliosis – a relapsing fever-like illness? Scand J Infect Dis Suppl 77:17-22
Burgdorfer W, Gage KL (1987) Susceptibility of the hispid cotton rat (*Sigmodon hispidus*) to the Lyme disease spirochete (*Borrelia burgdorferi*). Am J Trop Med Hyg 37:624-628
Burgdorfer W, Barbour AG, Hayes SF, Benach JL, Grunwaldt E, Davis JP (1982) Lyme disease – a tick-borne spirochetosis? Science 216:1317-1319
Burgdorfer W, Barbour AG, Hayes SF, Péter O, Aeschlimann A (1983) Erythema chronicum migrans – a tick-borne spirochetosis. Acta Trop (Basel) 40:17-83
Burgess EC (1986) Experimental inoculation of dogs with *Borrelia burgdorferi*. Zentralbl Bakteriol (A) 263:49-54
Cluss RG, Boothby JT (1990) Thermoregulation of protein synthesis in *Borrelia burgdorferi*. Infect Immun 58:1038-1042
Coleman JL, Benach JL, Beck G, Habicht GS (1986) Isolation of outer envelope from Borrelia burgdorferi. Zentralbl Bakteriol Hyg (A) 263:123-126
Elwell LP, Shipley PI (1980) Plasmid-mediated factors associated with virulence of bacteria to animals. Annu Rev Microbiol 34:465-496
Ferdows MS, Barbour AG (1989) Megabase-sized linear DNA in the bacterium *Borrelia burgdorferi*, the Lyme disease agent. Proc Natl Acad Sci USA 86:5969-5973
Fumarola D, Munno J, Marcuccio C, Miragliotta G (1986) Endotoxicity associated with the Lyme disease *Borrelia*: recent findings. Microbiologica 9:249-252
Garon CL, Dorward DW, Lubke LL, Corwin MD (1990) Ultrastructural detection of a DNA-containing compartment on the surface of *Borrelia burgdorferi* cells (poster). In: 4th international conference on Lyme borreliosis, Stockholm
Gassmann GS, Kramer M, Göbel UB, Wallich R (1989) Nucleotide sequenze of a gene encoding the *B. burgdorferi* flagellin. Nucleic Acids Res 17:3590
Goodman JL, Jurkovich P, Kramber JM, Johnson RC (1991) Molecular detection of persistent *Borrelia burgdorferi* in the urine of patients with active Lyme disease. Infect Immun 59:269-278

Habicht GS, Beck G, Benach JL, Coleman JL (1986) *Borrelia burgdorferi* lipopolysaccharide and its role in the pathogenesis of Lyme disease. Zentralbl Bakteriol Hyg (A) 263:137-141

Hansen K, Bangsborg JM, Fjordvang H, Pedersen NS, Hindersson P (1988) Immunochemical characterization and isolation of the gene for a *Borrelia burgdorferi* immunodominant 60-kilodalton antigen common to a wide range of bacteria. Infect Immun 56:2047-2053

Holt JG, Krieg NR (eds) (1984) Bergey's manual of systematic bacteriology, vol 1. Williams and Wilkins, Baltimore

Hyde FW, Johnson RC (1984) Genetic relationship of Lyme disease spirochetes to *Borrelia, Treponema*, and *Leptospira spp.* J Clin Microbiol 20:151-154

Hyde FW, Johnson RC (1988) Characterization of a circular plasmid from *Borrelia burgdorferi* etiologic agent of Lyme disease. J Clin Microbiol 26:2203-2205

Johnson RC, Hyde FW, Rumpel CM (1984a) Taxonomy of the Lyme disease spirochetes. Yale J Biol Med 57:79-87

Johnson RC, Schmid GP, Hyde FW, Steigerwalt AG, Brenner DJ (1984b) *Borrelia burgdorferi* sp.nov: etiologic agent of Lyme disease. Int J Syst Bacteriol 34:496-497

Johnson RC, Marek N, Kodner C (1984c) Infection of Syrian hamsters with Lyme disease spirochetes. J Clin Microbiol 20:1099-1101

Johnson RC, Kodner C, Russel M, Duray PH (1988) Experimental infection of the hamster with *Borrelia burgdorferi.* Ann NY Acad Sci 539:258-263

Kelly RT (1971) Cultivation of *Borrelia hermsii.* Science 173:443-444

Kelly RT (1976) Cultivation and physiology of relapsing fever borreliae. In: Johnson RC (ed) The biology of parasitic spirochetes. Academic, New York p 87

Kornblatt AN, Steere AC, Brownstein DG (1984) Experimental Lyme disease in rabbits: spirochetes found in erythema migrans and blood. Infect Immun 46:220-223

Kurtti TJ, Munderloh UG, Johnson RC, Ahlstrand GG (1987) Colony formation and morphology in *Borrelia burgdorferi.* J Clin Microbiol 25:2054-2058

Livermore BP, Bey RF, Johnson RC (1978) Lipid metabolism of *Borrelia hermsii.* Infect Immun 20:215-220

Persing DH, Telford SR, Spielman A, Barthold SW (1990) Detection of *Borrelia burgdorferi* infection in *Ixodes dammini* ticks with the polymerase chain reaction. J Clin Microbiol 28:566-572

Pillot Z (1965) Contribution à l' étude du genre Treponema: structure anatomique et antigenique. Thesis, Paris

Postic D, Edlinger C, Rauchaud C, Grimont F, Dufresne Y, Perolat P, Baranton G, Grimon PAD (1990) Two genomic species in *Borrelia burgdorferi.* Res Microbiol 141:465-475

Preac-Mursic V, Schierz G, Pfister-HW, Einhäupl K, Wilske B, Weber K (1984) Isolierung einer Spirochäte aus Liquor cerebrospinalis bei Meningoradikulitis Bannwarth. Munch Med Wochenschr 126:275-276

Preac-Mursic V, Wilske B, Schierz G (1986) European *Borrelia burgdorferi* isolated from humans and ticks: culture conditions and antibiotic susceptibility. Zentralbl Bakteriol Hyg (A) 263:112-118

Preac-Mursic V, Wilske B, Schierz G, Holmburger M, Süß E (1987) In vitro and in vivo susceptibility of *Borrelia burgdorferi.* Eur J Clin Mibrobiol 6:424-426

Preac-Mursic V, Patsouris E, Wilske B, Reinhardt S, Groß B (1990) Persistence of *Borrelia burgdorferi* and histopathological alterations in experimentally infected gerbils. Comparison with histopathological findings in human Lyme disease. Infection 18:1-10

Preac-Mursic V, Wilske B, Reinhardt S (1991) Culture of *Borrelia burgdorferi* on six solid media. Eur J Clin Microbiol 10:1076–1079

Rosa PA, Schwan TG (1991) A specific and sensitive assay for the Lyme disease spirochete *Borrelia burgdorferi* using polymerase chain reaction. J Infect Dis 160:1018-1029

Rosa PA, Hogan D, Schwan TG (1991) Polymerase chain reaction analyses indentify two distinct classes of Borrelia burgdorferi. J Clin Microbiol 29:524-532

Schaible UE, Kramer MD, Museteanu C, Zimmer G, Mossmann H, Simon MM (1989) The severe combined immunodeficiency (SCID) mouse. A laboratory model for analysis of Lyme arthritis and carditis. J Exp Med 170:1427-1431

Schmid GP, Steigerwalt AG, Johnson SE, Barbour AG, Steere AC, Robinson JM, Brenner DJ

(1984) DNA characterization of the spirochete that causes Lyme disease. J Clin Microbiol 20:155-158

Schwan TG, Barbour AG (1988) Efficacy of nucleic acid hybridization probes for the detection and identification of *Borrelia burgdorferi*. Ann NY Acad Sci 539:419-421

Schwan TG, Burgdorfer W (1987) Antigenic changes of *Borrelia burgdorferi* as a result of in vitro cultivation. J Infect Dis 156:852-853

Schwan TG, Burgdorfer W, Garon CF (1988a) Changes in infectivity and plasmid profile of the Lyme disease spirochete, *Borrelia burgdorferi,* as a result of in vitro cultivation. Infect Immun 56:1831-1836

Schwan TG, Burgdorfer W, Schrumpf ME, Karstens RH (1988b) The urinary bladder, a consistent source of *Borrelia burgdorferi* in experimentally infected white-footed mice (*Peromyscus leucopus*). J Clin Microbiol 26:893-895

Simpson WJ, Burgdorfer W, Schrumpf ME, Karstens RH, Schwan TG (1991) Antibody to a 39-Kilodalton *Borrelia burgdorferi* antigen (P39) as a marker for infection in experimentally and naturally inoculated animals. J Clin Microbiol 29:236-243

Stanek G, Burger I, Hirschl A, Wewalka G, Radda IA (1986) *Borrelia* transfer by ticks during their life cycle. Zentralbl Bakteriol Hyg (A) 263:29-33

Steere AC, Grodzicki RL, Kornblatt AN, Craft JE, Barbour AG, Burgdorfer W, Schmid GP, Johnson E, Malawista S (1983) The spirochetal etiology of Lyme disease. N Engl J Med 308:733-740

Stoenner HG (1974) Biology of *Borrelia hermsii* in Kelly medium. Appl Microbiol 28:540-543

Takayama K, Rothenberg RJ, Barbour AG (1987) Absence of lipopolysaccharide in the Lyme disease spirochete, *Borrelia burgdorferi*. Infect Immun 55:2311-2313

Wilske B, Preac-Mursic V, Schierz G (1985) Antigenic heterogeneity of European *Borrelia burgdorferi* strains isolated from patients and ticks. Lancet I: 1099

Wilske B, Preac-Mursic B, Schierz G, von Busch K (1986) Immunochemical and immunological analysis of European *Borrelia burgdorferi* strains. Zentral Bakteriol Hyg (A) 263:92-102

Wilske B, Preac-Mursic V, Schierz G, Kühbeck R, Barbour AG, Kramer M (1988) Antigenic variability of *Borrelia burgdorferi*. Ann NY Acad Sci 539:126-143

Wilske B, Preac-Mursic V, Fuchs R, Bruckbauer H, Hofmann A, Zumstein G, Jauris S, Soutschek E, Motz M (1990) Immundominant proteins of *Borrelia burgdorferi*; implications for improving serodiagnosis of Lyme borreliosis. In: Neu HC (eds) New antibacterial strategies. Churchill Livingstone, London

Wilske B, Anderson JF, Barbour AG, Hovind-Hougen K, Johnson RC, Preac-Mursic V (1991) Taxonomy of *Borrelia* spp. Scand J Infect Dis Suppl 77:108-129

III Ecology and Histopathology of Lyme Borreliosis

5 The Ecology of Lyme Borreliosis in Europe

L. Gern, W. Burgdorfer, A. Aeschlimann, and H.E. Krampitz

5.1 Introduction

The most common tick in Europe, the castor bean or sheep tick, *Ixodes ricinus* has been incriminated as a vector of the cause of the disease now referred to as Lyme borreliosis, ever since Dr. Arvid Afzelius in 1910 described, at the site of a tick bite on an elderly lady, the development of a skin lesion he referred to as erythema migrans. Since the etiology of this skin disorder was unknown, it was speculated that it might be a tick-associated toxin or infectious agent. The toxin hypothesis appeared unlikely because of the chronic nature of the disease. Lipschütz in 1923 suggested that the cause was associated with the tick bite and emphasized microscopic-bacteriologic investigations of the ticks' intestinal tract and salivary gland secretion. Similarly, Hellerström in 1950, referring to the demonstration of his colleague Carl Lennhoff of spirochete-like elements in skin sections of erythema migrans lesions, pointed to the possibility that this disease may be caused by tick-associated spirochetes.

Not only in erythema migrans, but also in other clinical manifestations now known to be related to Lyme borreliosis, ticks, particularly *I. ricinus,* have been implicated as the transmitting vectors. Thus, in 1922, Garin and Bujadoux described a case of tick paralysis in a patient bitten by *I. hexagonus*. The disorder was later recognized as lymphocytic meningoradiculitis or Bannwarth's syndrome (Bannwarth 1941) – a nervous system disorder commonly associated with the bite of *I. ricinus*. Similarly, numerous cases of acrodermatitis chronica atrophicans (ACA) and lymphadenosis benigna cutis were suspected to be associated with the bite(s) of *I. ricinus* even though many patients often no longer remembered having been bitten by ticks (Weber 1981; Rufli 1987).

The discovery in 1981 of a spirochete now known as *Borrelia burgdorferi* in the deer tick *I. dammini* in the United States (Burgdorfer et al. 1982), and in *I. ricinus* from Switzerland (Burgdorfer et al. 1982, 1983), which was found to be the agent of Lyme borreliosis, led to intensive epidemiological and ecological investigations on both sides of the Atlantic. This chapter is devoted to the natural history of *B. burgdorferi* in Europe and to the relationship(s) of this spirochete to its European tick vectors, primarily *I. ricinus*.

5.2 *Ixodes ricinus* – Principal Tick Vector of *B. burgdorferi* in Europe

5.2.1 Description and Geographic Distribution

The castor bean or sheep tick *I. ricinus* (Fig.1, see color plate I after page 118) is the most common tick in Europe where it has long been recognized as an important vector of various human and/or animal pathogens, including the virus of tick-borne encephalitis, *Babesia,* rickettsiae, trypanosomes, and microfilariae (Aeschlimann et al. 1979).

The literature on the biology of this tick is voluminous and has been referred to by Aeschlimann (1972, 1981). *I. ricinus* belongs to the genus *Ixodes* in the large family of ixodid or hard-shelled ticks – a family characterized by the chitinous scutum that, in male ticks, covers the entire dorsum, but in female ticks, only the anterior portion of the dorsum (Fig. 2). *I. ricinus* is a three-host tick whose developmental stages, i.e., the larva, the nymph, and the adult, parasitize different hosts (Fig. 3). In doing so, the fasting ticks leave their hiding places and climb surrounding vegetation such as grasses

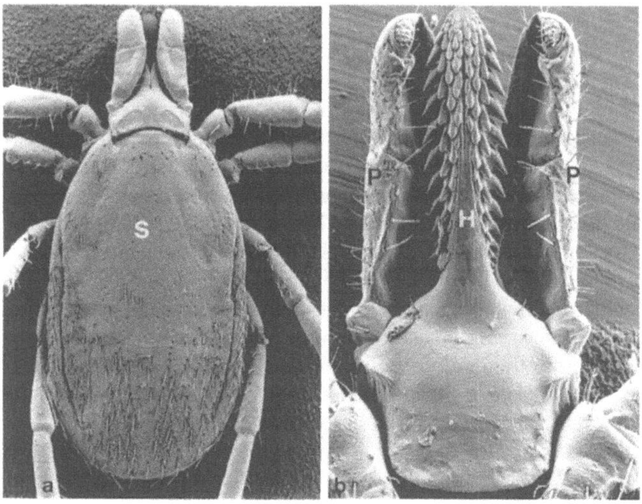

Fig. 2. a.b. *Ixodes ricinus*.
a Female, x 41.4 (*S*, scutum).
b Detail of a ventral view of the mouthpart, x 114.4 (*P*, pedipalps; *H*, hypostome). Scanning electron photographs (Courtesy of Tiago Cordas, Zoological Institute, Neuchâtel, Switzerland.

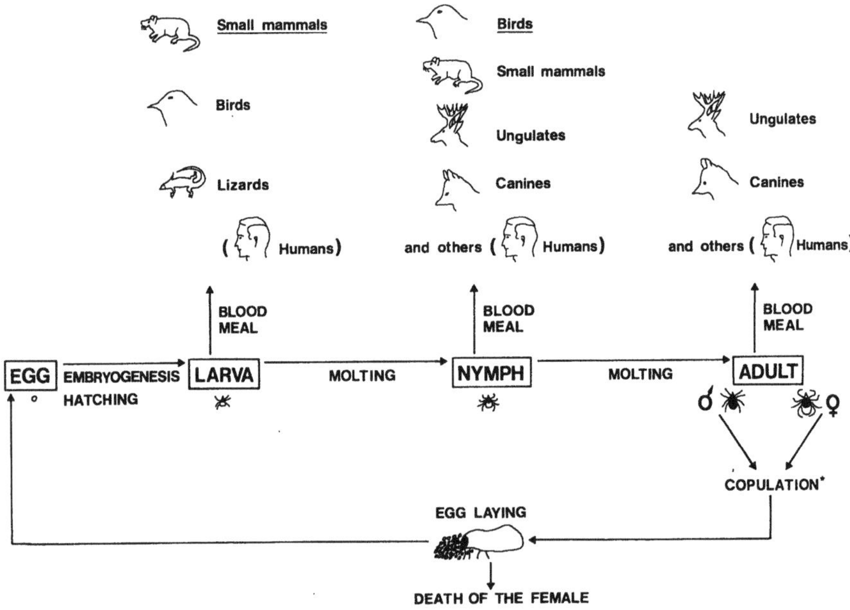

Fig. 3. Life cycle of *Ixodes ricinus* (modified from Aeschlimann et al. 1990)

and bushes where they lie in wait for a passing host. *I. ricinus* is notorious for the wide range of vertebrates it feeds on. In its immature stages, it attacks primarily small and medium-sized vertebrates such as lizards, ground-frequenting birds, rodents, insectivores, carnivores, wild pigs, and humans (Fig. 3). The adults prefer larger animals such as deer, cattle, dogs, and occasionally also humans.

I. ricinus is widely distributed throughout Europe and occurs in wooded and pasture areas from Ireland, Britain, and southern Scandinavia to Spain, Portugal, Algeria, Tunisia, and Morocco, much of southern Europe, central Europe, and European Russia to the Caspian Sea and northern Iran. Within this large distributional area, successful infestation by *I. ricinus* depends on fulfilling at least three biological and microclimatic requirements: (a) availability of suitable hosts, (b) temperature fluctuations between minus 10°C and 35°C with extreme values lasting for short periods of time only, and (c) constant relative humidity of at least 80% in the air and of near saturation in the soil. The ideal biotope of this tick, therefore, is a deciduous forest with damp soil covered by a thick layer and a rich undergrowth.

In central Europe, the tick's seasonal activity begins in March and ends in November. It is characterized by a bimodal pattern because of decreased activity during the hot summer months. At the end of November, the ticks disappear from hosts and vegetation to spend the winter months in the protective superficial layer of the soil. All developmental stages are active almost simultaneously, although nymphal and adult activity may occur 3–4 weeks before that of the larvae. Copulation of *I. ricinus* takes place during waiting periods on vegetation as well as on the host. Male ticks do attach, but ingest small quantities of blood only; they never engorge.

After 7–10 days of feeding, mated females drop to the ground where, after 8–30 days, they oviposit up to 3000 eggs, which develop after 4–8 weeks into the six-legged larval stage. Larval and nymphal *I. ricinus* that engorge in early summer molt to nymphs and adults, respectively, the same year. Engorged ticks or eggs oviposited during the fall period of activity go into a diapause, hibernate, and molt in early spring of the following year.

Although all stages of *I. ricinus* parasitize a large variety of animal hosts including humans, field studies in several European countries have shown that the larvae prefer small rodents, particularly the long-tailed and yellow-necked field mice (*Apodemus sylvaticus* and *A. flavicollis*, respectively) and the bank vole *(Clethrionomys glareolus)*. Nymphs, on the other hand, are apparently more frequently found on ground-frequenting birds and on larger animals, whereas adult *I. ricinus* parasitize roe deer *(Capreolus capreolus)* and red deer *(Cervus elaphus)*.

5.2.2 The Role of *Ixodes ricinus* as Vectors of *Borrelia burgdorferi* in Europe

The detection in *I. ricinus* from Switzerland of spirochetes identical with or similar to the causative agent of Lyme borreliosis in the United States has led to tick spirochete surveys in many countries including Austria, Germany, Switzerland, France, Czechoslovakia, the Scandinavian countries, the United Kingdom, and Ireland. According to the results summarized in Table 1, as many as 66% of the ticks examined were found to be infected with spirochetes, suggesting throughout large areas of Europe the existence of natural focid in which *B. burgdorferi* occurs in cycles between its tick vectors and animal hosts.

Recent tick/spirochete surveys in Germany and near Leningrad, Russia, revealed spirochetes also in other species of ticks such as *I. hexagonus* and *I. persulcatus* (see Table 1). The former parasitizes primarily hedgehogs and mustelidae, but is known to

Table 1. Prevalence of ticks infected with *Borrelia burgdorferi* in Europe

Country	Reference	Ticks examined			Infection rate		
		Species	n	Developmental stage	n	%	Range (%) *
Austria	Radda et al. 1986	*I. ricinus*	1207[a)]	adults and nymphs	27/90[b)]	20	2-40
Germany	Wilske et al. 1987	*I. ricinus*	1052	adults	220	21	11-34
			984	nymphs	104	11	3-26
			367	larvae	4	1	0-4
	Pelz et al. 1989	*I. ricinus*	129	adults	37	29	
			45	nymphs	3	7	
	Kahl et al. 1989	*I. ricinus*	173[a)]	adults	12/73[b)]	16	
			1365[a)]	nymphs	34/280[b)]	12	
			124	adults	10	8	
			49	nymphs	0		
Czechoslovakia	Kmeny et al. 1986	*I. ricinus*	550	adults	45	8	3-20
Sweden	Stiernstedt 1985	*I. ricinus*	132	adults	20	14	
	Mejlon et al. 1990	*I. ricinus*	159	adults	30	19	
			1032	nymphs	119	12	7-12
			388	larvae	0		
	Bergström et al. 1990	*I. ricinus*	88	adults	9	10	
Switzerland	Burgdorfer et al. 1983	*I. ricinus*	381	adults	112	29	
	Aeschlimann et al. 1986	*I. ricinus*	2500	nymphs		20	5-34

	Péter 1990	*I. ricinus*	508	adults	159	31	4-52
			119	nymphs	51	43	0-66
	Zhioua et al. 1988	*I. ricinus*	55	adults	8	15	
			258	nymphs	33	12	
			652	larvae	20	3	
	Miserez et al. 1990	*I. ricinus*	88	adults	22	25	
			94	nymphs	34	36	
			216	larvae	7	3	
United Kingdom	Muhlemann and Wright 1987	*I. ricinus*	130	adults	10	8	
			191	nymphs	15	8	
Germany	Liebisch et al. 1989	*I. hexagonus*	111	adults	13	12	
			178	nymphs	0		
			14	larvae	0		
	Liebisch and Olbrich 1990	*I. hexagonus*	610	larvae, nymphs and adults	52[c]	8	
Estonia	Barotov and Jögiste 1990	*I. persulcatus*	**	adults		12	≤35
		I. ricinus	**	adults		8	≤15
Russia	Korenberg (personal communication)	*I. persulcatus*					≤60
							≤30

*depending on the areas
[a] tested in pools
[b] infection rate of pools
[c] 2 larvae, 32 nymphs and 18 adults
**A total of 2638 *I. ricinus* and *I. persulcatus* were examined

attack also cats, dogs, and occasionally humans. Its role as a vector of *B. burgdorferi* remains to be evaluated. The taiga tick, *I. persulcatus,* on the other hand, is the principal vector of the Lyme borreliosis spirochete in Asian countries; its involvement in the ecology of *B. burgdorferi* in European Russia, where it coincides with *I. ricinus,* is not surprising.

5.2.3 Sources for Infecting *Ixodes ricinus* with *Borrelia burgdorferi*

Little information is available on the significance of animal hosts as sources for infecting *I. ricinus* with *B. burgdorferi.* The first isolation of spirochetes from rodents originated from Sweden (Hovmark et al. 1988), and additional isolations also from rodents were obtained (Aeschlimann et al. 1986) by feeding laboratory-reared ticks (xenodiagnosis) on field mice of the genus *Apodemus.* It is generally assumed that mice *(Apodemus* spp.*)* and voles *(Clethrionomys* spp*)* serve as spirochetal reservoirs. These rodents proved to be infective for ticks up to 1 year after their capture in nature (Vittoz et al. 1990). As yet, no data are available on whether mouse skin, rather than blood, may serve as a source of spirochetes as postulated by some investigators in the United States (Nakayama and Spielman 1989). Some species of ground-frequenting birds have also clearly demonstrated their reservoir competence for *B. burgdorferi* (Humair and Turrian-Vittoz 1990). Figure 4 illustrates the complexity of the ecology of Lyme borreliosis.

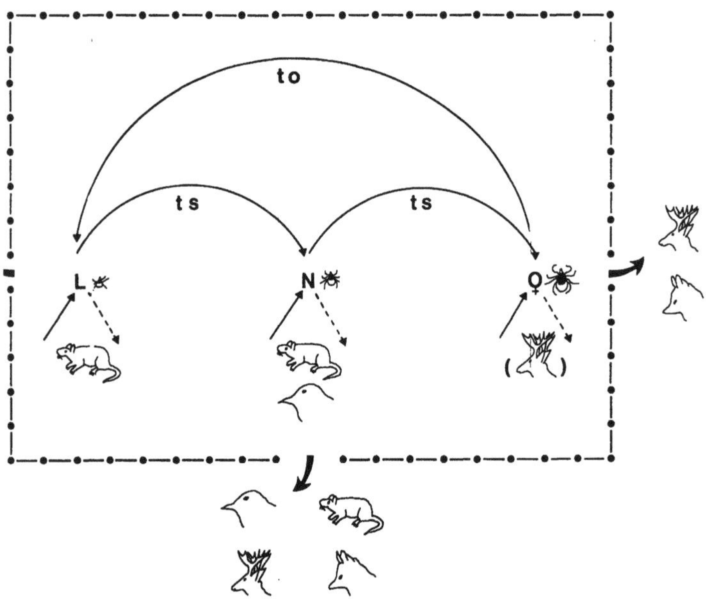

Fig. 4. Circulation of *Borrelia burgdorferi* in a Lyme borreliosis focus. A larva *(L)* can become infected on a small mammal and can transmit the infection to a nymph *(N)* and to an adult (here a female) transstadially *(ts).* This female can transmit the spirochetes to its progeny transovarially *(to).* The *arrows* (↘) indicate that every stage can transmit the infection to its host and can become infected (↗) on an infected host. The living area of *Ixodes ricinus* itself is confined to a small natural focus (—·—). Attached to the hosts, *Ixodes ricinus* may be dispersed and create new foci somewhere else (↓). (From Gern et al. 1991)

5.2.4 Relationship of *Borrelia burgdorferi* to Its Tick Vector *Ixodes ricinus*

The Lyme borreliosis spirochete exhibits, in its tick vectors, a behavior characteristically different from that of the hitherto investigated tick-borne spirochetes (Burgdorfer 1976) in that it causes a persistent infection in the ticks' midgut, from where it may penetrate during and after the ticks' engorgement to invade other tissues (Fig. 5). Of 112 infected adult *I. ricinus*, for example, only six had a generalized or systemic infection; all others had spirochetes in their midgut only (Burgdorfer et al. 1983).

Gut penetration and systemic infection by *B. burgdorferi* in American and European tick vectors have been the subjects of several investigations using conventional as well as electron microscopy (Benach et al. 1987; Ribeiro et al. 1987; Zung et al. 1989; Gern et al. 1990). Findings available so far suggest that *B. burgdorferi* persists in the midgut, where it aggregates between epithelial cells and becomes closely associated with the gut's basement membrane (Fig. 6). Penetration of the gut epithelium and dissemination into other tissues appears to take place during early and mid-feeding, i.e., 1–3 and 5–7 days after attachment, respectively. In *I. ricinus*, this phenomenon is said to take place as early as 3 days after tick attachment (Gern et al. 1990). Tick tissues other than the midgut are usually only mildly infected.

The presence and development of spirochetes in ovarian tissues has been demonstrated in all principal vectors of the Lyme borreliosis spirochete, and there is convincing evidence that transovarial transmission also occurs. This phenomenon, however, is

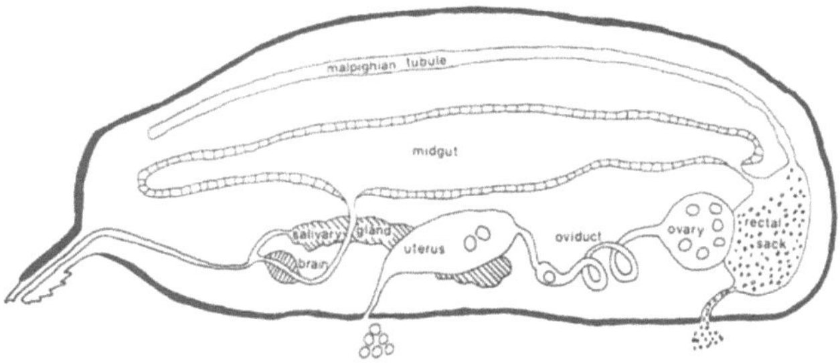

Fig. 5. Internal anatomy of *Ixodes ricinus*

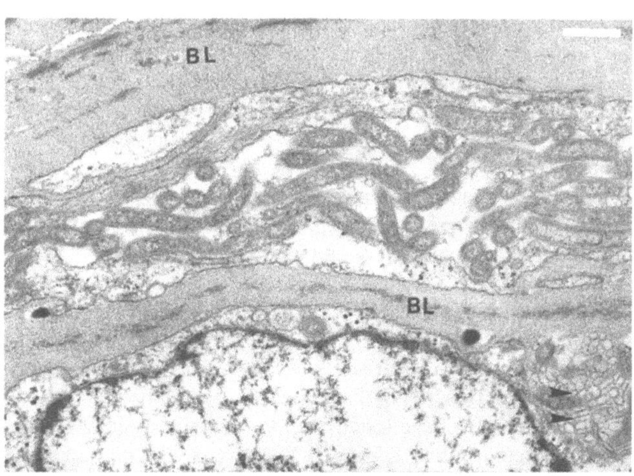

Fig. 6. *Borrelia burgdorferi* colonizing the basal lamina *(BL)* of *Ixodes ricinus* midgut. Spirochetes perturb the space of a gut wall muscle cell. *Arrows* denote colonization site at the base of a degenerating digestive cell. Transmission electron micrograph (*Bar*, 0.5 µm). Provided by one of us (W.B.) and S.F. Hayes, Rocky Mountain Laboratories, Hamilton, Montana, USA

not considered a major ecological factor for infecting *I. ricinus* with *B. burgdorferi* because filial infection rates established so far do not exceed 3% (Wilske et al. 1987; Zhioua et al. 1988). Oviposition of spirochete-infected eggs has been recorded for two *I. ricinus* from Switzerland (Burgdorfer et al. 1983). One produced 100% infected eggs, the other only 60%. The F_1 larvae were unfortunately not examined.

Investigations into the development of *B. burgdorferi* in the American tick vector *I. dammini* have also shown that massive spirochetal development in oogonia and oocytes may have an adverse effect on egg development (Burgdorfer et al. 1988; Burgdorfer and Hayes 1989). Spirochetes between the vitelline and oocyte membranes were invariably found to denude the oogonial surface and to interfere with the formation of the egg shell. Ticks with heavy systemic infections rarely produced viable eggs.

5.2.5 Transmission of *Borrelia burgdorferi* by *Ixodes ricinus*

Successful transmission of *B. burgdorferi* by ticks with spirochetal infections limited to the midgut led to the hypothesis that transmission may occur by regurgitation of gut material rather than by saliva (Burgdorfer 1984). This hypothesis, however, was questioned by the demonstration (Ribeiro et al. 1987) of spirochetes in experimentally obtained saliva as early as 3 days after tick attachment. Increasing numbers of spirochetes thereafter also suggested invasion by *B. burgdorferi* that had penetrated the gut epithelium during early feeding (see discussion above). Thus, if effective transmission via saliva does occur, it must take place during late feeding, i.e., from 3–8 days after tick attachment. Whether regurgitation of spirochete-containing gut material represents an additional mode of transmission awaits the results of ongoing investigations with *I. dammini* and *I. ricinus*.

5.3 Concluding Remarks

To understand the natural history of Lyme borreliosis in Europe, it is necessary to establish the parameters of interactions between the causative spirochete *B. burgdorferi*, its primary vector *I. ricinus*, and its animal hosts. Although valuable information has been obtained since the discovery of the spirochete 10 years ago, our understanding of how this agent interacts with the ecological components of its natural history is far from complete and appears to be more complex than initially assumed. *I. ricinus*, for instance, was considered the only tick vector until tick/spirochete surveys in Germany revealed *B. burgdorferi*-like spirochetes in the hedgehog tick *I. hexagonus*. If shown to serve as a competent vector of *B. burgdorferi* (see: Gern et al. 1991), this tick and possibly other species such as *I. trianguliceps* may contribute to the establishment of enzootic foci of Lyme borreliosis in areas where *I. ricinus* is rare or absent. Similarly, the finding (Stanek and Simeoni 1989) in Southern Tirol (Italy) of *B. burgdorferi* in the argasid tick *Argas persicus*, which is associated with pigeons, deserves our attention and suggests that certain species of birds may play a role in the ecology of this organism.

The question of species unity or plurality for the Lyme borreliosis spirochete has led to extensive efforts to recover spirochetes from patients, ticks, and tick hosts, for identification and comparison with isolates from the United States. Recent deoxyribonucleic acid (DNA)/DNA hybridization and ribosomal ribonucleic acid (rRNA) gene-restriction studies have identified two genomic species in *B. burgdorferi* (Postic et al. 1990). Whether either species is associated with certain ecological factors or specific clinical manifestations remains to be determined. Species plurality should be considered, and encounter, in other vertebrate and invertebrate hosts, of other species, such as *B. theileri,* the causative agent of bovine borreliosis, and of hitherto undescribed spirochetal agents should not be ruled out. Finally, according to Stanek (personal communication), the incidence of Lyme borreliosis in Europe, which has now been reported from at least 19 countries, will increase. Therefore, good surveillance of natural hosts and of the vector(s) of *B. burgdorferi* are essential in defining the boundaries of natural foci and the risk of human infection.

References

Aeschlimann A (1972) *Ixodes ricinus* Linné 1758 *(Ixodoidea, Ixodidae).* Essai préliminaire de synthèse sur la biologie de cette espèce en Suisse. Acta Trop (Basel) 29:321-340

Aeschlimann A (1981) The role of hosts and environment in the natural dissemination of ticks. Studies on a Swiss population of *Ixodes ricinus* L, 1758. Rev Adv Parasitol Warszawa 859-869

Aeschlimann A, Burgdorfer W, Matile H, Péter O, Wyler R (1979) Aspects nouveaux du rôle de vecteur joué par *Ixodes ricinus* L en Suisse. Acta Trop (Basel) 36:181-191

Aeschlimann A, Chamot E, Gigon F, Jeanneret JP, Kesseler D, Walther C (1986) *B. burgdorferi* in Switzerland Zentralbl Bakteriol Hyg (A) 263:450-458

Aeschlimann A, Brossard M, Haug T, Rutti B (1990) A survey of tick vaccines. Ann Res Dev 32:52-72

Afzelius A (1910) Verhandlungen der Dermatologischen Gesellschaft zu Stockholm. Arch Dermatol Syph 101:404-404

Bannwarth A (1941) Chronische lymphocytäre Meningitis, entzündliche Polyneuritis und "Rheumatismus". Arch Psychiatr Nervenkr 113:284-376

Barotov O, Jögiste A (1990) Studies of Lyme borreliosis in Estonia. 4th Int Conf Lyme Borreliosis, Stockholm, Sweden, June 18-21, abstract W/Th-9-2

Benach JL, Coleman JL, Skinner RA, Bosler EM (1987) Adult *Ixodes dammini* on rabbits. a hypothesis for the development and transmission of *Borrelia burgdorferi.* J Infect Dis 155:1300-1306

Bergström S, Burman N, Jaenson T, Jonsbo F, Jonsson M, Mejlon H, Olsén B (1990) Molecular characterization of *Borrelia burgdorferi* strains isolated from *Ixodes ricinus* in Northern Sweden. 4th international conference on Lyme borreliosis, Stockholm, Sweden, 18-21 June, abstract W/TH-P-17

Burgdorfer W (1976) The epidemiology of the relapsing fevers. In: Johnson RC (ed) The biology of parasitic spirochetes. Academic, New York, pp 191-200

Burgdorfer W (1984) Discovery of the Lyme disease spirochete and its relation to tick vectors. Yale J Biol Med 57:515-520

Burgdorfer W, Hayes SF (1989) Vector-spirochete relationships in louse-borne and tick-borne borreliosis with emphasis on Lyme disease. In: Harris KF (ed) Advances in disease vector research. Springer, Berlin Heidelberg New York, pp 127-150

Burgdorfer W, Barbour AG, Hayes SF, Benach JL, Grunwaldt E, Davis JP (1982) Lyme disease – a tick-borne spirochetosis? Science 216:1317-1319

Burgdorfer W, Barbour AG, Hayes SF, Péter O, Aeschlimann A (1983) Erythema chronicum migrans – a tickborne spirochetosis. Acta Trop (Basel) 40:79-83

Burgdorfer W, Hayes SF, Benach JL (1988) Development of *Borrelia burgdorferi* in ixodid tick vectors. Ann NY Acad Sci 539:172-179

Garin C, Bujadoux (1922) Paralysie par les tiques. J Med Lyon 71:765-767

Gern L, Zhu Z, Aeschlimann A (1990) Development of *Borrelia burgdorferi* in *Ixodes ricinus* females during blood feeding. Ann Parasitol Hum Comp 65:89-93

Gern L, de Marval F, Aeschlimann A (1991) Comparative considerations on the epidemiology of Lyme borreliosis and tick-borne encephalitis in Switzerland. In: Modern acarology Academic, The Hague/Academia, Prague, pp 249–254

Gern L, Toutoungi LN, Hu CM, Aeschlimann A (1991) *Ixodes (Pholeoixodes) hexagonus*, an efficient vector of *Borrelia burgdorferi* in the laboratory. Medical and Veterinary Entomology 5:431–435

Hellerström S (1950) Erythema chronicum migrans Afzelius with meningitis. South Med J 43:330-334

Hovmark A, Jeanson TGT, Asbrink E, Forsman A, Jansson F (1988) First isolates of *Borrelia burgdorferi* from rodents in Northern Europe. Acta Pathol Microbiol Immunol Scand 96:917-920

Humair PF, Turrian-Vittoz N (1990) Importance du rôle de réservoirs des micromammifères et des oiseaux dans la borréliose de Lyme. Travail de licence. Institut de Zoologie, Université de Neuchâtel.

Kahl O, Schmidt K, Schonberg A, Laukamm-Josten U, Knulle W, Bienzle U (1989) Prevalence of *Borrelia burgdorferi* in *Ixodes ricinus* ticks in Berlin (West). Zentralbl Bakteriol Hyg (A) 270:434-440

Kmety E, Rehacek J, Vyrostekova V (1986) Investigation of ticks for the presence of *Borrelia burgdorferi* in Czechoslovakia. Zentralbl Bakteriol Hyg (A) 263:468-470

Liebisch A, Olbrich S (1990) Hedgehog tick *Ixodes hexagonus* as vector of *Borrelia burgdorferi* in Europe. 8th international congress on acarology, Ceske Budejovice, Czechoslovakia, 6-11 Aug

Liebisch A, Olbrich S, Brand A, Liebisch G, Mourettou-Kunitz M (1989) Natürliche Infektionen der Zeckenart *Ixodes hexagonus* mit Borrelien *(Borrelia burgdorferi)*. Tierärztl. Umschau 44:809-810

Lipschütz B (1923) Weiterer Beitrag zur Kenntnis des "Erythema chronicum migrans". Arch Dermatol Syph 143:365-374

Mejlon HA, Jeanson TGT, Hovmark A, Asbrink E, Forsman A (1990) Ecology of Lyme borreliosis in Sweden. 4th international conference on Lyme borreliosis, Stockholm, Sweden, 18-21 June, abstract W/TH-P-6

Miserez V, Gern L, Aeschlimann A (1991) *Borrelia burgdorferi* in ticks of the canton Tessin (Switzerland). Parassitologia 32:293-299

Muhlemann MF, Wright DJM (1987) Emerging pattern of Lyme disease in the United Kingdom and Irish Republic. Lancet I:260-262

Nakayama Y, Spielman A (1989) Ingestion of Lyme disease spirochetes by ticks feeding on infected hosts. J Infect Dis 160:166-167

Pelz K, Wagner W, Vogt A (1989) *Ixodes ricinus* ticks as vectors of *Borrelia burgdorferi* in the Freiburg area. Zentralbl Bakteriol (Suppl) 18:35-39

Péter O (1990) Lyme borreliosis in the State of Valais, Switzerland. J Int Fed Clin Chem 2:121-124

Postic D, Edlinger C, Grimont F, Dufresne J, Perolat P, Baranton G, Grimont PAD (1990) Two genomic species in *Borrelia burgdorferi*. Res Microbiol 141:465-475

Radda A, Burger I, Stanek G, Wewalka G (1986) Austrian hard ticks as vectors of *Borrelia burgdorferi*. Zentralbl Bakteriol Hyg (A) 263:79-82

Ribeiro JMC, Mather TN, Piesman J, Spielman A (1987) Dissemination and salivary delivery of Lyme disease spirochetes in vector ticks (Acari: Ixodidae). J Med Entomol 24:201-205

Rufli Th (1987) Durch Zecken übertragene Erkrankungen. Schweiz Rundsch Med Prax 76:699-704

Stanek G, Simeoni J (1989) Are pigeons' ticks transmitters of *Borrelia burgdorferi* to humans? A preliminary report. Zentralbl Bakteriol (Suppl) 18:42-43

Stiernstedt G (1985) Tick-borne *Borrelia* infection in Sweden. Thesis, Danderyd Hospital

Vittoz N, Humair PF, Siegenthaler M, Aeschlimann A, Gern L (1990) Mammalian and avian reservoirs for *Borrelia burgdorferi* in a Lyme borreliosis focus in Switzerland. Rev Suisse Zool 97:783-783

Weber K (1981) Erkrankungen nach Zeckenbiss. Z Allgemeinmed 57:1158-1167

Wilske B, Steinhuber R, Bergmeister H, Fingerle V, Schierz G, Preac-Mursic V, Vanek E, Lorbeer B (1987) Lyme Borreliose in Süddeutschland. Dtsch Med Wochenschr 112:1730-1736

Zhioua E, Gern L, Monin R, Aeschlimann A (1988) Infection of free-living life-stages of *Ixodes ricinus* with *Borrelia burgdorferi* in Switzerland. Zentralbl Bakteriol Hyg (A) 306:293-293

Zung JL, Lewengrub S, Rudzinska MA, Spielman A, Telford SR, Piesman J (1989) Fine structural evidence for the penetration of the Lyme disease spirochete *Borrelia burgdorferi* through the gut and salivary tissues of *Ixodes dammini*. Can J Zool 67:1737-1748

6 Histopathology of Human Lyme Borreliosis

J. DE KONING and P.H. DURAY

6.1 General Observations

Lyme borreliosis is a disorder affecting many organs including the skin and reticuloendothelial, cardiovascular, nervous and musculoskeletal systems. The causative organism is *Borrelia burgdorferi*. After infection via a tick bite, a series of events follows in an attempt to destroy the microorganisms or limit its spread to other sites. The early skin lesion around the tickbite shows edema, vascular dilatation, extravasation of erythrocytes and an inflammatory infiltrate of neutrophils. If the causative organisms persist in the skin, the infiltrate shows predominantly lymphocytes and in many circumstances plasma cells.

After dissemination, *B. burgdorferi* appears to cause direct organ damage or to provoke immune mechanisms resulting in inflammation. The infiltrates in the early phase of the disease are histologically characterized by the presence of both B and T lymphocytes (helper and suppressor phenotype), macrophages, interdigitating reticulum cells, mast cells and plasma cells. During chronic disease, arterioles and venules may show concentric layers of perivascular cells, including fibroblasts. The blood vessels are sometimes narrowed with reduction or obliteration of the lumen, or they can be dilated. Hyalinization of the collagen may also be seen. These changes are particularly pronounced in the synovia in chronic Lyme arthritis and in the skin in acrodermatitis chronica atrophicans (ACA). Plasma cells predominate in this stage, but lymphocytes are also present. Fibrosis is a feature of late disease.

6.2 Histopathology

It is helpful to regard the disease as occurring in stages: an initial stage (stage I), early dissemination (stage II), and a chronic late phase (stage III).

6.2.1 Stage I

6.2.1.1 Skin

Skin involvement, as manifested by erythema migrans, is the characteristic lesion of early Lyme borreliosis (Weber et al. 1984). Borrelial lymphocytoma (BL) can be seen in both early and late disease, but early occurrence is more common. BL is typically seen in the ear lobe, areola mammae, and scrotum.

6.2.1.1.1 Erythema migrans

Biopsies from the center of erythema migrans lesions show slight subpapillary edema with a superficial and a deep perivascular and interstitital lymphohistiocytic infiltrate (Hauser 1965; Berger et al. 1983; Neubert et al. 1986; Büchner and Rufli 1987). Neutrophils may also be present around the blood vessels, especially in early lesions, but they are more often found in the interstitium between the collagen fibers of the dermis. Sometimes a lymphocytic infiltrate is located at the dermoepidermal junction with disruption of the basement membrane. Plasma cells may be seen in the perivascular region. Dilated lymph vessels are sometimes visible at the border of the infiltrate (Fig. 1). The erythematous border only shows a slight lymphohistiocytic infiltrate, occasionally mixed with plasma cells, and slightly dilated blood vessels (Fig. 2).

6.2.1.1.2 Localization of Spirochetes in the Skin

The *B. burgdorferi* microorganisms can be detected in the tissue by a modified Dieterle's stain (Duray et al. 1985) or a modified Steiner method (de Koning et al. 1987). They show up as coiled, undulating, or long elongated forms. In the skin, the spirochetes are located in the dermal papillae and especially in the reticular cutis, often adhering to collagen fibers (Fig. 3). They can sporadically be found in vessel walls and in the lumen of vessels (Fig. 4). Intracellular localizations of spirochetes have not been observed (de Koning and Hoogkamp-Korstanje 1986a). The preferential localization of the spirochetes between and in proximity to collagen fibers might be a reason for the destruction of these fibers, explaining the atrophy of the skin in chronic disease.

6.2.1.1.3 Borrelial Lymphocytoma

Pseudolymphomas are benign lymphocytic skin lesions of known or unknown etiology, showing a large variation in the composition of the infiltrate (Mach and Wilgram 1966; Caro and Helwig 1969; Burg and Braun-Falco 1983; Eckert et al. 1989). BL is a pseudolymphoma caused by *B. burgdorferi*. It is not entirely synonymous with lymphadenosis benigna cutis (LABC) (Weber et al. 1984, 1985), a term which was introduced by Bäfverstedt in 1943 and applies to lesions that may appear in a solitary or a disseminated form (LABC dispersa). Many of the solitary LABC lesions previously described may in fact be of borrelial origin, especially the lesions of the ear lobe, areola mammae, and scrotum. However, it is still undecided whether the disseminated lesions are *Borrelia*-induced. Thus, our knowledge of the histopathology of BL is based at present mostly on the unpublished observations of one uf us (de Koning) and previous experience of solitary LABC lesions (Bäfverstedt 1943; Paschoud 1954, 1957, 1958a,b,c; Asbrink and Hovmark 1988).

BL usually shows a diffuse, predominantly lymphocytic infiltrate with or without follicular structures resembling the germinal centers of lymph nodes (Lennert and Stein 1982). The infiltrate extends from the epidermis into the subcutaneous fat and consists of lymphocytes mixed with some centrocytes and centroblasts (Fig. 5). Follicular structures containing active germinal centers are composed of identical cell types as in the germinal centers of normal lymphatic tissue. Outside and in between the secondary follicles, the infiltrate consists of small lymphocytes, plasma cells, eosinophils, mast cells, and some dendritic reticulum cells and macrophages (Fig. 6,7). Solitary LABC lesions show a polyclonal proliferation of B and T cells (Braun-Falco and Burg 1975; Burg and Braun-Falco 1983). This holds true also for the BL lesions (de Koning unpublished data) (see Chap. 9).

The ultrastructure of the BL shows two main types of cells; namely lymphoid cells and a phagocytic reticulum cell (K. Weber, unpublished observations; de Koning, unpublished observations) and, therefore, strongly resembles the descriptions of LABC lesions presented by Schmoeckel et al. (1977) and Geerts and Kaiserling (1985). The phagocytic cells with their long cytoplasmatic processes have been thought to represent or to be closely related to macrophages (Fig. 8). In most cases of BL, spirochetes can be demonstrated, but sometimes serial sections have to be screened. Spirochetes are often seen in lesions with large active germinal centers (de Koning 1986b).

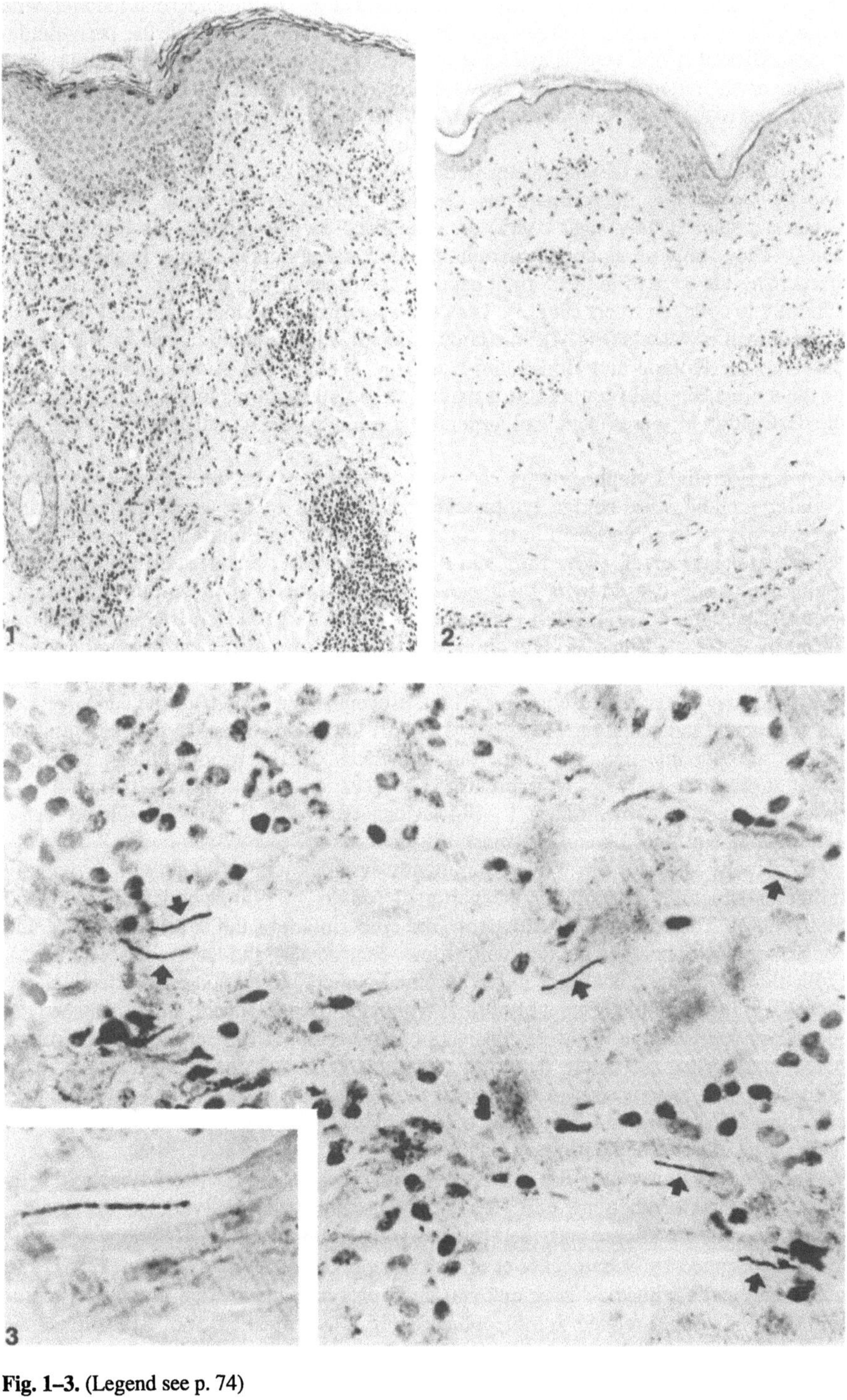

Fig. 1–3. (Legend see p. 74)

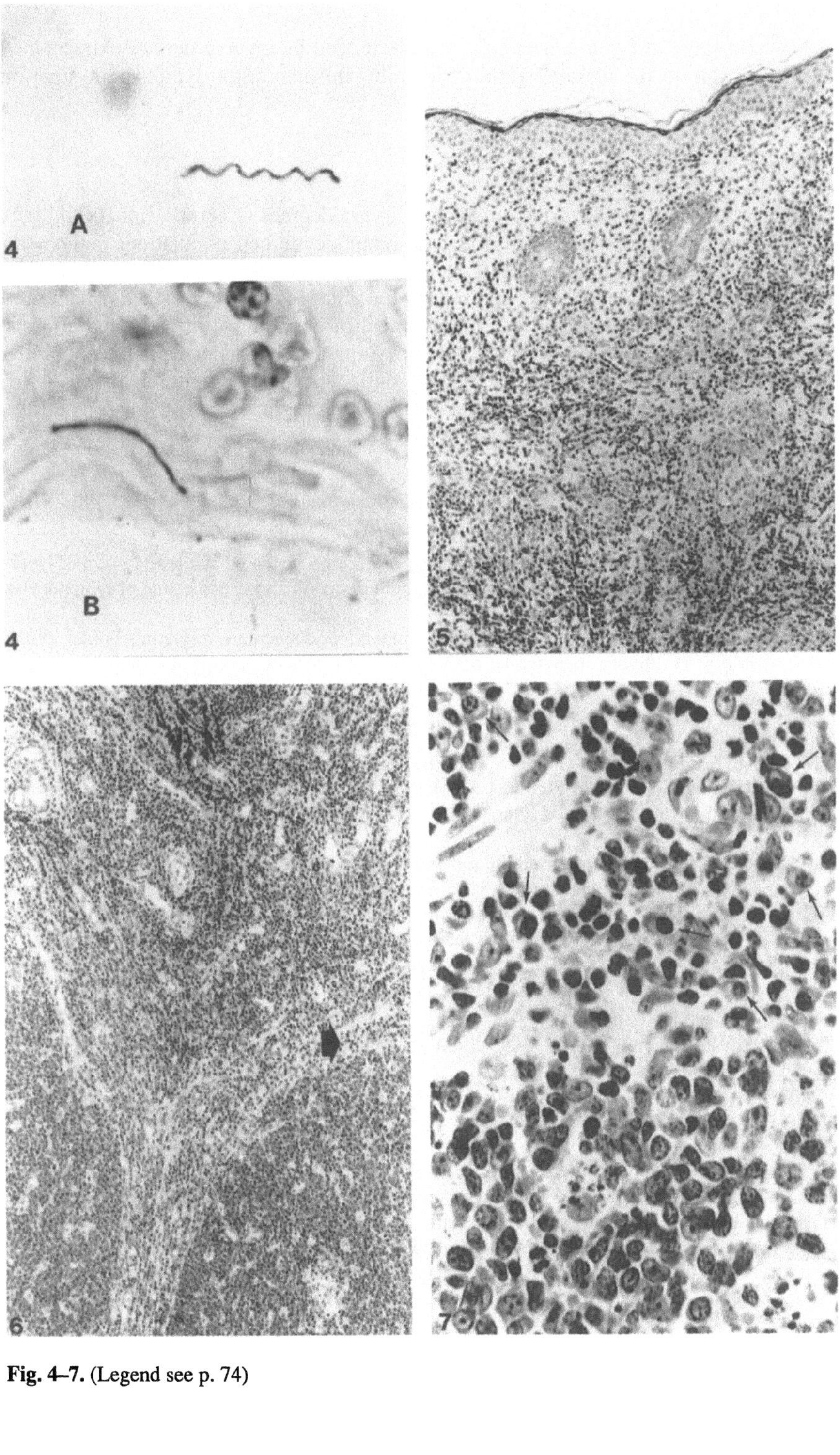

Fig. 4–7. (Legend see p. 74)

6.2.2 Stage II

The second stage of Lyme borreliosis is characterized by involvement of extracutaneous organs due to the dissemination of microorganisms via the lymphatic system or bloodstream.

6.2.2.1 Neurological Involvement

Patients with meningoradiculoneuritis have varying degrees of cerebrospinal fluid pleocytosis, consisting of a mixed plasmacytic-lymphocytic cell population. Some cells may show blast transformation (Chap. 8).

Lymphocytes may involve the cerebral leptomeninges during this stage (Meurers et al. 1990; Duray, unpublished data). Involvement of ganglia and afferent and efferent rootlets is also characterized by aggregates of lymphocytes. Plasma cells are not prominent. We have not seen spirochetes in the ganglia, but the assumption is that they are there. The central nervous system may show focal microgliosis during this stage of the disease (Duray 1989a,b; Meurers et al. 1990).

◁

Fig. 1. Erythema migrans. Biopsy taken from the center of a lesion on the left leg of a 41-year-old woman. Superficial and deep perivascular and interstitial infiltrate consisting of lymphocytes and histiocytes, H & E, x 60
Fig. 2 Erythema migrans. Biopsy from the peripheral erythematous migrating border (same lesion as in Fig. 1). Slight subepidermal perivascular infiltrates composed of lymphocytes, H & E, x 40.
Fig. 3. Erythema migrans of the upper arm of a 34-year-old woman. Biopsy taken from the center of the lesion. The spirochetes *(arrows)* stain black, modified Steiner method, x 400. Insert shows a high magnification of another area of the same section (polarized light), x 1250.
Fig. 4 *A,B*. Same biopsy as in Fig. 3. *A* Spirochetes located in the lumen of a lymph vessel. *B* Spirochete in the wall of a dilated lymph vessel. (Semi-thin section, modified Steiner method, x 1000)
Fig. 5. Borrelial lymphocytoma from the right ear lobe of a 12-year-old boy. A diffuse lymphoid infiltrate is seen throughout the entire cutis. A free subepidermal zone is not present, H & E, x 40
Fig. 6. Borrelial lymphocytoma from the scrotum of a 4-year-old-boy. Follicular structures are present on both sides of the lower part of the picture. The infiltrate between the follicular structures is predominantly composed of small lymphocytes, with plasma cells and some eosinophils. The *arrow* indicates the area detailed in Fig. 7, H & E, x 60
Fig. 7. Borrelial lymphocytoma (detail of Fig. 6). The follicular structure in the lower part shows some phagocytic cells. The infiltrate in the upper part contains many plasma cells *(arrows)*, H & E, x 300

▷

Fig. 8. Borrelial lymphocytoma from the nipple of a 5-year-old girl. The infiltrate is mainly composed of lymphoid cells *(L)* and phagocytic reticulum cells *(R)* and their processes. Reticulum cells contain many cytoplasmic processes with ribosomes (*) and sometimes also with phagosomes (**). (Electron micrograph kindly supplied by K. Weber)
Fig. 9 A - D. Carditis. Biopsies of right ventricle of a 24-year-woman. *A* The band-like subendocardial infiltrate consists of lymphocytes and plasma cells, H & E, x 125. *Arrow* indicates the area detailed in B. *B* Higher magnification of the subendocardial infiltrate with plasma cells, H & E, x 400. *C, D* Spirochetes in the interstitium at the border of the muscle fibers, Bosma-Steiner method, x 650

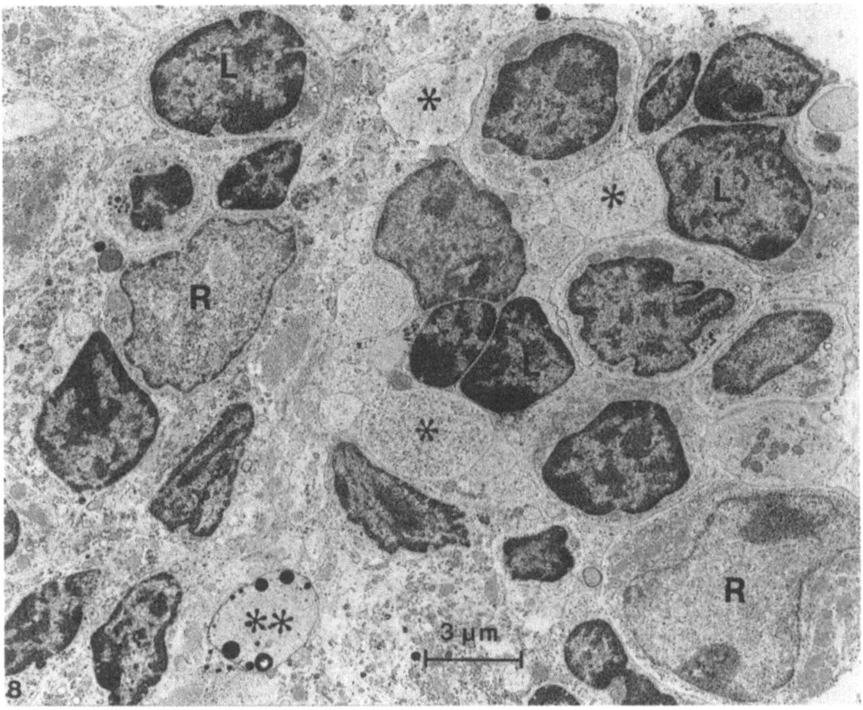
8
L
L
L
R
R
*
*
*
**
3 µm

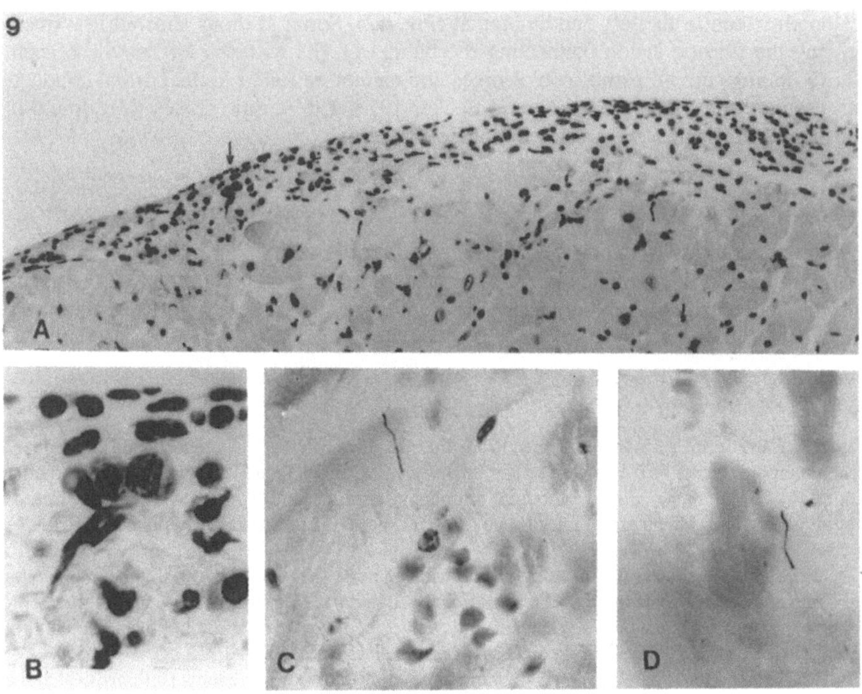
9
A
B
C
D

Peripheral nerve biopsies show heavy infiltrations around epineural vasa nervorum and slight infiltrations around endoneural capillaries. These infiltrates are composed of lymphocytes, histiocytes, and plasma cells. Necrotizing changes of blood vessel walls are not present, but thrombosis and recanalization can be seen in some epineurial vessels. The nerves may show loss of myelinated axons due to axon degeneration (Vallat et al. 1987; Meier et al. 1989; Meurers et al. 1990) (see Chapt. 11). Spirochetes have not been seen in the peripheral nerves so far, but have been demonstrated in smears of the cerebrospinal fluid with silver staining methods (Wokke et al. 1987a).

6.2.2.2 Heart

The histopathologic findings in cardiac biopsies include an interstitial infiltrate of lymphocytes and plasma cells involving the myocardium, pericardium, and endocardium. Aggregates of lymphocytes may be seen in the myocardium. In general, the muscle fibers are intact. Sporadically, individual myocardial fibers show infiltration by lymphocytes. The endocardium shows band-like infiltrates of lymphocytes and plasma cells (Marcus et al. 1985; Reznick et al. 1986; Vacher et al. 1987; van der Linde et al. 1990) (Fig. 9). Neutrophils and eosinophils are absent. Obliterative microangiopathy cannot be demonstrated. Occasionally, spirochetes with the typical shape of *B. burgdorferi* may be found near and in the infiltrates, between the muscular fibers, and in the endocardium (Marcus et al. 1985; de Koning et al. 1989) (Fig. 9).

Recently, we were able to study the conducting system of the heart of a 38 year-old man (de Koning, Hoogkamp-Korstanje, and Koster, unpublished data). The sinoatrial and the atrioventricular nodes showed slight lymphocytic infiltration and localized edema (Fig. 10,11). Fibers with contraction band necrosis were found in an edematous area of the sinotrial node. Focal edema was present in the bundle of His, and a fibrotic lesion was seen in the left bundle branch (Fig. 12). Serial sections showed few fibers passing the fibrotic lesion connecting the bundle of His with the left bundle branch. These findings are all pathologic changes and cannot be found in the normal conducting tissue of the heart (Anderson and Becker 1978, 1980). Spirochetes were found in the bundle of His near the areas with localized edema.

6.2.2.3 Lymph Nodes

Lymph nodes show follicular hyperplasia, as does the white pulp of the spleen. There is a proliferation of plasma cell precursors and an increase in mature plasma cells. The plasma cell precursors are large, appear atypical, and may resemble Reed-Sternberg

▷

Fig. 10. Carditis in a 38-year-old man who died 21 days after cardiac arrest (and subsequent severe cerebral ischemia). The patient had received intravenous penicillin for 10 days. Sinoatrial node *(Sn)* shows edema and an interstitial lymphocytic infiltrate. The artery of the node *(star)* is not involved. The subepicardial fatty tissue *(asterisk)* contains infiltrates of lymphocytes, H & E, x 25

Fig. 11. Same patient as in Fig. 10. Atrioventricular node *(asterisk)* shows slight lymphocytic infiltrate and edema in the zone of the transitional cells *(stars)*. The fibrous tissue *(f)* of the membranous septum (located between atrium and ventricle) reveals no abnormalities. The septum interventriculorum *(S)* is seen in the lower part of the picture, H & E, x 25

Fig. 12. Same patient as in Fig. 10. The branching bundle shows focal edema *(thin arrows)* and a fibrous lesion of the left bundle branch *(thick arrow)*, H & E, x 25

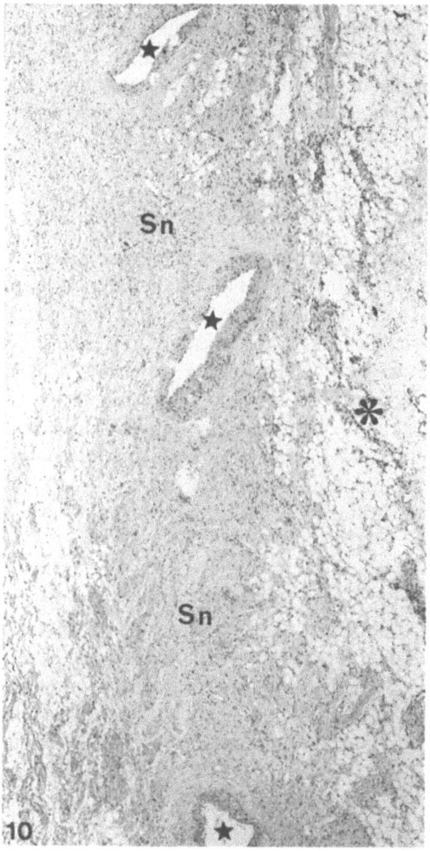
Sn
Sn
10

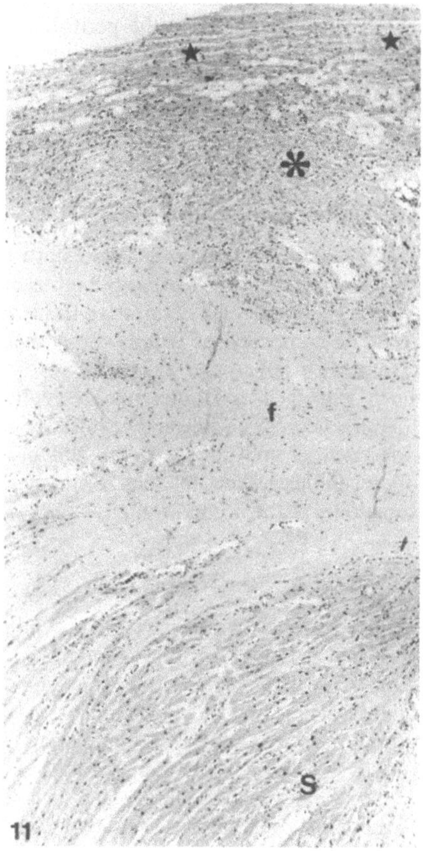
f
S
11

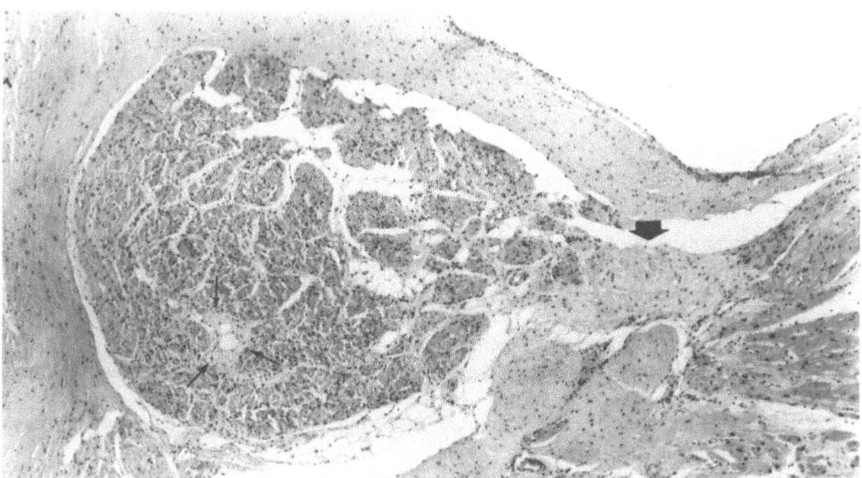
12

cells. Other cells look like typical immunoblasts. Neutrophils and clustered macrophages may be found in the perifollicular area. Degeneration with karyorrhexis and nuclear debris of lymphoid elements have been seen in a cervical lymph node of a patient with repeated high fever and marked discomfort in the neck nodes (Duray and Steere 1988). Spirochetes can be detected in the spleen (Cimmino et al. 1988) and in the perifollicular area of the lymph nodes (Fig. 13).

6.2.2.4 Muscular System

Involvement of muscular tissue in stage II Lyme borreliosis shows focal nodular myositis and interstitial myositis (Schmutzard et al. 1986; Atlas et al. 1988; Reimers et al. 1989). The perimysial connective tissue may show small diffuse or focal inflammatory infiltrate composed of lymphocytes, macrophages, and some plasma cells. There are also small perivascular aggregates of lymphocytes and histiocytes (Fig. 14). Muscle fibers near these infiltrates may show hyaline, granular, and vacuolar fiber degeneration and necrosis, with phagocytosis of some fibers. There are also signs of regeneration such as splitting of fibers and proliferation of cells. Vasculitis or obliterative microangiopathy are not seen. Spirochetes are located between the muscle fibers and sometimes at the border of the lymphocytic infiltrate (Fig. 15).

6.2.2.5 Other Internal Organs

Liver involvement can vary from a mild lymphocytic portal triaditis all the way to liver cell derangement that simulates acute hepatitis (Duray 1987; Duray and Steere 1988). The liver cells at this stage show degeneration with swollen clear cytoplasm, microvesicles, and fat. Plasma cells may be present throughout the sinusoids and the portal tracts. Neutrophils are numerous in the sinusoids in acute hepatitis-like lesions (Duray and Steere 1986; Duray 1989a).

The kidney shows hardly any histological changes; spirochetes are occasionally found in the renal tubules overlaying the tubular epithelial cells (Duray 1989a).
Uterine involvement in stage II Lyme borreliosis is a significant factor in pregnancy since it may result in transplacental transmission of *B. burgdorferi* to the fetus. The amount of Hofbauer cells may be increased in the villi of the placenta. An inflammatory infiltrate may or may not be present. One of us has recently encountered cases of

▷

Fig. 13. Lymphnode in the left groin of a 16-year-old boy with arthralgia in the left knee a few weeks after a tick bite and erythema migrans on the left leg. A spirochete *(long arrow)* is seen in association with a macrophage. The *small arrow* indicates a blast-like lymphoid cell, modified Steiner method, x 1250

Fig. 14. Right musculus quadriceps femoris of a 61-year-old man suffering from myositis. The infiltrate is composed of lymphocytes and histiocytes and is located between the muscular fibers, H & E, x 300

Fig. 15. Same patient as in Fig. 14. Spirochete in the interstitium between the muscular fibers, modified Steiner method, x 1250

Fig. 16. Acrodermatitis chronica atrophicans, infiltrative phase. Biopsy from the right leg of a 63-year-old woman. Pronounced subepidermal band-like infiltrate and slight perivascular and sparse interstitial infiltrate throughout the entire cutis composed of plasma cells and lymphocytes, H & E, x 60

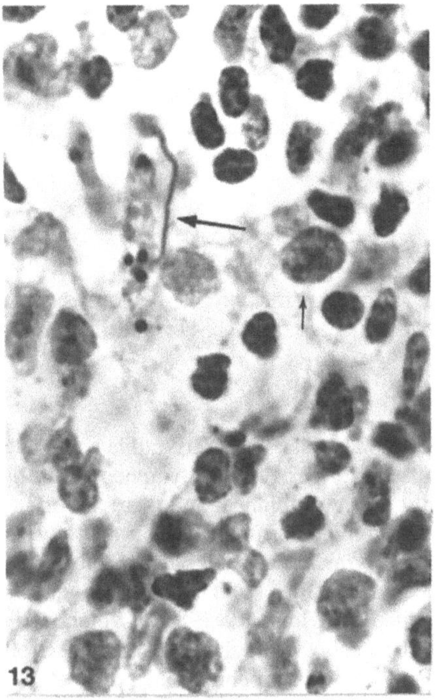
13

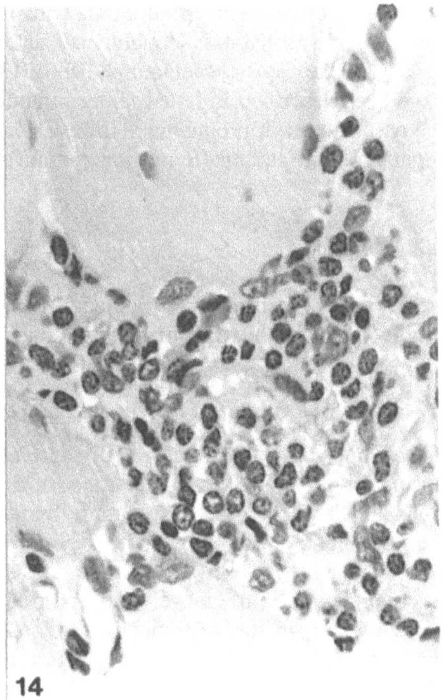
14

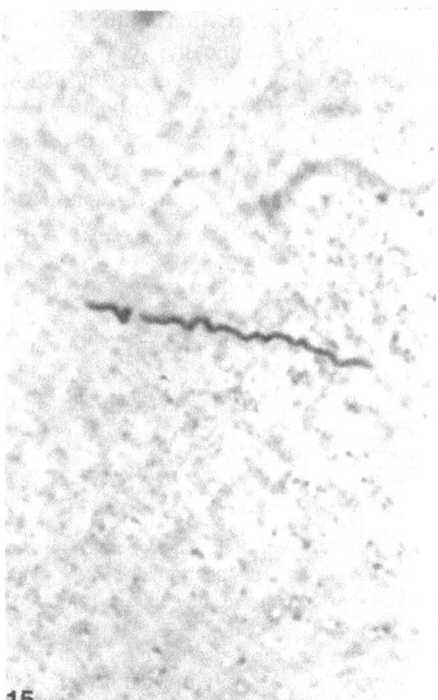
15

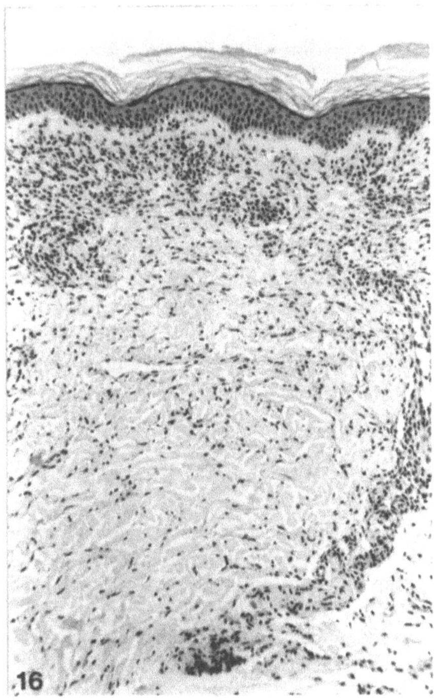
16

decidual necrosis with inflammation in patients with intrauterine infection due to *B. burgdorferi* (de Koning, unpublished data). Spirochetes may occasionally be seen in organs of the fetus (MacDonald 1986). Cortical blindness and cardiac abnormalities have been described in fetuses after infection of the mother with *B. burgdorferi* during the early stages of pregnancy. However, no inflammatory infiltrates were found in the organs of these infants (Schlesinger et al. 1985; Weber et al. 1988).

6.2.3 Stage III

Acrodermatitis chronica atrophicans (ACA), arthritis, and chronic nervous system involvement are the main manifestations of the third stage of Lyme borreliosis.

6.2.3.1 Acrodermatitis Chronica Atrophicans

The skin lesions in ACA can be divided into two phases: an early infiltrative and a late atrophic phase. The disease starts with a nonspecific perivascular lymphocytic infiltrate. In early lesions, the epidermis frequently becomes thinned, and there are orthokeratosis and a loss of rete ridges. The upper and middle portion of the dermis shows a band-like and perivascular infiltrate consisting of lymphocytes and plasma cells, often in combination with a more or less pronounced edema (Fig. 16). Dilated blood vessels can be found in the superficial dermis (Asbrink and Hovmark 1988). Lymphocytic infiltrates surround blood vessels, hair follicles, and sweat glands. Under electron microscopy, macrophages show phagocytosis of dense osmiophilic material (Fig. 17) and some vacuoles filled with osmiophilic substance, presumably fat (Fig. 18).

Juxta-articular fibroid nodules may be present in the deeper portions of the reticular cutis extending into the subcutaneous fat (Hauser 1965). Clinically, they resemble rheumatoid nodules, but they have a different histological structure. The nodules show a homogeneous eosinophilic center surrounded by irregular fascicles of collagen fibers (Fig. 19,20). The latter are typically arranged in an onion-like, concentric fashion (Fig. 19). Fibrosis is pronounced, and degeneration of fibrils of the collagen may be found (Fig. 20). Necrosis of the collagen with palisading granulomas as seen in rheumatoid nodules or granuloma annulare are not present (Hardmeier 1968). Perivascular infiltrates of plasma cells and lymphocytes are seen surrounding the center of the fibroid nodule. The inflammatory cells extend along the blood vessels into the homogeneous material, decreasing in density towards the center. The blood vessels may be surrounded by concentric cell layers. The origin of these cells is still unclear. The overlying epidermis is usually atrophic, with some orthokeratosis. In addition, plaque-like lesions may be present. They resemble the nodular lesions, but the infiltrate is usually

▷

Fig. 17. Acrodermatitis chronica atrophicans (same patient as in Fig. 16). Macrophages *(M)* are located between patchy irregularly distributed collagen *(C)* fibrils. The macrophages show phagocytic activity of osmium-dense material. Fibroblast *(F)* with indentated nucleus (eletron micrograph, x 8000)

Fig. 18. Acrodermatitis chronica atrophicans (same patient as in Fig. 16). The normal structure of the collagen *(C)* fibers has partly disappeared, the fibrils show a focally irregular pattern. The infiltrate shows lymphocytes *(L)* and macrophages *(M)* with phagocytosis of osmium-dense material and small vacuoles in the cytoplasma filled with osmiophilic material, probably fat *(F)*. An elastic fiber is seen in the right corner *(E)* (electron micrograph, x 5000)

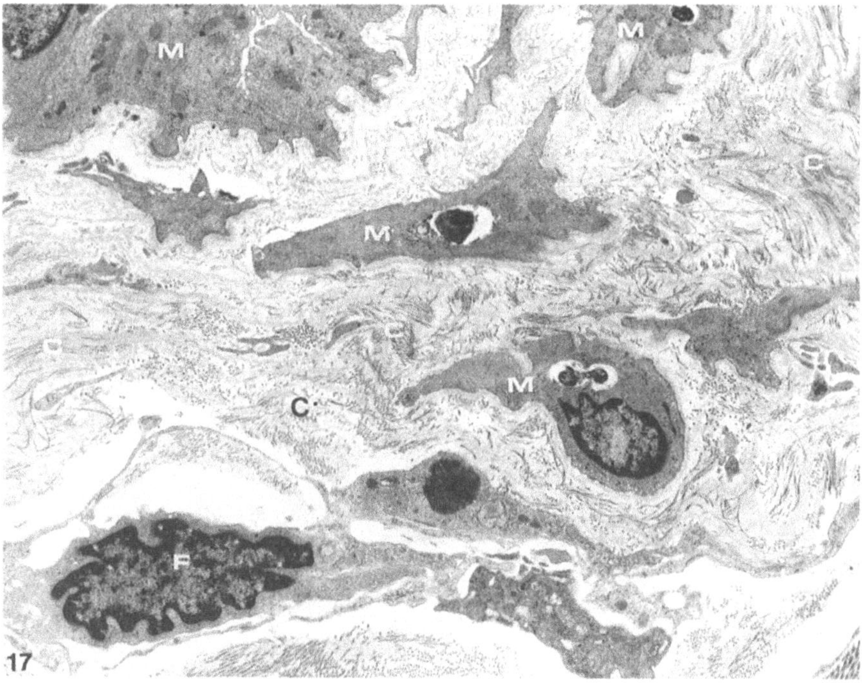

17

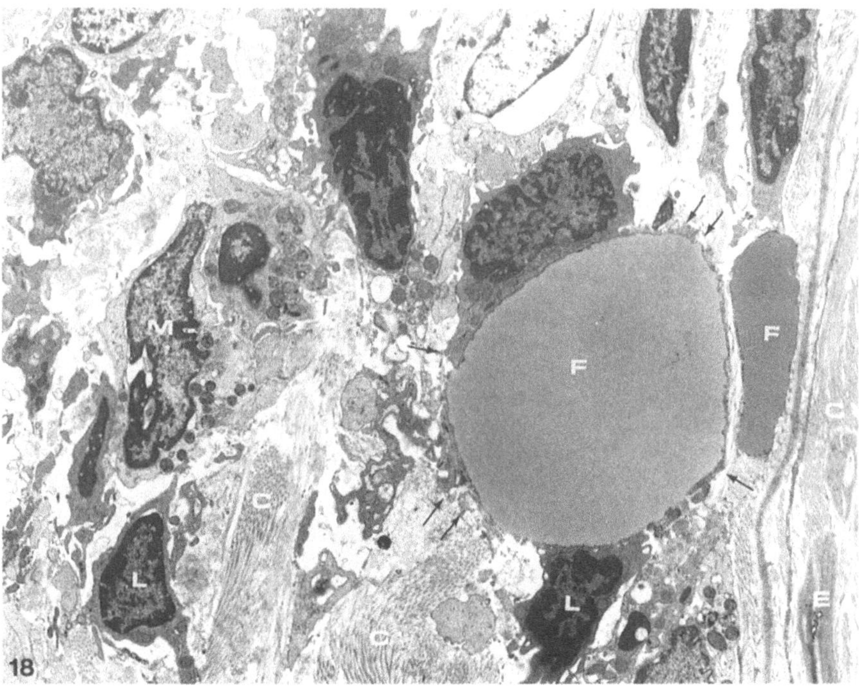

18

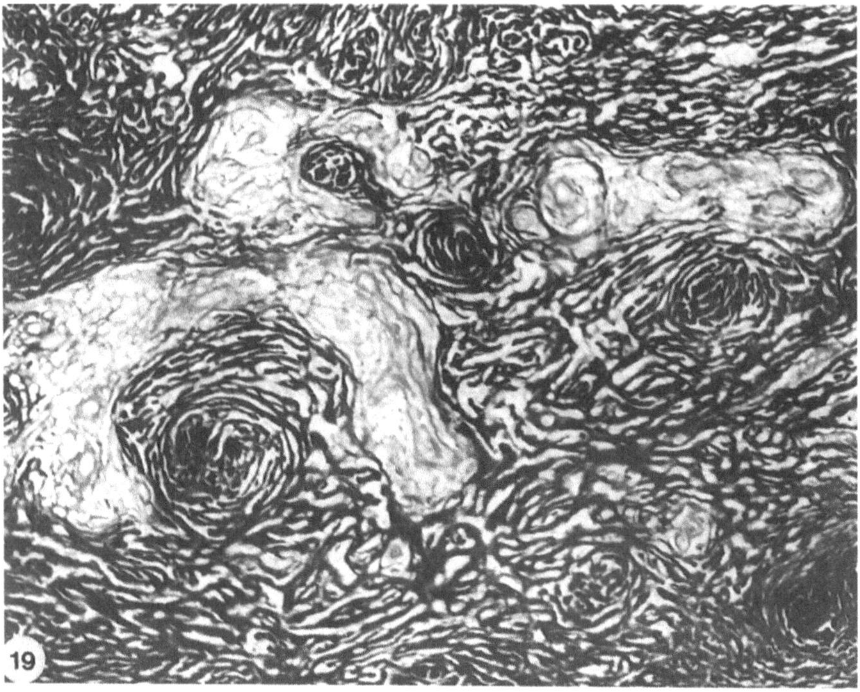
19

20

more abundant and there is less fibrosis (Ehrmann and Falkenstein 1925; Hauser 1965). It was this type of infiltrate that reminded Ehrmann and Falkenstein (1925) of the histological changes seen in syphilis, which led them to assume an infectious etiology of ACA.

In the late stage of ACA, cutaneous atrophy is a characteristic feature, as is a more or less pronounced inflammation. Rete ridges have disappeared in this stage. The epidermis is often only a few cell layers thick, sometimes with liquefaction and degeneration of basal cells (Fig. 21). Dilated blood vessels surrounded by lymphocytes and plasma cells can be found in the superficial cutis (Hauser 1965; Asbrink 1985; Asbrink et al. 1986). In very late 'burned-out' atrophic lesions, the inflammatory infiltrates are sparse or may even be lacking (Hauser 1965). The collagen fibers are strongly reduced in number, and are sometimes very thin with split fibrils. A particularly characteristic finding of this late stage is the degeneration of the elastic fibers, especially in the papillary dermis (Fig. 22). Elastic fibers are irregular and may be surrounded by macrophages (Fig. 23). If still present, atrophic sweat glands are infiltrated by lymphocytes. Small fat vacuoles and fat cells can be demonstrated (Hauser 1965).

In the early stage of ACA, spirochetes are usually located interstitially between or adhering to the collagen fibers. Only a few microorganisms are present in the late stage, often with fragmentation or shortened forms of the spirochete. Most spirochetes can be found in the early stage and then particularly in the fibroid nodules next to the homogeneous center.

6.2.3.2 Arthritis

The pathologic alterations in arthritis are those of a nonspecific synovitis. Chronically inflamed hypertrophic synovial villi (Fig. 24) with the deposition of fibrinaceous, eosinophilic material in the synovia are seen in specimens of synovectomized patients. The synovia is rarely scarred, unlike in other chronic nonspecific inflammations of the joints. In rheumatoid arthritis, reticulin fibers are visible in villous structures, but not so in the fibrin-filled villi in Lyme arthritis (Fig. 25).

Obliterative arteritis of medium-sized and sometimes smaller vessels can be seen in nearly all patients with chronic joint disease of other causes. However, small blood vessel obstruction with synovial fibrin deposition is more often seen in Lyme borreliosis (Johnston et al. 1985; de Koning and Hoogkamp-Korstanje 1986a; Duray and Steere 1986; Herzer 1989).

The inflammatory infiltrate shows predominantly lymphocytes, often in follicular structures with incomplete germinal centers, and plasma cells. In some cases, the plasma cells are as abundant as in rheumatoid arthritis. Mast cells can easily be found in the areas of increased vascularization. Eosinophils and neutrophils cannot usually be

◁

Fig. 19. Fibroid nodule in association with acrodermatitis chronica atrophicans. Biopsy from the left leg of a 69-year-old man. Multiple collagen fibers are arranged in a concentric, onion-like manner; the interstitium between the collagen fibers is not stained. The larger lighter areas represent the perivascular infiltrate, azan, x 40

Fig. 20. Fibroid nodule in association with acrodermatitis chronica atrophicans (same patient as in Fig. 19). The center of the nodule shows a collagen fiber surrounded by fibrinoid material. Degeneration of fibrils with loss of periodic banding is present in the fibrinoid material (electron micrograph, x 8500)

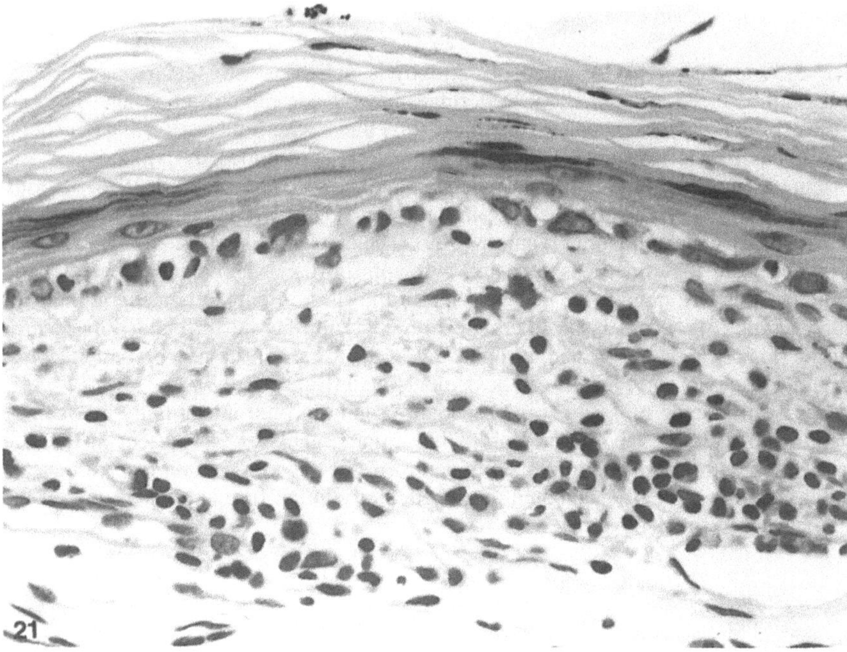
21

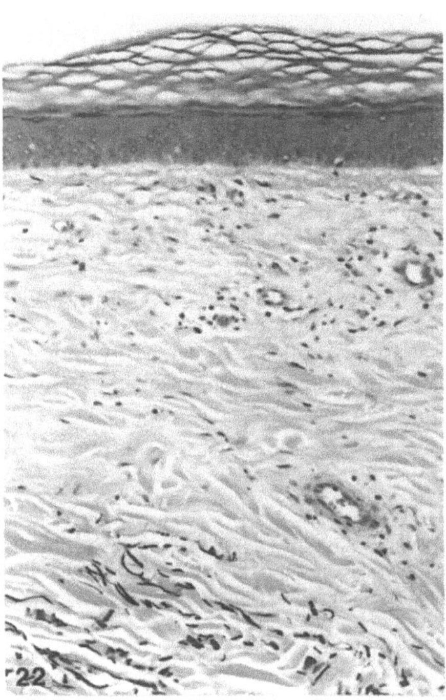
22

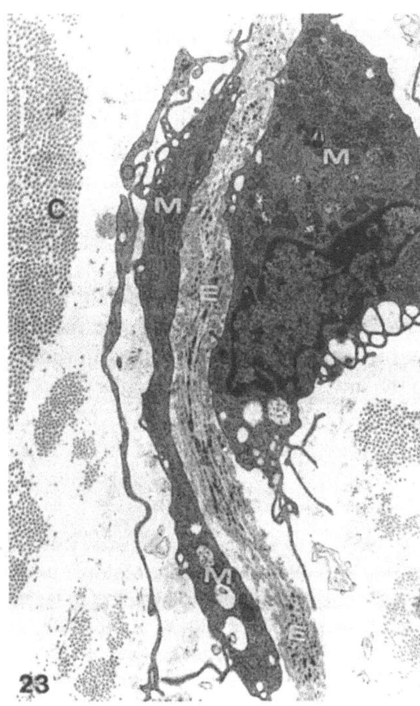
C
M
M
E
M
E
23

demonstrated, although neutrophils seem to occur in Lyme arthritis of recent onset and in acute exacerbations of Lyme arthritis. Granulomatous lesions or abcesses are not present (Johnston et al. 1985; Duray 1989b).

6.2.3.3 Late Neurological Involvement

The most common neurological manifestation in human chronic Lyme borreliosis consists of peripheral neuropathy, which may be accompanied by ACA (Kristoferitsch et al. 1988) (Chap.16). Lymphocytes and plasma cells are demonstrated around the blood vessels in the perineurium, with occasional sparse lymphocytes in the vessel wall (Fig. 26). The infiltrates are similar to the ones found in stage II, but they persist for many years; the vessel walls show no signs of necrosis. With time, the vessels outside the nerve sheath become thickened and obliterated, as do synovial vessels (Duray 1989b). Thrombosis of perineural vessels can be found (Camponovo and Meier 1986). Fibers within the nerve eventually lose myelin. However, the most striking sequela of peripheral nerve involvement is axonal degeneration (Kristoferitsch et al. 1988). Spirochetes have not been seen in the peripheral nerves so far.

Another late manifestation of Lyme borreliosis is encephalitis. Patients with stupor and coma-like conditions that have been biopsied showed mild, generalized cerebral edema, focal oligodendrocytic hyperplasia, focal increase of microglial cells, and perivascular lymphocytic infiltrate (Duray and Steere 1988; Pachner et al. 1989). Focal cerebral microglial reaction may correspond to the multiple lesions seen on magnetic resonance imaging, but future studies are necessary to confirm this (Wokke et al. 1987b; Halperin et al. 1989).

6.2.3.4 Lymph Node and Bone Marrow Involvement in Chronic Disease

The histological changes of the enlarged regional lymph nodes seen in patients with ACA reveal a nonspecific reactive hyperplasia (Hauser 1965). Large germinal centers are present with a pronounced number of plasma cells and histiocytes in the interfollicular area, hilus, sinus, and capsule. Russell bodies may be present; neutrophils or eosinophils are not found (Hauser 1955, 1965). The spleen may also show a reactive hyperplasia (Cimmino et al. 1988).

The bone marrow involvement of late disease is characterized by plasmocytosis. Plasmoblasts with mitoses and sometimes multinucleation of plasma cells may be present. The lymphoid aggregates may demonstrate follicular center cells and a few immunoblasts. There is an increased number of eosinophils and lymphoid cells. Mast

◁

Fig. 21. Acrodermatitis chronica atrophicans, atrophic phase. Biopsy from the left leg of a 61-year-old woman. The atrophic epidermis is only 3 to 4 cell layers thick; the basal cells show vacuolization. Slight subepidermal lymphocytic infiltrate surrounds dilated blood vessels, H & E, x 400

Fig. 22. Acrodermatitis chronica atrophicans, atrophic phase. Biopsy from the left foot of a 51-year-old woman. The elastic fibers are absent in the upper part of the cutis. In the lower part, the elastic fibers are irregular, show fragmentation and are diminished in number, Verhoef staining, x 80

Fig. 23. Acrodermatitis chronica atrophicans, atrophic phase. Same patient as in Fig. 22. Elastic fiber *(E)* engulfed by a macrophage *(M)*; *C*, collagen fiber; electron micrograph, x 6000

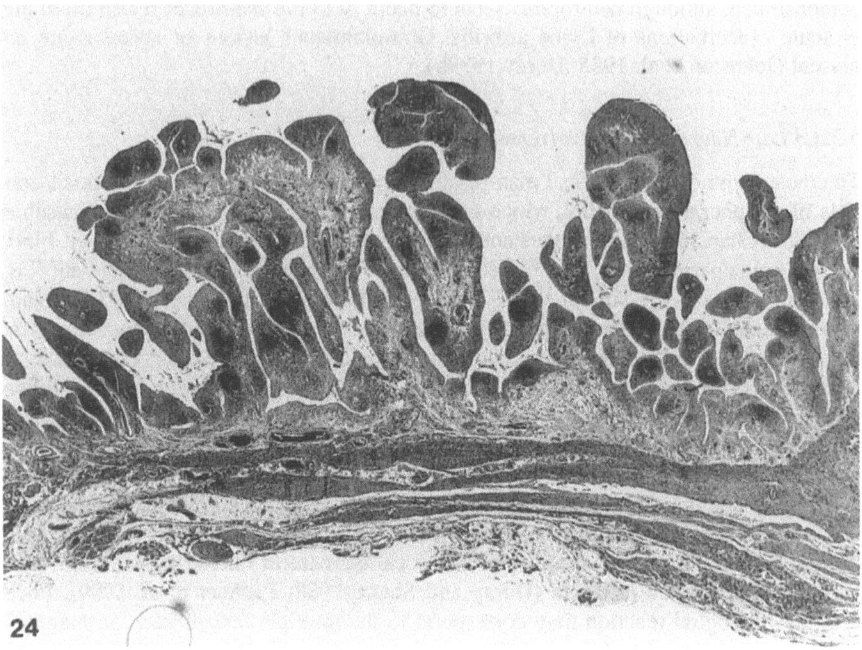

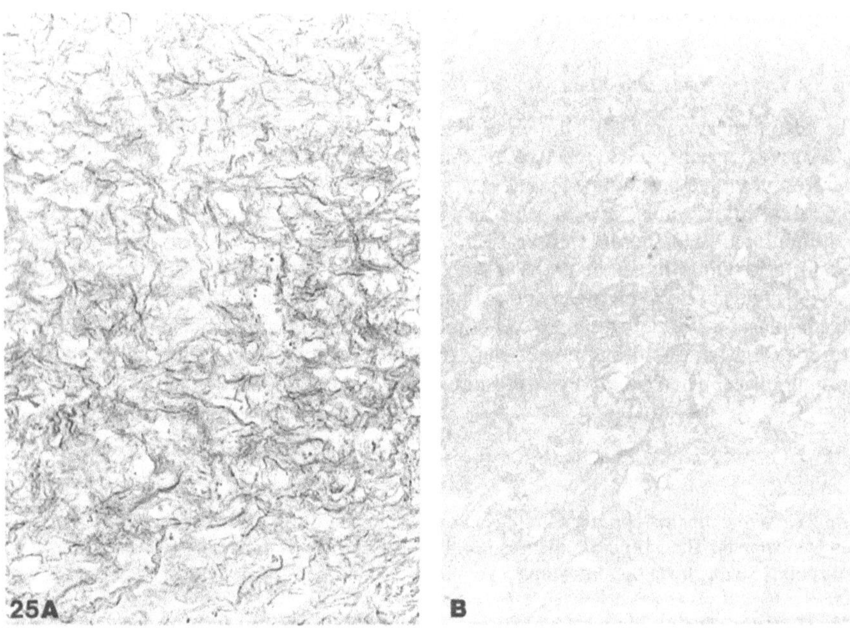

Fig. 24. Lyme arthritis with synovial villous hypertrophy, synovial cell hyperplasia and lymph follicle-like structures. (HE, x 10)

Fig. 25 A, B. Lyme arthritis. *A* Reticulin fibers are present in rheumatoid arthritis. *B* Reticulin fibers are absent in the villous stromal fibrin of the synovia of Lyme borreliosis, Foote's reticulin stain, x 400

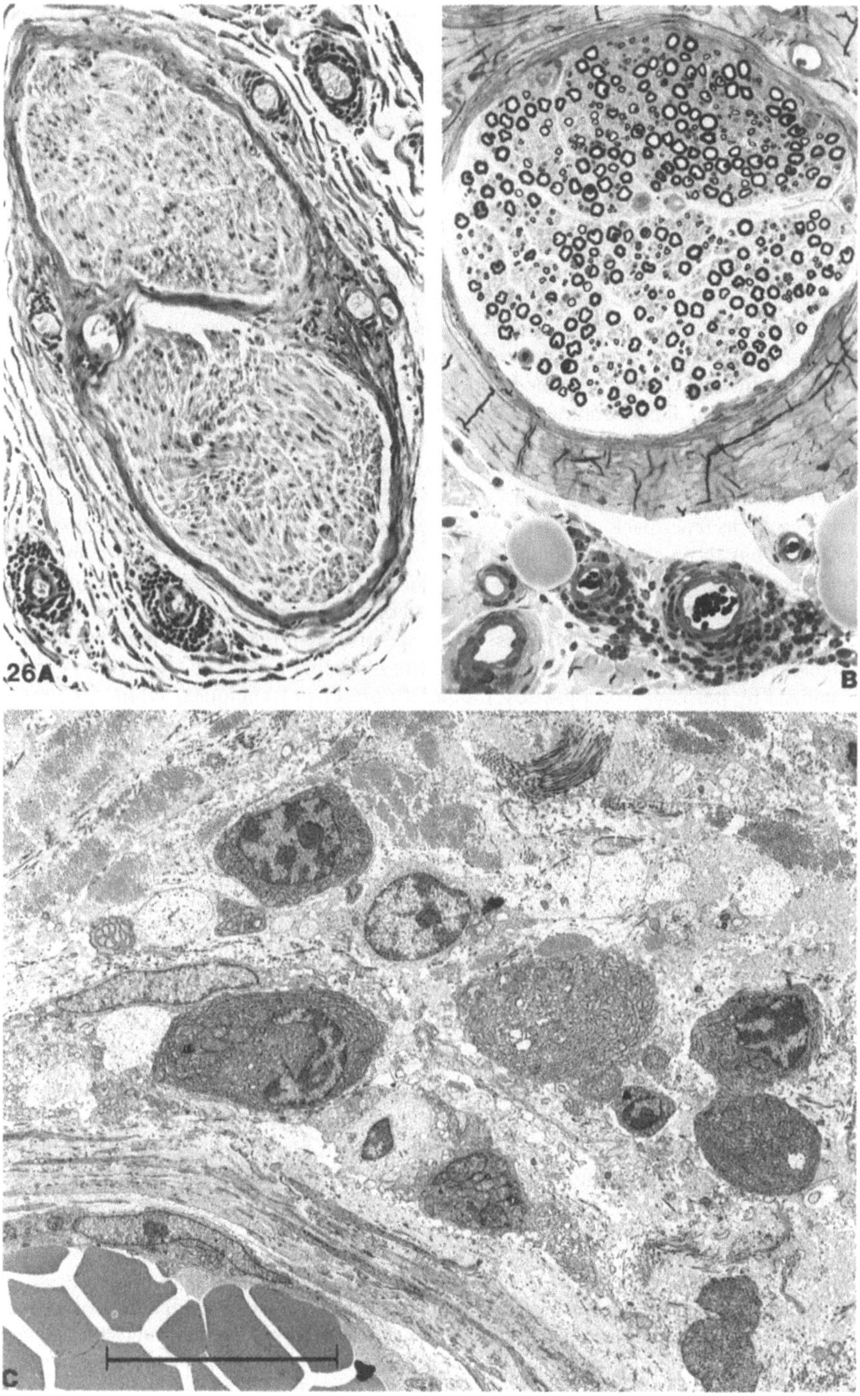

Fig. 26 A–C. (Legend see p. 88)

cells, which are usually not present in the bone marrow, can be demonstrated in small numbers (Hauser 1955, 1965). However, all these signs are not specific for ACA because they can also be seen in other (chronic) diseases.

6.3 Staining Methods

B. burgdorferi can be demonstrated in tissue using silver impregnation methods. However, these staining methods are not specific, because they stain all types of spirochetes (Thompson 1966). Experience shows that *B. burgdorferi* is not usually present in such high numbers in tissue as *Treponema pallidum* or *B. hermsii*. Finding a few spirochetes with both short and long forms in a given tissue section is quite suggestive of *B. burgdorferi*, and this implies that a good silver stain, such as the Bosma-Steiner or the modified Dieterle's method may indeed be useful for the investigation of a biopsy specimen. Good contrast with a bright yellow background and well-stained microorganisms are important in order to recognize spirochetes when screening histological slides. The modified Dieterle's method (Duray et al. 1985) and modified Steiner technique (de Koning et al. 1987) yield good results because the spirochetes stain black or blue-black against a clean yellow to yellow-orange background. All glassware has to be clean in silver staining techniques. It is important to prepare fresh reagents on the day of staining. The color development should be continuously watched by a technician in order to produce an optimal stain.

Immunohistochemical techniques show specific staining of the microorganism. Polyclonal and monoclonal antibodies are currently being used. The spirochetes can be visualized by peroxidase conjugates. immunofluoresence techniques, avidin- or streptavidin-biotin labeling and immunogold-silverstaining (Holgate et al. 1983; Hacker et al. 1985, 1988; Roth et al. 1989). Frozen sections are preferable when monoclonal and polyclonal antibodies are used. Spirochetes stain poorly with monoclonal antibodies in skin material which was fixed in formalin and embedded in paraffin. Polyclonal antibodies can be used with good results. Monoclonal antibodies reacting with different antigens of *B. burgdorferi* have been produced in the United States and in Europe. (Chap. 21).

The Bosma-Steiner silver stain shows good contrast, and the histological structures can be reasonably well recognized. Table 1 compares the Bosma-Steiner silver stain with the other techniques mentioned above. An advantage of the immunofluorescence method is the strong contrast between the sharply demarcated spirochetes and the tissue structures. The streptavidin-biotin technique shows excellent preservation of clusters of spirochetes, but detection of solitary microorganisms may be difficult.

With the immunogold-silver technique, spirochetes are labeled with gold and in the final step metallic silver reacts with the gold, resulting in enhancement with a stronger reaction. The background can be counterstained with hematoxylin-eosine. Two light sources are used in screening immunogold-silver-stained sections: the first light source

◁

Fig. 26 A - C. Late peripheral neuropathy. Biopsy of nervus suralis from a 65-year-old woman with acrodermatitis chronica atrophicans and sensory disturbances of the legs and feet. (Supplied by Dr. C Meier, Bern). *A* Perivasculitis of the epineural vasa nervorum, H & E, semi-thin section, x 140. *B* Several epineural small arteries and arterioles surrounded by a perivascular infiltrate, methylene-blue stain, semi-thin section, x 220. *C* The perivascular infiltrate shows several plasma cells (electron micrograph; *bar*, 10 mm)

Table 1. A comparison of staining methods

Method	Contrast	Tissue preservation	Specificity
Bosma-Steiner silver stain	+	+	–
Immunofluorescence	++	–	++
Streptavidine-biotin	+	++	++
Immunogold-silver	++	++	++

(like immunofluorescence) demonstrates the spirochetes by reflected light with a dark background. This is called epipolarization and generates a good constrast, which is especially helpful in screening slides for spirochetes. The second light source is used as in ordinary light microscopy. The hematoxylin-eosin-stained histological structures are easily visualized. By combining both light sources, the spirochetes become visible (by the reflected light) and can be related with the histological structures.

A strong cross-reaction is found with peripheral nerve structures when using monoclonal antibodies against the 41 kDa antigen (Sigal and Tatum 1988). In the immunoperoxidase method, suppression of endogenous peroxidase activity through incubation in methanol with 1% H^2O^2 is recommended. Nonspecific immunoglobulin (Ig) staining is suppressed by incubation with a 1:10 dilution of normal horse or goat serum. Incubation with the primary monoclonal antibody is followed by biotinylated horse anti-mouse IgG and avidin-biotin conjugate reagent. Washing with phosphate buffer has to be done after each step, especially before the final chromogen substrate step. The last step involves immersion in 1% diaminobenzidine, activated by one drop of 30% H^2O^2 for 6 min. The background should be controlled by microscopy.

References

Anderson RH, Becker AE (1978) Anatomy of conducting tissues revisited. Brit Heart J 40 (Suppl):2-16

Anderson RH, Becker AE (1980) The conducting system. In Anderson RH, Becker AE (eds) cardiac anatomy, vol 1. Churchill Livingstone, New York, pp 615-630

Asbrink E (1985) Erythema chronicum migrans Afzelius and acrodermatitis chronica atrophicans. Early and late manifestations of *Ixodes ricinus*-borne Borrelia spirochetes. Akademisk Avhandling, Stockholm

Asbrink E, Hovmark A (1988) Early and late cutaneous manifestations in *Ixodes*-borne borreliosis (Erythema migrans borreliosis, Lyme borreliosis). Ann N Y Acad Sci 539:4-15

Asbrink E, Brehmer-Andersson E, Hovmark A (1986) Acrodermatitis chronica atrophicans - a spirochetosis. Clinical and histopathological picture based on 32 patients. Am J Dermtatopathol 8:209-219

Atlas E, Stuart ST, Novak N, Duray PH, Steere AC (1988) Lyme myositis: muscle invasion by *Borrelia burgdorferi*. Ann Intern Med 109:245-246

Berger BW, Clemmensen OJ, Ackermann AB (1983) Lyme disease is a spirochetosis. Am J Dermatopathol 5:111-124

Bäfverstedt B (1943) Über Lymphadenosis benigna cutis. Acta Derm Venerol (Stockh) (Suppl 11) 24:1-202

Braun-Falco O, Burg G (1975) Lymphoreticulare Proliferationen in der Haut. Cytochemische und immunocytologische Untersuchungen bei Lymphadenosis benigna cutis. Hautarzt 26:124-132

Büchner SA, Rufli T (1987) Erythema chronicum migrans: evidence for cellular immune reaction in the skin lesions. Dermatologica 174:144-149

Burg G, Braun-Falco O (1983) Cutaneous lymphomas, pseudolymphomas, and related disorders. Springer Berlin Heidelberg New York, pp 415-441

Burg G, Braun-Falco O, Hoffmann-Fezer G, Schmoeckel C (1982) Differentiation between pseudolymphomas and malignant B-cell lymphomas of the skin. In: Goos U, Christophers E (eds) Lymphoproliferative diseases of the skin. Springer, Berlin Heidelberg New York

Camponovo F, Meier C (1986) Neuropathy of vasculitic origin in a case of Garin-Bujadoux-Bannwarth syndrome with positive antibody response. J Neurol 233:69-72

Caro WA, Helwig EB (1969) Cutaneous lymphoid hyperplasia. Cancer 24:487-502

Cimmino MA, Azzolini A, Tobia F, Pesce CM (1988) Spirochetes in the spleen of a patient with chronic Lyme disease. Am J Clin Pathol 91:95-97

De Koning J, Hoogkamp-Korstanje JAA (1986a) Demonstration of spirochetes in biopsies. Zentral Bacteriol Hyg (A) 263:179-188

De Koning J, Hoogkamp-Korstanje JAA (1986b) Significance of spirochetes in lymphadenosis benigna cutis. IXth international congress of infections and parasitic disease. Munich, 20-26 July abstract 1260

De Koning J, Bosma RB, Hoogkamp-Korstanje JAA (1987) Demonstration of spirochaetes in patients with Lyme disease with a modified silver stain. J Med Microbiol 23:261-267

De Koning J, Hoogkamp-Korstanje JAA, Linde MR van der, Crijns HJGM (1989) Demonstration of spirochetes in cardial biopsies of patients with Lyme disease. J Infect Dis 106:150-153

Duray PH (1987) The surgical pathology of human Lyme disease: an enlarging picture. Am J Surg Pathol 11 (suppl 1):47-60

Duray PH (1989a) Visceral histopathology in Lyme borreliosis: New observations. Zentralbl Bacteriol (Suppl) 18:116-125

Duray PH (1989b) Clinical pathologic correlations of Lyme disease. Rev Infect Dis 11 (Suppl 6):1487-1493

Duray PH, Steere AC (1986) The spectrum of organ and systems pathology in human Lyme disease. Zentral Bacteriol Hyg (A) 263:169-178

Duray PH, Steere AC (1988) Clinical pathologic correlations of Lyme disease by stage. Ann N Y Acad Sci 539:65-79

Duray PH, Kusnitz A, Ryan J (1985) Demonstration of the Lyme disease spirochete by a modification of the Dieterle stain. Lab Med 16:685-687

Eckert F, Schmidt U, Kaudewitz P, Burg G, Braun-Falco O (1989) Follicular lymphoid hyperplasia of the skin with high content of Ki-1 positive lymphocytes. Am J Dermatopathol 11:345-352

Ehrmann S, Falkenstein S (1925) Über dermatitis atrophicans und ihre pseudo-sklero-dermatischen Formen. Arch Dermatol Syph 149:142-175

Geerts MJ, Kaiserling E (1985) A morphologic study of lymphadenosis benigna cutis. Dermatologica 170:121-127

Hacker GW, Springall DR, Van Noorden S et al. (1985) The immunogold-silver staining method. A powerful tool in histopathology. Virchows Arch (A) 406:449-461

Hacker GW, Grimelius L, Dancher G, Bernatzky G et al (1988) Silver acetate autometallography: an alternative enhancement technique for immunogold-silver staining (IGSS) and silver amplification of gold, silver, mercury and zink in tissue. J Histotechnol 11:213-221

Halperin JJ, Luft BJ, Anand AK (1989) Lyme neuroborreliosis: central nervous system manifestations. Neurology 39:753-759

Hardmeier Th (1968) Zur Histopathologie der fibroiden Knoten bei Acrodermatitis chronica atrophicans. Arch Klin Exp Derm 232:373-383

Hauser W (1955) Zur Kenntnis der Acrodermatitis chronica atrophicans. Arch Dermatol Syph 199:350-393

Hauser W (1965) Wahrscheinliche Infektionskrankheiten der Haut. In: Jadassohn J (ed) Handbuch der Haut- und Geschlechtskrankheiten, vol IV, part 1A Springer, Berlin Heidelberg New York, pp 556-629

Herzer P (1989) Lyme-Borreliose. Steinkopff, Darmstadt

Holgate CS, Jackson P, Cowen PN, Bird CC (1983) Immunogold-silver staining: new method of immunostaining with enhanced sensitivity. J Histochem Cytochem 31:177-186

Johnston YE, Duray PH, Steere AC, Kashgarian M, Buza J, Malawista SE, Askenase PW (1985) Lyme arthritis spirochaetes found in synovial microangiopathic lesions. Am J Pathol 118:26-34

Kristoferitsch W, Sluga E, Graf M, Partsch H, Neumann R, Stanek G, Budka H (1988) Neuropathy associated with acrodermatitis chronica atrophicans. Ann N Y Acad Sci 539:35-45

Lennert K, Stein H (1982) The germinal center: morphology, histochemistry and immunohistology. In: Goos U, Christophers E (eds) Lymphoproliferative disease of the skin. Springer, Berlin Heidelberg New York

MacDonald AB (1986) Human fetal borreliosis, toxemia of pregnancy, and fetal death. Zentrabl Bacteriol Hyg (A) 263:189-200

Mach KW, Wilgram GF (1966) Characteristic histopathology of cutaneous lymphoplasia (lymphocytoma). Arch Dermatol 94:26-32

Marcus LC, Steere AC, Duray PH, Anderson AE, Mahoney EB (1985) Fatal pancarditis in a patient with coexistent Lyme disease and babesiosis. Ann Intern Med 103:374-376

Meier C, Grahmann F, Engelhardt A, Dumas M (1989) Peripheral nerve disorders in Lyme borreliosis. Nerve biopsy studies from eight cases. Acta Neuropathol Berl 79:271-278

Meurers B, Kohlhepp W, Gold R, Rohrback E, Mertens HG (1990) Histopathological findings in the central and peripheral nervous system in neuroborreliosis. A report of three cases. J Neurol 237:113-116

Neubert U, Krampitz HE, Engl H (1986) Microbiological findings in erythema (chronicum) migrans and related disorders. Zentralbl Bacteriol Hyg (A) 263:237-252

Pachner AR, Duray PH, Steere AC (1989) Central nervous system manifestations of Lyme disease. Arch Neurol 46:790-795

Paschoud JM (1954) Lymphocytom nach Zeckenbiss. Dermatologica 108:434-437

Paschoud JM (1957) Die Lymphadenosis benigna cutis als übertragbare Infektionskrankheit. Hautarzt 8:197-211

Paschoud JM (1958a) Die Lymphadenosis benigna cutis als übertragbare Infektionskrankheit. Hautarzt 9:153-165

Paschoud JM (1958b) Die Lymphadenosis benigna cutis als übertragbare Infektionskrankheit. Hautarzt 9:263-269

Paschoud JM (1958c) Die Lymphadenosis benigna cutis als übertragbare Infektionskrankheit. Hautarzt 9:311-315

Reimers CD, Pongratz DE, Neubert U, Pilz A, Hubner G, Neagele M, Wilske B, Duray PH, De Koning J (1989) Myositis caused by *Borrelia burgdorferi*: report of four cases. J Neurol Sci 91:215-226

Reznick JW, Braunstein DB, Walsh RL, Wolfson PM, Gierke LW, Gorelkin L, Chandler FW (1986) Lyme carditis. Am J Med 81:923-927

Roth J, Taatjes DJ, Warthol MJ (1989) Prevention of nonspecific interactions of gold-labeled reagents on tissue sections. Histochemistry 92:47-56

Schlesinger PA, Duray PH, Burke BA, Steere AC, Stillman MT (1985) Maternal-fetal transmission of the Lyme disease spirochete, *Borrelia burgdorferi*. Ann Intern Med 103:67-68

Schmoeckel C, Burg G, Wolf HH, Braun-Falco O (1977) The Ultrastructure of lymphadenosis benigna cutis (pseudolymphoma cutis). Dermatol Res 258:161-167

Schmutzhard E, Willeit J, Gerstenbrand F (1986) Meningopolyneuritis Bannwarth with focal nodular myositis. Klin Wochenschr 64:1204-1208

Sigal LH, Tatum AH (1988) Lyme disease patients' serum contains IgM antibodies to *B. burgdorferi* that cross-react with neural antigens. Neurology 38:1439-1442

Thompson SW (1966) Modified Warthin-Starry method for the demonstration of spirochetes. In: Selected histochemical and histopathological methods. Thomas, Springfield, pp 1048-1049

Vacher D, Lusson JR, Peycelon P, Kantellip P, Veyere A (1987) La myocardite de Lyme: Intérèt des examens scintigraphiques (Gallium 67 et Thallium 201) et de la biopsie endomyocardique (à propos de deux cas). Coeur 18:374-379

Van der Linde MR, Crijns HJGM, De Koning J, Hoogkamp-Korstanje JAA, De Graaf JJ, Piers DA, Van de Galien A, Lie KJ (1990) Range of atrioventricular conducting disturbances in Lyme borreliosis: a report of four cases and review of other published reports. Br Heart J 63:162-168

Vallat JM, Hugon J, Lubeau M, Leboutet MJ, Dumas M, Desproges-Gutteron R (1987) Tick-bite

meningoradiculoneuritis: clinical, electrophysiologic, and histologic findings in 10 cases. Neurology 37:749-753
Weber K, Schierz G, Wilske B, Preac-Mursic V (1984) European erythema migrans disease and related disorders. Yale J Biol Med 57:463-471
Weber K, Schierz G, Wilske B, Preac-Mursic V (1985) Das Lymphozytom - eine Borreliose? Z Hautkr 60:1585-1598
Weber K, Bratzke HJ, Neubert U, Wilske B, Duray PH (1988) *Borrelia burgdorferi* in a newborn despite oral penicillin for Lyme borreliosis during pregnancy. Pediatr Infect Dis J 7:286-289
Wokke JHJ, De Koning J, Stanek G, Jennekens FGI (1987a) Chronic muscle weakness caused by *Borrelia burgdorferi* meningoradiculitis. Ann Neurol 22:389-392
Wokke JHJ, Gijn J van, Elderson A, Stanek G (1987b) Chronic forms of *Borrelia burgdorferi* infection of the nervous system. Neurology 37:1031-1034

IV Clinical Features of Lyme Borreliosis

7 Clinical Overview

K. WEBER, H.-W. PFISTER, and C.D. REIMERS

7.1 Introduction

As with syphilis, Lyme borreliosis develops in three stages (Steere et al. 1984, 1986a; Weber et al. 1984b; Asbrink 1985; Pfister and Einhäupl 1986; Weber 1986, 1988, 1989a, b; Asbrink and Hovmark 1988; Herzer 1989; Steere 1989). The first two stages, which occur within a few weeks or months after infection due to *Borrelia burgdorferi,* represent the early phase of the disease (Asbrink and Hovmark 1988; Weber 1988; Steere 1989). The third stage arises after several months or years and is, therefore, considered to be the late phase of the disease (Table 1).

The development in three stages is not readily apparent in individual patients, since patients usually experience only a part of the spectrum of the disease. Observation of a large number of patients, however, led to the concept of the development in stages.

This chapter presents a clinical overview and discusses in somewhat more detail items which are not dealt with elsewhere in this volume. Selected, primarily European references are included.

7.2 Vectors and Time of Onset

7.2.1 Vectors

Lyme borreliosis is transmitted mainly by ticks of the *Ixodes ricinus* complex (Fig. 1, 2). However, only about 30-50% of the infected patients remember a tick bite (Weber

Table 1. Main clinical features of Lyme borreliosis

Stage	Clinical features
I	Erythema migrans
	Constitutional symptoms
II	Meningoradiculoneuritis (Bannwarth's syndrome)
	Meningitis / meningoencephalitis / cerebral vasculitis
	Arthralgia / (arthritis)
	Myalgia (myositis)
	Carditis
	Borrelial lymphocytoma
	Multiple erythema migrans lesions
III	Acrodermatitis chronica atrophicans
	Arthritis
	Encephalomyelitis / cerebral vasculitis
	Peripheral neuropathy

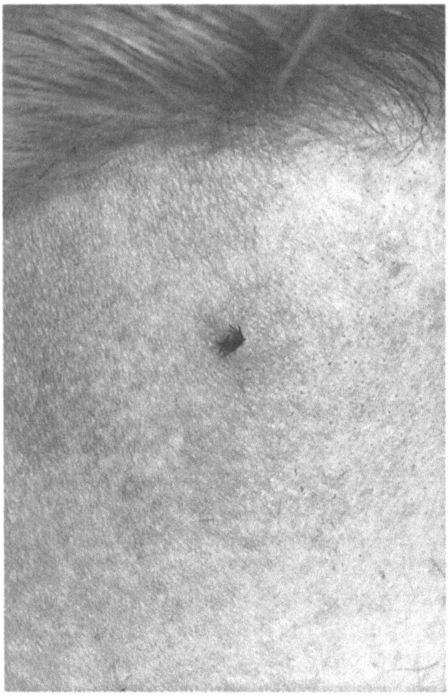

Fig. 1. Ixodid tick attached to neck skin with little engorgement and almost no surrounding erythema

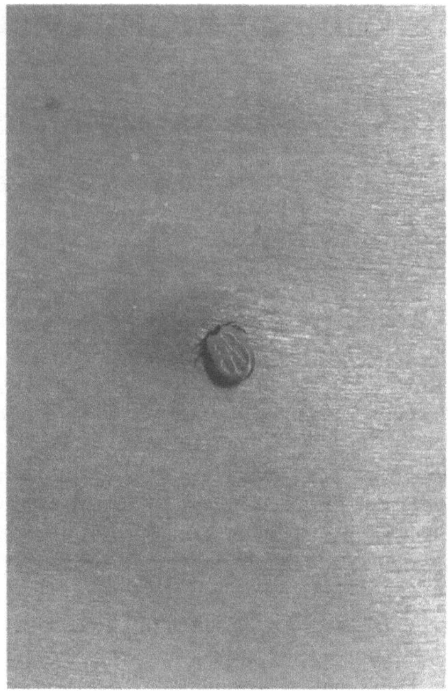

Fig. 2. Well-engorged ixodid tick attached to skin of the back with small, nonspecific surrounding erythema; the tick has been present for at least 2 days

et al. 1983; Asbrink and Olsson 1985; Weber and Neubert 1986). It is currently assumed that most of the remaining patients experience a tick bite, too, although they do not notice it. Many patients who claim to have had an insect bite cannot really remember that they were actually stung by a flying insect; careful questioning reveals in most instances that these patients confuse the red spot that they have seen with an insect bite. Nevertheless, horse flies were implicated as vectors in about 10% of cases in one series (Schaltenbrand 1967), and the fly *Stomoxys calcitrans* was identified by a veterinarian in another instance (Herzer et al. 1983).

7.2.2 Time of Onset

Lyme borreliosis in Europe starts in most instances between May and November, during the time of vector activity. Figure 3 shows when a group of German patients with erythema migrans experienced their tick bite and when the disease started. The time interval between tick bite and onset of the erythema migrans in these 42 patients was a median of 1.5 weeks (range 0.5–16 weeks). The early neurological manifestations and the other signs and symptoms of the first two stages of the disease start at about the same time or shortly later. The time interval between tick bite and the beginning of early neurological involvement is a median of approximately 3 (range 1 - 18) weeks (see Chap. 11). The time interval between tick bite or onset of erythema migrans and

Fig. 3. Tick bite (striped boxes) in 42 patients and onset of erythema migrans (striped and unshaded boxes) in 103 patients with early Lyme borreliosis (from Weber and Neubert 1986). See also Fig. 2 in Chap. 29, which shows fewer cases in October

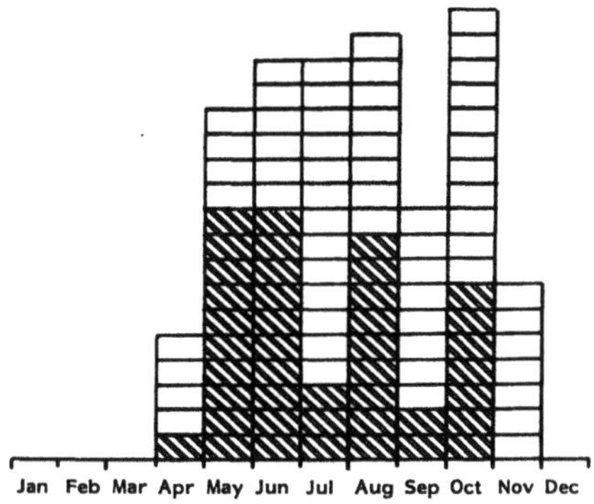

early arthralgia has been found to be a median of 8 (> 0 - 32) weeks in 23 erythema migrans patients (Weber and Neubert 1986).

The onset of the third stage manifestations does not follow a seasonal pattern. Even if a patient with third stage manifestations of the disease remembers a tick bite, it remains a matter of speculation whether this particular tick bite led to initiation of the disease.

7.3 Overall Description of the Disease by Stages

7.3.1 Stage I

The hallmark of stage I and of the disease in general is a skin condition called erythema migrans. The erythema migrans in its typical appearance is a unique sign not seen in any other disease. Erythema migrans may be associated with flu-like constitutional symptoms (Weber and Neubert 1986; Asbrink and Hovmark 1987). These symptoms may also occur without erythema migrans (see Chap. 8). In rare instances, a borrelial lymphocytoma may be seen as the first sign of Lyme borreliosis.

7.3.2 Stage II

A few weeks, usually not more than 3 months after infection, manifestations of the nervous system, the musculoskeletal apparatus, the heart, the skin, and occasionally of other organs can arise due to dissemination of *B. burgdorferi* (disseminated infection). The erythema migrans may still be present while these manifestations develop. However, the erythema migrans disappears soon spontaneously in the majoritiy of these patients within a few weeks.

A considerable number of patients develop meningoradiculoneuritis (Bannwarth's syndrome) and some acquire plain meningitis without peripheral neuropathy, meningoencephalitis or cerebral vasculitis. Meningoradiculoneuritis is the most common neurological manifestation of the second stage in Europe. It is usually associated with sometimes excruciating pain and lacks the typical signs of a meningitis such as severe headaches, nausea, or vomiting. However, frank meningitis without peripheral signs and symptoms may also occur, especially in children. The meningoencephalitis is predominantly quite mild in nature (Hörstrup and Ackermann 1973; Pfister et al. 1986a).

Arthralgia and myalgia are the most common symptoms of early musculoskeletal involvement. Frank arthritis can also occasionally be observed in the first few months of the disease (patients 1 and 2 in Fig. 4) (Steere et al. 1977; Weber et al. 1983; Herzer et al. 1986; Weber and Neubert 1986; Steere et al. 1987; Herzer 1989). Myositis has been encountered rarely so far (Schmutzhard et al. 1986; Reimers et al. 1989).

In patients with carditis, the most common finding is a conduction disturbance. Rhythm disturbances, a more diffuse endomyocarditis, and/or a pericarditis may occasionally be seen (Steere et al. 1980) (Chap. 10).

Borrelial lymphocytoma can often be observed at particular sites such as the ear lobe or the nipple. Borrelial lymphocytoma presents usually as a small solitary bluish-red plaque or nodule. Multiple lesions may occur, too. Borrelial lymphocytoma represents a specific type of a cutaneous pseudolymphoma caused by *B. burgdorferi* (Weber et al. 1984b, 1985; Asbrink et al. 1989). It appears to be a sign of all stages of Lyme borreliosis, but may be most commonly encountered during the second stage (Weber 1989a). Multiple erythema migrans-like lesions are another skin manifestation of the second stage.

In addition to the regional lymphadenopathy which can be found during stage I in association with the erythema migrans, patients with stage II disease may develop a generalized lymphadenopathy (Steere et al. 1983; Weber and Neubert 1986). The enlarged lymph nodes are usually not tender. Splenomegaly may be observed in early Lyme borreliosis, too (Steere et al. 1983), but pronounced splenomegaly which is detectable by palpitation is uncommon, at least in European erythema migrans patients (K. Weber, unpublished observation in 40 patients).

Some patients have slightly abnormal liver function tests, especially elevated transaminase values (Weber et al. 1983; Weber and Neubert 1986). Acute hepatitis has been observed in a patient with meningoradiculoneuritis (Killmann et al. 1987). Patients with preexisiting liver damage or concomitant drug therapy may be prone to develop frank hepatitis, as has been shown in a patient reported by Prinz et al. (1986). Hepatomegaly and mildly to moderately abnormal liver function tests (such as elevation of transaminases and lactate dehydrogenase) have been found in a variety of patients in the United States of America (Steere et al. 1983).

Occasional patients experience dry cough (Steere et al. 1983; Weber et al. 1983; Shresta et al. 1985; Weber and Neubert 1986), but frank pneumonia has not been observed.

In Europe, the main ophthalmological manifestation seems to be a chorioiditis, but other ocular signs and symptoms may also occur (see Chap. 19).

Lyme borreliosis affects the ear, nose, and throat in several ways. Sore throat and hoarseness have been noted in 9% and 3% of patients, respectively, and rhinitis and tracheolaryngitis in individual patients (Weber et al. 1983; Weber and Neubert 1986). Borrelial lymphocytoma is often seen on the ear lobe. Cranial nerve involvement may lead to facial palsy (Hörstrup and Ackermann 1973; Pfister et al. 1986a).

Testicular swelling indicating orchitis has been noted in two American patients (Steere et al. 1983), but not so far in Europe. Various constitutional symptoms may accompany the manifestations of the early phase of the disease (see Chap. 8).

7.3.3 Stage III

Months to years after the manifestations of the early phase have disappeared, or 6 to 12 months (an exact date has not been generally agreed upon) after onset of the disease, the third stage or late phase begins.

Here again, a unique dermatological manifestation, acrodermatitis chronica atrophicans (ACA), can be encountered. ACA starts with inflammatory changes which will sooner or later be followed by atrophy in the affected areas. If untreated, ACA usually develops over years and decades. Hundreds of cases have been documented in Europe so far (Hauser 1965) (see Chap. 14).

Late involvement of the nervous system consists of chronic encephalitits/encephalomyelitis, cerebral vasculitis or peripheral neuropathy. The late manifestations of the nervous system tend to run a chronic course. Patients with encephalomyelitis may experience spastic pareses and psychic changes (Ackermann et al. 1985; Kohlhepp et al. 1987). Cerebral vasculitis may lead to cerebral infarction, presenting clinically as stroke (May and Jabbari 1990). The late peripheral neuropathy is usually associated with an ACA, which provides clues to the correct diagnosis (Hopf 1966; Kristoferitsch 1989) see (Chaps. 15, 16).

Arthritis is more commonly found during the late than during the early phase of the disease. The large joints, especially the knee, are particularly often affected. The course is typically intermittent. Arthritis can become chronic. Chronic gonarthritis may lead to a stiff knee. In addition, arthralgia may accompany the arthritis or the other late manifestations, or arthralgia may be present alone for years, even in patients treated with antibiotics during early disease (Steere et al. 1977, 1987; Herzer et al. 1986; Weber and Neubert 1986; Weber et al. 1986a; Herzer 1989).

A few cases of late myositis have been observed, especially in association with an ACA (Chap. 13). A dermatomyositis-like syndrome has also been noticed (Chap. 20). Chronic heart involvement may exist, since *B. burgdorferi* has been isolated from a patient with cardiomyopathy (Stanek et al. 1990). Chronic lymphadenopathy is often present in patients with ACA, even in those treated with certain antibiotics (Hauser 1955, 1965; Weber et al. 1984a, b and 1988). It has also been known for many years that there may be a reactive hyperplasia of the bone marrow in patients with ACA (Hauser 1955, 1965). A few patients with keratitis have been observed in the USA (see Chap. 19) (Steere 1989).

7.3.4 Latent Phase

There is considerable evidence that patients undergo a latent phase of the disease. These patients lack clinical signs and symptoms, but have elevated antibody titers, usually of the immunoglobulin G (IgG) type. Three points support the notion of latency in Lyme borreliosis. First, the history of patients with certain third stage manifestations such as ACA or arthritis reveals that several years may elapse without symptoms between manifestations of the early and the late phase of the disease (see examples in Fig. 4).

Second, *B. burgdorferi* could be isolated in two patients who lacked typical signs and symptoms at the time when material for laboratory investigation was taken. In one, a long-standing erythema migrans had completely disappeared from the site of the biopsy material, which yielded growth of *B. burgdorferi* (Weber et al. 1990), and an elevated IgG antibody titer of 1:64 was found (K. Weber and B. Wilske, unpublished data). The second case involved a boy with multiple tick bites, elevated serum IgG antibody titer, and normal cerebrospinal fluid (regarding cell count and protein content); *B. burgdorferi* could be isolated from his cerebrospinal fluid (Pfister et al. 1989a).

Third, epidemiological studies have shown that some forest workers with elevated antibody titers against *B. burgdorferi*, usually of the IgG type, apparently lack symp-

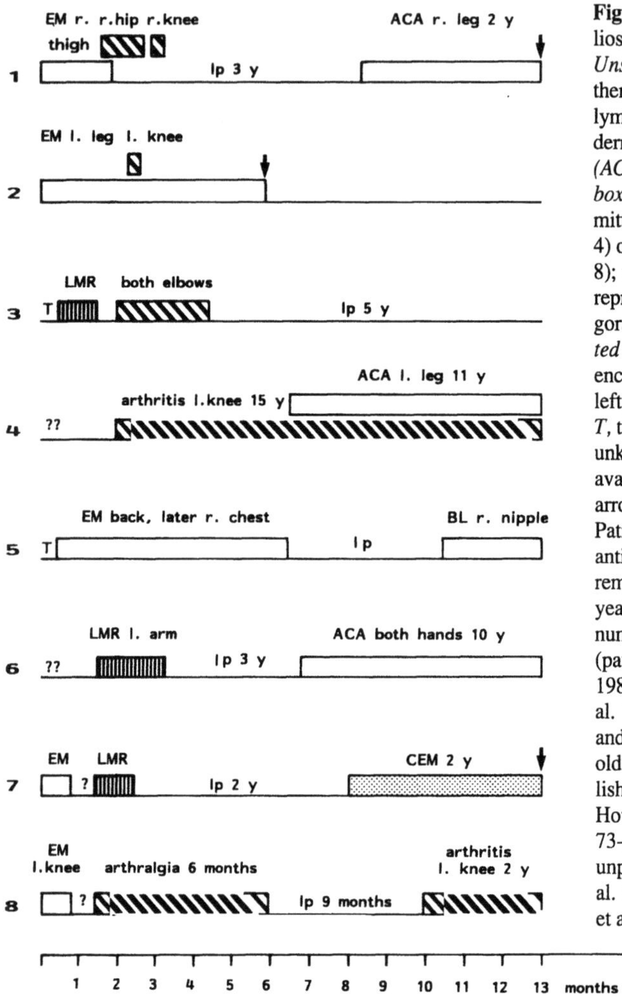

Fig. 4. Course of Lyme borreliosis in eight patients. *Unshaded boxes* represent erythema migrans *(EM)*, borrelial lymphocytoma *(BL)*, or acrodermatitis chronica atrophicans *(ACA)*; *diagonally striped boxes* represent arthritis (intermittent type in patients 3 and 4) or arthralgia (only in patient 8); *vertically striped boxes* represent lymphocytic meningoradiculoneuritis *(LMR)*; *dotted box* represents chronic encephalomyelitis *(CEM)*, (*l*, left; *lp*, latency period; *r*, right; *T*, tick bite; *y*, year(s); ?, unknown duration; ??, no data available about initial phase; arrow, antibiotic treatment). Patients 3, 4, and 6 refused antibiotic therapy; patient 3 remained asymptomatic for 5 years. References for patient numbers: *1*, Weber et al. 1984b (patient DH); *2*, Weber et al. 1983 (patient P); *3*, Weber et al. 1983 (patient L) and Weber and Neubert 1986; *4*, 79-year-old man (K. Weber, unpublished); *5*, Asbrink and Hovmark 1988 (patient 1); *6*, 73-year-old woman (K. Weber, unpublished); *7*, Ackermann et al. 1985 (patient 1); *8*, Herzer et al. 1986 (patient E)

toms typical for Lyme borreliosis (Wilske et al. 1985; Neubert et al. 1988). A North American prospective study has also shown that there are asymptomatic patients with elevated (IgG) antibody titers persisting for years (Steere et al. 1986b). It appears that there is latent Lyme borreliosis just as there is latent syphilis.

7.4. Course of the Disease

7.4.1 Survey

It is characteristic of the disease that most patients develop one or usually a few of the aforementioned manifestations. Figure 4 presents a few examples of the course of the disease. It becomes apparent from these examples and observations in other patients that different manifestations may occur simultaneously (patients 1,2, and 4 in Fig. 4) or consecutively, whereby they are separated by time intervals varying from days to years.

Meningoradiculoneuritis, facial palsy, borrelial lymphocytoma, arthritis and ACA occur more or less frequently in close proximity to the preceding erythema migrans (Fig. 4).

Manifestations of the early phase of the disease are usually self-limited, whereas manifestations of the late phase usually take a chronic course. However, erythema (chronicum) migrans and early arthritis may occasionally remain for a few years before they disappear (see Chaps. 8, 12) representing a continuous transition from the early into the late phase; but still, this means self-limitation after a prolonged course. Late arthritis often subsides spontaneously, too, thus behaving differently from the other late manifestations of the disease.

Certain manifestations of Lyme borreliosis such as arthritis and some of the early signs and symptoms have a more or less pronounced tendency to relapse. This may result in an intermittent course of the disease. The relapsing course is reminiscent of another borrelial disease, relapsing fever. Some other signs and symptoms have a tendency to wax and wane, for instance, the arthralgia in some patients or the erythema migrans when it is about to disappear. However, manifestations such as ACA usually have a definite tendency to remain constant over months or years.

7.4.2 Natural Course

7.4.2.1 Natural Course in USA and Europe

The natural course of Lyme borreliosis is difficult to assess. In the United States, 55 untreated patients with erythema migrans have been followed for several years. Of these patients, 80% developed arthritis (62%) or joint, periarticular, or musculoskeletal pain (18%) (Steere et al. 1987), 9% experienced recurrent erythema migrans (Steere et al. 1983), about 8% developed neurological and 4% cardiac abnormalities (Steere et al. 1980b), and a total of 20% did not develop any consecutive signs and symptoms (Steere et al. 1987). Steere (1989) recently estimated, however, that the occurrence of frank neurological involvement was about 15%–20%, and cardiac involvement amounted to 4%–8% of North American patients with Lyme borreliosis.

In Europe, knowledge of the natural course of the disease has to be deduced from the body of information provided by previous and recent investigations on different manifestations of Lyme borreliosis. This has been a tedious process because the whole spectrum of Lyme borreliosis was not recognized in the preantibiotic era, and thereafter, prospective studies in untreated patients have not been performed. A few patients with erythema migrans who refused therapy and are often difficult to follow up have developed meningoradiculoneuritis, arthritis, and symptoms suggestive of cardiac involvement (Weber et al. 1983; Asbrink and Olsson 1985; Weber and Neubert 1986). It should, however, be realized that patients with erythema migrans already represent a selection, although a reliable one, among the total number of patients infected with *B. burgdorferi* (Weber 1989b). The possibly prospective study by Bannwarth (1941) and the retrospective analyses by Krüger et al. (1989) and Pfister et al. (1989b) have provided valuable information on the natural course of patients with meningoradiculoneuritis. To get more information, retrospective evaluation of patients with later manifestations of the disease, epidemiological, and other data have to be used. Nevertheless, despite our uncertainty about exact prognostic data, patients with early Lyme borreliosis are at a considerable risk of developing more severe disease unless they are promptly treated with effective antibiotics.

7.4.2.2 Clinical Differences Between European and North American Lyme Borreliosis

Experiences based on various studies have provided hints that Lyme borreliosis may be somewhat different in Europe and the United States. Presently it appears that early disease is less severe in Europe and that borrelial lymphocytoma, ACA, and possibly meningoradiculoneuritis occur more frequently in Europe. However, bias may at least in part account for these differences (Weber 1989b).

7.4.3 Duration of the Diseases

The duration of Lyme borreliosis appears to be quite variable. Chronic manifestations last for years if untreated. For instance, patients with ACA can be ill for several decades, even up to 50 years. It seems as if some untreated patients with erythema migrans may experience a self-limitation of the disease (Steere et al. 1987). However, many years may elapse before late manifestations become apparent. Not enough data are presently available to make definite statements regarding the question of spontaneous cure of the disease.

7.5 Reinfection

There have now been well-documented cases of reinfection in patients who previously had erythema migrans or lymphocytic meningoradiculoneuritis. Reinfection can be assumed when primary and secondary erythema migrans occur at different sites, when there is a relatively long time interval between primary infection and reinfection, often exceeding the interval that can be seen with a relapse, and when a tick bite precedes reinfection, especially if the site of the tick bite is in the center of a second erythema migrans. Experience in several patients has shown that the clinical picture occurring after reinfection is similar to the one seen following the primary infection (Hollström 1958; Weber et al. 1985, 1986b; Pfister et al. 1986b).

Antibody titers may or may not be elevated during reinfection. In a few well-documented patients with erythema migrans, the antibody titers remained negative during primary infection and reinfection, but antibiotic treatment had been administered relatively early on both occasions (Weber et al. 1986b). On the other hand, two of our patients showed an elevation of the IgG antibody titers after reinfection, and, in one instance, there was a significant rise after the decline of the IgG antibody titer following treatment for lymphocytic meningoradiculoneuritis. In both patients, *B. burgdorferi* was isolated from the erythema migrans after reinfection when the IgG antibody titer was elevated (Pfister et al. 1986b; Weber et al. 1990; K. Weber, V. Preac-Mursic, B. Wilske, unpublished). Reinfection has also been observed in the United States (AC Steere, personal communication).

Superinfection also seems to be possible since individual cases of an erythema migrans occurring in ACA patients have been observed (Hauser 1965; Asbrink et al. 1984).

7.6 Congenital Infection

B. burgdorferi can apparently be transmitted via the placenta to the fetus. Only a few reports documenting congenital infection have been published in the United States and in Europe so far (Schlesinger et al. 1985; McDonald et al. 1987; Weber et al. 1988). In the case seen in Germany, *B. burgdorferi* could be demonstrated in the brain and liver of a newborn despite therapy of the mother with oral penicillin during early pregnancy (Weber et al. 1988). However, in none of these cases has a causal relationship between borrelial fetal invasion and unfavorable outcome been unequivocally shown.

7.7 Diagnosis

The diagnosis of Lyme borreliosis can easily be made when the characteristic skin lesions are present, i.e., erythema migrans, borrelial lymphocytoma at typical sites (nipple or ear lobe), or ACA. The recent history of a typical erythema migrans and significantly rising specific antibody titers are diagnostic, too. The history of a tick bite may be supportive, but most tick bites have no sequelae. The presence of a lymphocytic meningoradiculoneuritis, especially with its sometimes excruciating pain, in an endemic area during the summer and fall is quite suggestive of Lyme borreliosis, but needs confirmation at least by a positive serological blood test, preferably by a test for the specific intrathecal antibody production.

The diagnosis of Lyme borreliosis on the basis of the presence of other manifestations is usually less indicative, but may be suggested or confirmed by the demonstration of elevated antibody titers and/or by the morphological or cultural identification of *B. burgdorferi* in biopsied tissue or body fluids. The cultural identification of *B. burgdorferi* confirms the diagnosis, but the procedure is difficult and is therefore confined to a few laboratories. An elevated (IgG) antibody titer against *B. burgdorferi* alone is by no means proof of the diagnosis; it may indicate latent disease or can be found after sufficient antibiotic treatment, even for other reasons; a serological test can also be false positive.

7.8 Prognosis

Lyme borreliosis seems to have a relatively good prognosis. However, the morbidity in individual cases might be considerable, especially when patients experience syncopes or other signs of more severe carditis, the sometimes excruciating pain or pareses due to meningoradiculoneuritis, the sequelae of chronic nervous system involvement, permanent impairment of joint function due to chronic arthritis, or malignancy due to long-standing ACA.

Mortality of the disease is low. One fatality did not seem to be due to the *B. burgdorferi* infection alone, since this North American patient who died of pancarditis was found to have a dual infection with *B. burgdorferi* and *Babesia microti*. Nevertheless, *B. burgdorferi* was detected in the heart indicating that this spirochete was instrumental in eliciting the carditis (Marcus et al. 1985). In Europe, two fatalities can be indirectly attributed to Lyme borreliosis, namely, the cardiac arrest due to Lyme carditis with subsequent severe brain damage and fatal pneumonia in a 38-year-old man (De Koning et al., unpublished) (see Chap. 6) and the fatal paralytic ileus in a 75-year-old woman with severe paraparesis due to meningoradiculoneuritis and a history of a recent erythema migrans (Meurers et al. 1990). Suicide has been attempted by two patients with meningoradiculoneuritis (mentioned by Wolf 1970).

7.9 Nomenclatural Aspects

Although the main dermatological and the early neurological manifestations of the disease were already well known in Europe by 1976 and although a relationship between them had been recognized, a generally accepted nomenclatural proposal had not been made. After the description of Lyme arthritis (Steere et al. 1977) and Lyme disease (Steere et al. 1980a, b) in the USA, several terms have been used in Europe. Weber and coworkers (1983, 1984b, 1986), referring to an earlier nomenclatural proposal (Weber 1974), have used the term "erythema migrans disease", while Ackermann et al. (1984) preferred "erythema chronicum migrans disease" and later "erythema migrans borreliosis" (Ackermann et al. 1988), and Asbrink and Hovmark (1988) have spoken of an *Ixodes*-borne borreliosis. From the European standpoint, there is no ideal solution to this nomenclatural problem. The majority of the European contributors to this book have now voted to accept the term Lyme borreliosis, which was proposed at the Second International Lyme Symposium in Wien in 1985 (Bozsik et al. 1986; Stanek et al. 1987).

References

Ackermann R, Gollmer E, Rehse-Küpper B (1985) Progressive Borrelien-Enzephalomyelitis. Dtsch Med Wochenschr 110:1039-1042
Ackermann R, Rehse-Küpper B, Gollmer E, Schmidt R (1988) Chronic neurologic manifestations of erythema migrans borreliosis. Ann NY Acad Sci 539:16-23
Asbrink E, Hovmark A (1987) Cutaneous manifestations in Ixodes-borne borrelia spirochetosis. Int J Dermatol 26:215-223
Asbrink E, Hovmark A (1988) Early and late cutaneous manifestations in Ixodes-borne borreliosis (erythema migrans borreliosis, Lyme borreliosis). Ann NY Acad Sci 539:4-15
Asbrink E, Olsson I (1985) Clinical manifestations of erythema chronicum migrans Afzelius in 161 patients. Acta Derm Venereol (Stockh) 65:43-52
Asbrink E, Hovmark A, Hederstedt B (1984) The spirochetal etiology of acrodermatitis chronica atrophicans Herxheimer. Acta Derm Venereol (Stockh) 64:506-512
Asbrink E, Hovmark A, Olsson I (1989) Lymphadenosis benigna cutis solitaria-borrelia lymphocytoma in Sweden. Zentralbl Bakteriol (Suppl) 18:156-163
Bannwarth A (1941) Chronische lymphocytäre Meningitis, entzündliche Polyneuritis and "Rheumatismus". Arch Psychiatr Nervenkr 113:284-376
Bozsik BP, Lakos A, Budai J, Telegdy L, Ambrozy G (1986) Occurrence of Lyme borreliosis in Hungary. Zentralbl Bakteriol Hyg (A) 263:466-467
Hauser W (1955) Zur Kenntnis der Akrodermatitis chronica atrophicans. Arch Dermatol Syph 199:350-393
Hauser W (1965) Wahrscheinliche Infektionskrankheiten der Haut. Springer, Berlin Heidelberg New York, pp 556-629 (Handbuch der Haut- und Geschlechtskrankheiten, vol IV, part 1A)
Herzer P (1989) Lyme-Borreliose. Steinkopf: Darmstadt
Herzer P, Schattenkirchner M, Zöllner N (1983) Lyme-Arthritis-eine zu selten bedachte Diagnose? Verh Dtsch Ges Inn Med 89:299-302
Herzer P, Wilske B, Preac-Mursic V, Schierz G, Schattenkirchner M, Zöllner N (1986) Lyme arthritis: clinical features, serological and radiographic findings of cases in Germany. Klin Wochenschr 64:206-215
Hörstrup P, Ackermann R (1973) Durch Zecken übertragene Meningopolyneuritis (Garin-Bujadoux, Bannwarth). Fortschr Neurol Psychiatr 41:583-606
Hollström E (1958) Penicillin treatment of erythema chronicum migrans Afzelius. Acta Derm Venereol (Stockh) 38:285-289
Hopf HC (1966) Acrodermatitis chronica atrophicans (Herxheimer) und Nervensystem. Springer, Berlin Heidelberg New York (Monographien aus dem Gesamtgebiet der Neurologie und Psychiatrie, vol 114)

Killmann H, Lind P, Stanek G (1987) Akute Hepatitis bei Lyme-Borreliose. Wien Med Wochenschr 137:343-346

Kohlhepp W, Mertens HG, Oschmann P, Rohrbach E (1987) Akute und chronische Erkrankung bei zeckenvermittelter Borreliose. Nervenarzt 58:557-563

Kristoferitsch W (1989) Neuropathien bei Lyme-Borreliose. Springer, Vienna New York

Krüger H, Reuss K, Pulz M, Rohrbach E, Pflughaupt KW, Martin R, Mertens HG (1989) Meningoradiculitis and encephalomyelitis due to *Borrelia burgdorferi:* a follow-up study of 72 patients over 27 years. J Neurol 236:322-328

MacDonald AB, Benach JL, Burgdorfer W (1987) Stillbirth following maternal Lyme disease. NY J State Med 87:615-616

Marcus LC, Steere AC, Duray PH, Anderson AE, Mahoney EB (1985) Fatal pancarditis in a patient with coexistent Lyme disease and babesiosis. Ann Intern Med 103:374-376

May EF, Jabbari B (1990) Stroke in neuroborreliosis. Stroke 21:1232-1235

Meurers B, Kohlhepp W, Gold R, Rohrbach E, Mertens HG (1990) Histopathological findings in the central and peripheral nervous system in neuroborreliosis. J Neurol 237:113-116

Neubert U, Münchhoff P, Völker B, Reimers CD, Pflüger KH (1988) *Borrelia burgdorferi* infections in Bavarian forest workers. Ann NY Aca Sci 539:476-479

Pfister HW, Einhäupl KM (1986) Lyme Borreliosis - eine durch *Borrelia burgdorferi* hervorgerufene Multisystemerkrankung. Internist Welt 3:76-84

Pfister HW, Einhäupl KM, Wilske B, Preac-Mursic V (1986a) Bannwarth's syndrome and the enlarged neurological spectrum of arthropod-borne borreliosis. Zentralbl Bakteriol Hyg (A) 263:343-347

Pfister HW, Neubert U, Wilske B, Preac-Mursic V, Einhäupl KM, Borasio GD (1986b) Reinfection with *Borrelia burgdorferi* (letter). Lancet II:984-985

Pfister HW, Preac-Mursic V, Wilske B, Einhäupl KM, Weinberger K (1989a) Latent Lyme neuroborreliosis: presence of *Borrelia burgdorferi* in the cerebrospinal fluid without concurrent inflammatory signs. Neurology 39:1118-1120

Pfister HW, Wilske B, Preac-Mursic V, Einhäupl KM (1989b) Clinical and serological follow up of patients with Bannwarth's syndrome: comparison of patients with and without penicillin treatment. Zentralbl Bakteriol (Suppl) 18:276-279

Prinz A, Weiss P, Stanek G (1986) Generalized exanthema, acute hepatitis with porphyrinuria and eosinophilia. Zentralbl Bakteriol Hyg (A) 263:389-391

Reimers CD, Pongratz DE, Neubert U, Pilz A, Hübner G, Naegele M, Wilske B, Duray PH, DeKoning J (1989) Myositis caused by *Borrelia burgdorferi.* Report of 4 cases. J Neurol Sci 91:215-226

Schaltenbrand G (1967) Durch Arthophoden übertragene Erkrankung der Haut und des Nervensystems. Verh Dtsch Ges Inn Med 72:975-1005

Schlesinger PA, Duray PH, Burke BA, Steere AC, Stillmann T (1985) Maternal-fetal transmission of the Lyme disease spirochete, *Borrelia burgdorferi.* Ann Intern Med 103:67-68

Schmutzhard E, Willeit J, Gerstenbrand F (1986) Meningopolyneuritis Bannwarth with focal nodular myositis. Klin Wochenschr 64:1204-1208

Shresta M, Grodzicki RL, Steere AC (1985) Diagnosing early Lyme disease. Am J Med 78:235-240

Stanek G, Flamm H, Barbour AG, Burgdorfer W (eds) (1987) Lyme borreliosis. Fischer, Stuttgart

Stanek G, Klein J, Bittner R, Glogar D (1990) Isolation of *Borrelia burgdorferi* from the myocardium of a patient with longlasting cardiomyopathy. N Engl J Med 322:249-252

Steere AC (1989) Lyme disease. N Engl J Med 321:586-596

Steere AC, Malawista SE, Hardin JA, Ruddy S, Askenase PW, Andimann WA (1977) Erythema chronicum migrans and Lyme arthritis: the enlarging clinical spectrum. Ann Intern Med 86:685-698

Steere AC, Malawista SE, Newman JH, Spieler PN, Bartenhagen NH (1980a) Antibiotic therapy in Lyme disease. Ann Intern Med 93:1-8

Steere AC, Batsford WP, Weinberg M, Alexander J, Berger HJ, Wolfson S, Malawista SE (1980b) Lyme carditis: cardiac abnormalities of Lyme disease. Ann Intern Med 93:8-16

Steere AC, Bartenhagen NH, Craft JE, Hutchinson GJ, Newman JH, Rahn DW, Sigal LH,

Spieler PN, Stenn KS, Malawista SE (1983) The early clinical manifestations of Lyme disease. Ann Intern Med 99:76-82
Steere AC, Malawista SE, Bartenhagen NH, Spieler PN, Newman JH, Rahn DW, Hutchinson GJ, Green J, Snydman DR, Taylor E (1984) The clinical spectrum and treatment of Lyme disease. Yale J Biol Med 57:453-461
Steere AC, Bartenhagen NH, Craft JE, Hutchinson GJ, Newman JH, Pachner AR, Rahn DW, Sigal LH, Taylor E, Malawista SE (1986a) Clinical manifestation of Lyme disease. Zentralbl Bakteriol Hyg (A) 263:201-205
Steere AC, Taylor E, Wilson ML, Levine JF, Spielman A (1986b) Longitudinal assessment of the clinical and epidemiological features of Lyme disease in a defined population. J Infect Dis 154:295-300
Steere AC, Schoen RT, Taylor E (1987) The clinical evolution of Lyme arthritis. Ann Intern Med 107:725-731
Weber K (1974) Erythema-chronicum-migrans-Meningitis - eine bakterielle Infektionskrankheit. Munch Med Wochenschr 116:1993-1998
Weber K (1986) Die Lyme-Borreliose. Hautarzt 37:583-586
Weber K (1988) Klinik der Erythema-migrans- (Lyme-) Borreliose. Z Hautkr 63:314-319
Weber K (1989a) Borrelien-bedingte Dermatosen. Berichtsband der Medicinale XIX, Iserlohn
Weber K (1989b) Clinical differences between European and North-American Lyme borreliosis - a review. Zentralbl Bakteriol (Suppl) 118:146-155
Weber K, Neubert U (1986) Clinical features of early erythema migrans disease and related disorders. Zentralbl Bakteriol Hyg (A) 263:209-228
Weber K, Puzik A, Becker T (1983) Erythema-migrans-Krankheit: Beitrag zur Klinik und Beziehung zur Lyme-Krankheit. Dtsch Med Wochenschr 108:1182-1190
Weber K, Schierz G, Wilske B, Preac-Mursic V (1984a) Zur Klinik und Ätiologie der Acrodermatitis chronica atrophicans. Hautarzt 35:571-577
Weber K, Schierz G, Wilske B, Preac-Mursic V (1984b) European erythema migrans disease and related disorders. Yale J Biol Med 57:463-471
Weber K, Schierz G, Wilske B, Preac-Mursic V (1985) Das Lymphozytom - eine Borreliose? Z Hautkr 60:1585-1598
Weber K, Neubert U, Thurmayr R (1986a) Antibiotic therapy in early erythema migrans disease and related disorders. Zentralbl Bakteriol Hyg (A) 263:377-388
Weber K, Schierz G, Wilske B, Neubert U, Krampitz HE, Barbour AG, Burgdorfer W (1986b) Reinfection in erythema migrans disease. Infection 14:32-35 (Abstract: Zentralbl Haut Geschlechtskr (1985) 150:628-629)
Weber K, Preac-Mursic V, Neubert U, Thurmayr R, Herzer P, Wilske B, Schierz G, Marget W (1988) Antibiotic therapy of early European Lyme borreliosis and acrodermatitis chronica atrophicans. Ann NY Acad Sci 539:324-345
Weber K, Preac-Mursic V, Wilske B, Thurmayr R, Neubert U, Scherwitz C (1990) A randomized trial of ceftriaxone versus oral penicillin for the treatment of early European Lyme borreliosis. Infection 18:91-96
Wilske B, Münchhoff P, Schierz G, Preac-Mursic V, Roggendorf M, Zoulek G (1985) Zur Epidemiologie der *Borrelia-burgdorferi*-Infektion bei Waldarbeitern aus Oberbayern. Munch Med Wochenschr 127:171-172
Wolf G (1970) Über die chronische lymphozytäre Meningitis unter dem Bilde der Polyneuritis (Bannwarth). Fortschr Neurol Psychiatr 38:221-234

8 Erythema Migrans and Early Signs and Symptoms

K. WEBER, U. NEUBERT, and S.A. BÜCHNER

8.1 Erythema Migrans

8.1.1 Clinical Findings

8.1.1.1 Definition

Erythema migrans is a homogeneous or ring-shaped erythematous and usually peripherally expanding skin lesion caused by *Borrelia burgdorferi.*

8.1.1.2 Age

The erythema migrans can occur almost at any age. The median age among 434 European patients described in the dermatological literature was 41 years (range 1-88) (Hollström 1958; Sedlacek 1960; Neubert et al. 1986; Neumann et al. 1986; Stiernstedt et al. 1986; Weber and Neubert 1986), while it was 43 years (1-89) in 560 patients reported by Stanek et al. (1988). It appears that the median age of 52 years found in a series of 231 patients from a certain area of Stockholm, Sweden (Asbrink et al. 1986), represents a selection of a somewhat older group of patients. The median age of 314 American patients with erythema migrans reported by Steere et al. (1983) was 28 years (range 2-88), clearly lower than in Europe.

8.1.1.3 Sex Ratio

The male/female ratio of 719 European erythema migrans patients diagnosed by dermatologists was 0.53 (250 men, 469 women) in the studies of Sedlacek (1960), Hauser (1965), Asbrink and Olsson (1985), Stiernstedt et al. (1986), Neubert et al. (1986), Neumann et al. (1986) and Weber and Neubert (1986) and 0.56 in the study of Stanek et al. (1988). The male/female ratio of the above-mentioned 314 Americans was 1.1 (164 men and 150 women), again considerably different from the sex ratio found in Europe.

We assume that the differences in age and sex ratio among European and American patients with erythema migrans may be due to a different selection of patients and possibly different social habits rather than biomedical discrepancies. For instance, two American epidemiologic studies on patients with Lyme disease of whom 93% in one study and 75% in the other study had erythema migrans, have indicated a rather high percentage of children and a considerable predominance of boys over girls (Bowen et al. 1984; Williams et al. 1986). In one study, adult men clearly predominated over adult women, too (Bowen et al. 1984). Similarly, a male/female ratio of 1.1 (19 boys, 17 girls) has also been observed in a recent Swiss investigation on children with Lyme borreliosis (Schaad et al. 1986).

8.1.1.4 Clinical Description

An erythema migrans may arise at the site of the inoculation of *B. burgdorferi* in the skin (Figs. 1-8; Figs. 1, 2 and 4 see color plate I between pp 118–119). The incubation period (time interval between tick bite and onset of the erythema migrans) varies between about 3 days and 4 months, but is commonly 1–2 weeks. According to the

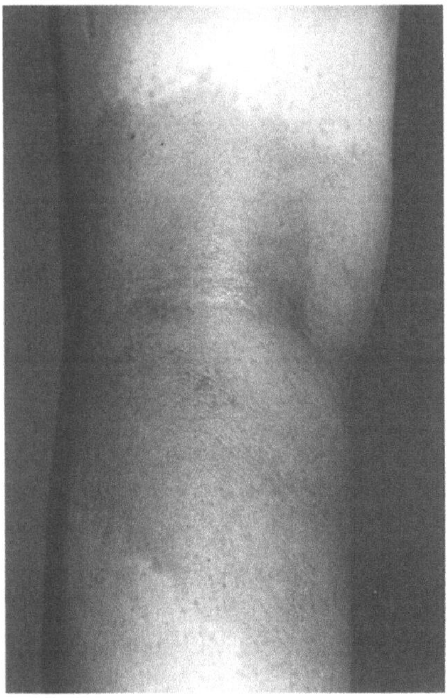

Fig. 3. Homogeneous erythema migrans of 10 days duration on the left knee of a 46-year-old woman. A tick was removed 17 days previously. *B. burgdorferi* was isolated from the border of the erythema migrans

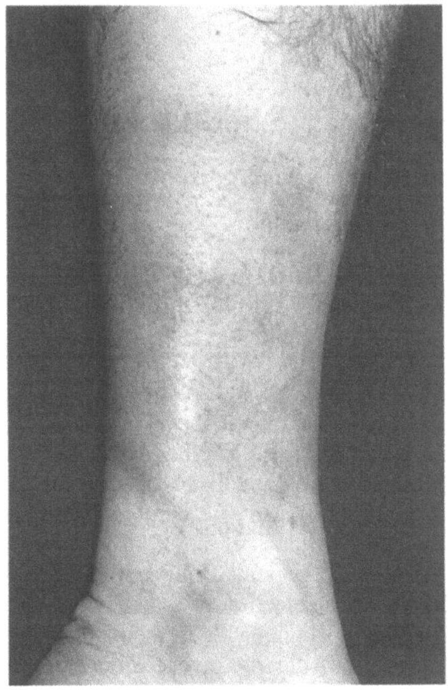

Fig. 5. Ring-shaped erythema migrans of at least 8 weeks duration on the left lower leg of a 46-year-old man. No tick was noted. *B. burgdorferi* was isolated from the border of the erythema migrans

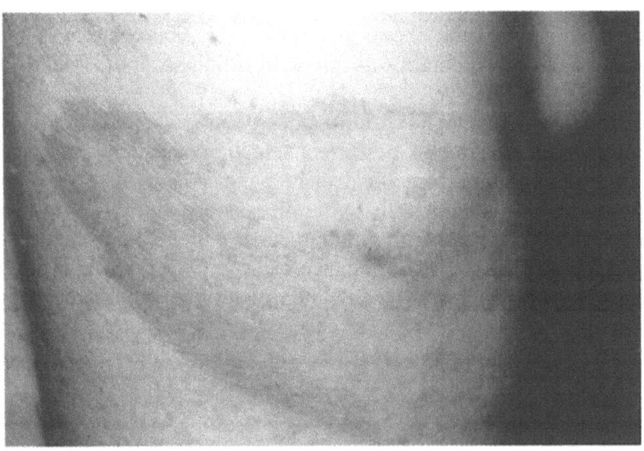

Fig. 6. Ring-shaped erythema migrans of at least 6 weeks duration with central lymphocytoma-like lesion on the trunk in a 48-year-old woman (from Weber et al. 1984a)

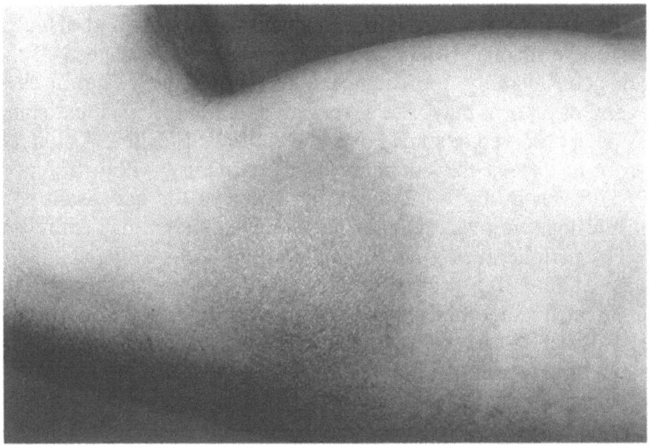

Fig. 7. Pale homogeneous erythema migrans of 4 weeks duration with pronounced central lymphocytoma-like plaque on the right upper arm of a 52-year-old man. No tick was noted. *B. burgdorferi* was isolated from the border of the central plaque

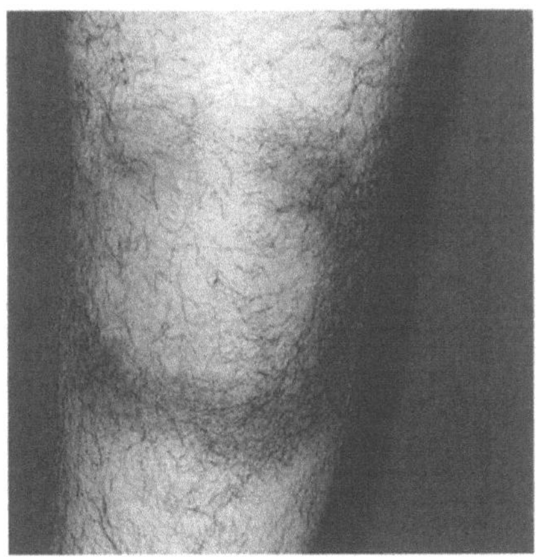

Fig. 8. Ring-shaped, hemorrhagic erythema migrans with central bluish-red discoloration of 4 days duration on the right dorsal thigh of a 35-year-old man. He developed ten associated signs and symptoms besides his erythema migrans. No tick was noted

observations of the affected patients, the erythema migrans is preceded in about 30%–50% of the cases by a tick bite and possibly in a few instances by a sting of a flying insect (Hauser 1965; Weber et al. 1983; Weber and Neubert 1986; Asbrink and Hovmark 1988). It must be assumed that in instances, in which neither a tick bite nor an insect sting was observed by the patient, a tick bite, probably from a tiny nymph, took place unnoticed. The tick bite leads to a small central erythematous macule or papule. The macule or papule is often still present when the erythema migrans develops.

The initial erythema migrans is usually a homogeneous lesion. This lesion may either be evenly bright red (Figs. 2-4; Figs. 2 and 4 see color plate I between pp 118–119) or it may turn bluish-red quite early (Fig. 8), but a bright red border often remains. In some patients, the homogeneous character of the lesion may not change until the patient is treated or until the lesion disappears spontaneously (Hauser 1965; Weber and Neubert 1986; Asbrink and Hovmark 1988).

The majority of lesions will eventually fade or totally clear in the center leaving a peripheral bright red erythematous ring as a remnant (Figs. 4-8). As this ring-shaped erythema expands, the lesion becomes bigger and bigger (Lipschütz 1913; Afzelius 1921). If the erythematous ring spreads around an extremity, the leading parts will merge, and the lesion may then expand proximally and distally on the extremity in a sleeve-like manner (Riehl 1913; Hauser 1965). The speed of the peripheral spreading varies considerably; for instance, a lesion of about 20 cm in diameter may arise within several days, or it may take several weeks. In individual patients, the erythema migrans has been observed to spread over the entire body within 4 months (Hellerström 1930) or hardly to exceed the boundary of an extremity within 2 years (Bruder 1950; Sonck 1965). Even in a given patient, the speed of the expansion is often different, so that most lesions are oval or somewhat elongated rather than entirely round.

The peripheral ring-like erythema usually measures about 1–2 cm in width, with a range of about 0.5–4 cm. The outer border of the erythema migrans is usually sharply demarcated and with a slightly irregular outline. The size of the erythema migrans varies between 2 cm and several dozens centimeters. The median size is about 20 cm (mean of the two maximum diameters) (Weber and Neubert 1986).

Approximately 21% of erythema migrans lesions contain a more or less pronounced central efflorescence, sometimes resembling a borrelial lymphocytoma. This central efflorescence may consist of a small macule or papule of a few millimeters in diameter or of more intensely stained erythemas or slightly raised bluish-red plaques of about 1 cm to a few centimeters, occasionally up to 12 cm in diameter. The central efflorescence can be found in homogeneous and more often in ring-shaped erythema migrans lesions (Weber and Neubert 1986).

Sometimes, atypical erythema migrans lesions can be observed. Such a lesion may be unusually small with a diameter of 2 to about 4 cm (Fig. 9, see color plate I between pp 118–119) and they hardly expand, if at all. *B. burgdorferi* was isolated from three such lesions, seen by one of us (K.W.), in the laboratory of V. Preac-Mursic (Weber et al. 1988 and unpublished). Strongly elongated lesions may also be seen. Other lesions have pronounced irregular, and sometimes bizarre outlines or are not truly annular, since they only have a few erythemas of about 1–2 cm in size arranged in a

Table 1. Erythema migrans characteristics in 103 patients (Weber and Neubert 1986)

Characteristics	*n*
Duration	
≥ 4 weeks	84[a]
< 4 weeks	19
Symptoms	
Asymptomatic	44
Pruritus	33
Pain, burning, heat	26
Appearance ($n = 100$)	
Ring-shaped	79
Homogeneous	19
Hemorrhagic	2
Central lesions	21
Multiple lesions	4
Spontaneous clearing	9

[a] 61 before therapy

somewhat annular fashion. Exceptionally, lesions may be hemorrhagic (Fig. 8). Central ulceration, blistering, or central scaling are also rare and may be a nonspecific reaction due to the tick bite or due to scratching of the initial tick bite papule (Weber and Neubert 1986; Asbrink and Hovmark 1988).

The duration of the erythema migrans varies approximately from 1 week to 1 year, and in rare cases up to 2 years. Lesions of less than 4 weeks duration are often followed by neurologic or other more severe manifestations, but this is not invariably the case. Most erythema migrans lesions in Europe probably last longer than 4 weeks (Table 1). They usually disappear spontaneously within a few months (Weber et al. 1983; Asbrink and Olsson 1985; Weber and Neubert 1986). An erythema chronicum migrans lasting 2 years has been observed in two instances (Bruder 1950; Sonck 1965).

An erythema migrans can occur on almost any part of the body, most frequently on the legs and, in children, quite often on the head. The localization of the primary erythema migrans depends on the site of the tick bite (Afzelius 1921; Hauser 1965; Weber et al. 1983; Weber and Neubert 1986; Asbrink and Hovmark 1988).

Erythema migrans lesions may wax and wane, especially prior to their disappearance. Some lesions become visible only if the patient takes a hot bath or is exposed to heat in another way. However, an erythema migrans can sometimes recur after its disappearance, usually within about 1 year. The relapse takes place at the site of the original erythema migrans lesion (Hollström 1951; Weber et al. 1983; Asbrink and Olsson 1985).

Reappearance of an erythema migrans due to reinfection is a different phenomenon. In reinfection, primary and secondary erythema migrans lesions occur at different sites and are usually separated by time intervals of more than 1 year (Weber et al. 1986b). Yet another phenomenon is superinfection: an erythema migrans has been observed to appear when an acrodermatitis chronica atrophicans was already present (Hauser 1965; Asbrink et al. 1984).

Erythema migrans lesions may be asymptomatic or pruritic or the patient may experience a feeling of pain, burning, or heat (Table 1). The pruritus is usually mild or moderate in intensity. It has been shown in one study that the feeling of pain, burning, or heat is significantly more often associated with special manifestations of the disease than if the lesions are asymptomatic or pruritic (Weber and Neubert 1986).

Some patients experience several tick bites at the same time, but only occasionally do these patients develop more than one primary erythema migrans lesion. It is more common for the patients to acquire a single erythema migrans lesion and develop multiple erythema migrans-like lesions later. In Europe, about 4%–8% of the patients are found to have multiple lesions (Weber et al. 1983; Asbrink and Olsson 1985; Stiernstedt et al. 1986; Weber and Neubert 1986). The incidence of multiple lesions in the United States is between 25% and 50% (Steere et al. 1983; Berger 1984). Since the majority of the multiple erythema migrans-like lesions are of the secondary type, similar to the secondary lesions in syphilis, these lesions are now regarded as part of the secondary stage of the disease (WHO Case Definition Committee 1989). The secondary erythema migrans-like lesions can be located anywhere on the skin. In Europe, most patients do not have more than ten lesions, whereas in the United States up to 100 lesions have been encountered (Steere et al 1983). However, a German patient who received irradiation for cancer developed about 70 lesions (Vaillant-Rieder and Hey 1987). The secondary erythema migrans-like lesions resemble the primary erythema migrans in that they are either homogeneous or annular erythemas (Fig. 10). Their diameter varies mostly between 5 and 15 cm. *B. burgdorferi* has been isolated from these lesions (K. Weber and V. Preac-Mursic; U. Neubert, unpublished). Important erythema migrans characteristics are listed in Table 1.

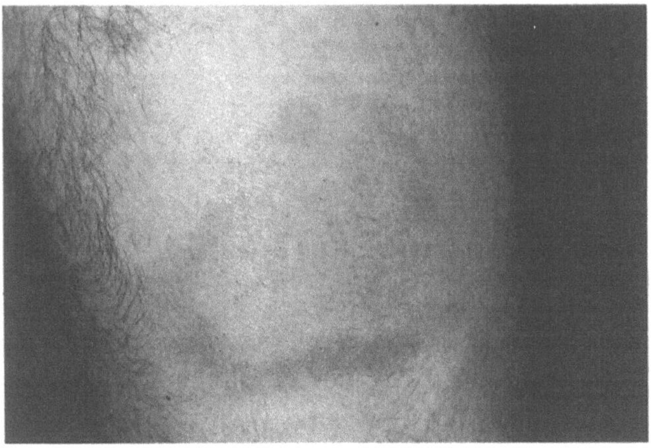

Fig. 10. Secondary erythema-migrans-like lesion of a few days duration on the right dorsal thigh of a 37-year-old man appearing 4 weeks after a primary erythema migrans elsewhere. *B. burgdorferi* was isolated from the secondary and primary lesions

8.1.1.5 Nomenclature

General usage of the term "erythema chronicum migrans", which implies a long-lasting nature of this skin manifestation, appears to be a mistake (Weber 1986) because it does not take into account that some lesions are present only for a short period of time (Hörstrup and Ackermann 1973; Steere et al. 1977; Weber and Neubert 1986). One of us (K.W.) has recently tried to reconcile the historical terms "erythema migrans" (Afzelius 1910) and "erythema chronicum migrans" (Lipschütz 1913) with our present knowledge of the manifestation (Weber et al. 1983; Weber 1986). It seems that the shorter term "erythema migrans" should be the general designation and the word "chronicum" is usually not needed. However, in order to preserve the historical term "erythema chronicum migrans", its usage must be restricted to long-lasting lesions. We now suggest that only lesions lasting 6 months and more may be called "erythema chronicum migrans". Our original suggestion (Weber et al. 1983) to separate "erythema migrans" from "erythema chronicum migrans" on the basis of a duration of 4 weeks has now been abandoned. Our present suggestion is in agreement with the case report of B. Lipschütz, who described a lesion of more than 7 months duration as representing an "erythema chronicum migrans" (1913).

In our opinion, the term "localized erythema migrans" which has recently been used (Steere 1989; Kuiper et al. 1990) ought to be avoided. Erythema migrans is a localized phenomenon in itself, and this fact does not need special emphasis. However, early dissemination of *B. burgdorferi* may already occur when the erythema migrans is present and may lead to disseminated infection, even though symptoms and other signs may be lacking (for instance, in an asymptomatic meningitis). "Localized" erythema migrans might imply that dissemination of *B. burgdorferi* does not take place, and this can lead to misinterpretations.

8.1.2 Histology and Immunohistology

8.1.2.1 Histologic Findings

8.1.2.1.1 Erythema Migrans

Histologic exmination of the peripheral and central parts of the erythema migrans reveals a superficial perivascular infiltrate composed of lymphocytes and some histio-

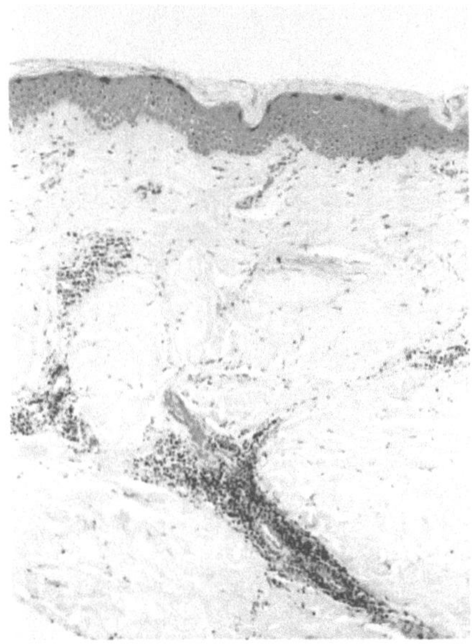

Fig. 11. Erythema migrans, peripheral border; perivascular lymphohistiocytic infiltrate in the dermis, H & E, x32

cytes; eosinophils and plasma cells can sometimes be seen and rarely also neutrophils. Occasionally, there is an interstitial infiltrate and, particularly in the central part of the erythema migrans, also a deep perivascular infiltrate (Fig. 11). The epidermis is usually unaffected, although a slight acanthosis may be noted in some areas, and a few lymphocytes may be present in the lower parts of the epidermis. A slight edema of the papillary dermis can sometimes be observed. The blood vessels may show endothelial swelling and deposition of fibrinoid material. Extravasation of erythrocytes is frequently found (Lipschütz 1913; Hauser 1965; Berger 1984; de Koning and Hoogkamp-Korstanje 1986; Neubert et al. 1986; Büchner and Rufli 1987).

8.1.2.1.2 Tick Bite Reaction

The area of a tick bite shows circumscribed necrotic changes of the skin and dermal edema and a dense inflammatory infiltrate that may extend into the subcutaneous tissue. The epidermis shows pseudoepitheliomatous hyperplasia and hyperkeratosis. The inflammatory infiltrate appears more polymorphic and consists of lymphocytes, numerous eosinophils, and clusters of plasma cells. In rare instances, multinucleated giant cells are present. The infiltrating lymphoid cells spread between collagen bundles. Occasionally, parts of a tick are present in the dermis. The persistent tick-bite reaction shows a dense lymphohistiocytic infiltrate, usually with follicle formation, and may be indistinguishable from lymphocytoma (Wegelin 1947; Allen 1948; Burg and Braun-Falco 1983).

8.1.2.2 Immunohistology

8.1.2.2.1 Immunohistochemical Findings

An in situ analysis of the inflammatory cell infiltrate in skin biopsy specimens taken from the periphery of the erythema migrans shows many T lymphocytes in the dermis (Fig. 12). Subset analysis reveals a slight preponderance of the helper /inducer T cell subset. Suppressor/cytotoxic T cells are also present at the dermoepidermal interface

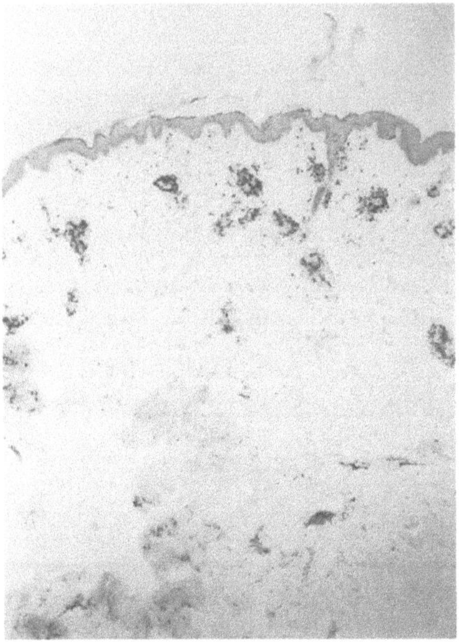

Fig. 12. Erythema migrans, peripheral border; LEU4-reactive (Pan) T lymphocytes around the blood vessels in the upper dermis, immunoperoxidase, x32

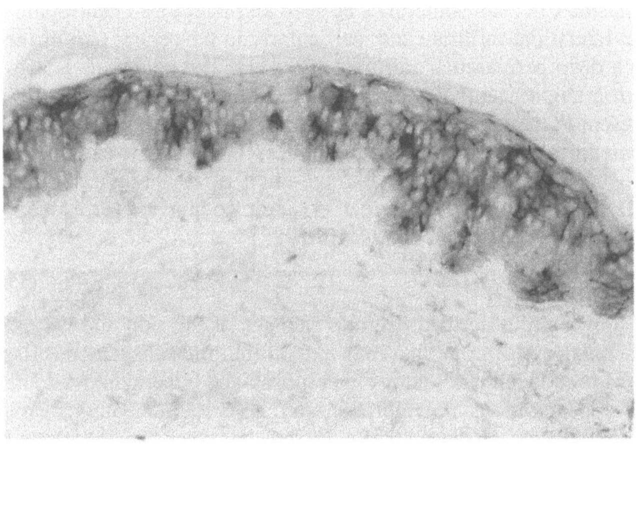

Fig. 13. Erythema migrans, peripheral border; numerous LEU6-reactive dendritic Langerhans cells in the epidermis, immunoperoxidase, x32

and within the epidermis. B lymphocytes are found only in small numbers in the dermal infiltrate. Numerous Langerhans cells are seen in the epidermis and in the dermal infiltration (Fig. 13). In the center of erythema migrans lesions, T cells were found to be present in much smaller numbers. In addition, high numbers of Langerhans cells could be identified in the epidermis of the clinically noninflamed skin outside the erythematous portion of the lesions.

The presence of T cells and increased numbers of Langerhans cells suggests that cell-mediated immune mechanisms are initially involved in the host response to *B. burgdorferi* (Büchner and Rufli 1987). These observations are in agreement with other studies on the cell-mediated immune response in the early phase of the disease (Moffat et al. 1984; Dattwyler et al. 1988).

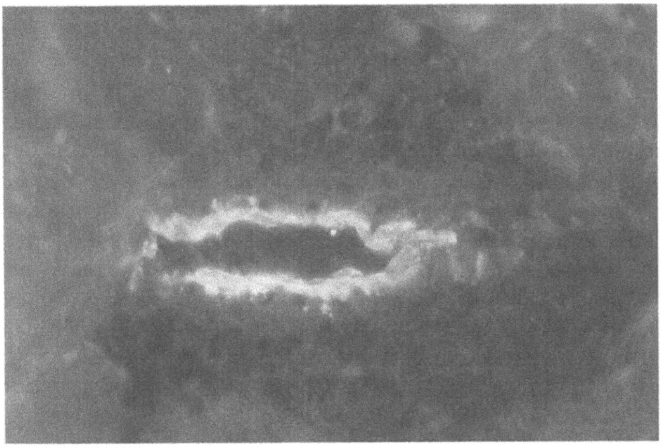

Fig. 14. Erythema migrans, central papule; pronounced deposition of complement C3 in the wall of a dermal blood vessel (see Weber et al. 1984b).

8.1.2.2.2 Immunohistologic Findings

By direct immunofluorescence, deposition of complement factor C3 (Fig. 14) and weak deposition of immunoglobulin M (IgM) in and around blood vessels has been found in three out of six patients with erythema migrans (Weber et al. 1984b). A recent, more detailed evaluation (U. Neubert and M. Meurer, unpublished observations) has shown that C3, C4d, fibrin, fibrinogen, IgM, and IgG are deposited in the blood vessels, and sometimes even at the dermoepidermal junction, in varying degrees in the central and to a lesser extent in the peripheral part of the erythema migrans.

8.1.3 Identification of *B. burgdorferi*

8.1.3.1 Cultural Identification of B. burgdorferi

B. burgdorferi is always present in erythema migrans lesions. Although the isolation of *B. burgdorferi* is regarded to be difficult (Steere 1989), comparatively high success rates have recently been achieved by V. Preac-Mursic who, for instance, isolated *B. burgdorferi* in 29 (52%) of 56 patients with erythema migrans (Weber et al. 1990; see also Chap. 21).

8.1.3.2 Morphological Identification of B. burgdorferi

B. burgdorferi can be identified in the skin using monoclonal antibody methods or, somewhat less reliably various silver stains (See Chap. 6).

8.1.4 Diagnosis and Differential Diagnosis

8.1.4.1 Diagnosis

In most instances, the diagnosis of an erythema migrans can be made on the basis of its clinical appearance alone. Particularly the ring-shaped lesion is pathognomonic, and the peripheral spreading is also quite typical. Homogeneous lesions are more difficult to identify as erythema migrans. The history of a tick bite is helpful for the diagnosis, but by no means a prerequisite for it. Contrary to many nonspecific insect bite reactions, the erythema migrans needs usually at least 2 or 3 days to arise. Moreover, most erythema migrans lesions start to develop between May and November.

The histologic picture is usually nonspecific, but the presence of plasma cells might be supportive diagnostically. The morphological identification of *B. burgdorferi* is tedious. The cultural identification of *B. burgdorferi* proves the diagnosis, but this procedure usually takes a few weeks and has so far only been accomplished in special laboratories. A positive serological test is found only in a portion of the cases; and although this supports the diagnosis, it does not prove it, because it could be the result of a previous infection with *B. burgdorferi*. However, a significant increase in antibody titers against *B. burgdorferi* is diagnostically very helpful.

Atypical and early homogeneous lesions may be so difficult to diagnose that the cultural or morphological identification of *B. burgdorferi* is required for the diagnosis. The history of a tick bite, the appearance of the lesion during the summer months, and/or a positive serological test support the diagnosis in such instances.

8.1.4.2 Differential Diagnosis

Among the most important differential diagnoses are insect bite reaction, erysipelas, and tinea. An insect bite reaction usually starts within 24 h after the insect bite, commonly lasts for a few days, and gradually subsides, whereas an erythema migrans typically continues to spread peripherally and does not start as early as an insect bite. Erysipelas is frequently located on the foot or on the face near to the port of entry of the causative organism (interdigital areas of the toes and the nose), while the erythema migrans is only seldom seen at those localizations. In addition, erysipelas is frequently associated with an abrupt rise in body temperature, a considerable elevation of the erythrocyte sedimentation rate (ESR) and leukocytosis. The last two abnormalities are found rather uncommonly in association with an uncomplicated erythema migrans (Weber and Braun-Falco 1974; Weber and Neubert 1986). Tinea corporis or tinea inguinalis are characterized by a scaling erythema (scaling is very unusual in an erythema migrans) and often by papulopustules, which are predominantly found at the periphery of the lesion. Positive fungal cultures confirm the diagnosis of a tinea. Of minor importance are fixed drug eruptions, eczema, and erythema annulare centrifugum.

8.2 Early Signs and Symptoms

8.2.1 Symptoms

During the early phase of the disease, when the erythema migrans is present, a variety of symptoms can occur (Table 2). A recent evaluation in Germany has shown that patients with erythema migrans had zero to ten symptoms besides the erythema migrans (Weber et al. 1983, 1984a; Weber and Neubert 1986). Other authors have made similar observations (Asbrink and Olsson 1985; Stiernstedt et al. 1986).

Two types of symptoms can be distinguished although this distinction is not clear-cut and somewhat speculative. First, there are constitutional symptoms such as mild headache, fatigue, fever, chills, sore throat, profuse sweating, diarrhea, and hoarseness. The other type of symptoms point rather to an involvement of specific organs such as severe headache, arthralgia, myalgia, sensory disturbances, dizziness, stiff neck, palpitations, and concentration problems.

The spectrum of symptoms as presented in the study of Weber and Neubert (1986) is depicted in Table 3. Eleven patients had five or more associated symptoms as well as erythema migrans, and one other patient had five symptoms without an erythema migrans. These patients showed a prolonged course, and most of them developed special manifestations. Other patients had one to four associated symptoms, and yet another group of patients lacked symptoms other than erythema migrans. A comparison between the severity of initial illness (as documented by the number of associated symptoms) and outcome revealed that a more severe initial disease was a risk factor

Table 2. Symptoms in 104 patients with early Lyme borreliosis (Weber and Neubert 1986)

Symptoms	*n*
Headaches	39
Fatigue, malaise	33
Lesional pruritus	33
Lesional pain, heat, burning	26
Arthralgia	19
Myalgia	18
Fever	17
Sensory disturbances	10
Chills	10
Sore throat	9
Dizziness	9
Stiff neck	8
Profuse sweating	5
Palpitations	5
Sleeplessness	5[a]
Diarrhea	3
Hoarseness	3
Difficulty concentrating	3[a]
Double vision	2
Cough	2
Weight loss	2[a]
Vomiting	2
Nausea	2
Chest pain	1
Eye pressure	1

[a] Sometimes after antibiotic therapy; 103 patients had an erythema migrans; one additional patient had mainly dysesthesia, arthralgia, and a more than four-fold increase of the IgG antibody titer.

Table 3. Spectrum of symptoms in 103 patients (Weber and Neubert 1986)

Symptoms	*n*
EM alone	32
EM and 1 symptom	14
EM and 2 symptoms	23
EM and 3 symptoms	16
EM and 4 symptoms	7
EM and 5–10 symptoms	11

EM, erythema migrans

for the development of later symptoms (Weber and Neubert 1986; Weber et al. 1986b). This was in remarkable agreement with a similar detailed study in the United States (Steere et al. 1983), although the American patients had more symptoms than the German patients.

Some of the symptoms are present only for a rather short period of time. For instance, palpitations may be noted only for a few minutes, and patients may have only one to five attacks of palpitations. Headache may be present for only 1-3 days. Some of the symptoms occur intermittently, and patients tend to forget them quite quickly. All this may explain why symptoms were noted in only a few European patients with erythema migrans (Stadelmann 1934; Sonck 1965; Weber and Braun-Falco 1974) prior to the description of Lyme disease in the United States (Steere et al. 1977). Even so, symptoms need to be asked for specifically, otherwise they might easily be missed.

8.2.2 Signs

Signs besides the erythema migrans are not common and often nonspecific in the early phase of the disease (Table 4). The most common extracutaneous findings are regional lymphadenopathy and elevated temperature. The regional lymph nodes are usually about 0.5–1 cm in diameter, and they are mostly not tender (Hauser 1965; Weber and Neubert 1986). Some of the other signs shown in Table 4 might already reflect an involvement of particular organs. Although 40 patients with erythema migrans underwent a general physical examination by one of us (K.W.), hepatomegaly or splenomegaly were not observed in any of these patients. Malar rash or a maculopapular eruption have been noted in only very few European patients (Asbrink and Olsson 1985), but have been noted in a variety of American patients (Steere et al. 1977, 1983).

An important point in evaluating patients with erythema migrans is the question of whether they have meningitis. Most patients with erythema migrans do not have meningitis (Putkonen et al. 1962; Kuiper et al. 1990), but Kuiper et al. (1990) have recently

Table 4. Findings in 104 patients with early Lyme borreliosis (Weber and Neubert 1986)

Findings	*n*
Erythema migrans	100
Regional lymphadenopathy	20
Elevated temperature	17
Tender joints on pressure	8
Pharyngitis	6
Extraregional lymphadenopathy	6
Heart beat 80-100/min	4
Multiple erythema migrans-like lesions	4
Muscle tenderness	3
Extrasystoles[a]	3
Hoarseness	2
Tracheolaryngitis	1
Conjunctivitis	1
Hypesthesia[b]	1
Pain on neck flexion	1
Bradycardia[b]	1

[a] In one man after therapy
[b] After therapy; three patients had a history of a recent erythema migrans.

found inflammatory changes in the cerebrospinal fluid in a few erythema migrans patients with and without dysesthesia. At present, it is not clear which patients are at risk for developing early meningitis. One may assume that erythema migrans patients with early neurologic symptoms such as severe headache, stiff neck, and dysesthesia, and possibly those with a greater number of associated non-neurologic signs and symptoms are more prone to having or developing meningitis.

8.3 Early Laboratory Changes

8.3.1 Serology

Only a certain proportion of the patients with erythema migrans have a positive serological test, depending on the method used. During the first 6 weeks, IgM antibody titers against *B. burgdorferi,* and, after 6 weeks, IgG antibody titers against *B. burgdorferi* are more frequently elevated (Wilske et al. 1984, 1989) (see Chap. 21). For example, antibody titers against *B. burgdorferi* were negative in all patients shown in Figs. 1–9. However, despite prompt antibiotic therapy, IgM antibody titers were found to be positive in patients 1,2,3,5, and 8 three to five weeks after the pictures were taken, whereas the IgG and IgM antibody titers remained negative in the others; all patients were seen by one of us (K.W.), and the serological methods used were according to Wilske et al. (1984, 1989).

8.3.2 Inflammatory Parameters

Besides the specific serological tests, most if not all laboratory changes in patients with erythema migrans are nonspecific and usually mild in character. They are found only in a certain proportion of the patients.

8.3.2.1 Erythrocyte Sedimentation Rate

The ESR is elevated in only a minority of the patients with erythema migrans (Weber et al. 1983, 1984a; Asbrink and Olsson 1985; Stiernstedt et al. 1986; Weber and Neubert 1986). The median ESR in 72 patients in one series was 6 mm/h (range 2-27), while the median of abnormal values was 19 mm/h (11-27) (Weber and Neubert 1986). It has been reported that 4%–8% of patients have values of more than 20 mm/h (Asbrink and Olsson 1985; Stiernstedt et al. 1986; Weber and Neubert 1986).

8.3.2.2 Serum Immunoglobulins

About 10% (range 7%–13%) of European patients with erythema migrans have an elevated serum IgM level (Weber et al. 1983; Asbrink et al. 1986; Weber and Neubert 1986). This is considerably less than the 33% found in American patients with Lyme disease (Steere et al. 1983). The median of the serum IgM levels determined in 89 patients of one series was 134 mg%, while the median of abnormal serum IgM values in 9 (10%) of these 89 patients was 283 mg% (252–395; normal value up to 250 mg%) (Weber and Neubert 1986). An American study has provided evidence that an elevated serum IgM level predisposes to the development of a subsequent arthritis in untreated patients (Steere et al. 1979); such a correlation has not been observed in European patients so far, but most erythema migrans patients in Europe who visited a physician were promptly treated with antibiotics so that there is not enough experience to prove that point. A correlation between elevated serum IgM levels and subsequent arthralgia has not been found either (Weber and Neubert 1986).

Only 3% of patients had a serum IgG level of more than 1500 mg% in both an European and an American study and 4% and 7% of the patients in these two studies, respectively, had an elevated serum IgA level (Steere et al. 1983; Weber and Neubert 1986). There seems to be no correlation between an elevation of the serum IgG and IgA level and the development of special manifestations.

8.3.2.3 Immune Complexes and Cryoglobulins

Systematic investigations on immune complexes and cryoglobulins have not yet been published from Europe. Detailed reports from the United States have shown that immune complexes and cryoglobulins are frequently formed early in the illness and tend to persist in patients with a complicated course (Hardin et al. 1979; Steere et al. 1979). A small amount of cryoglobulins could be detected in one patient (12.6 mg/dl) (Weber et al. 1984a), but not in any of the other 52 European patients examined (Weber et al. 1983; Asbrink and Olsson 1985). Immune complexes have been found to be elevated only in a few patients (Asbrink and Olsson 1985; Stiernstedt et al. 1986; U. Neubert and W. Samtleben, unpublished).

8.3.2.4 Leukocytes

European studies on patients with erythema migrans have shown that there is usually no leukocytosis during the early phase of the disease (Weber and Braun-Falco 1974; Weber et al. 1983; Stiernstedt et al. 1986; Weber and Neubert 1986). On the other side, leukocytosis seems to be associated with a more complicated course, for instance with meningoradiculoneuritis (Chap. 11).

8.3.2.5 Differential Blood Count

Eosinophilia of up to 6%, monocytosis of up to 13%, 2% basophiles, and changes in lymphocyte count and morphology have been occasionally noted in patients with erythema migrans. In the first few weeks of the illness, some patients may have an absolute lymphogenia, while somewhat later patients may show a relative lymphocytosis and plasma cell-like lymphocytes (Weber and Braun-Falco 1974; Weber et al. 1983; Weber and Neubert 1986). An early absolute lymphopenia has also been observed in American patients (Shresta et al. 1985).

8.3.3 Other Laboratory Changes

8.3.3.1 Urine Analysis

Significant early changes found in urine analysis have not been reported from Europe so far. Microscopic hematuria and sometimes mild proteinuria have been noted occasionally in American patients (Steere et al. 1983).

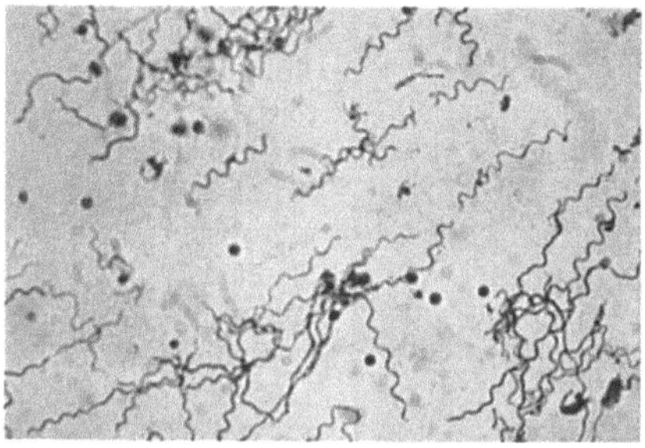

Fig. 1, Chap. 4. *Borrelia burgdorferi* in culture, Fuchsin, x 1600

Fig. 1, Chap. 5. *Ixodes ricinus* larva *(below left)*, nymph *(above right)*, adult male *(below right)*, and adult female *(above left)*. One *square* corresponds to 1 mm^2.

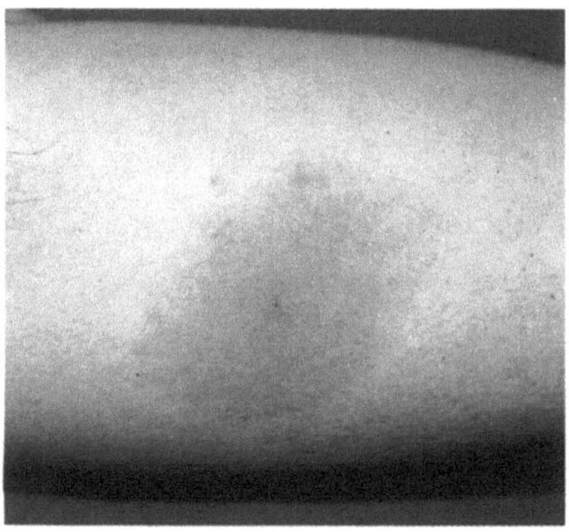

Fig. 1, Chap. 8. Homogeneous erythema migrans of 4 days duration on the left upper arm of a 56-year-old woman. A tick had been removed 2 weeks previously from the center of the lesion. *B. burgdorferi* was isolated from both the tick and the border of the erythema migrans

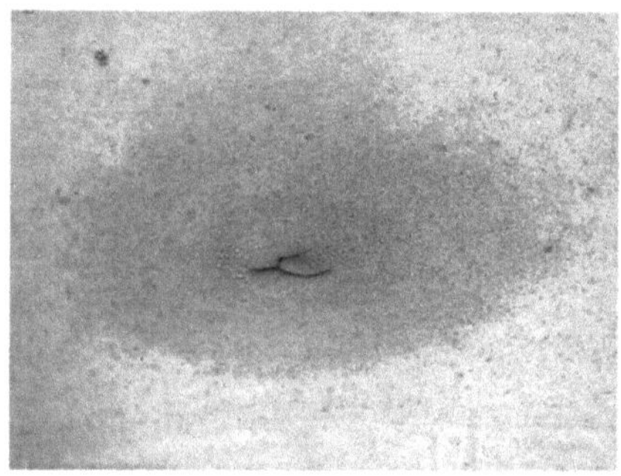

Fig. 2, Chap. 8. Homogeneous erythema migrans of 10 days duration on the navel of a 54-year-old woman. A tick had been removed 11 days previously. *B. burgdorferi* was isolated from the border of the erythema migrans

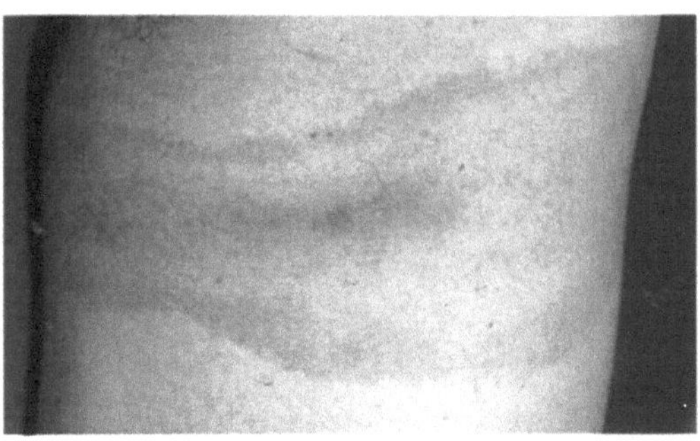

Fig. 4, Chap. 8. Ring-shaped erythema migrans of 2 weeks duration on the left thigh of a 60-year-old woman. No tick was noted. *B. burgdorferi* was isolated from the border of the erythema migrans

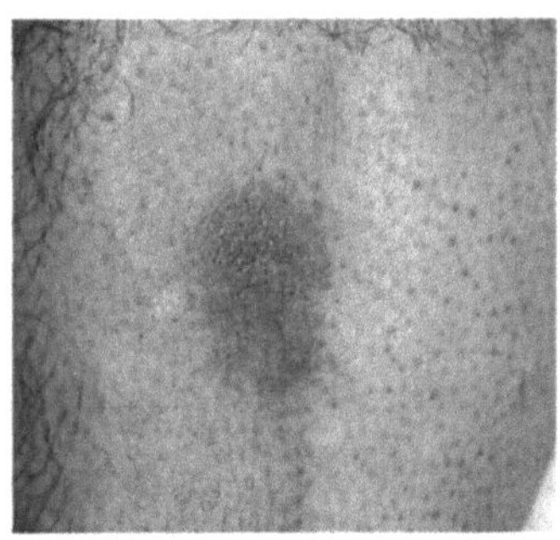

Fig. 9, Chap. 8. Atypical stationary erythema "migrans" measuring 4x2.5 cm in diameter of 3 days duration on the right lower leg of a 43-year-old man. No tick was noted. *B. burgdorferi* was isolated from the lesion

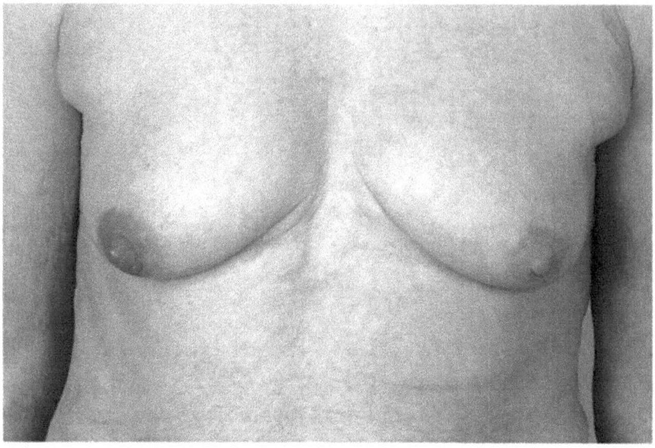

Fig. 1, Chap. 9. Borrelial lymphocytoma on the left nipple and areola mammae region in a 71-year-old woman. Duration 2.5 months. *B. burgdorferi* was cultivated from this lesion

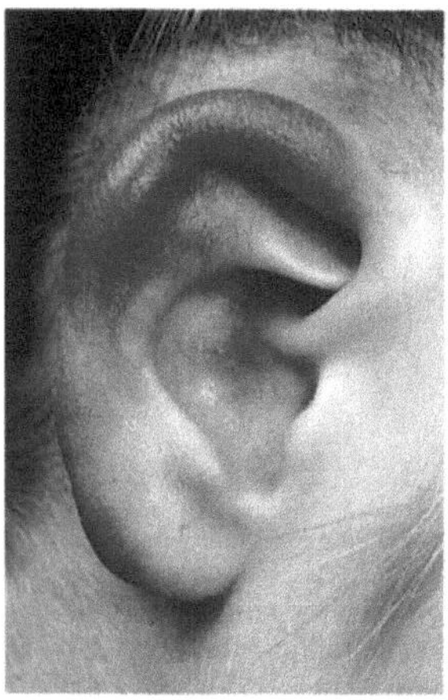

Fig. 3, Chap. 9. Borrelial lymphocytoma on the helix in a 9-year-old girl. Duration 2 months. The patient also had a facial palsy. Elevated antibody titers to *B. burgdorferi* were found both in the serum and cerebrospinal fluid

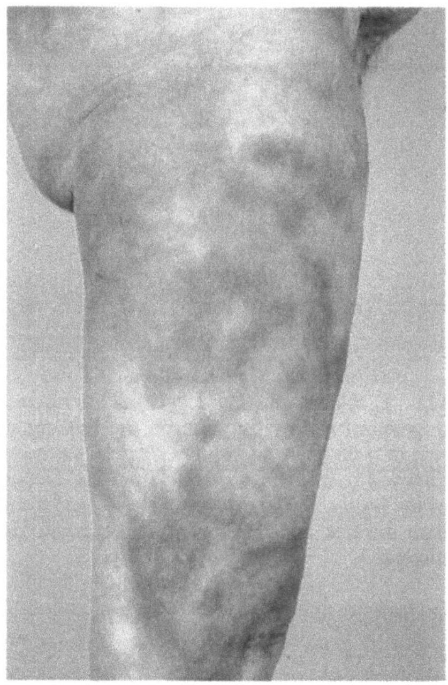

Fig. 1, Chap. 14. Acrodermatitis chronica atrophicans; patchy and cyanotic erythema on the thigh

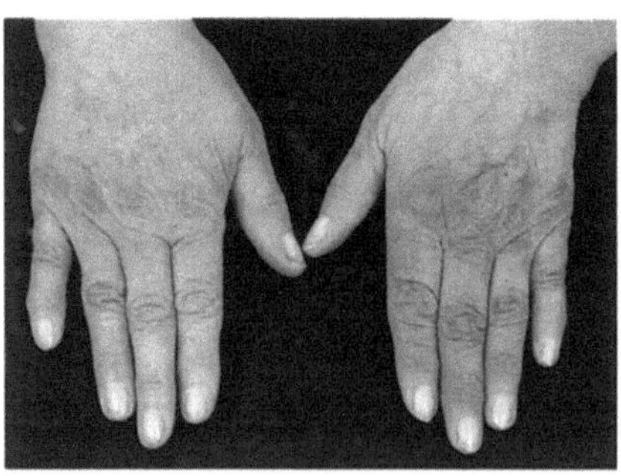

Fig. 2, Chap. 14. Acrodermatitis chronica atrophicans; bluish–red discoloration of one of the hands

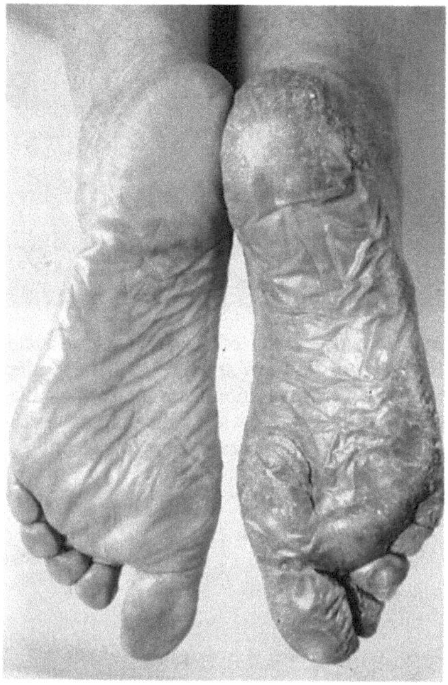

Fig. 5, Chap. 14. Acrodermatitis chronica atrophicans; bluish-red discoloration and atrophy of the sole of one foot

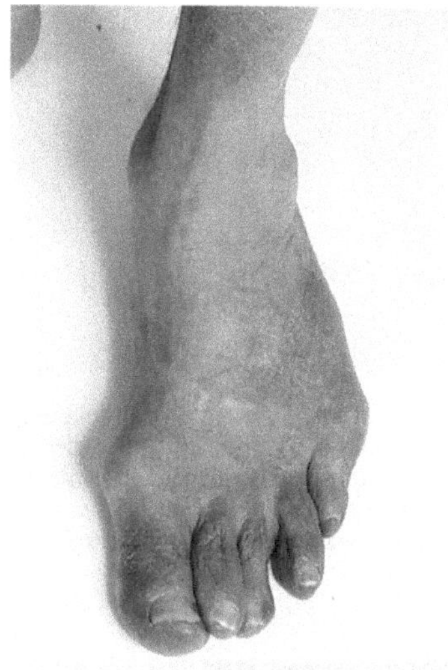

Fig. 8, Chap. 14. Acrodermatitis chronica atrophicans; enlarged, deformed forefoot with luxation of some of the toe joints

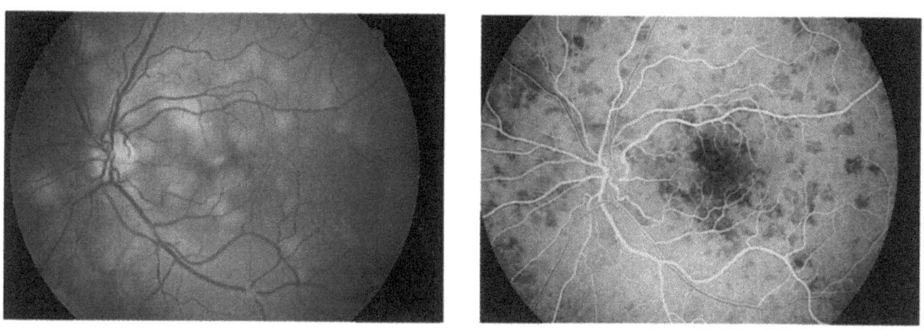

a b

Fig. 1 a, b, Chap. 19. Multifocal choroiditis in the left eye of patient 5. Elevated IgG and IgM antibody titers against *B. burgdorferi* with simultaneous bilateral gonarthritis and lymphadenopathy. The visual acuity was 0.1. Prompt response to doxycycline with subsequent recovery of full visual acuity.
a Fundus. Multifocal whitish foci in the level of the choroid. **b** Standard fluorescein angiography 19 s after injection of sodium fluorescein in a cubital vein. Hypofluorescence of the choroid in the affected areas indicating the involvement of the choriocapillaries. The retinal vessels are not involved

Fig. 2 a, b, Chap. 19. Neuroretinitis (papillitis and peripapillary retinal vasculitis) in the right eye of patient 8. Elevated IgG and IgM antibody titers against *B. burgdorferi*. The visual acuity was 0.4. Prompt response to doxycycline with subsequent recovery of full visual acuity.

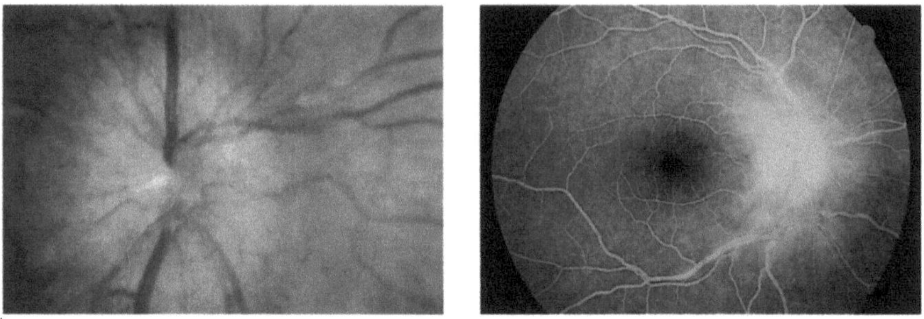

a b

Fig. 2, Chap. 19 (continued). **a** Optic nerve head in detail. Edema of the optic nerve head. Dilatation of the optic nerve head capillaries. **b** Standard fluorescein angiography 103 s after injection of sodium fluorescein in a cubital vein. Fluorescein extravasation from the disk capillaries and the peripapillary retinal vessels and staining of the retinal vessel walls indicating the breakdown of the inner blood-retina barrier located in the endothelium of the vessels due to vasculitis

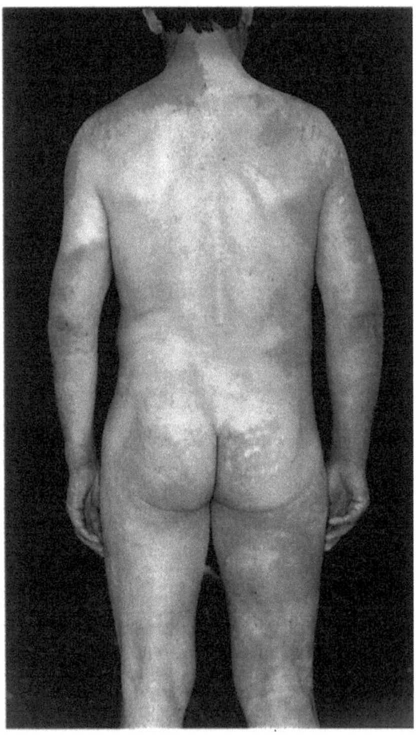

Fig. 1, Chap. 20. Borrelial dermatomyositis-like syndrome. Patient 1. Diffuse erythemas on the trunk and extremities

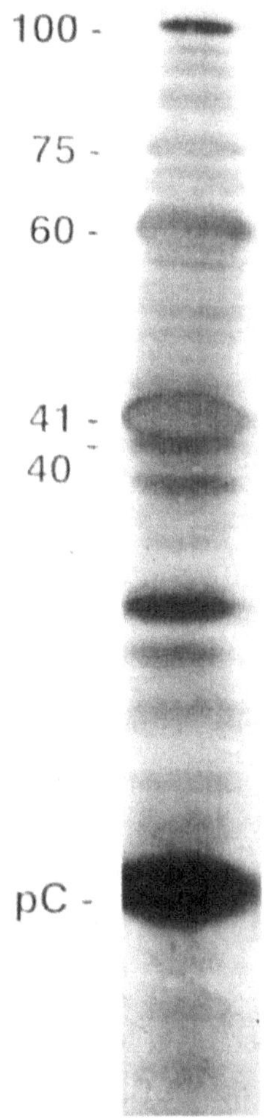

Fig. 8, Chap. 21. *Borrelia burgdorferi* proteins in a two-color blot; specific proteins are stained blue with immune serum against *Borrelia burgdorferi* and non specific proteins are stained brown with immune serum against *Treponema phagedenis*

8.3.3.2 Transaminases

A usually mild elevation of the serum glutamic-oxaloacetic transaminase (SGOT) (ALA) and/or of the serum glutamate pyruvate transaminase (SGPT) (ASA) levels has been observed in a few patients with erythema migrans (Weber et al. 1983, 1984a; Weber and Neubert 1986). In the United States, 15%–19% of patients have sometimes pronounced elevation of the SGOT and SGPT, and somtimes also of the lactate dehydrogenase (LDH) level (Steere et al. 1983).

8.4 Comparison Between Initial Illness in Europe and the United States

A comparison between prospective studies of several hundred erythema migrans patients observed over a number of years in Europe (Weber et al. 1983; Asbrink and Olsson 1985; Weber and Neubert 1986) and the United States (Steere et al. 1977, 1983) and other data seem to indicate that the initial illness is more severe in Americans than in Europeans, but this statement may be influenced by bias in patient selection (Weber 1989).

8.5 Concluding Remarks

In conclusion, erythema migrans and its associated signs and symptoms represent the first stage of Lyme borreliosis, but they may extend into the second stage of the disease. Erythema chronicum migrans and the secondary erythema migrans-like lesions are examples of this extension. Erythema migrans is a unique sign. It is the hallmark of Lyme borreliosis and found only during the early phase of the disease.

Acknowledgements. We are indebted to Dr. Vera Preac-Mursic for isolating *B. burgdorferi* (see figures) and to Dr. W. Maciejewski for providing Fig. 14.

References

Afzelius A (1910) Verhandlungen der Dermatologischen Gesellschaft zu Stockholm, 28 Oct 1909. Arch Dermatol Syph 101:404

Afzelius A (1921) Erythema chronicum migrans. Acta Derm Venereol (Stockh) 2:120-125

Allen C (1948) Persistent "insect bites" (dermal eosinophilic granulomas) simulating lymphoblastomas, histiocytoses and squamous cell carcinomas. Am J Pathol 24:367-375

Asbrink E, Hovmark A (1988) Early and late cutaneous manifestations in *Ixodes*-borne borreliosis (erythema migrans borreliosis, Lyme borreliosis). Ann NY Acad Sci 539:4-15

Asbrink E, Olsson I (1985) Clinical manifestations of erythema chronicum migrans Afzelius in 161 patients. Acta Derm Venereol (Stockh) 65:43-52

Asbrink E, Hovmark A, Hederstedt B (1984) The spirochetal etiology of acrodermatitis chronica atrophicans Herxheimer. Acta Derm Venereol (Stockh) 64:506-512

Asbrink E, Olsson I, Hovmark A (1986) Erythema chronicum migrans Afzelius in Sweden. A study on 231 patients. Zentralbl Bakteriol Mikrobiol Hyg (A) 263:229-236

Berger BW (1984) Erythema chronicum migrans of Lyme disease. Arch Dermatol 120:1017-1021

Bowen GS, Griffin M, Hayne C, Slade J, Schulze TL, Parkin W (1984) Clinical manifestations and descriptive epidemiology of Lyme disease in New Jersey, 1978 to 1982. JAMA 251:2236-2240

Bruder K (1950) Zur Kenntnis des Erythema chronicum migrans (Lipschütz). Dermatol Wochenschr 121:337-344
Büchner SA, Rufli T (1987) Erythema chronicum migrans: evidence for cellular immune reaction in the skin lesion. Dermatologica 174:144-149
Burg G, Braun-Falco O (1983) Cutaneous lymphomas, pseudolymphomas and related disorders. Springer, Berlin Heidelberg New York
Dattwyler RJ, Volkman DJ, Halperin JJ, Luft BJ, Thomas JA, Golightly MG (1988) Specific immune responses in Lyme borreliosis: characterization of T cell and B cell responses to *Borrelia burgdorferi*. Ann NY Acad Sci 539:93-102
De Koning J, Hoogkamp-Korstanje JAA (1986) Diagnosis of Lyme disease by demonstration of spirochetes in tissue biopsies. Zentralbl Bakteriol Mikrobiol Hyg (A) 263:179-188
Hardin JA, Steere AC, Malawista SE (1979) Immune complexes and the evolution of Lyme arthritis: dissemination and localization of abnormal Clq binding activity. N Engl J Med 301:1358-1363
Hauser W (1965) Wahrscheinliche Infektionskrankheiten der Haut. In: Marchionini A, Götz H (eds) Infektionskrankheiten der Haut I. Springer, Berlin Heidelberg New York, pp 555-629 (Handbuch der Haut- und Geschlechtskrankheiten, vol IV, part 1 A)
Hellerström S (1930) Erythema chronicum migrans Afzelii. Acta Derm Venereol (Stockh) 11:315-321
Hollström E (1951) Successful treatment of erythema migrans Afzelius. Acta Derm Venereol (Stockh) 31:235-243
Hollström E (1958) Penicillin treatment of erythema chronicum migrans Afzelius. Acta Derm Venereol (Stockh) 38:285-289
Kuiper H, de Jongh B, Ramselaar T, Spanjaard L (1990) Evidence of asymptomatic nervous system involvement in localized erythema migrans (abstract). IVth international conference on Lyme borreliosis, Stockholm, June 1990
Lipschütz B (1913) Über eine seltene Erythemform (Erythema chronicum migrans). Arch Dermatol Syph 118:349-356
Moffat CM, Sigal LH, Steere AC, Freeman DH, Dwyer JM (1984) Cellular immune findings in Lyme disease: correlation with serum IgM and disease activity. Am J Med 77:625-632
Neubert U, Krampitz HE, Engl H (1986) Microbiological findings in erythema chronicum migrans, acrodermatitis chronica atrophicans and lymphadenosis cutis benigna. Zentralbl Bakteriol Mikrobiol Hyg (A) 263:237-252
Neumann R, Aberer E, Stanek G (1986) Treatment and course of erythema chronicum migrans. Zentralbl Bakteriol Mikrobiol Hyg (A) 263:372-376
Putkonen T, Mustakallio KK, Salminen A (1962) Erythema chronicum migrans with meningitis: a rare coincidence of two tick-borne diseases? Dermatologica 125:184-188
Riehl G (1913) Arch Dermatol Syph 115:401
Schaad UB, Flüeler U, Schaub H et al. (1986) Durch *Ixodes-ricinus*-Spirochäten (*Borrelia burgdorferi*) verursachte Krankheitsbilder (Lyme-Krankheit) bei pädiatrischen Patienten in der Schweiz. Schweiz Med Wochenschr 116:1426-1430
Sedlacek V (1960) Erythema chronicum migrans - poznamky ke klinice, etiopatogenezi a zarazeni. Cs Dermatol 35:386-399
Shresta M, Grodzicki RL, Steere AC (1985) Diagnosing early Lyme disease. Am J Med 78:235-240
Sonck CE (1965) Erythema chronicum migrans with multiple lesions. Acta Derm Venerol (Stockh) 45:34-36
Stadelmann R (1934) Ein Beitrag zum Krankheitsbild des Erythema chronicum migrans Lipschütz. Dissertation, University of Marburg
Stanek G, Pletschette M, Flamm H et al. (1988) European Lyme borreliosis. Ann NY Acad Sci 539:274-282
Steere AC (1989) Lyme disease. N Engl J Med 321:586-596
Steere AC, Malawista SE, Hardin JA, Ruddy S, Askenase PW, Andiman WA (1977) Erythema chronicum migrans and Lyme arthritis: the enlarging clinical spectrum. Ann Intern Med 86:685-698
Steere AC, Hardin JA, Ruddy S, Mummaw JG, Malawista SE (1979) Lyme arthritis: correlation

of serum and cryoglobulin IgM with activity, and serum IgG with remission. Arthritis Rheum 22:471-483
Steere AC, Bartenhagen NH, Craft JE et al. (1983) The early clinical manifestations of Lyme disease. Ann Intern Med 99:76-82
Stiernstedt G, Eriksson W, Enfors H et al. (1986) Erythema chronicum migrans in Sweden. Scand J Infect Dis 18:217–224
Vaillant-Rieder L, Hey D (1987) Atypischer Verlauf einer Erythema-migrans-Krankheit. Dtsch Med Wochenschr 112:1680-1681
Weber K (1986) Remarks on the infectious disease caused by *Borrelia burgdorferi*. Zentralbl Bakteriol Mikrobiol Hyg (A) 263:206-208
Weber K (1989) Clinical differences between European and North-American Lyme borreliosis - a review. Zentralbl Bakteriol (Suppl) 18:146-155
Weber K, Braun-Falco O (1974) Blutbildveränderungen bei Erythema chronicum migrans. Hautarzt 25:611-613
Weber K, Neubert U (1986) Clinical features of early erythema migrans disease and related disorders. Zentralbl Bakteriol Mikrobiol Hyg (A) 263:209-228
Weber K, Puzik A, Becker T (1983) Erythema-migrans-Krankheit: Beitrag zur Klinik und Beziehung zur Lyme-Krankheit. Dtsch Med Wochenschr 108:1182-1190
Weber K, Schierz G, Wilske B, Preac-Mursic V (1984a) European erythema migrans disease and related disorders. Yale J Biol Med 57:463-471
Weber K, Schierz G, Wilske B, Preac-Mursic V (1984b) Zur Klinik und Ätiologie der Acrodermatitis chronica atrophicans. Hautarzt 35:571-577
Weber K, Neubert U, Thurmayr R (1986a) Antibiotic therapy in early erythema migrans disease and related disorders. Zentralbl Bakteriol Mikrobiol Hyg (A) 263:377-388
Weber K, Schierz G, Wilske B, Neubert U, Krampitz HE, Barbour AG, Burgdorfer W (1986b) Reinfection in erythema migrans disease. Infection 14:32-35
Weber K, Preac-Mursic V, Neubert U, Thurmayr R, Herzer P, Wilske B, Schierz G, Marget W (1988) Antibiotic therapy of early European Lyme borreliosis and acrodermatitis chronica atrophicans. Ann NY Acad Sci 539:324-345
Weber K, Preac-Mursic V, Wilske B, Thurmayr R, Neubert U, Scherwitz C (1990) A randomized trial of ceftriaxone versus oral penicillin for the treatment of early European Lyme borreliosis. Infection 18:91-96
Wegelin C (1947) Zur Histologie des Zeckenstiches. Dermatologica 94:368-376
WHO Case Definition Committee (Asbrink E, Hovmark A, Weber K) (1989) Cutaneous manifestations of Lyme borreliosis. Zentralblatt Bakt (Suppl) 18:350
Williams CL, Curran AS, Lee AC, Sousa VO (1986) Lyme disease: epidemiologic characteristics of an outbreak in Westchester county, NY. Amer J Publ Health 76:62-65
Wilske B, Schierz G, Preac-Mursic V, Weber K, Pfister HW, Einhäupl K (1984) Serological diagnosis of erythema migrans disease and related disorders. Infection 12:331-337
Wilske B, Schierz G, Preac-Mursic V, Pfister HW, Weber K, von Busch K, Baruschke A (1989) IgM- and IgG-immune response to *Borrelia burgdorferi* in erythema migrans and neuroborreliosis. Zentralbl Bakteriol (Suppl) 18:290-298

9 Borrelial Lymphocytoma

A. HOVMARK, E. ASBRINK, K. WEBER, and P. KAUDEWITZ

9.1 Introduction

Lymphocytoma as a diagnostic term has been used since 1923 to designate cutaneous proliferations of lymphoid cells without detectable extracutaneous involvement; it was first used by Biberstein following a suggestion of Kaufmann-Wolf. Burckhardt (1911) and Kaufmann-Wolf (1921) reported on several cases of what was then categorized as hyperplasia of cutaneous lymphoid tissue. Lymphocytoma and lymphadenosis benigna cutis (LABC), often used synonymously, belong to the group of pseudolymphomas, which are lymphoreticular disorders with a benign course (Burg and Braun-Falco 1983).

The term LABC has been in use since 1943, when it was coined by Bäfverstedt. It has been used to designate benign lymphoreticular proliferations of the skin having a tendency to form germinal centers. Bäfverstedt classified the disorder into two main forms, localized LABC (usually a solitary lesion, but sometimes with a few regionally circumscribed ones) and the disseminated form, LABC dispersa. The former was found to be most common and to occur in both children and adults at predilection sites such as the ear lobes, the nose, the nipples and areola mammae, and the scrotum. The rarer form, LABC dispersa, was found only in adults and had no predilection sites. In LABC dispersa, there are multiple, disseminated lesions. These disseminated lymphoreticular proliferations can be very long-lasting - years to decades - and are sometimes entirely subcutaneous.

The pathogenesis of pseudolymphomas remained unclear. An association with tick bites was reported in some patients (Strandberg 1920; Bäfverstedt 1943; Paschoud 1954; Mach 1965) and the term "Zeckenbiss-Lymphocytom" (tick-bite lymphocytoma) was used by Jordan and Holtschmidt (1951). These observations, the good results achieved with penicillin treatment (Bianchi 1950), and the finding that LABC could develop after injection of skin homogenates from patients with tick-bite-associated LABC (Paschoud 1957, 1958a, 1958b) all suggested that LABC might be induced by a transmissible, infective, tick-borne agent. After *Borrelia burgdorferi* was found to be the causative organism of Lyme disease in the United States (Burgdorfer et al. 1982), this agent also became a candidate for causing LABC.

Clinical observations and results of serologic and histopathologic tests during the last few years have indicated that, at least in areas endemic for Lyme borreliosis in Europe, most cases of typical solitary LABC lesions are of borrelial etiology (Weber et al. 1984, 1985; Asbrink 1985; Hovmark et al. 1986; de Koning and Hoogkamp-Korstanje 1986; Weber and Neubert 1986; Asbrink et al. 1989). Similar investigations in patients with the rare LABC dispersa are still lacking.

The first successful cultivation of *B. burgdorferi* from a cutaneous LABC lesion was described in 1986 (Hovmark et al. 1986), and ultimately proved that this organism can cause LABC. In his experiments with transmission of LABC and erythema migrans, Paschoud (1957, 1958a, 1958b) found that the development of an LABC lesion was more likely to occur if the skin homogenate was injected deep into the dermis, and an erythema migrans more likely to occur after a superficial inoculation.

Although the *B. burgdorferi* infection may give rise to pseudolymphomas, antigen injection (Bernstein et al. 1974), acupuncture (Bork 1983), and tattooing (Blumental et al. 1982; Rijlaarsdam et al. 1988) may induce these conditions, too. These observations indicate that different kinds of persistent subepidermal antigenic stimulation may result in the development of lymphocytomas. In order to determine which cases are caused by *B. burgdorferi*, Weber et al. introduced the term "borrelial lymphocytoma" (BL) in 1984 and 1985. The etiology in cases of pseudolymphomas seen before a *B. burgdorferi* infection could be diagnosed, therefore, have to be regarded with caution unless they were associated with an erythema migrans.

9.2 Definition

BL can be defined as a bluish-red-tumor-like skin infiltrate with lymphoreticular proliferation in the dermis and/or subcutis caused by *B. burgdorferi*.

9.3 Prevalence and Incidence

Most cases of lymphocytoma have been described from central, eastern and northern Europe. There are also American case reports (Mach and Wilgram 1966; Self et al. 1969), but in the United States cutaneous lymphocytomas have not yet been described in connection with Lyme disease. Whether this is due to differences in the clinical spectra of Lyme borreliosis on each side of the Atlantic Ocean caused by dissimilarities of the spirochetes involved or other factors, or merely reflects varying awareness of this condition, is still not known.

At the Department of Dermatology of Södersjukhuset, Stockholm, Sweden, 22 cases of solitary BL were diagnosed over a period of 2 years (Asbrink et al. 1989). During the same period, 130 cases of erythema migrans and 43 cases of the late borrelial manifestation acrodermatitis chronica atrophicans (ACA) were seen at the same clinic. This indicates that although BL is not a very rare condition in an area highly endemic for Lyme borreliosis, it is less common than erythema migrans and ACA.

9.4 Clinical Features

9.4.1 Clinical Description of BL

The literature of LABC/lymphocytoma cutis written before the discovery of *B. burgdorferi* must be viewed with caution, as all these different clinical descriptions may not be relevant for lymphocytomas caused by borreliae. As the etiology of BL was only recently revealed, clinical descriptions of patients with lymphocytomas of proven borrelial etiology are still sparse, and the full clinical spectrum of BL has yet to be elucidated.

As is the case in other manifestations of Lyme borreliosis, many, but far from all, patients with BL are aware of a preceding tick bite. The lesion may start within a few weeks after the causative tick bite, but sometimes it will take as long as 10 months before a BL develops (Asbrink et al. 1989). In the cases of successful transmission of solitary LABC lesions from the ear lobe of one individual to the ear lobe or to the back of another, the LABC usually started 3–7 weeks after the injection of the skin homogenates (Paschoud 1957, 1958b).

A typical BL presents as a tumor-like and generally bluish-red nodule or plaque, usually 1–5 cm in diameter. A BL may reach its maximum size within weeks, but it may also continue to grow for several months, The lesion may be slightly tender, and the patient often has regional lymphadenopathy. Predilection sites for solitary BL are the nipples and areolae mammae (Fig. 1, see color plate II between pp 118–119) and - particularly in children - the ear lobes (Fig. 2) (Bäfverstedt 1943; Paschoud 1958a; Mach 1965; Hauser 1965; Weber et al. 1984; Asbrink 1985; Hovmark et al. 1986; Asbrink et al. 1989). The scrotum and the nose have also been regarded as predilection sites for solitary lymphocytomas (Bäfverstedt 1943; Hauser 1965).

In cases of long-lasting lesions, patients have also not seldom had lesions which are typical for ACA (Mulzer and Keining 1929; Gottron 1938; Bäfverstedt 1943, 1953; Herzberg 1952; Matras 1953; Hauser 1955; Mach 1965). In these cases the borrelial etiology of the lymphocytomas seems to be obvious. Some of the cases described as lymphocytomas have been located around elbows in patients with ACA (Bäfverstedt 1943), and some of these lesions may rather be classified as fibrotic nodules, which may appear in patients with ACA.

Besides typical solitary lesions many clinical variants of lymphocytomas have been described. Such variants include groups of papules and small, nodular, disseminated lesions, but also large, multiple lesions and flat forms lying below the skin. Further studies are needed to answer the question of whether these more unusual forms of lymphocytomas may be of borrelial etiology or not.

In many cases there will be no or only local symptoms, but patients with BL may sometimes suffer from constitutional symptoms such as headache, profound fatigue, arthralgia/myalgia, and/or fever (Asbrink et al. 1989). The lesion may sometimes be preceded by or occur concomitantly with an erythema migrans (Strandberg 1920; Bäfverstedt 1943; Jordan and Holtschmidt 1951; Weber et al. 1984, 1985; Asbrink 1985; Hovmark et al. 1986). In one study, 10 of 22 patients with BL had a history of a spontaneously healing or a concomitant erythema migrans near the site of BL development (Asbrink et al. 1989). One of these 22 patients also developed a sclerotic, morphea-like or lichen-sclerosus-et-atropicus-like lesion, which healed completely after antibiotic treatment. Such sclerotic lesions have previously been described in associa-

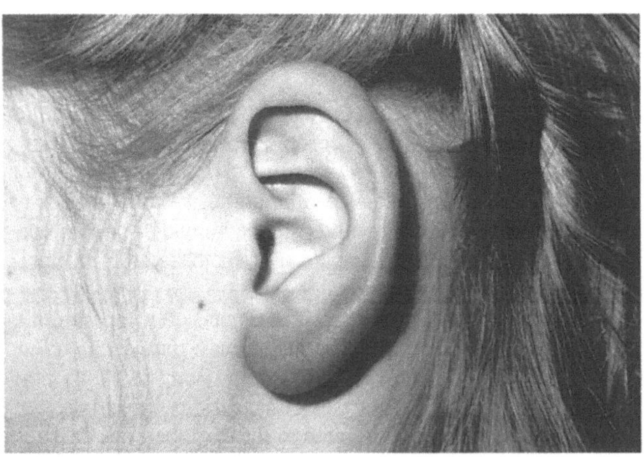

Fig. 2. Borrelial lymphocytoma of 8 weeks duration in a 10-year-old girl. A tick bite occurred 6 weeks prior to the beginning of the lymphocytoma. Elevated IgG antibody titer to *B. burgdorferi* (from Weber et al. 1984)

tion with lymphocytomas (Bäfverstedt 1943) and also occur in about 10% of patients with ACA (Asbrink 1985). In 1954, Paschoud reported on meningoradiculoneuritis occurring in a patient with a lymphocytoma cutis. Of 22 patients with BL, three developed meningitis with facial palsy (Fig. 3, see color plate II between pp 118–119), one trigeminal neuralgia and choroiditis, and one arthritis (Asbrink et al. 1989).

Children appear to be more prone to develop BL than adults (Paschoud 1958a). Among seven patients documented by Weber and Neubert (1986), five were children, and in another 22 patients the median age was 9 years (range 1.5-71 years) (Asbrink et al. 1989).

9.4.2 Course of BL

BL often has a more prolonged course than erythema migrans. It will usually take several months and sometimes more than a year for BL to heal without treatment. Lesions lasting more than 6 months are not uncommon, and BL with a duration of 15 months has been described (Asbrink et al. 1989). A typical case of solitary lymphocytoma on the areola mammae which lasted 2 years has been reported (Bäfverstedt 1943). Bäfverstedt (1962) has also described a large solitary scrotal lymphocytoma, which had been present for about 20 years, and this infiltrate disappeared after penicillin treatment. It is still an open question whether lesions which have persisted for many years and have been described as LABC dispersa (Bäfverstedt 1943; Bianchi 1950) are of borrelial origin.

BL usually heals without any residual sign. In the same way as syphilis, Lyme borreliosis has been divided into early (stages one and two) and late (stage three) manifestations. Symptoms or signs of different stages may, however, sometimes overlap, and the clinical borders between the different stages are not always strict. As previously mentioned, a BL may appear simultaneously with the primary stage manifestation erythema migrans and during the secondary stage together with a meningoradiculoneuritis or an early arthritis; there have also been case reports of solitary lymphocytomas occurring in patients with the tertiary borrelial manifestation ACA. In a follow-up study after 17 years in nine of his previously examined patients with LABC, Bäfverstedt found that all LABC lesions had healed, either spontaneously or after radiotherapy, but one of the patients had developed ACA (Bäfverstedt 1960). This indicates that an early untreated BL may in some patients result in the development of late/chronic Lyme borreliosis.

It may be discussed whether an early, solitary BL should be classified as a stage one or a stage two manifestation of a *B. burgdorferi* infection. As in the primary stage manifestation, erythema migrans, a BL may appear at or very near the site of a tick bite - the site of the spirochetal inoculation. However, it may sometimes follow an untreated, spontaneously healing erythema migrans and may occasionally develop far away from the site of the causative tick bite. It may occasionally take as long as 6–10 months after the start of an erythema migrans before a BL develops. Thus, it has been found that erythema migrans lesions which have developed after tick bites on the back, the abdomen, or in the umbilical region have been followed several months later by the development of BL in the areola mammae region (Bäfverstedt 1943; Hovmark et al. 1986; Asbrink et al. 1989). In these cases, the lymphocytomas developed after the erythema migrans lesions had reached the front of the chest and disappeared. Lymphocytomas may also appear simultaneously with ACA. Thus, according to our present knowledge, BL may be a manifestation seen in all stages of the disease.

9.5 Histopathologic Findings

The histopathologic picture of pseudolymphomas has been investigated by several authors (Burckhardt 1911; Bäfverstedt 1943; Mach and Wilgram 1966; Wantzin et al. 1982). A number of histopathologic features have been elaborated (Burg and Braun-Falco 1983). In conjunction with immuno-histochemical results, these form the basis for a modern classification of pseudolymphomas (Kerl 1989).

BL may histologically present either as a diffuse, dense dermal lymphocytic infiltrate or as a follicular lesion with germinal centers (Fig. 4). There is an admixture of plasma cells, macrophages, and sometimes also eosinophils. The infiltrate is often interspaced by small proliferating vessels. Usually the epidermis is unaffected, but slight reactive hyperplasia may be seen. Some of the lymphocytes may be large, and there is often mitotic activity in the germinal centers. Thus, on histological grounds, BL may not differ essentially from B cell pseudolymphomas of nonborrelial origin, and the histopathologic features may sometimes simulate a highly differentiated malignant lymphoma. Further insight into the cellular composition of BL is provided by immuno-histochemical investigations. In typical cases, these demonstrate the presence of a germinal center of B cell origin (Fig. 5a) with a dense network of dendritic reticulum cells (Fig. 5b) and a high proliferative activity (Fig. 5c). As in other types of follicular cutaneous lymphoid hyperplasia, these may be surrounded by a B cell mantle zone and a perifollicular T cell area (Fig. 5d). Immunohistochemical investigation further facilitates differentiation from malignant B cell lymphoma (Hovmark et al. 1986; Smolle et al. 1987). In diffuse monomorphous B cell infiltrates, monoclonal (Gk kappa) or (Gk lambda) light chain expression is a hallmark of malignant lymphoma.

Besides BL in the skin, pseudolymphomatous lesions with similar histopathologic features have been found in oral mucous membranes (Schneider 1964), in the ocular conjunctiva (Epstein 1936), in the orbit (Bäfverstedt et al. 1956), and in the vulvar and

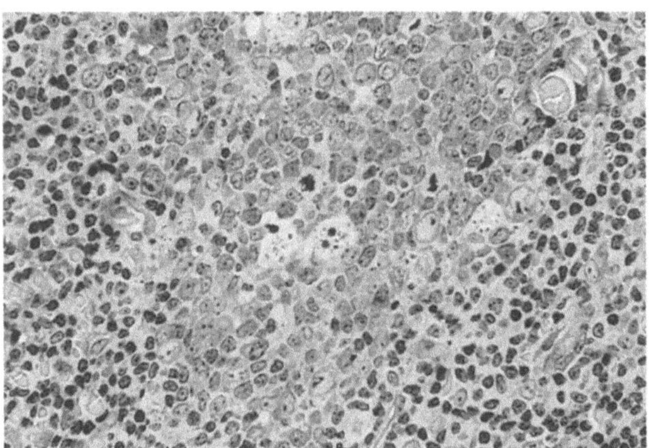

Fig. 4. Borrelial lymphocytoma on the right lower leg in a 51-year-old male (paraffin section, Giemsa staining). The germinal center region of a borrelial lymphocytoma is characterized by the typical components of a reactive cutaneous lymphoid hyperplasia centroblasts, starry sky macrophages, and many mitotic figures

▷

Fig. 5a-d. Same patient as in Fig. 4. Cryostat sections of a snap-frozen unfixed biopsy for demonstration of leukocyte antigens using the alkaline phosphatase anti-alkaline phosphatase (APAAP) technique. Reactivity for a given antigen is shown by the presence of a reaction product (– red in the original color pictures –) on the cell surface (CD22, R 4/23, CD3) or in the nucleus (Ki-67) (see arrows). **a** Staining of B cells with anti-CD22 antibody. B cells are arranged

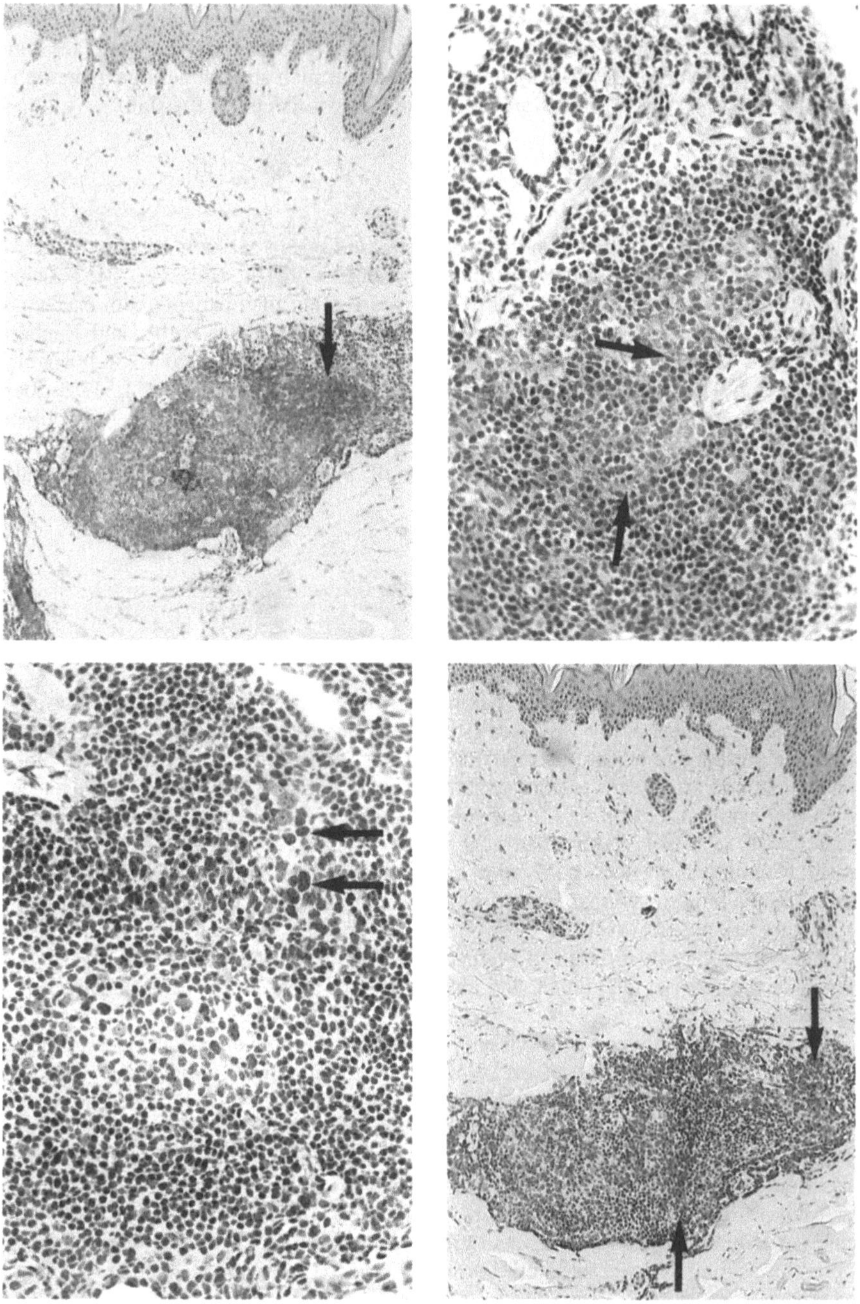

a b c d

in a follicular pattern. **b** Demonstration of follicular dendritic reticulum cells by antibody R 4/23 (arrow, see also Fig. 8, Chap. 6, p. 75). Dendritic cells are a typical component of a follicular borrelial lymphocytoma and are easily visualized by immunostaining, whereas on paraffin section they are hardly detectable in routine stainings. **c** Detection of proliferating cells using monoclonal antibody Ki-67. These cells are concentrated in the germinal center. **d** Perifollicular and interfollicular T cells as detected by anti-CD3 antibody rectal mucosa. The future will show whether such lesions may also be of borrelial etiology. In the inflamed synovia of patients with Lyme arthritis, lymphoid follicles may also be found (Steere 1988). It may be speculated that such an immunologic reaction is a feature of *B. burgdorferi* infection which is not only restricted to the skin.

9.6 Serologic Tests

With the currently used serologic methods, increased serum antibody titers to *B. burgdorferi* are often, but not always, found in patients with BL. However, BL patients seem to have increased serum antibody titers more often than patients with erythema migrans. Using an immunofluorescence test Weber et al. (1984, Weber and Neubert 1986) found elevated immunoglobulin G (IgG) antibody titers of up to 1:256 in six out of seven German patients with BL and elevated IgM antibodies of up to 1:512 in four out of seven. In a Swedish study, in which sonicated spirochetes were used as antigen in enzyme-linked immunosorbent assay (ELISA) tests, 14 of 19 patients with BL had serum IgG titers above the 95th percentile of that found in healthy individuals (Asbrink et al. 1989).

9.7 Diagnosis and Differential Diagnosis

9.7.1 Diagnosis

A bluish-red nodule or plaque on the skin, especially if present on the ear lobe or nipple and areola mammae, is suggestive of BL. A positive spirochetal cultivation can prove the presence of a borrelial infection, but this procedure is complicated and time-consuming, and often leads to false-negative results. In many cases of BL, for instance, when the lesion is located on the ear lobe of a child, sufficiently large biopsy specimens for the recovery of spirochetes cannot be obtained. Thus, with the methods currently available, spirochetal cultivation cannot be used as a routine diagnostic procedure. The diagnosis may also be confirmed by morphologic identification of borreliae in tissue sections. However, as these procedures are not routine diagnostic methods and since they may be difficult to evaluate, the diagnosis often has to be based on the history of the disease, the clinical and histopathologic findings, and/or a positive serologic test.

The presence or a history of erythema migrans or the previous occurrence of a tick bite in the vicinity of the lesion are helpful clues to the correct diagnosis of BL. The histopathologic picture may be suggestive of the diagnosis, but is not specific. If the lesion does not resolve after adequate antibiotic treatment, the diagnosis of a BL ought to be reconsidered. It must, however, be kept in mind that after treatment it will often take more than a month before the lesion has disappeared completely. A lesion which has been present for a very long time may also need a prolonged treatment period. In one case of lymphocytoma, which had been present for 5 years, penicillin was administered for as long as 7 weeks before healing occurred (Lasthein-Andersen et al. 1982).

9.7.2 Differential Diagnosis

The differential diagnosis in BL must include malignant lymphoma and leukemic infiltrates. In lesions without reaction centers, the differentation may be difficult.

Immunohistochemical characterization of the cell infiltrate may be of diagnostic help in distinguishing between the polyclonal infiltrate of a BL and the more monoclonal infiltrate of a malignant lymphoma (Smolle et al. 1987). In Jessner's lymphocytic infiltrate of the skin, the lesions characteristically wax and wane in severity. Jessner's lesion composed mainly of T cells (Willemeze et al. 1984) can be separated immunohistochemically from the BL infiltrates, in which there is a predominance of B cells.

As is the case in many other borrelial manifestations, BL has parallels with syphilis. Pseudolymphomatous skin lesions may appear in the second stage of syphilis (Hodak et al. 1987). Secondary syphilis can, however, be excluded by negative serologic tests (Wassermann and TPI). Histopathologic investigations may be decisive in distinguishing a BL from granulomatous disorders such as granuloma faciale, granuloma annulare, sarcoidosis, and lupus vulgaris. Histological examination may also help in differentiating BL from lupus erythematosus, polymorphous light eruption, keloids, and arthropod-bite granuloma.

References

Asbrink E (1985) Erythema chronicum migrans Afzelius and acrodermatitis chronica atrophicans, Early and late manifestations of *Ixodes ricinus*-borne *Borrelia* spirochetes. Acta Derm Venereol (Stockh) (Suppl) 118:1-63

Asbrink E, Hovmark A, Olsson I (1989) Lymphadenosis benigna cutis solitaria-borrelial lymphocytoma in Sweden. Zentralbl Bakteriol (Suppl) 18:156-163

Bäfverstedt B (1943) Über lymphadenosis benigna cutis. Eine klinische und pathologisch-anatomische Studie. Norstedt, Stockholm, pp 1-202

Bäfverstedt B (1953) Lymphadenosis benigna cutis as a symptom of malignant tumours. Acta Derm Venereol (Stockh) 33:171-180

Bäfverstedt B (1960) Lymphadenosis benigna cutis (LABC) its nature, course and prognosis. Acta Derm Venereol (Stockh) 40:10-18

Bäfverstedt B (1962) Unusual forms of Lymphadenosis benigna cutis (LABC). Acta Derm Venereol (Stockh) 42:3-10

Bäfverstedt B, Lundmark C, Mossberg H, Stenbeck A (1956) Lymphadenosis benigna orbitae. Acta Ophtalmol (Copenh) 34:367-376

Bernstein H, Schupack J, Ackerman AB (1974) Cutaneous pseudolymphoma resulting from antigen injections. Arch Dermatol 110:756-757

Bianchi G (1950) Die Penicillinbehandlung der Lymphozytome. Dermatologica 100:270-273

Biberstein H (1923) Lymphocytome. Zentralbl Hautkr 6:70-71

Blumental G, Okun MR, Ponitch JA (1982) Cutaneous pseudolymphoma resulting from a tattoo. J Am Acad Dermatol 6:485-488

Bork K (1983) Komplikation einer Akupunkturbehandlung. Hautarzt 34:496-499

Burckhardt JL (1911) Zur Frage der Follikel- und Keimzentrenbildung in der Haut. Frankf Z Pathol 6:352-359

Burg G, Braun-Falco O (1983) Cutaneous lymphomas, pseudolymphomas, and related disorders. Springer, Berlin Heidelberg New York, pp 415-441

Burgdorfer W, Barbour AG, Hayes SF, Benach JL, Grunwaldt E, Davis JP (1982) Lyme disease - a tickborne spirochetosis? Science 216:1317-1319

De Koning J, Hoogkamp-Korstanje A (1986) Diagnosis of Lyme disease by demonstration of spirochetes in tissue biopsies. Zentralbl Bakteriol Hyg (A) 263:179-188

Epstein S (1936) Lymphocytome der Haut mit Beteiligung der Conjunctiva Bulbi: Beitrag zur Pathogenese der Lymphocytome. Arch Dermatol Syph 173:181-195

Gottron H (1938) Lymphadenosis cutis circumscripta im Bereich der Mamille bei gleichzeitiger Acrodermatitis chron. atrophicans der Extremitäten. Zentralbl Hautkr 59:633-712

Hauser W (1955) Zur Kenntnis der Acrodermatitis atrophicans arthropathica. Arch Dermatol 199:350-393

Hauser W (1965) Wahrscheinliche Infektionskrankheiten der Haut. In: Marchionini A, Götz H (eds) Infektionskrankheiten der Haut I. Springer, Berlin Heidelberg New York, pp 556-629 (Handbuch der Haut- und Geschlechtskrankheiten vol IV, part 1 A)
Herzberg J (1952) Retikulosarcomatöse Umwandlung multipler subcutaner Lymphocytome bei Acrodermatitis chronica atrophicans. Dermatol Wochenschr 125:422-429
Hodak E, David M, Rothem A, Bialowance M, Sandbank M (1987) Nodular secondary syphilis mimicking cutaneous lymphoreticular process. J Am Acad Dermatol 17:914-917
Hovmark A, Asbrink E, Olsson I (1986) The spirochetal etiology of lymphadenosis benigna cutis solitaria. Acta Derm Venereol (Stockh) 66:479-484
Jordan P, Holtschmidt J (1951) Traumatisches Zeckenbiss-Lymphocytom und Erythema chronicum migrans. Hautarzt 2:397-401
Kaufmann-Wolf M (1921) Über gutartige lymphocytäre Neubildungen der Scrotalhaut des Kindes. Arch Dermatol Syph 130:425-435
Kerl H (1989) Classification of cutaneous pseudolymphomas. 10th international dermatopathology colloquium, Charleston, USA (abstract)
Lasthein-Andersen B, Brandrup F, Petri J (1982), Lymphocytoma cutis: a pseudomalignancy treated with penicillin. Acta Derm Venereol (Stockh) 62:83-85
Mach K (1965) Die gutartigen Lymphoplasien der Haut. Arch Klin Exp Dermatol 222:325-349
Mach K, Wilgram G (1966) Characterstic histopathology of cutaneous lymphoplasia (lymphocytoma). Arch Dermatol 94:26-32
Matras A (1953) Tumorförmige Lymphocytombildung in atrophischer Haut. Hautarzt 4:85-87
Mulzer P, Keining E (1929) Über miliare Lymphozytome der Haut. Dermatol Wochenschr 8:293-300
Paschoud JM (1954) Lymphocytom nach Zeckenbiss. Dermatologica 108:435-437
Paschoud JM (1957) Die Lymphadenosis benigna cutis als übertragbare Infektionskrankheit. I. Hautarzt 8:197-211
Paschoud JM (1958a) Die Lymphadenosis benigna cutis als übertragbare Infektionskrankheit. III. Hautarzt 9:263-269
Paschoud JM (1958b) Die Lymphadenosis benigna cutis als übertragbare Infektionskrankheit. IV. Hautarzt 9:311-315
Rijlaarsdam JU, Brunynzeel DP, Wos W, Meijer C, Willemze R (1988) Immunohistochemical studies of lymphadenosis benigna cutis occurring in a tattoo. Am J Dermatopatol 10:518-523
Schneider W (1964) Lymphadenosis benigna cutis der Mundschleimhaut. Z Haut Geschlechtskr 36:17-24
Self S, Carter V, Noojin R (1969) Disseminated lymphocytoma cutis. Arch Dermatol 100:459-467
Smolle J, Kaudewitz P, Aberer E, Torne R, Burg G, Kerl H (1987) Immunhistochemische Klassifizierung von kutanen Lymphomen und Pseudolymphomen. Hautarzt 38:461-466
Steere (1988) Pathogenesis of Lyme arthritis. Ann NY Acad Sci 539:87-92
Strandberg J (1920) Regarding an unusual form of migratory erythema caused by tick bites. Acta Derm Venereol (Stockh) 1:422-427
Wantzin G, Hou-Jensen K, Nielsen M (1982) Cutaneous lymphocytomas: clinical and histological aspects. Acta Derm Venereol (Stockh) 62:119-124
Weber K, Neubert U (1986) Clinical features of early erythema migrans disease and related disorders. Zentralbl Bakteriol Hyg (A) 263:209-228
Weber K, Schierz G, Wilske B, Preac-Mursic V (1984) European erythema migrans disease and related disorders. Yale J Biol Med 57:463-471
Weber K, Schierz G, Wilske B, Preac-Mursic V (1985) Das Lymphozytom - eine Borreliose? Z Hautkr 69:1585-1598
Willemze R, Dijkstra A, Meijer C (1984) Lymphocytic infiltration of the skin (Jessner): a T-cell lymphoproliferative disease. Br J Dermatol 110:523-529

10 Lyme Carditis

M.R. VAN DER LINDE and P.E. BALLMER

10.1 Introduction

Lyme carditis can present as conduction or rhythm disturbances, (endo)myocarditis and pericarditis (with or without heart failure) (Steere et al. 1980; Bedell et al. 1981; Goldings and Jericho 1986; Hansen and Madsen 1986; Kapusta et al. 1986; Meyer and Swenson 1987; Hassler et al. 1987, 1988a,b; Lorcerie et al. 1987; Lefevre 1988; van der Linde et al. 1989). Lyme carditis may occur at any age and in either sex, although most European patients have been men (Table 2). Lyme borreliosis as an entity was described in 1977 by Steere et al., and, in 1980, Steere et al. characterized the cardiac manifestations of this disease. The first specific articles on Lyme carditis in Europe were reported in 1984 (Cornuau et al. 1984a,b; Houwerzijl et al. 1984). Many authors, however, had previously related cardiac complaints and abnormalities to other manifestations of Lyme borreliosis (Bannwarth 1941; Hopf 1966; Weber et al. 1983).

In 1973 a French patient presented with erythema migrans, arthralgias, parästhesias, and electrocardiographic (ECG) evidence of Wenckebach atrioventricular (AV) block and perimyocarditis; a probable rickettsiosis was diagnosed because of the clinical pattern, although serologic markers were negative. The patient recovered within 5 weeks, without specific therapy. This seems to be the first documented case of Lyme carditis in Europe, since serologic tests performed retrospectively in 1985 proved to be positive for *Borrelia burgdorferi* (immunofluorescence assay, IFA, 1:512; Table 2) (Schott 1985). In 1987, Vacher et al. published the first report on a European patient with Lyme carditis with biopsy-proven myocarditis, but they failed to demonstrate spirochetes. It was only in 1989 that de Koning et al. and van der Linde et al. reported the first biopsy-proven case of Lyme carditis in Europe. These workers demonstrated spirochetes in the endomyocardial tissue (de Koning et al. 1989), thus confirming the observations in North-American patients by Marcus et al. (1985) and Reznick et al. (1986).

In recent years, the number of publications on Lyme carditis has increased rapidly. One of the explanations for the late "establishment" of Lyme carditis is the difficulty of relating cardiac abnormalities to the infection in any given patient. In the North American literature, the incidence of Lyme carditits is estimated as occurring in 8% of patients with Lyme borreliosis (Steere et al. 1980). The frequency of cardiac manifestations of Lyme borreliosis in Europe has not yet been established. In 1985, Schmidt et al. reported a frequency of 1.6% for Lyme carditis in a population of 817 German patients with Lyme borreliosis. In 1987, Dournon et al. reported on 272 French patients with Lyme borreliosis nine of whom (3.3%) had second- or third- degree AV block. Difficulties in confirming Lyme carditis and other factors, however, may lead to over- or underestimation of the frequency of this manifestation.

In this chapter, the clinical features of Lyme carditis, the differential diagnosis, and the investigative process are described. Furthermore, the documented cases of Lyme carditis in Europe are reviewed. The treatment of Lyme carditis is discussed in Chap. 26.

10.2 Clinical features, Investigational Procedures and Diagnosis

Lyme carditis may present as:
- Conduction disturbances (disturbances of impulse formation and conduction)
- Rhythm disturbances (extrasystoles, brady- and tachyarrhythmias)
- Pericarditis and/or pericardial effusion
- (Endo)myocarditis and/or heart failure

Intra-atrial, AV, and intraventricular conduction disturbances may all occur (Allal et al. 1986; Kapusta et al. 1986; van der Linde et al. 1989). A fluctuating degree of AV block is, however, the most frequent sign of Lyme carditis (Steere et al. 1980; Olson et al. 1986; Roelli et al. 1989) (Section 10.7). Rapid changes in degree of AV block may be present (Kapusta et al. 1986; van der Linde et al. 1990). The block may be restricted to the supra-hisian (i.e., in the AV node) or infra-hisian areas (i.e., in the common His bundle and bundle branches) or diffuse, in which case the AV node, common His bundle, bundle branches and the Purkinje system may be all affected, with a concomitant delay of intra-atrial conduction (Kapusta et al. 1986; van der Linde et al. 1989). Rhythm disturbances in Lyme carditis seem to be less frequent than conduction disturbances (Steere et al. 1980; Ballmer and Hany 1988), but may include atrial fibrillation or flutter, paroxysmal atrial tachycardia, premature supraventricular and ventricular beats and (sinus)bradycardia (Table 2). Ventricular tachycardias have not been reported. Clinically significant heart failure is a rare event (Table 2). Pericarditis and pericardial effusion in Lyme carditis are common (Table 2). The time interval between the tick bite and the occurrence of cardiac manifestations may be as short as 10 days or as long as 120 days (Vieyres et al. 1987a).

10.2.1 Symptoms

The specific cardiac symptoms depend on the nature of the underlying disturbance of conduction, rhythm, or contractility. In advanced AV conduction disturbances, the patient may experience fatigue, dizziness on exertion, or even Stokes-Adams attacks. Rhythm disturbances most often present as palpitations, diminished exercise tolerance, and chest discomfort. Peri- and myocarditis can mimic all degrees of right-sided and left-sided heart failure, but most often the patient complains of shortness of breath.

10.2.2 Physical Examination

There may be no abnormal findings in cases of AV conduction disturbances. Complete AV block is indicated by the presence of cannon waves (abnormal and pronounced retrograde pulsations) in the jugular veins and varying intensitiy of the first heart sound. Rhythm disturbances such as atrial fibrillation induce an irregular pulse, a pulse deficit during rapid ventricular rates, and varying intensity of the heart sounds. Pericarditis may be manifested by friction rub and pericardial effusion, sometimes leading to tamponade. As with other causes of cardiomyopathy, heart failure may be accompanied by tachycardia, elevated central venous pressure (CVP), a third heart sound, pulmonary rales, and low blood pressure.

10.2.3 Blood Tests

10.2.3.1 Routine Laboratory Tests

There may be an increased erythrocyte sedimentation rate (ESR), a leukocytosis and raised levels of immunoglobulins (Ig) (Goldings and Jericho 1986). If Lyme carditis is suspected, the creatinephosphokinase (CPK) and CPK-myocardial band (MB) levels should be determined over several days.

10.2.3.2 Specific Tests

IgM and IgG antibody titers against *B. burgdorferi* should be evaluated. In Europe, no patients with carditis have been reported in whom neither elevated IgM nor elevated IgG antibody titers have been detected (Table 2), but the diagnosis of Lyme carditis is hard to prove in patients with normal antibody titers.

10.2.4 Electrocardiogram

Fluctuating degrees of AV block are most commonly seen in ECG (Fig. 1). Escape foci may be stable or unstable (Fig. 2) (Allal et al. 1986; van der Linde et al. 1990) and originate from different levels of the conduction system (AV node, common His bundle, and bundle branches). Intra-atrial (prolonged duration and abnormal configuration of the P wave) and intraventricular conduction delay (right bundle branch block, left

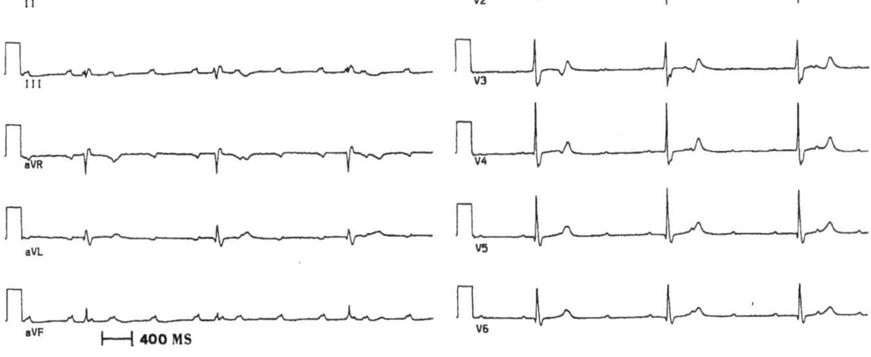

Fig. 1. Electrocardiogram of a 40-year old patient with complete AV block due to (biopsy proven) Lyme carditis. The duration and configuration of the escape complexes suggest that they originate in the left bundle branch. The duration and configuration of the P waves indicate intra-atrial conduction delay. Sinus frequency 100/min, escape frequency 35/min, QRS duration 130 ms, and P wave duration 120ms.

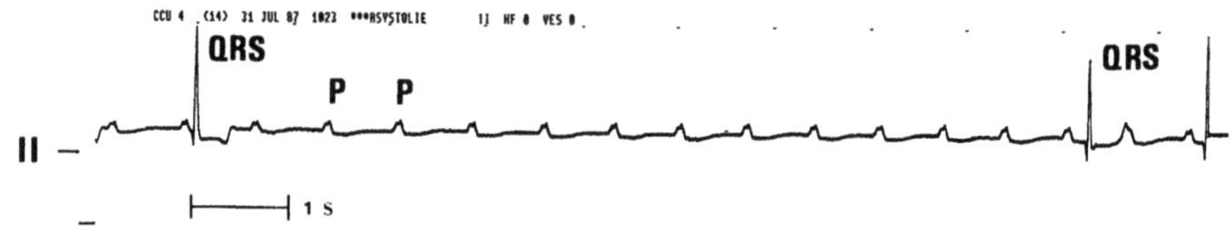

Fig. 2. Rhythm strip of the same patient as in fig. 1 (lead II). Complete AV block: the instability of the escape focus in this patient with Lyme carditis is demonstrated by the ventricular standstill of approximately 9 seconds (despite infusion of isoproterenol). A temporary pacemaker was inserted.

bundle branch block, and nonspecific prolongation of the QRS complex) may be present. Atrial flutter, (paroxysmal) atrial fibrillation, atrial tachycardia, (sinus) bradycardia, and premature supraventricular and ventricular complexes may also be seen (Table 2). Furthermore, nonspecific ST segment changes and T wave inversion may be present as a sign of diffuse cardiac involvement (Goldings and Jericho 1986). Sequential ECG changes may reveal the evolution of the carditis in an individual patient.

10.2.5 Holter-ECG-Monitoring

Ambulatory continuous ECG monitoring for a variable time period is an important tool for detecting paroxysmal rhythm disturbances. In patients with rapidly variable AV block, unstable escape foci, or hemodynamically significant rhythm disturbances, hospital admission for continuous monitoring is necessary.

10.2.6 Chest X-Ray

There are no specific radiological abnormalities in Lyme carditis which distinguish it from other forms of heart disease.

10.2.7 Echocardiogram

Pericardial effusion may be detected and its hemodynamic significance assessed by echocardiography. There are no findings which are specific for Lyme endomyocarditis, but as the examination is easy to perform it may reveal unsuspected involvement of the heart.

10.2.8 Scintigraphic Investigations

10.2.8.1 Gallium-67 Scintigraphy

Gallium-67 is taken up in inflammatory lesions, making it a nonspecific tool for the detection of Lyme carditis. It can detect more or less diffuse (active) endomyocarditis, even if this is not clinically suspected (O'Connell et al. 1984). The test can play an important role in assessing the likelihood of the diagnosis Lyme carditis in the case of positive *B. burgdorferi* serology (Jacobs et al. 1984; Alpert et al. 1985; Rienzo et al. 1987) and is helpful in distinguishing active myocarditis from primary cardiomyopa-

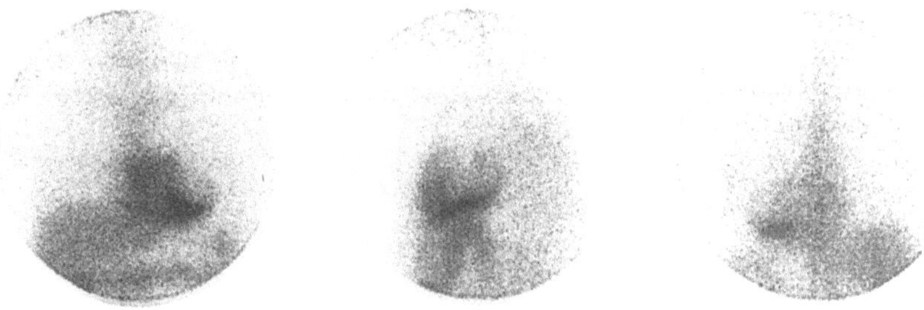

Fig. 3. Gallium-67 scintigraphy of a 40-year old patient with Lyme carditis 48 hours after injection of 100 MBq gallium-67. A pronounced diffuse uptake of gallium-67 in the cardiac wall is demonstrated in all three projections. (left panel, anterior view; mid panel, left lateral view; right panel, posterior view of the chest).

thies (Ponsonnaille et al. 1986) and other noninflammatory myocardial disorders (Fig. 3). Studies may be made between 24 and 96 h after the injection of gallium-67 (Jacobs et al. 1984; Alpert et al. 1985; van der Linde et al. 1990). A normal gallium-67 scan, however, does not exclude (Lyme) myocarditis (O'Connell et al. 1984; van der Linde et al. 1990) and may even enhance the need for endomyocardial biopsies in case of diagnostic difficulties.

10.2.8.2 Thallium-201 Scintigraphy

Thallium-201 scintigraphy has not often been used in cases of Lyme carditis and is not specific for the condition. Vacher et al. (1987) and Ponsonaille et al. (1986) performed thallium-201 scans in three patients; in two of them a heterogeneous distribution of thallium in the cardiac wall was observed. The authors concluded that this heterogeneous distribution was due to noncoronary heart disease or cardiomyopathy (coronary angiography was normal). Again, this is not specific for Lyme carditis. Therefore, thallium-201 scintigraphy seems to be an irrelevant investigation for Lyme carditis, unless there is reason to believe that additional coronary artery disease plays an important role in a particular patient.

10.2.8.3 Multigated Bloodpool Scintigraphy

Multigated bloodpool scintigraphy with red blood cells labeled with technetium-99m is used to determine the left and right ventricular ejection fractions (LVEF and RVEF) and to detect abnormal cardiac wall motion. Multigated bloodpool scintigraphy has been used to monitor improvement in cardiac function after treatment of Lyme carditis (Steere et al. 1980). Some preliminary data suggest possible late development of cardiomyopathy in Lyme carditis (Stanek et al. 1990). Sequential assessment of the LVEF and RVEF by multigated bloodpool scintigraphy may provide more information on this issue.

10.2.9 Endomyocardial Biopsies

Using an invasive technique, it is possible to obtain endomyocardial biopsies of patients in order to reveal specimens of the affected cardiac tissue. Endomyocarditis may be demonstrated in these biopsies by routine staining techniques (Duray and

Steere 1986; O'Connell et al. 1984). It is possible to differentiate between acute endomyocarditis and a more chronic inflammation. The presence of spirochetes in endomyocardial tissue may be demonstrated by special staining methods (de Koning et al. 1987, 1989) (see also Chap. 6).

10.2.10 Electrophysiologic Studies

Electrophysiologic studies can be useful in Lyme carditis, although they are not fundamental for diagnosis and treatment. Electrophysiologic studies in Lyme carditis may be restricted to patients in whom it is already necessary to introduce a temporary pacemaker and those in whom myocardial biopsies are performed. For research purposes, it may be performed in patients who have given special informed consent. In combination with the surface ECG, location and extension of the conduction disturbances (intra-atrial, AV-nodal, in the common His bundle and/or the bundle branches, or intraventricular) can be determined by measuring time intervals (e.g., sinoauricular, AV-nodal-His, and His-ventricular intervals) and by recording the presence of His spikes (Fig. 4) (Dunica et al. 1986; Fauchier et al. 1988; van der Linde et al. 1990). The Wenckebach point and retrograde conduction can be demonstrated. Inducibility of atrial and ventricular rhythm disturbances can be tested with programmed stimulation.

10.2.11 Bicycle Exercise Tests

Bicycle exercise tests are not specific for Lyme carditis, but may, however, be useful in predicting the likelihood of (hemodynamic) problems in outpatients and may also indicate the coexistence of ischemic heart disease.

10.3 Proposal for Uniform Criteria for the Diagnosis of Lyme Carditis

Diagnostic features of Lyme carditis are (see also Fig. 5):

1. (Recent) history or presence of typical manifestations of Lyme borreliosis
2. Positive serology
3. ECG changes (conduction disturbances, rhythm disturbances, T wave inversion or ST changes)
4. Increased uptake of gallium-67 in the cardiac wall in gallium-67 scintigraphy
5. (Endo)myocarditis in tissue biopsies, with or without demonstration of spirochetes
6. Cardiac abnormalities with ECG changes which cannot otherwise be explained

We propose that Lyme carditis is:

1. *Virtually proven* if the combination of the above features 1, 2 and 5, and possibly 3 and 4 is present
2. *Highly suggested* if the combination of the following above features is present:
   - 1, 2, and 4 (and possibly 3)
   - 1, 2, and 6 (and possibly 3)
   - 2, 3 (or 6), and 5
3. *Suggested* if the combination of the following above features is present:
   - 2, (3), and 4
   - 2 and 6

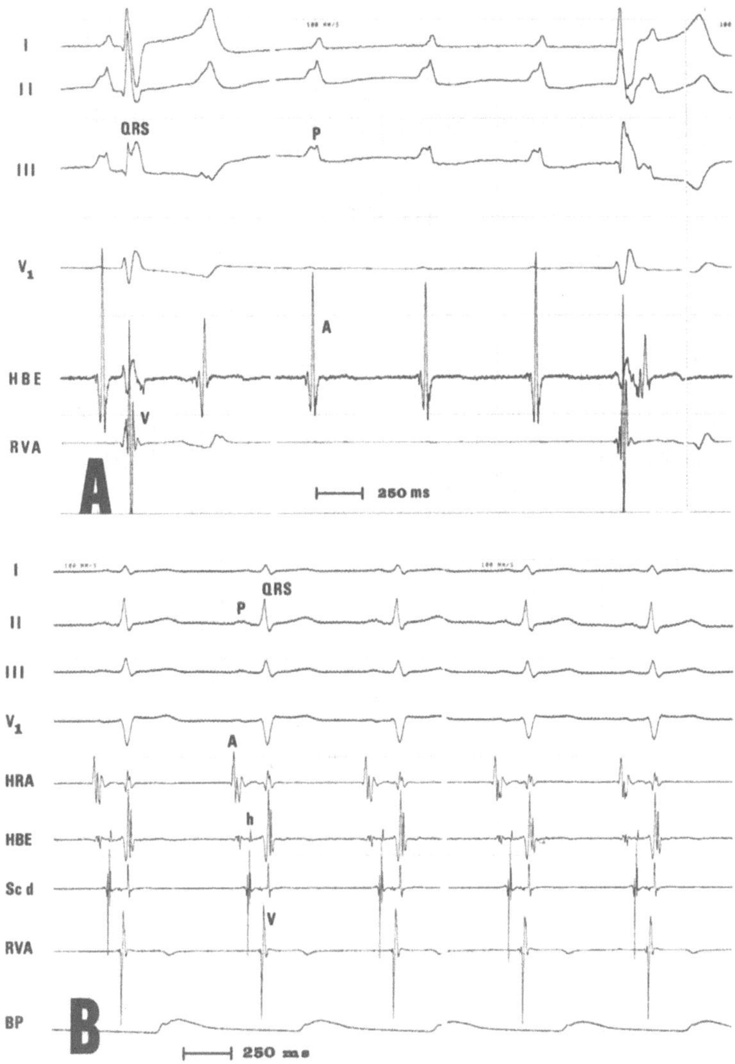

Fig. 4. Electrophysiological recordings, using transvenously introduced catheters in various positions.
(panel **A**): In a 40-year old patient with complete AV block due to (biopsy proven) Lyme carditis. No His spike is present in the HBE recordings, neither following the atrial activity, nor preceding the ventricular activity. This suggests absence of atrioventricular conduction, with location of the block in the common His bundle and/or the AV node. The surface electrocardiographic recordings (I, II, III and V1) show abnormal configuration and duration of the (unstable) escape focus and delay of intra-atrial conduction.
(panel **B**): In a normal patient. A clear His spike is visible in the HBE recordings, following the atrial and preceding the ventricular activity. Time intervals are normal (AH 110 ms, HV 45 ms, heart rate 81/min). The surface electrocardiographic recordings show a normal pattern.
I, II, III and V1, surface electrocardiogram leads; A, atrial electric activity; BP, blood pressure; h, His spike; HBE, His bundle electrogram; HRA, electric catheter in high right atrium position; RVA, catheter in the right ventricular apex; Scd, catheter in the sinus coronarius dextra; V, ventricular electric activity

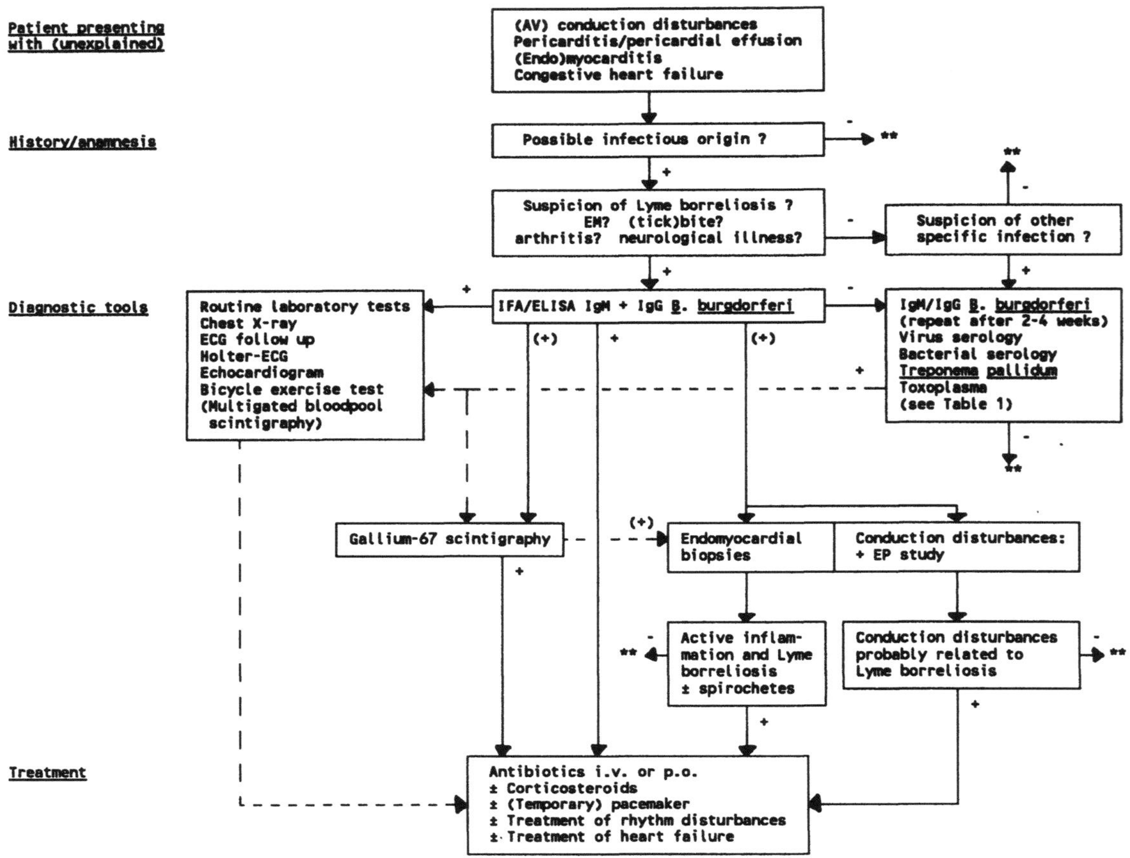

Fig. 5. Flow chart for the management of Lyme carditis. AV = atrioventricular, ELISA = enzyme-linked immunosorbent assay, EM = erythema migrans, EP-study = electrophysiologic study, IFA = immunofluorescence assay. ** = appropriate management of this exceeds the intention of the flow chart, — = negative, + = positive.

As Lyme carditis may be the only expression of Lyme borreliosis, a recent history of this infection cannot unconditionally be a necessary criterion for diagnosis. Resolution of the (cardiac) manifestations with antibiotic therapy supports the diagnosis of Lyme carditis, especially if a Jarisch-Herxheimer reaction occurs. Although seronegative Lyme borreliosis has been reported, we currently think that positive serology is still required for the diagnosis of Lyme carditis.

10.4 Differential Diagnosis

All diseases which can cause conduction and rhythm disturbances, endomyocarditis and heart failure, pericarditis, and pericardial effusion must be included in the differential diagnosis of Lyme carditis (Table 1). Most problematic is the attribution of rhythm disturbances to the infection because of the large number of potential alternative causes. The clinical manifestations of other diseases, specific laboratory tests, and general information such as age and state of health at the onset of the illness can help in differentiating from Lyme carditis.

Table 1. Differential diagnosis of Lyme carditis

Manifestation	I. Atrioventricular conduction disturbances	II. Pericarditis	III. Endomyocarditis
A. Infectious	1. Viral coxsackie A and B echo type 6 and 8 adeno type 3 influenza A hepatitis B Epstein-Barr mumps poliomyelitis varicella 2. Bacterial diphtheria *Yersinia enterocolica* *Treponema pallidum* 3. Parasitic toxoplasmosis Chagas' disease 4. Rickettsial infections	1. Viral (see I.A.1) 2. Meningococcal 3. Tuberculous 4. Syphilitic 5. Mycotic 6. Parasitic(see I.A.3)	Myocarditis 1. Viral: (see I.A.1) 2. Bacterial diphtheritic meningococcal 3. Parasitic (see I.A.3) leptospirosis trichinosis Endocarditis 1. Bacterial *Streptococcus viridans* 2. Mycotic *Candida* Aspergillus 3. Other (Q-fever)
B. Non infectious	1. Coronary artery disease 2. Post cardiac surgery 3. Calcified aortic stenosis 4. Congenital heart disease 5. Drug-induced (digoxine, betablockers, verapamil, diltiazem, tricyclic antidepressants, etc)	1. Acute myocardial infarction 2. Post myocardial infarction (Dressler's syndrome) 3. Postpericardiotomy 4. Uremia 5. Neoplasma 6. Post radiation	Myocarditis 1. Post radiation 2. Chemical poisons paraquat carbon monoxide 3. Immunosuppressive drugs Adriamycin daunorubicin

Table 1. Continued

Manifestation	I. Atrioventricular conduction disturbances	II. Pericarditis	III. Endomyocarditis
	6. Other acute rheumatic fever polyarthritis systemic lupus erythematosus amyloidosis sarcoidosis myxedema idiopathic sclerodegenerative disease (Lenègre's disease) fibrocalcareous encroachment (Lev's disease)	7. Trauma 8. Drug-induced (procainamide) 9. Other rheumatic fever myxedema sarcoidosis collagen vascular disease	4. Other acute rheumatic fever postinfarction Fiedler's myocarditis idiopathic giant cell myocarditis

10.5 Long-term Follow-up

Although chronic arthritic and neurological sequelae of Lyme borreliosis are well known, follow-up of Lyme carditis has received little attention. Complete or almost complete resolution of Lyme carditis within 2 weeks is generally reported, although in a few patients persistent major cardiac problems have been described (Steere et al. 1980; Marcus et al. 1985; Mayer et al. 1989; van der Linde et al. 1990). Furthermore, there are some preliminary data suggesting that Lyme borreliosis may be associated with dilated cardiomyopathy (Stanek et el. 1990). There is clearly a need for long-term follow-up studies.

10.6 Pathogenesis of Lyme Carditis

It is conceivable that Lyme carditis is due to the deposition of immune complexes in the heart (Hardin et al. 1979) or to the release of lipopolysaccharides from the outer wall of *B. burgdorferi* itself (Fumarolo et al. 1986; Habicht et al. 1986). Patients with Lyme borreliosis treated with antibiotics may develop a Jarisch-Herxheimer reaction due to the release of spirochetal lipopolysaccharides which trigger an increased production of interleukin-1 (Habicht et al. 1987), an important mediator in the pathogenesis (Beck et al. 1986). It is still uncertain, however, how these mechanisms relate to the histopathological changes observed in Lyme carditis. In a fatal case of Lyme borreliosis, a transmural myocarditis including the epicardium and the endocardium was found with lymphocytic and plasma cell infiltration, but neutrophils, giant cells, and eosinophils were lacking (Duray and Steere 1986). De Koning has observed previously unreported necrosis of isolated myocardial cells surrounded by lymphocytes in a patient with Lyme carditis (Chap. 6).

10.7 Discussion of the Documented European Cases of Lyme Carditis

We have now collected data on 66 patients with highly probable Lyme carditis in Europe by reviewing the literature and through personal communications and personal observations (see Table 2). In only five patients were endomyocardial biopsies performed (Vacher et al. 1987; Fauchier et al. 1988; de Koning et al. 1989; van der Linde et al. 1990). Four of the patients showed evidence of endomyocarditis (Vacher et al. 1987; de Koning et al. 1989; van der Linde et al. 1990) and spirochetes were demonstrated in the (endo)myocardium of three patients (de Koning et al. 1989; van der Linde et al. 1990). There were 50 men and 16 women (ratio 3:1). The mean age of the patients was 43.6 years (range 19-84 years) in the men and 46.4 years (range 23-58 years) in the women. Cases have been reported in most European countries. In Denmark 2 cases were reported, in France 24, in Germany 25, in The Netherlands 7, and in Switzlerand 8, and in addition patients have been described in Italy (Trevison et al. 1986), England (Muhlemann and Wright 1987), and Austria (Stanek et al. 1986) for whom it was not possible to obtain detailed clinical data.

10.7.1 General Clinical and Laboratory Data

IgG antibody titers against *B. burgdorferi* tested using IFA or enzyme-linked immunosorbent assay (ELISA), were reported to be elevated in 98% (65 patients), while IgM antibody titers were elevated in only 21% (14 patients). As many as 47% of all patients recalled a (tick)bite. In addition to Lyme carditis, erythema migrans was observed in 58% of the patients, arthritis in 45%, and a neurological illness (mostly meningoradiculitis) in 26%.

10.7.2 Clinical Manifestations of Carditis

The most frequent clinical sign of Lyme carditis in these 66 patients was temporary complete AV block (48%), which corresponds well with the data of Steere et al. (1980). The maximal observed AV conduction disturbance was second degree in 18% and first degree in 8% of the patients. These figures are not directly comparable to those of Steere et al. (1980), because of differences in classification, but if they are first categorized according to Steere et al. (1980), very similar figures are obtained; 48% of our patients had complete AV block, 66% had (only or also) second degree AV block and 74% had (only or also) first degree AV block. In some patients, rapid transition in AV block or an unstable escape focus was reported (Allal et al. 1986; Kapusta et al. 1986; Hassler et al. 1987; van der Linde et al. 1989), but in many patients with complete AV block but without these objectivated features, a history of collapse was present. These features are possibly more often present than generally is assumed. In 9% of all patients a bundle branch block was present; in four patients this was a right bundle branch block, and in two a left bundle branch block, and in five of these six patients this was found in combination with an AV block. Two patients had a combined sinoauricular and AV block (Kapusta et al. 1986; Ponsonnaille et al. 1986). In 15% of the patients rhythm disturbances were present, mainly of supraventricular origin (three patients with atrial fibrillation, two with atrial flutter, one with atrial tachycardia, one with AV nodal tachycardia, two with sinusbradycardia and one with premature ventricular beats). In all but one patient (persisting sinusbradycardia), rhythm disturbances disappeared after antibiotic treatment. In ten patients (15%) peri-

Table 2. Data of the European patients with documented Lyme carditis

Country Author (year)	Pa-tients (n)	IFA/Elisa IgM	IFA/Elisa IgG	Recall (tick) bite	Ery-thema migrans	Arthri-tis	Neurolo-gical Illness	Maximal AV-Block 1°	2°	3°	Rhythm Distur-bances	Heart-failure	Echo-cardio-graphy	Chest X-ray	Gallium-67 scan	Electro-physiologic conduction disturbances I-A	AH	HV	Endomyo-cardial biopsies Imflam-mation	Spiro-chetes	Treat-ment	Course	Special Features
Denmark																							
Hansen and Madsen (1986)	2	1	2	1	1	0	0	0	0	2	0	0	0/2	-	-	-	-	-	-	-	1 P, 1 T	2 CR	-
France																							
Cornuau et al (2 unpublished cases, 1984, 1986)	4	-	4	3	4	3	3	0	2	2	0	0	1/4	0/4	-	0/1	0/1	1/1	-	-	2 A, 1 none 1 NSAID 1 TPM 1 DPM	4 CR	2 pericarditis 1 effussion 1 BBB
Lavaud et al. (1985)	1	1	-	1	1	1	-	0	0	1	0	0	1/1	-	-	-	1/1	0/1	-	-	none; after 10 days; T	SCR	pericarditis effussion
Schott (unpublished case, 1985)	1	-	1	1	1	1	1	0	1	0	0	0	-	0/1	-	-	-	-	-	-	none	SCR	pericarditis
Allal et al (1986); Vieyres et al. (1987b)	3	-	3	3	3	3	2	2	0	1	0	-	0/2	-	-	-	-	-	-	-	2 P, 1 T 1 NSAID 1 TPM	3 CR	1 myopericarditis 1 RBBB 1 LBBB 1 unstable escape focus
Dunica et al. (1986)	1	-	1	1	1	1	0	0	0	1	0	0	0/1	-	-	0/1	1/1	0/1	-	-	T, TPM	CR	0

Kapusta et al. (1 unpublished case, 1986)	3	-	3	-	3	0	0	0	1	2	0	0	0/3	0/3	-	1/3	3/3	0/3	-	-	3 T 2 NSAID 2 TPM	3 CR	1 SAB 1 RBBB
Ponsonnaille et al. (1986)	1	-	1	1	1	1	1	1	0	0	0	0	0/1	0/1	1/1	1/1	1/1	0/1	-	-	none	SCR	1 SAB coronary angiography normal
Lorcerie et al. (1987)	2	-	2	-	-	0	0	1	0	0	0	1	2/2	-	-	-	-	-	-	-	2 P 2 NSAID	2 CR	2 pericarditis + 2 effusion 1 RHF + 1 persistent T wave inversion
Vacher et al. (1987)	2	0	2	2	1	1	0	0	0	2	0	1	1/2	1/2	2/2	0/1	1/1	0/1	1/1	0/1	1 C 1 NSAID 1 TPM	2 CR	1 pericarditis + RHF 1 hypokinetic con traction pattern 1 coronary angiography normal
Veyssier et al. (1987)	2	-	2	-	2	2	0	0	0	0	0	1	2/2	2/2	-	-	-	-	-	-	2 P 2 NSAID	2 CR	1 pericarditis + effusion 1 myoperi-carditis + effusion + RHF
Vieyres et al. (1987; see also Allal et al. 1986)	1	-	1	1	1	1	1	0	1	0	0	0	0/1	-	-	0/1	1/1	1/1	-	-	T	possibly persistent 1° AVB	-

Country Author (year)	Pa-tients (n)	IFA/Elisa IgM	IgG	Recall (tick) bite	Ery-thema migrans	Arthri-tis	Neurolo-gical Illness	Maximal AV-Block 1°	2°	3°	Rhythm Distur-bances	Heart-failure	Echo-cardio-graphy	Chest X-ray	Gallium-67 scan	Electro-physiologic conduction disturbances I-A	AH	HV	Endomyo-cardial biopsies Imflam-mation	Spiro-chetes	Treat-ment	Course	Special Features
Fauchier et al. (1988)	1	-	1	1	1	0	0	0	0	0	1	1	1/1	1/1	-	0/1	0/1	1/1	0/1	-	T	CR	atrial flutter + LBBB LHF + RHF hypokinetic contraction pattern coronary angiography normal
Lefèvre (1988)	1	-	1	1	1	0	0	0	0	1	0	1	-	0/1	-	-	-	-	-	-	T, A NSAID TPM	CR	RBBB
Dournon (1989, unpublished case)	1	1	1	1	0	1	0	0	0	1	0	0	-	-	-	-	-	-	-	-	none; later CE	SCR	-
Germany Hassler et al. (1988a)	11	2	11	4	2	4	4	1	0	1	5	6	4/7	6/8	-	-	-	-	-	-	6 P, 8 CE 2 T 1 TPM	11 CR	6 LHF 2 pericarditis + STT + 1 effusion 2 AF, 1 atrial flutter 1 ANVT 1 PVB Pwave prolongation 3 hypokinetic contraction

Mayer et al. (1989)	2	1	2	1	1	0	1	0	1	1	0	0	-	-	-	-	-	-	-	-	2 CE	1 CR 1 persi-stent 1° AVB	1 T wave inversion
Wunderlich et al. (2 unpub-lished cases, 1988, 1989)	11	-	11	5	5	6	1	0	2	7	2	0	0/6	0/11	-	-	-	-	-	-	10 P, 1 A 4 C 6 NSAID 4 TPM	10 CR 1 persistent sinusbrady-cardia STT	2 myope-ricarditis 2 sinusbrady-cardia
Borgmann et al. (1989)	1	1	1	-	-	1	0	0	0	1	0	0	-	0/1	-	-	-	-	-	-	P, C	CR	0
The Netherlands Houwerzijl et al. (1984)	1	1	1	1	1	0	0	0	0	1	0	0	0/1	0/1	-	-	-	-	-	-	P, T	CR	0
Blaauw et al. (1988)	1	1	1	1	1	1	1	0	1	0	0	0	-	0/1	-	-	-	-	-	-	none; later T	CR	0
Kuiper et al. (1988)	1	1	1	-	1	-	1	0	0	1	0	0	0/1	0/1	-	-	-	-	-	-	DPM; later P	CR	pacemaker later removed
Van der Linde et al. (1989)	4	3	4	-	1	2	0	0	0	4	0	0	1/4	0/4	1/2	1/3	2/3	2/3	3/3	3/3	3 P, 1 T 1 CE, 1 C 1 later P 1 TPM 1 DPM	2 CR 1 SCR 1 persistent 3° AVB	1 hypokinetic contraction pattern 1 unstable escape focus
Switzerland Laperrouza (1985)	1	-	1	-	1	-	-	0	1	0	-	-	-	-	-	-	-	-	-	-	P	CR	-

Country Author (year)	Pa-tients (n)	IFA/Elisa IgM	IFA/Elisa IgG	Recall (tick) bite	Ery-thema migrans	Arthri-tis	Neurolo-gical Illness	Maximal AV-Block 1°	2°	3°	Rhythm Distur-bances	Heart-failure	Echo-cardio-graphy	Chest X-ray	Gallium-67 scan	Electro-physiologic conduction disturbances I-A	AH	HV	Endomyo-cardial biopsies Inflam-mation	Spiro-chetes	Treat-ment	Course	Special Features
Satz et al. (1986)	1	-	1	-	-	0	-	0	0	0	0	0	-	0/1	-	-	-	-	-	-	P	CR	myoperi-carditis T wave inversion
Steurer et al. (1987)	1	-	1	1	1	-	0	0	0	1	0	0	-	0/1	-	-	-	-	-	-	P	CR	-
Ballmer and Hany (1988)	1	-	1	-	1	0	0	0	1	0	1	0	-	0/1	0/1	-	-	-	-	-	T	CR	"slow" AF
Hany and Ballmer (1988)	1	1	1	-	1	0	1	0	0	0	1	0	0/1	0/1	0/1	-	-	-	-	-	P, BB	CR	AT
Roelli et al. (1989)	3	0	3	1	2	1	0	0	1	2	0	0	0/3	0/3	0/2	-	-	-	-	-	2 P 1 NSAID	1 CR 1 SCR 1 persi-stent 1° AVB	0
Total	66	14	65	31	38	30	17	5	12	32	10	11	13/45	10/49	4/9	3/12	10/13	5/13	4/5	3/4	36 P,4 A 16 T,11 CE 1 BB, 7 C 17 NSAID 13 TPM, 3 DPM	54 CR 7 SCR 5 NCR (see text)	(see text)

A, ampicillin; AF, atrial fibrillation; AT, atrial tachycardia; AVB, atrioventricular block; AVNT, AV, nodal tachycardia; BB, beta-blocker; BBB, bundle branch block, unspecified; C, corticosteroids; CE, cephalosporin; CR, complete resolution; DPM, definitive pacemaker; ELISA, enzyme-linked immunosorbent assay; I-A, intra-atrial; IFA, immunofluorescence assay; LBBB, left bundle branch block; LHF, left-sided heart failure; NSAID, nonsteroidal antiinflammatory drug; P, penicillin; PVB, premature ventricular beats; RBBB, right bundle branch block; RHF, right-sided heart failure; SAB, sinoauricular block; SCR, spontaneous complete resolution; STT, STT abnormalities; T, tetracycline; TPM, temporary pacemaker; 0, negative or normal; - (minus), unknown, absent, or not performed; 1°, first-degree; 2°, second degree; 3°, third-degree

Unpublished reports:
- Dournon E (1989) A case of Lyme carditis. (Department of infectious diseases, Hôpital R. Poincare, Garches, France). Unpublished correspondence.
- Hany A, Ballmer PE (1988) Atrial tachycardia in a patient with Lyme borreliosis. (Department of Internal Medicine, University of Bern, Inselspital, Switzerland). Unpublished correspondence.

carditis was diagnosed; in six of these there was an echocardiographically proven pericardial effusion and in two right-sided heart failure. In the other patients, the diagnosis pericarditis was made because of the presence of a friction rub in combination with typical ECG changes. Five other patients (8%) had a myopericarditis (pericarditis associated with enzyme changes without evidence of myocardial infarction), and, in one of these, this was complicated by pericardial effusion and right-sided heart failure. One patient had right-sided heart failure without pericarditis and seven patients had left-sided heart failure (combined with right-sided heart failure in one patient). Overall there were 15 patients (23%) with (myo)pericarditis and 11 patients (17%) with clinical heart failure. This compares with the observations of Steere et al. (1980) of 5% (one patient) of his patients with pericarditis and 15% (three patients) with heart failure.

10.7.3 Investigations

In 49 patients (74%) a chest X-ray was performed. In ten of these patients (20%) the chest X-ray was abnormal. In almost all cases a temporary cardiomegaly was present, in some patients with accompanying pulmonary venous congestion. Almost all these patients had clinically clear signs of heart failure. An echocardiogram was performed in 45 patients (68%); in 13 of them (29%) abnormalities were found (seven cases of pericardial effusion and six cases of hypokinesia of one or more parts of the left ventricle). In four of the seven patients with pericardial effusion, this was not clinically suspected. In five patients with an initially abnormal echocardiogram, findings reverted to normal some weeks after the antibiotic treatment.

A gallium-67 scan was performed in nine patients (14%); in four patients (44%) it was abnormal, with extensive or diffuse uptake of gallium-67 in the cardiac wall suggesting localized or diffuse myocarditis. Endomyocardial biopsies were taken in five patients (8%) (Vacher et al. 1987; Fauchier et al. 1988; de Koning et al. 1989; van der Linde et al. 1990), in at least two of them these were performed as a result of the findings of the gallium-67 scan. In four of these patients (80%) signs of an (endo)myocarditis were found and in three patients spirochetes were demonstrated in the endomyocardium (de Koning et al. 1989; van der Linde et al. 1990). In agreement with the results obtained by Marcus et al. (1985) and Reznick et al. (1986), the findings in these last patients support the hypothesis that the presence of *B. burgdorferi* is one of the triggering factors for inflammation in cardiac tissue.

To study the characteristics of the conduction disturbances, electrophysiologic studies were performed in thirteen patients (20%). In ten patients a prolonged AH interval was found, suggesting a proximal location of the AV block. In five patients a prolonged HV interval was found. This suggests a more distal location of the AV block. Two patients had a prolonged AH and HV interval, suggesting very diffuse injury of the AV conduction system. Three patients had combined intra-atrial and AV conduction disturbances.
Treatment and course are described in Chap. 26.

10.8 Summary

Lyme carditis includes AV, intra-atrial and intraventricular conduction disturbances, rhythm disturbances, (endo)myocarditis and pericarditis. Carditis may even be the only manifestation of Lyme borreliosis. The diagnosis of Lyme carditis can be difficult, since the clinical pattern can be very heterogeneous and has to be differentiated from many other diseases. A recent history of erythema migrans accompanied by positive serology in a patient with cardiac abnormalities is highly suggestive, but not conclusive proof of Lyme carditis. Futher investigations such as negative serology for other causative agents of carditis, isotope scans, and endomyocardial biopsies may be necessary. With special staining or culture methods, it is possible to demonstrate spirochetes in endomyocardial tissue. The combination of positive serology and the demonstration of spirochetes in inflamed endomyocardial tissue is a virtual proof of the diagnosis.

The criteria presented in Fig. 5, Table 1, and Section 10.3 in this chapter may help in standardizing the diagnosis of this condition.

Since 1984, 66 patients with Lyme carditis have been documented in Europe. Of these, 50 were men (mean age 43.6 years) and 16 women (mean age 46.4 years) (male/female ratio 3:1). In 98% of these patients, the IgG antibody titer against *B. burgdorferi* was elevated, while elevated IgM antibody titers were found in only 21%. An erythema migrans was observed in 58% of the patients, with 47% recalling a (tick) bite. Arthritis was present in 45% and neurological illness in 26% of the patients. The most frequent cardiac sign was AV conduction disturbances: 74% of all patients had a certain degree of (temporary) AV block (48% maximal third degree, 18% maximal second degree, and 8% maximal first degree). In 9% of the patients, a temporary bundle branch block was present, mostly in combination with an AV block. In 3% of the patients, a combined transient sinoauricular and AV block was found. In 15% of the patients, temporary supraventricular rhythm disturbances were found (atrial fibrillation and flutter, atrial tachycardia, AV nodal tachycardia and sinusbradycardia), in 23% transient (myo)pericarditis (with pericardial effusion in 50%), and in 17% of the patients resolving clinical heart failure was found.

Although the frequency of Lyme carditis in Europe is not yet well established, it does not seem to be very rare. We are aware that despite an extensive literature search and personal correspondence with many colleagues throughout Europe, we have been unable to collect data on all the diagnosed cases of Lyme carditis in Europe. We hope that anyone who has seen patients with this condition will contact us in order that we may collate the data from an expanding series of patients for forthcoming editions of this book.

Acknowledgement. The authors thank Dr. D. Hassler (General Practitioner, Kraichtal, Germany), Dr. E. Dournon (Hôpital R. Poincare, Garches, France), Dr. E. Wunderlich (Bezirkskrankenhaus Dresden-Friedrichstadt, Dresden, Germany) and all other physicians who unrestrictedly provided us with all the information needed on the Lyme carditis cases in this chapter and the useful suggestions they conferred.

In addition, the authors thank Prof. R.L. Himsworth, Aberdeen Royal Infirmary, Scotland, United Kingdom for his help in the final preparation of this manuscript.

References

Allal J, Coisne D, Thomas P, Vieures C, Gallimard JF, Becq-Giraudon B, Foullon P, Barraine R (1986) Manifestations cardiaque de la maladie de Lyme. Ann Med Interne (Paris) 137:372-374

Alpert LI, Welch P, Fisher N (1985) Gallium-positive Lyme disease myocarditis. Clin Nucl Med 10:617

Ballmer PE, Hany A (1988) Lyme-Karditis. Schweiz Med Wochenschr 118:358-362

Bannwarth A (1941) Chronische lymphocytäre Meningitis, entzündliche Polyneuritis und Rheumatismus. Arch Psychiatr Nervenkr 113:284-376

Beck G, Habicht GS, Benach JL, Coleman JL, Lysik RM, O'Brien RF (1986) A role for interleukin-1 in the pathogenesis of Lyme disease. Zentralbl Bakteriol Hyg (A) 263:133-136
Bedell SE, Pastor BM, Cohen SI (1981) Symptomatic high grade heart block in Lyme disease. Chest 79:236-237
Blaauw AAM, Braat S, van Santen-Hoeufft HM, van der Linden SJ (1988) Cardiologische afwijkingen bij patienten met Lyme-borreliose. Ned Tijdschr Geneeskd 132:2111-2114
Borgman H, Kumpf HP, Thess G, Schulze HJ (1989) Ein Fall von herzschrittmacherpflichtiger Lyme-Karditis. Z Gesamte Inn Med 44:247-248
Cornuau C, Bardet M, Baudoin P, Daumas PL, Oblet B, Poirot G, Valois M (1984a) Bloc auriculo-ventriculaire aigu, syncopal, au cours de la maladie de Lyme (letter). Presse Med 13:888
Cornuau C, Bernard M, Daumas PL, Oblet B, Poirot G, Valois M (1984b) Les manifestations cardiaques de la maladie de Lyme. A propos de deux observations. Ann Cardiol Angiol (Paris) 33:395-399
De Koning J, Bosma RB, Hoogkamp-Korstanje JAA (1987) Demonstration of spirochetes in patients with Lyme disease with a modified silver stain. J Med Microbiol 23:261-267
De Koning J, Hoogkamp-Korstanje JAA, van der Linde MR, Crijns HJGM (1989) Demonstration of spirochetes in cardiac biopsies of patients with Lyme disease. J Infect Dis 160:150-153
Dournon E, Assous M, Fourcade C (1987) Clinical features of Lyme disease in France. In: Abstract book of the 27th interscience conference on antimicrobial agents and chemotherapy (ICAAC), New York, p 319, 1274
Dunica S, Piette JC, Nassar N, Beaufils P (1986) Une nouvelle cause de bloc auriculo-ventriculaire aigu transitoire: la maladie de Lyme. Arch Mal Coeur 79:1251-1255
Duray PH, Steere AC (1986) The spectrum of organ and systems pathology in human Lyme disease. Zentralbl Bakteriol Hyg (A) 263:169-178
Fauchier JP, Cosnay P, Sirinelli A, Moquet B, Rabut H (1988) Myocardite de Lyme sans trouble conductifs auriculo-ventriculaires. Press Med 17:2036-2037
Fumarolo D, Munno I, Marcuccio C, Miragliotta G (1986) Endotoxin-like activity associated with Lyme disease borrelia. Zentralbl Bakteriol Hyg (A) 263:142-145
Goldings EA, Jericho J (1986) Lyme disease. Clin Rheum Dis 12:343-367
Habicht GS, Beck G, Benach JL, Coleman JL (1986) *Borrelia burgdorferi* lipopolysaccaride and its role in the pathogenesis of Lyme disease. Zentralbl Bakteriol Hyg (A) 263:137-141
Habicht GS, Beck G, Benach JL (1987) Lyme disease. Sci Am 257:60-65
Hansen K, Madsen JK (1986) Myocarditis associated with tickborne *Borrelia burgdorferi* infection. Lancet I:1323-1324
Hardin JA, Steere AC, Malawista SE (1979) Immune complexes and the evolution of Lyme arthritis: dissemination and localization of abnormal Clq binding activity. N Engl J Med 301:1358-1363
Hassler D, Zipperle G, Ackermann R, Lembke U, Heinrich F (1987) Kardiale Beteiligung auch bei der Europaeischen Erythema-migrans-Borreliose? Dtsch Med Wochenschr 112:1506-1508
Hassler D, Geiselhardt B, Zipperle G, Zöller L, Sonntag HG, Heinrich F, Lang H, Wetzel H, Ackermann R (1988a) Internistische Aspekte der Lyme-Borreliose. Therapiewoche 38:2408-2413
Hassler D, Zipperle G, Ackermann R, Heinrich F (1988b) Die Borreliose. Z Allg Med 64:175-183
Hopf HC (1966) Acrodermatitis Chronica Atrophicans (Herxheimer) und Nervensystem. Springer, Berlin Heidelberg New York, pp 70-71 (Monographien aus dem Gesamtgebiete der Neurologie und Psychiatrie, vol 114)
Houwerzijl J, Root JJ, Hoogkamp-Korstanje JAA (1984) A case of Lyme disease with cardiac involvement in the Netherlands (letter). Infection 12:358
Jacobs JC, Rosen JM, Szer IS (1984) Lyme myocarditis diagnosed by gallium scan. J Pediatr 105:950-952
Kapusta P, Fauchier JP, Cosnau P, Huguet R, Grezard O, Rouesnel P (1986) Troubles conductifs sino-auriculaires et auriculo-ventriculaires de la maladie de Lyme. A propos de deux observations. Arch Mal Coeur 79:1361-1366
Kuiper H, De Jongh BM, Senden PJ (1988) Pacemaker-implantatie wegens totaal atrioventriculair block bij Lyme-borreliose. Ned Tijdschr Geneeskd 132:2109-2111

Laperrouza C (1985) La borréliose de Burgdorfer ou maladie de Lyme: expérience en zone endémique. Med Hyg 43:2885-2896

Lavaud P, Etienne J, Chamot E, Atallah G, Moleur P, Brun Y, Touboul P (1985) Bloc auriculo-ventriculaire aigu associé une maladie de Lyme. Un cas confirmé par sérologie (letter). Presse Med 14:2020

Lefèvre T (1988) Quel est votre diagnostic? Concours Med 110:2473-2475

Lorcerie B, Boutron MC, Portier H, Beuriat P, Ravisy J, Martin F (1987) Manifestations pericardique de la maladie de Lyme. Ann Med Interne (Paris) 138:601-603

Marcus LC, Steere AC, Duray PH, Anderson AE, Mahoney EB (1985) Fatal pancarditis in a patient with coexistent Lyme disease and babesiosis: demonstation of spirochetes in the myocardium. Ann Intern Med 103:374-376

Mayer W, Kleber FX, Wilske B, Preac-Mursic V, Maciejewski W, Sigl H, Holzer E, Doering W (1989) Persistent atrioventricular block in Lyme disease. Klin Wochenschr 68:431-435

Meyer LK, Swenson DB (1987) Lyme carditis: high-grade heart block in Lyme disease. Minn Med 70:345-346

Muhlemann MF, Wright DJ (1987) Emerging pattern of Lyme disease in the United Kingdom and Irish Republic. Lancet I:260-262

O'Connell JB, Henkin RE, Robinson JA, Subramanian R, Path MRC, Scanlon PJ, Gunnar RM (1984) Gallium-67 imaging in patients with dilated cardiomyopathy and biopsy-proven myocarditis. Circulation 70:58-62

Olson LJ, Okafor EC, Clements IP (1986) Cardiac involvement in Lyme disease: manifestations and management. Mayo Clin Proc 61:745-749

Ponsonnaille J, Citron B, Karsenty B, D'Agrosa MC, Peyselon P, Veyre A, Gras H (1986) Myocardite aigue au cours d'un syndrome de Lyme. Intérét de la scintigraphie myocardique au gallium 67. Arch Mal Coeur 79:1946-1950

Reznick JW, Braunstein DB, Walsh RL, Smith CR, Wolfson PM, Gierke LW, Gorelkin L, Chandler FW (1986) Lyme carditis. Electrophysiologic and histopathologic study. Am J Med 81:923-927

Rienzo RJ, Morel DE, Prager D, Barron L, Post R (1987) Gallium avid Lyme myocarditis. Clin Nucl Med 12:475-476

Roelli H, Angehrn W, Reutter FW (1988) Atrio-ventrivuläre Blockierung bei Lyme-Karditis. Schweiz Med Wochenschr 118 (Suppl 23):6

Roelli H, Angehrn W, Reutter FW (1988) Atrio-ventriculäre Blockierung bei Lyme-Karditis. Schweiz Med Wochenschr 119:81-83

Satz N, Ott A, Knoblauch M (1986) Die Erythema-migrans-Krankheit. Das klinische Spektrum der Infektion mit *Borrelia burgdorferi*. Schweiz Med Wochenschr 116:763-769

Schmidt R, Kabatzki J, Hartung S, Ackermann R (1985) Erythema-migrans-Borreliose in der Bundesrepublik Deutschland: Epidemiologie und klinisches Bild. Dtsch Med Wochenschr 110:1803-1807

Schott B (1985) Erytheme de Lipschütz avec atteinte cardiaque. A propos d'un cas: étiology méconne de BAV paroxystique. Synthèse et rapport avec la maladie de Lyme aux USA. Thesis, Louis Pasteur University, Strasbourg, France

Stanek G, Flamm H, Groh V, Hirschl A, Kristoferitsch W, Neumann R, Schmutzhard E, Wewalka G (1986) Epidemiology of Borrelia infections in Austria. Zentralbl Bakteriol Hyg (A) 263:442-449

Stanek G, Klein J, Bittner R, Glogar D (1990) Isolation of *Borrelia burgdorferi* from the myocardium of a patient with longstanding cardiomyopathy. N Engl J Med 322:249-252

Steere AC, Malawista SE, Snydman DR, Shope RE, Andiman WA, Ross MR, Steele FM (1977) Lyme arthritis: an epidemic of oligoarticular arthritis in children and adults in three Connecticut communities. Arthritis Rheum 20:7-17

Steere AC, Batsford WP, Weinberg M, Alexander J, Berger HJ, Wolfson S, Malawista SE (1980) Lyme carditis: cardiac abnormalities of Lyme disease. Ann Intern Med 93:8-16

Steurer J, Bansky G, Greminger P, Siegenthaler W (1987) Acrodermatitis chronica atrophicans und Lymesche Erkrankung. Internist 28:693-697

Trevison G, Crovato F, Marcuccio C, Fumarola D, Scarpa C (1986) Lyme disease in Italy. Zentralbl Bakteriol Hyg (A) 263:459-463

Vacher D, Lusson JR, Peyselon P, Kantelip B, Veyre A, Cassagnes J (1987) La myocardite de Lyme: intérét des examens scintigraphiques (Gallium 67 et Thallium 201) et de la biopsie endomyocardique (à propos de deux cas). Coeur 18:374-379

Van der Linde MR, Crijns HJGM, Lie KI (1989) Transient complete atrioventricular block in Lyme disease. Chest 96:219-221

Van der Linde MR, Crijns HJGM, de Koning J, Hoogkamp-Korstanje JAA, de Graaf JJ, Piers DA, van der Galien A, Lie KI (1990) Range of atrioventricular conduction disturbances in Lyme borreliosis: a report of 4 cases and review of other published reports. Br Heart J 63:162-168

Veyssier P, Davous N, Kalousstian E, Maitre B, Lallement PY, Serret A (1987) Atteintes cardiaques au cours de la maladie de Lyme. Deux observations. Rev Med Interne 8:357-360

Vieyres C, Allal J, Coisne D, Amiel A, Barraine R (1987a) Manifestations cardiaques de la maladie de Lyme. Mises Jour Cardiol 16:13-21

Vieyres C, Allal J, Coisne D, Thomas Ph, Neau JP, Breux JP (1987b) Aspects européens de la maladie de Lyme. Huit cas. Press Med 16:59-62

Weber K, Puzik A, Becker T (1983) Erythema-migrans-Krankheit. Beitrag zur Klinik und Beziehung zur Lyme-Krankheit. Dtsch Med Wochenschr 108:1182-1190

Wunderlich E, Thess G, Hetze A, Witzleb W, Schmidt PKH (1988) Lyme-Borreliose - eine mögliche Ursache von atrioventrikulären (AV-)Blockierungen. Z Kardiol 77:256-258

11 Early Neurological Involvement (Bannwarth's Syndrome)

H.W. Pfister, W. Kristoferitsch, and C. Meier

11.1 Introduction

The nomenclature for the neurological manifestations of *Borrelia burgdorferi* infections is confusing and not uniformly used. In the literature, the terms used include Bannwarth's syndrome (Meyer-Rienecker and Hitzschke 1977), meningopolyneuritis Garin-Bujadoux-Bannwarth (Hörstrup and Ackermann 1973), Borrelia meningitis (Hansen et al. 1986), erythema chronicum migrans meningitis (Weber 1974), radiculomyelomeningitis after tick bite (Schaltenbrand 1962), meningomyeloradiculitis after tick bite with erythema (Bammer and Schenk 1965), neuroborreliosis (Stiernstedt et al. 1988), and Lyme neuroborreliosis (Halperin et al. 1989). In addition, these terms confuse the neurological manifestations during the early disseminated infection and those in chronic late infection. All these terms have disadvantages; for example, Bannwarth's syndrome is characterized by painful meningoradiculoneuritis (Bannwarth 1941). Use of the word "meningitis" is not advisable because, despite the inflammatory cerebrospinal fluid (CSF) syndrome, patients in Europe usually do not suffer from the typical signs or symptoms of meningitis such as headache, stiff neck, fever, vomiting, and photophobia. The term "neuroborreliosis" is nonspecific because other borrelial infections such as relapsing fever may also cause neurological manifestations. Thus, the term Lyme neuroborreliosis seems to be most suitable to describe the neurological manifestations of *B. burgdorferi* infection including early and late manifestations.

This chapter focuses attention on the spectrum of early Lyme neuroborreliosis, the neurological manifestations of early disseminated infection as observed in Europe. The spectrum of Lyme neuroborreliosis was markedly enlarged when serodiagnostic tests became available. However, the association between elevated serum antibody titers against *B. burgdorferi* and neurological symptoms in a given patient does not prove a causal relationship. In considering atypical cases, we have to distinguish carefully between *definite* cases and cases of *suspected* Lyme neuroborreliosis.

In the following chapter we distinguish between:
- Typical Bannwarth's syndrome
- Definite cases in patients with an atypical clinical picture such as predominant encephalitis, myelitis, meningitis, or myositis
- Suspected cases

11.2 Typical Bannwarth's Syndrome

11.2.1 Definition

Bannwarth's syndrome is characterized by painful meningoradiculoneuritis with or without peripheral paresis, with or without cranial nerve affection – in most cases facial nerve palsy – , and lymphocytic pleocytosis in the CSF caused by infection with *B. burgdorferi.*

11.2.2 Demographic Data

Bannwarth's syndrome has been reported from all European countries where Lyme borreliosis exists (Chapter 29). Patients may be affected at all ages, although painful radiculoneuritis as a neurological manifestation of early disseminated Lyme borreliosis is a rare event in children (see Chapter 17). In the majority of patients, Bannwarth's syndrome occurs over the age of 40 years. In some studies there has been a female predominance (Stiernstedt et al. 1985; Kristoferitsch 1989), in others a male predominance (Hörstrup and Ackermann 1973; Wenig 1975; Schmidt and Ackermann 1985). One study found that 78 % of adults with early Lyme neuroborreliosis had typical Bannwarth's syndrome (Pfister et al. 1986). Similar results have been found in other reports (Stiernstedt et al. 1985; Kohlhepp et al. 1987).

11.2.3 Seasonal Distribution

Bannwarth's syndrome in central Europe usually starts in summer or fall. In one study 86 % of all patients had the onset of the disease between July and November, (Kristoferitsch 1989) (Fig.1). In one Scandinavian study, all patients had fallen ill between June and November (Stiernstedt et al. 1985). This seasonal distribution was also mentioned in the first systematic studies on Bannwarth's syndrome (Schaltenbrand 1967; Erbslöh and Kohlmeyer 1968; Reisner 1969; Hörstrup and Ackermann 1973).

11.2.4 Tick Bite

The frequency of patients who remember arthropod bites preceding Bannwarth's syndrome varies in different reports between 39 % and 46 % (Stiernstedt et al. 1985; Pfister et al. 1986; Kristoferitsch 1989). In one study two–thirds of the patients reported on tick bites and one–third on bites from flying insects, which were followed by marked erythema (Kristoferitsch 1989). Some evidence can be drawn from these and former European reports (Schaltenbrand 1967; Reisner 1969) that apart from ticks flying insects may also transmit *B. burgdorferi.* The interval between arthropod bite

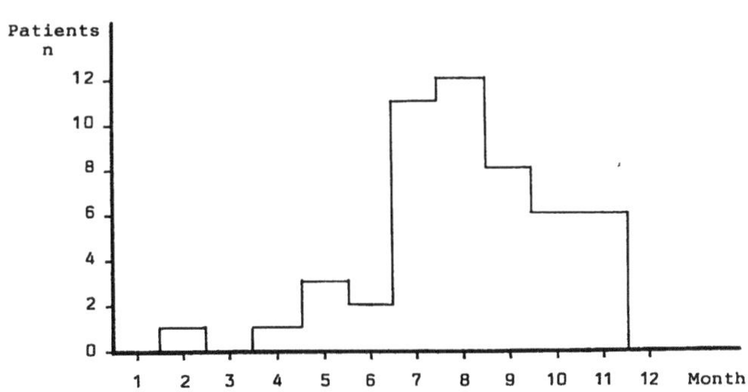

Figure 1. Seasonal distribution of onset of Bannwarth's syndrome in 50 patients

and the onset of neurological symptoms is a median of 3 weeks (range 1 – 18 weeks) (Kristoferitsch 1989). Long incubation periods have to be interpreted with some scepticism since the patient may have had additional but unrecognized tick bites. Nevertheless, an incubation period of 18 weeks without the possibility of additional reinfection has been observed (Kristoferitsch 1989).

11.2.5 Erythema Migrans

It has been reported that 34 % – 64 % of patients with Bannwarth's syndrome noticed an erythema migrans prior or concomitant to Bannwarth's syndrome (Stiernstedt et al. 1985; Schmidt and Ackermann 1985; Pfister et al. 1986; Kristoferitsch 1989). From the dermatological point of view, 17 % of patients with erythema migrans have been reported to develop neurological symptoms, 3 % proven meningoradiculoneuritis (Weber and Neubert 1986). Multiple erythemata or chronic erythema migrans occur only rarely in European patients with Bannwarth's syndrome (Schmidt and Ackermann 1985; Pfister et al. 1986; Kristoferitsch 1989). Patients with long–lasting erythema migrans seldom develop neurological disease (Asbrink and Olsson 1985; Weber and Neubert 1986). In four–fifths of patients, erythema migrans has already disappeared by the time of onset of neurological symptoms. Bammer and Schenk (1965) have demonstrated in detail the close topical association between the cutaneous region involved by erythema migrans and the initial radicular lesion in Bannwarth's syndrome. This close association could also be demonstrated in larger groups of European patients (Schaltenbrand 1967; Erbslöh and Kohlmeyer 1968; Hörstrup and Ackermann 1973; Kristoferitsch et al. 1983; Pfister et al. 1986), but not in American patients (Pachner and Steere 1985). This observation may indicate that besides hematogenous spreading of borreliae spreading along peripheral nerves may also occur.

11.2.6 Clinical Picture of Bannwarth's Syndrome

11.2.6.1 Symptoms

11.2.6.1.1 Pain

Pain due to radiculoneuritis is a frequent symptom in early Lyme borreliosis. It has been described as severe, burning, tearing, migrating pain with characteristic exacerbations at night. At the site involved by pain, the skin may be dysesthetic and hyperpathic. Slight contact as from the pressure of clothing may cause severe burning pain. Pain in Bannwarth's syndrome will not respond to the commonly used analgesics. However, it will decrease dramatically within a few days with antibiotic treatment in the majority of patients. Pain without additional pareses has frequently been interpreted as aggravation, and there have even been reports of patients who have attempted to commit suicide (Wolf 1970). Meningism, fibromyalgia, or arthralgia may also cause pain in early Lyme borreliosis (Weber and Neubert 1986; Kristoferitsch 1989). In most patients with Bannwarth's syndrome, these symptoms cannot be differentiated from the overwhelming radicular neuritic pain. Pain as the sole clinical manifestation of Bannwarth's syndrome has been reported in about 20 % of patients without antibiotic treatment (Kristoferitsch et al. 1983).

11.2.6.1.2 Neuropsychological Symptoms

Agitation, depression, anxiety, and restlessness are frequent symptoms in Bannwarth's syndrome and have been found in more than one–third of patients. Three–quarters of patients suffer from sleeplessness. The state of agitation and restlessness has not exclusively been related to the state of severe pain, but has also been interpreted as a symptom of mild encephalitis or encephalopathy (Kristoferitsch et al. 1983). Severe fatigue and lethargy in Bannwarth's syndrome have not been as frequently observed in Europe as in the USA (Pachner and Steere 1985). Patients with Bannwarth's syndrome and organic psychosis have been described on several occasions (Hellerström 1930; Erbslöh and Kohlmeyer 1968), yet this has not been studied systematically so far.

11.2.6.2 Clinical Signs

11.2.6.2.1 Radiculitis

Pareses of extremities occur in about 50 % – 60 % of patients without antibiotic treatment (Hörstrup and Ackermann 1973; Kristoferitsch et al. 1983). In most patients, one single lower extremity is affected and lower extremities have been found to be more frequently involved than upper extremities. Usually the paresis starts at the extremity affected by the tick bite or erythema migrans. It frequently spreads to the contralateral extremity and later to other extremities. The pattern of motor involvement ranges from an oligoradiculitic type with involvement of a few adjacent nerve roots to a polytopical asymmetric radiculitis. Symmetric polyradiculoneuritis does not occur frequently in Bannwarth's syndrome. In one–third of patients, pareses are proximally accentuated. Pareses of abdominal muscles can be observed in about 5 % of patients without antibiotic treatment. On rare occasions one–sided or bilateral phrenicus pareses have been reported. This has been related to radiculitis of the fourth cervical segment (Pfister et al. 1985; Kristoferitsch 1989).

While pain and hyperesthesia are common symptoms in Bannwarth's syndrome, hypoesthesia or sensory loss is not so prominent and has been found in about 30 % of patients. Its distribution only roughly follows a clear dermatomal pattern (Kristoferitsch et al. 1983). Distally to the nerve–roots, localized nerve involvement such as plexitis or mononeuritis multiplex (Erbslöh and Kohlmeyer 1968) is generally overlapped by the dominating radiculitis, although distal nerve involvement has been consistently found in nerve biopsies (Vallat et al. 1984, 1987; Camponovo and Meier 1986; Meier and Grehl 1988; Engelhardt et al. 1989). The severity of the pareses ranges from mild weakness to paralysis.

11.2.6.2.2 Vegetative Symptoms

Vegetative symptoms as manifestations of the autonomous peripheral nervous system (PNS) have been occasionally seen. Symptoms such as hyperhydrosis, obstipation, and disturbance of micturation have been connected with involvement of the autonomous nervous system (Kristoferitsch 1989). A systematic study of these phenomena is lacking in Bannwarth's syndrome. On rare occasions, sympathetic reflex dystrophy develops as sequel of Bannwarth's syndrome (Kohler and Thoden 1987).

11.2.6.2.3 Cranial Nerve Affection

Cranial nerve affection has been found in 47 % – 82 % of patients without antibiotic treatment (Erbslöh and Kohlmeyer 1968; Hörstrup and Ackermann 1973; Kristoferitsch et al. 1983; Krüger et al. 1989). In 72 % – 92 % of these patients, the facial nerve is affected. Apart from the facial nerve, the VIth, Vth, IIIrd, and VIIIth cranial nerves seem to be next frequently affected. Few patients with optic neuropathy and Bannwarth's syndrome have been described (Erbslöh and Kohlmeyer 1968; Schmutzhard et al. 1988). In 29 % (Weber et al. 1987) to 71 % (Wenig 1975) of patients with facial palsy, bilateral involvement has been reported. Multiple cranial neuritis occurs in combination with uni– or bilateral facial nerve palsy and has been found in the spontaneous course of 25 % – 47 % of patients with Bannwarth's syndrome (Erbslöh and Kohlmeyer 1968; Hörstrup and Ackermann 1973).

11.2.6.2.4 Other Neurological Signs
Meningism has been observed in about 4% – 21 % of patients with typical Bannwarth's syndrome (Erbslöh and Kohlmeyer 1968; Hörstrup and Ackermann 1973; Pfister et al. 1986; Kristoferitsch 1989). Usually the symptoms of meningism are mild. Meningitis as the sole neurological manifestation of early Lyme disease has also been reported. These patients do not fit the criteria of Bannwarth's syndrome and will be discussed later. Clinical signs of additional encephalitis or myelitis have been detected in about 10 % of patients with Bannwarth's syndrome (Pfister et al. 1986; Krüger et al. 1989). Prominent or exclusive meningitis or central nervous system (CNS) manifestations in early Lyme borreliosis do not fulfill the criteria for Bannwarth's syndrome and will be discussed later in this chapter.

11.2.7 Clinical Course of Bannwarth's Syndrome

11.2.7.1 Spontaneous Clinical Course

Hörstrup and Ackermann (1973) have stated that the course of Bannwarth's syndrome follows a characteristic order; tick bite and erythema migrans are followed by migratory pain, lymphocytic CSF pleocytosis, and, in the majority of patients, facial palsy and/or peripheral asymmetric pareses. Pain is the initial neurological symptom in all adult patients, while pareses may follow days to several weeks later (Kristoferitsch 1989). Sometimes the course is fluctuating and, while the primary involved region is already improving, other regions become affected. The disease duration without antibiotic treatment has been estimated to be at least 3 months (Hörstrup and Ackermann 1973) or up to 6 months (Kristoferitsch et al. 1983), 2 – 6 months (Krüger et al. 1990), 2 – 12 months (Bammer 1981), or several months to 1,5 years (Schaltenbrand, cited by Wenig 1975). Improvement of pain was seen in 32 % of patients without antibiotic or steroid therapy, improvement of cranial nerve pareses in 44 % of patients and improvement of radicular pareses in only 8 % of patients 4 weeks after hospital admission. Four month after the onset of neurological disease improvement of pain was found in 79 % of patients, improvement of cranial nerve pareses in 78 %, and improvement of radicular pareses in only 11 %. Overall, 63 % of patients have been classified as improved 4 months after onset of neurological symptoms (Kristoferitsch 1989).

11.2.7.2 Residual Symptoms

Bannwarth's syndrome is a self–limiting manifestation in the majority of patients. The risk of developing chronic progressive encephalomyelitis must be judged as low. Five of 48 patients with chronic progressive encephalomyelitis had a previous history consistent with Bannwarth's syndrome (Ackermann et al. 1988). Two similar patients have been reported by Kohler and Thoden (1987). In 16 % of patients with acrodermatitis chronica atrophicans (ACA) (Asbrink et al. 1986) and 18 % of patients with ACA–associated neuropathy (Kristoferitsch 1989) a history of neurological symptoms was found, which could be connected with early Lyme neuroborreliosis. Krüger et al. (1989), however, could not detect chronic progressive encephalomyelitis or recurrent *B. burgdorferi* infections in a follow–up examination of 72 patients who had Bannwarth's syndrome or Bannwarth's syndrome combined with encephalomyelitis or encephalitis 5 – 27 years previously. None of these patients had antibiotic treatment, all of them were in good health at the follow–up examination and were able to return to their former professions. Mild residual symptoms have been observed such as temporary radicular pain in 37 % of patients, mild residual pareses in 32 %, decreased visual power in 10 %, or disturbances in coordination in 8 %. Similar results have been found in three smaller surveys in 15 patients 4,5 – 7,5 years, in 11 patients 2 – 6 years, and in 9 patients 5 – 16 years after antibiotically untreated Bannwarth's syndrome (Pfister et al. 1989; Kristoferitsch 1989; Gollmer et al. 1989). In two of these studies, new neurological disorders (pheripheral neuropathy in two patients and facial palsy in one) had been observed, but they could not be unequivocally linked to the previous Bannwarth's syndrome. The clinical long–term outcome did not seem to differ compared with patients who had antibiotic treatment (Gollmer et al. 1989; Kristoferitsch 1989; Krüger et al. 1989; Pfister et al. 1989).

11.2.8 Cerebrospinal Fluid Findings

Abnormalities have been uniformly detected in the CSF of patients with Bannwarth's syndrome. Characteristic findings are lymphocytic pleocytosis with a high number of plasma cells (with an average of 9 % to a maximum of 30 %) (Fig. 2) (Kristoferitsch 1989) and elevated total protein content. The latter is due to the blood–brain barrier damage, wich is measurable by the elevated albumin ratio, or to an intrathecal immunoglobulin (Ig) synthesis, demonstrated by elevated IgG, IgM, and IgA index values, or oligoclonal bands. Oligoclonal bands consist of *B. burgdorferi*–specific and nonspecific IgG (Krüger et al. 1981; Felgenhauer 1982; Kristoferitsch et al. 1983; Stiernstedt et al. 1984; Murray et al. 1986; Henriksson et al. 1986; Martin et al. 1988; Hansen et al. 1990). A significant decrease in cell count and total protein content could be demonstrated between the 4th and the 10th week of the disease, while the amount of intrathecally synthesized IgG or IgM did not decline. Especially IgM showed a tendency to increase throughout the first months of the disease. This has been interpreted as marker for the disease activity (Kristoferitsch et al. 1983; Maida et al. 1986). By means of isoelectrofocusing of the CSF, oligoclonal bands could still be demonstrated in almost one–third of patients 5 – 17 years after the occurrence of Bannwarth's syndrome (Krüger et al. 1989, 1990). Oligoclonal bands could also be found in long-term follow up studies of antibiotically treated patients (Sköldenberg et al. 1983).

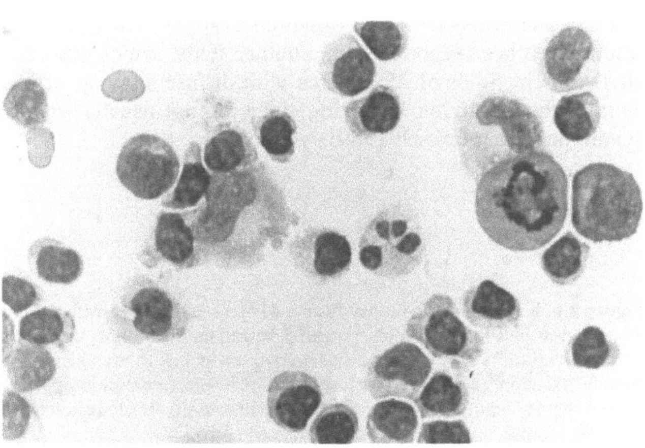

Figure 2. Cerebrospinal fluid cytology in a patient with Bannwarth's syndrome. Note the high grade of stimulation characterized by lymphoid cells, plasma cells and mitosis.

11.2.9 Electroneurography and Electromyography

Only a few electrophysiological studies on Bannwarth's syndrome have been published, and their results or interpretations show some controversies. The first systematic electromyographic (EMG) study was performed by Rohmer et al. (1974) on 36 patients with meningoradiculitis – 16 of them had had a tick bite. The predominant alterations were signs of severe denervation while nerve conduction velocity was found to be normal. These results were interpreted "to confirm the radicular involvement without neuritic participation". Wulff et al. (1983), performing conduction studies of the facial nerve in four patients with Bannwarth's syndrome observed long latencies in two patients, indicating either segmental demyelination or regeneration after Wallerian degeneration. On the basis of the retrospectively analyzed data of 29 patients with Bannwarth's syndrome, Graf et al. (1986) found prolonged nerve conduction velocities and prolonged distal latencies to be the prominent electroneurographic pattern. Hänny and Häuselmann (1987) could detect signs of acute denervation in EMG investigations of seven patients. In addition, they found "mild slowing of nerve conduction velocity" in 4 of 15 patients, which was interpreted to be compatible with peripheral neuritis of demyelinating nature. The electrophysiological studies of Vallat et al. (1987), however, provided "provocative" hints for axonal degeneration in ten patients with Bannwarth's syndrome. These authors reported on an association of denervation potentials with normal or subnormal conduction velocities and slight prolongation of distal latencies and a moderate slowing of sensory conduction velocities, sometimes with reduced amplitudes. They also described increased F wave latencies in two out of ten investigated patients. In another study, six out of seven investigated patients showed pathological F wave latencies, and in four investigated patients H reflexes were found to be prolonged, indicating nerve root involvement (Kristoferitsch 1989). Similar results of studies reported from the United States were compatible either with demyelinating neuropathy (Sterman et al. 1982), or mild axonal neuropathy (Pachner and Steere 1985)

11.2.10 Electroencephalography

Systematical electroencephalography (EEG) studies in Bannwarth's syndrome have rarely been conducted. Schmidt and Ackermann (1985) mentioned a diffuse slowing of basic rhythm or slight dysrhythmia in 66 % of 41 patients investigated. The remission of these alterations could be observed in nearly all follow–up investigations. Similar results have been reported from another study, which showed slightly abnormal EEG alterations in 67 % of 25 patients with diffuse slowing of basic rhythm or increased delta or theta activity; localizing signs are not usually found in typical Bannwarth's syndrome (Kristoferitsch 1989).

▷

Figure 3 a, b. Sural nerve biopsy from a 41–year–old patient with meningoradiculoneuritis. **a** Heavy mononuclear cell infiltration of epineurial vessels in the vicinity of a nerve fascicle, H & E x 230. **b** The perineurium is also heavily infiltrated. *Arrow* points to myelinated axons undergoing break down (semithin section, methylene blue). **c** Detail of **b**. Electronmicroscopy reveals destruction of the perineurial sheath (*small arrows*). Large *arrow* points to myelinated axon undergoing Wallerian degeneration (bar, 5 μm).

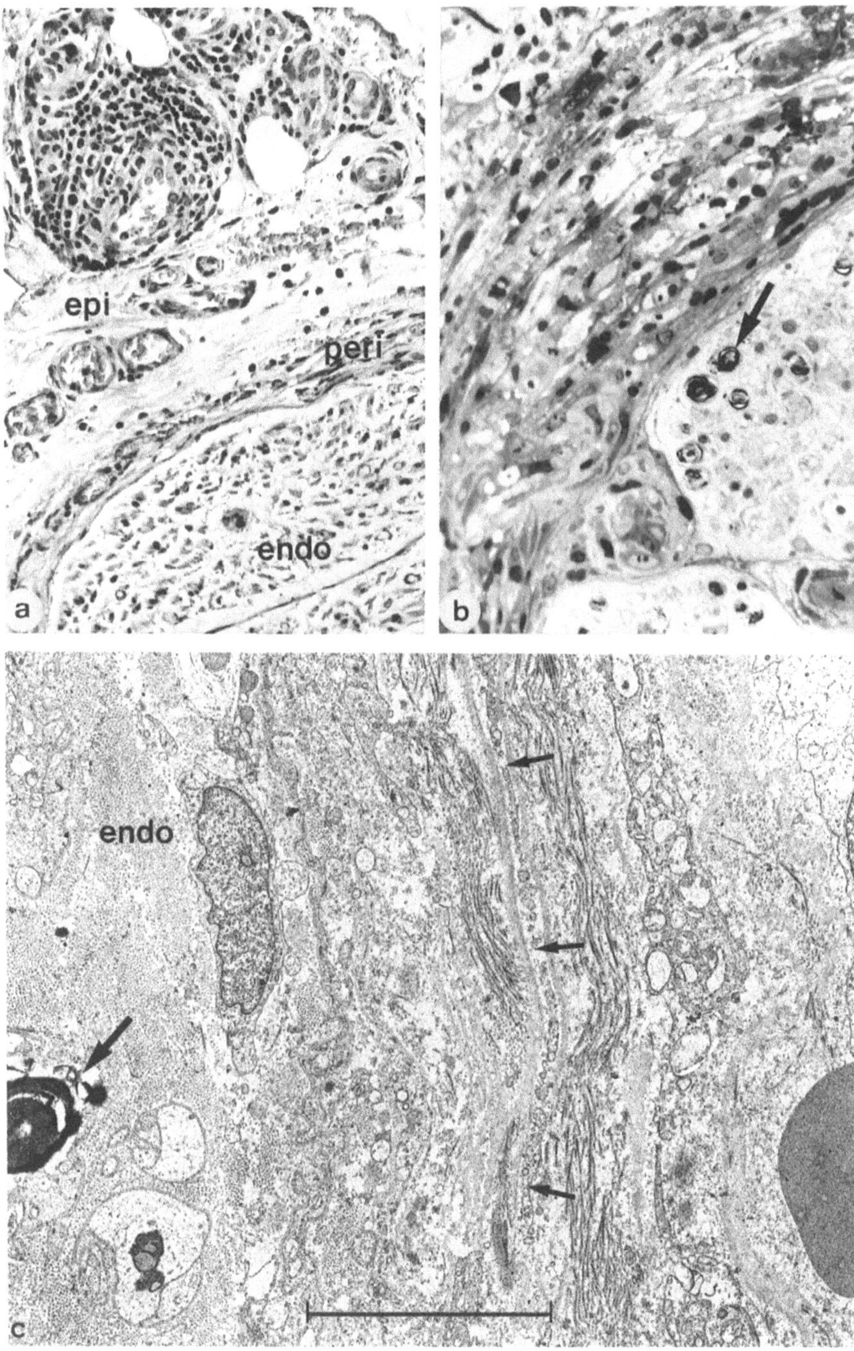

(Fig. 3 d see page 160)

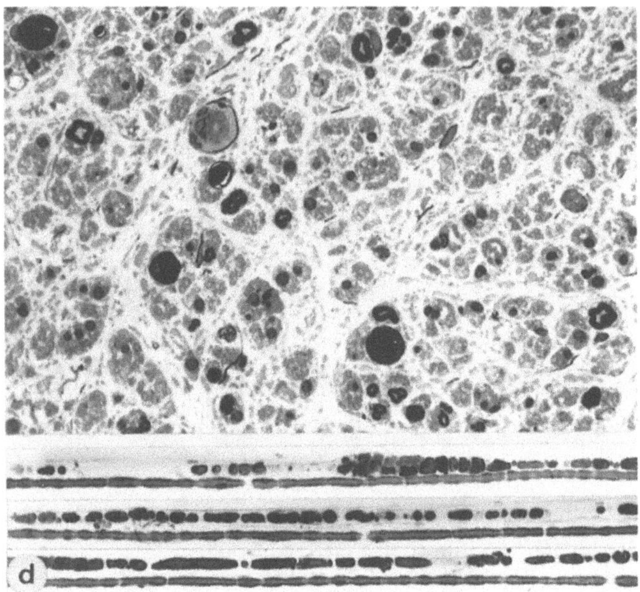

Fig. 3 d. Heavy destruction of many myelinated axons shown in cross sections (methylene blue x 450) and teased fiber preparation (x130)

11.2.11 Histological Findings and Pathogenesis

Sural nerve biopsy findings from patients with PNS–involvement of Lyme borreliosis have been reported by Camponovo and Meier (1986), Vallat et al. (1987) and Meier et al. (1989). These authors observed perivascular infiltrations around the vasa nervorum of the epi-, peri- and endoneurium. The infiltrates consisted of lymphocytes, histiocytes, and plasma cells (Figs. 3,4). Some of the vessels in the epineurium showed evidence of thrombotic occlusion or recanalization. The authors emphasized that necrotizing lesions of the vessel walls were never observed. The nerve fascicles exhibited fiber loss of variable degree due to axonal degeneration. Occasionally segmental demyelination was found next to axonal degeneration. These observations suggest that vasculitis is an important mechanism in the pathogenesis of PNS involvement in Lyme borreliosis. *B. burgdorferi* was not observed in the vasculitic lesions or elsewhere in the biopsied nervous tissue. Various theories have been advanced to explain the extensive tissue damage found in the PNS in spite of the absence of the spirochetes. The first theory implies that immune complexes consisting of *Borrelia* antigen, antibodies and complement accumulate in the vessel walls, inducing an immunecomplex vasculitis. However, the immunohistochemical studies performed so far in the tissue samples from the PNS of patients with Lyme borreliosis have not confirmed such a mechanism (Vallat et al. 1987; Meier et al. 1989). Another theory suggests, that lipopolysaccharides from the gram–negative outer surface of *B. burgdorferi* may induce the release of interleukin–1, wich in turn would exert a local amplifying immune response (Habicht et al. 1988). The third theory is based on the observation, that the sera from patients with Lyme borreliosis contain antibodies to *Borrelia burgdorferi* that crossreact with nervous tissue antigens (Suchanek et al. 1986; Sigal and Tatum 1988; Garcia-Monco et al. 1988; Aberer et al. 1989). These observations have to be confirmed and carefully evaluated. Scepticism has been nourished by the current work of Dumas et al. (1989), who have shown that both seropositive and seronegative sera were equally affected by immunoadsorption with neuroaxonal and *B. burgdorferi* antigens.

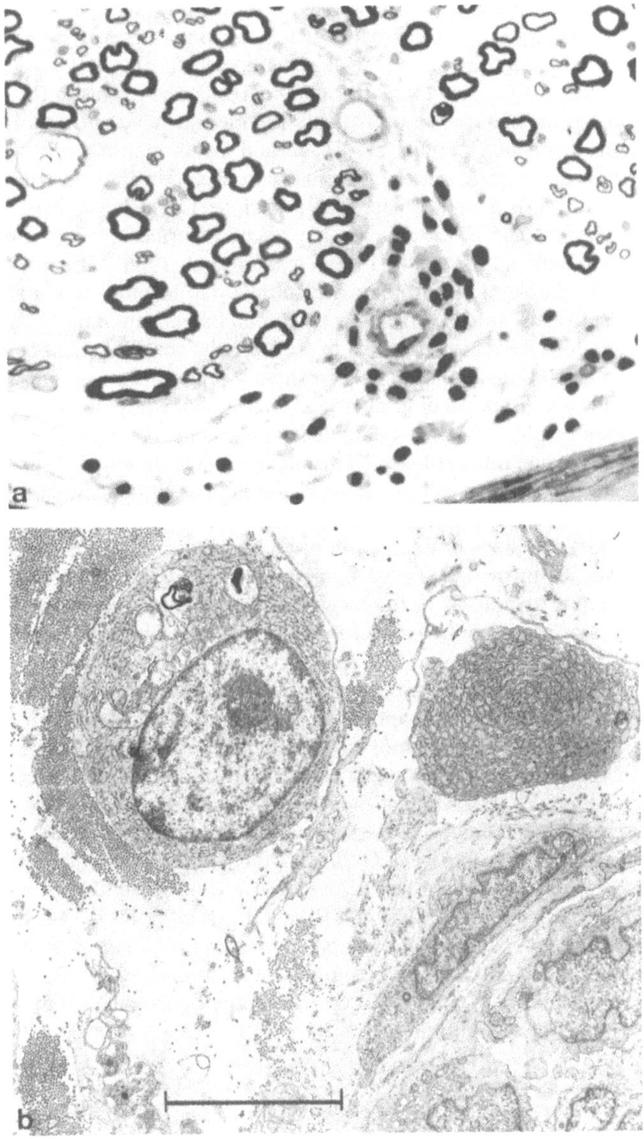

Figure 4 a, b. Sural nerve biopsy from a 28–year–old patient with meningoradiculo-neuritis.
a Small pericapillary infiltration in the endoneurium (methylene blue–stained semithin section, x520).
b Electronmicroscopy reveals that the infiltration contains plasma cells (bar, 5 μm)

11.2.12 Diagnosis

Diagnostic criteria are outlined in Sects. II.2.1 and II.3.1

11.2.13 Differential Diagnosis

In the differential diagnosis of Bannwarth's syndrome, other tick–transmitted disorders have to be considered as well as diseases with predominant radicular or migrating pain, with radicular lesions or with cranial nerve pareses. Apart from Lyme borreliosis central European encephalitis or tick–borne encephalitis (TBE) is the only tick–transmitted neurological disorder which is of importance in Europe. Especially TBE with clinical predominance of myeloradiculoneuritis (Radl et al. 1983) or meningoradiculoneuritis (Queisser 1962; Reisner 1981) may be mistaken for Bannwarth's syndrome. On the other hand, some patients reported to have suffered from TBE with atypical long incubation periods may have had Lyme neuroborreliosis (Queisser 1962). The diagnosis of TBE, which can reliably be ascertained by serodiagnostic tests (Hoffmann et al. 1983), does not necessarily exclude Lyme borreliosis. Double infections with TBE virus and *B. burgdorferi* have been reported in a few patients (Baumhackl et al. 1986; Kristoferitsch et al 1986; Neubert et al. 1986). This possibility may cause diagnostic problems. Neurological disorders due to other tick–transmitted infections like rickettsiosis or babesiosis are of minor interest in Europe. On rare occasions, such infections have been detected in persons who acquired them outside Europe.

Disk protrusion has been a frequent primary diagnosis in patients with Bannwarth's syndrome (Stiernstedt et al. 1988), and in one study 30 % of such patients underwent myelography or spinal computed tomography (CT) prior to the diagnosis of Bannwarth's syndrome (Pfister and Einhäupl 1988). In another study, 10 out of 17 patients were primarily diagnosed as having a lumbar or cervical root compression syndrome (Dieterle et al. 1989). Especially monoradiculitis may mimic disk herniation, and the correct diagnosis of Bannwarth's syndrome was established in some patients only after unsuccessful neurosurgical intervention (Meier et al. 1989). The absence of typical vertebral symptoms in Bannwarth's syndrome as well as the frequent migrating character of pain may be helpful in the differential diagnosis and CSF analysis will prove the correct diagnosis.

Leukemic or neoplastic meningoradiculopathy must also be considered in the differential diagnosis. In lymphoma, the CSF cell differentiation may appear similar to that in Bannwarth's syndrome, which may cause additional diagnostic problems (Razavi–Encha et al. 1987; Stiernstedt et al. 1988). Amongst viral diseases, herpes zoster radiculitis may be mistaken for Bannwarth's syndrome, but the vesicular eruptions will lead to the correct diagnosis. Another viral disorder which causes severe pain is Bornholm disease. The diagnosis of Guillain–Barre' syndrome, which in some patients may also start with severe pain and hyperesthesia, can be confirmed by the typical CSF albuminocytologic dissociation. Bilateral facial palsy, which is characteristic of Bannwarth's syndrome, can also be found in Guillain–Barre' syndrome as well as in sarcoidosis, tuberculous meningitis, and meningeal carcinosis. In unclear cases, the demonstration of intrathecal specific antibody synthesis may be helpful in confirming Bannwarth's syndrome.

11.3 Definite Atypical Early Lyme Neuroborreliosis

To confirm the diagnosis in patients with an atypical clinical picture, at least two of the following criteria should be fulfilled, provided that other diseases have been excluded:

1. The cultural demonstration of the etiologic agent in the affected organ
2. The intrathecal antibody production against *B. burgdorferi*
3. Previous erythema migrans 3 months or less prior to the onset of the neurological symptoms
4. Typical pathomorphological findings in biopsies, for example, peripheral nerve biopsy or muscle biopsy
5. Clinical response to antibiotic therapy

Stiernstedt et al. (1986) found signs due to CNS involvement in seven out of 46 patients (15%) with acute neurological manifestations of Lyme borreliosis. These signs, which were due to encephalitis or myelitis, included hemiparesis, spastic paraparesis, ataxia, and somnolence. Similar observations were made by Baumhackl et al. (1986), who reported myelitis in three and encephalitis in four of 50 patients with early Lyme neuroborreliosis. Pfister et al. (1986) found encephalitic symptoms in seven out of 80 patients with early Lyme neuroborreliosis. The signs and symptoms included central vestibular dysfunction, dysarthria, emotional lability, and difficulty in concentrating. Three out of 80 patients had clinical signs of involvement of the spinal cord and presented with spastic paraparesis or sensibility disorders due to transverse myelitis. Recently, Pfister et al. (1991) reported a patient with acute encephalitis and marked catatonic symptoms who was initially admitted at the department of psychiatry; diagnosis of Lyme encephalitis was based upon lymphocytic pleocytosis in the CSF, marked intrathecal *B. burgdorferi*–specific antibodies and isolation of *B. burgdorferi* from the CSF; the clinical symptoms completely resolved after intravenous ceftriaxone therapy. Hansen et al. (1986) observed one out of 23 patients with early neurological manifestations of Lyme borreliosis who had acute focal encephalitis with slight hemiparesis. Kohlhepp et al. (1987) reported on encephalitis in four out of 58 (7%) patients with acute neurological manifestations of Lyme borreliosis.

Two patients had acute Parkinson's syndrome associated with *B. burgdorferi* infection. The diagnosis was based on positive CSF IgM antibody titers against *B. burgdorferi*, lymphocytic pleocytosis and clinical improvement after antibiotic therapy. Several case reports showed that cerebral vasculitis with focal neurological deficits may occur during the second stage of the disease (Hänny and Häuselmann 1987; Schmutzhard et al. 1987; Veenendal-Hilbers et al. 1988; May and Jabbari 1990; Kuntzer et al. 1991).

Typical symptoms due to lymphocytic meningitis were seen in 15%–25% of patients with early Lyme neuroborreliosis (Stiernstedt et al. 1986; Pfister et al. 1986; Baumhackl et al. 1986). The main symptoms included headache (27%–54%), stiff neck (13%–18%), and photophobia (3%) (Stiernstedt et al. 1986; Pfister et al. 1985, 1986; Baumhackl et al. 1986). These typical meningitic symptoms were often combined with painful radiculitis. European (Schmutzhard et al. 1986; Schoenen et al. 1989; Reimers et al. 1989) and American (Atlas et al. 1988) authors have reported on patients with myositis associated with *B. burgdorferi* infection. This topic is discussed in chap. 13.

11.4 Suspected Early Lyme Neuroborreliosis

In a Swedish study performed by Hanner et al. (1989), the authors suggested that there is a causal relationship between hearing impairment and *B. burgdorferi* infection. In 17 (17%) out of a total of 98 patients, who had different patterns of hearing dysfunction, elevated serum antibody titers against *B. burgdorferi* were found. The hearing dysfunction included sudden hearing loss, disorders similar to Menière's disease, or hearing loss combined with facial palsy or vertigo. The diagnosis in these patients is doubtful because CSF findings were not mentioned in this study.

The significance of the case reports of Schroeter et al. (1988) and Bendig and Ogilvie (1987) on paralysis of the recurrent nerve and severe encephalopathy, respectively, remains questionable because in the first patient CSF findings were not mentioned, and in the second patient inflammatory CSF changes were found, but there was no intrathecal *B. burgdorferi* antibody production. The diagnosis of further unusual patients reported from the United States by Raucher et al. (1985) and Glauser et al. (1989) on pseudotumor cerebri and Horner's syndrome associated with *B. burgdorferi* infection, respectively, was also only doubtful because neither inflammatory CSF changes nor an intrathecal specific antibody production could be documented.

Schmutzhard et al. (1985) reported on three patients with polyneuritis cranialis associated with *B. burgdorferi* infection. All patients had elevated IgG and IgM antibody titers agains *B. burgdorferi*, but not painful radiculitis or encephalitis.

References

Aberer E, Brunner C, Suchanek G, Klade H, Barbour AG, Stanek G, Lassmann H (1989) Molecular mimicry and Lyme borreliosis: a shared antigenic determinant between *Borrelia burgdorferi* and human tissue. Ann Neurol 26: 732–737

Ackermann R, Rehse–Küpper B, Gollmer E, Schmidt R (1988) Chronic neurologic manifestations of erythema migrans borreliosis. Ann NY Acad Sci 539: 16–23

Asbrink E, Olsson I (1985) Clinical manifestations of erythema chronicum migrans Afzelius in 161 patients. Acta Derm Venereol Stockholm 65: 43–52

Asbrink E, Hovmark A, Olsson I (1986) Clinical manifestations of acrodermatitis chronica atrophicans in 50 Swedish patients. Zentralbl Bakteriol Hyg A 263: 253–261

Atlas E, Novak SN, Duray PH, Steere AC (1988) Lyme myositis: muscle invasion by *Borrelia burgdorferi*. Ann Intern Med 109: 245–246

Bammer HG (1981) Meningomyeloradikulitis nach Zeckenbiß mit Erythem. In: Hopf HC, Poeck K, Schliack H (eds), Neurologie in Praxis und Klinik, Vol 2. Thieme, Stuttgart, New York, pp 4.233–4.235

Bammer H, Schenk K (1965) Meningo–Myelo–Radikulitis nach Zeckenbiß mit Erythem. Dtsch Z Nervenheilkunde 187: 25–34

Bannwarth A (1941) Chronische lymphozytäre Meningitis, entzündliche Polyneuritis und Rheumatismus. Arch Psychiatr Nervenkr 113: 284–376

Baumhackl U, Kristoferitsch W, Sluga E, Stanek G (1986) Neurological manifestation of *Borrelia burgdorferi*–infections: the enlarging clinical spectrum. Zentralbl Bakteriol Hyg A 263: 334–336

Bendig JWA, Ogilvie D (1987) Severe encephalopathy associated with Lyme disease. Lancet I: 681–682

Camponovo F, Meier C (1986) Neuropathy of vasculitic origin in a case of Garin-Bujadoux-Bannwarth syndrome with positive borrelial antibody response. J Neurol 233: 69–72

Dieterle L, Kubina FG, Staudacher T, Büdingen HJ (1989) Neuroborreliose oder Bandscheibenvorfall. Dtsch MedWochenschr 114: 1602–1606

Dumas M (1989) Antibodies against nervous tissue antigens in sera of patients with positive Lyme-serology. 9th Peripheral Nerve Study Group Meeting, Padova (abstract)

Engelhardt A, Grahmann F, Neundörfer B (1989) Vasculitic neuropathy in a case of Bannwarth's syndrome. Zentralbl Bakteriol (Suppl) 18: 136–140

Erbslöh F, Kohlmeyer K (1968) Über polytope Erkrankungen des peripheren Nervensystems bei lymphozytärer Meningitis. Fortschr Neurol Psychiatr 36: 321–342

Felgenhauer K (1982) Differentiation of the humoral immune response in inflammatory diseases of the central nervous system. J Neurol 228: 223–237

Garcia-Monco JC, Coleman JL, Benach JL (1988) Antibodies to myelin basic protein in Lyme disease. J Infect Dis 158:667 (letter)

Glauser TA, Brennan PJ, Galetta SL (1989) Reversible Horner's syndrome and Lyme disease. J Clin Neuro ophthalmol 9: 225–228

Gollmer E, Schmidt R, Kabatzki J, Hartung S, Rehse–Küpper B, Ackermann R (1989) Borrelia meningopolyneuritis Garin–Bujadoux–Bannwarth, catamnesis of 25 cases. Zentralbl Bakteriol (Suppl) 18: 206–211

Graf M, Kristoferitsch W, Baumhackl U, Zeitlhofer J (1986) Electrophysiologic findings in meningopolyneuritis of Garin–Bujadoux–Bannwarth. Zentralbl Bakteriol Hyg (A) 263: 324–327

Habicht GS, Beck G, Benach JL (1988) The role of interleukin-1 in the pathogenesis of Lyme disease. Ann N Y Acad Sci 539:80–86

Hänny PE, Häuselmann HJ (1987) Die Lyme-Krankheit aus der Sicht des Neurologen. Schweiz Med Wochenschr 117:901–915

Halperin JJ, Luft BJ, Anand AK, Roque CT, Alvarez O, Volkman DJ, Dattwyler RJ (1989) Lyme neuroborreliosis: central nervous system manifestations. Neurology 39: 753–759

Hanner P, Rosenhall U, Edström S, Kaijser B (1989) Hearing impairment in patients with antibody production against *Borrelia burgdorferi* antigen. Lancet I: 13–15

Hansen K, Cruz M, Link H (1990) Oligoclonal *Borrelia burgdorferi* – specific IgG antibodies in cerebrospinal fluid in Lyme neuroborreliosis. J Infect Dis 161:1194–1202

Hansen K, Rechnitzer C, Pedersen NS, Arpi M, Jessen O (1986) Borrelia meningitis in Denmark. Zentralbl Bakteriol Hyg (A) 263:348–350

Hellerström S (1930) Erythema chronicum migrans Afzelii. Acta Derm Venereol Stockholm 11: 315–321

Henriksson A, Link H, Cruz M, Stiernstedt G (1986) Immunglobulin abnormalities in cerebrospinal fluid and blood over the course of lymphocytic meningoradiculitis (Bannwarth's syndrome). Ann Neurol 20: 337–345

Hoffmann H, Heinz FX, Dippe H (1983) ELISA for IgM and IgG antibodies against tick–borne encephalitis virus: quantification and standardization of results. Zentral Bakteriol Hyg A 255: 448–455

Hörstrup P, Ackermann R (1973) Durch Zecken übertragene Meningopolyneuritis (Garin–Bujadoux, Bannwarth). Fortschr Neurol Psychiatr 41: 583–606

Kohler I, Thoden U (1987) Schmerzsyndrom bei Zeckenübertragenen Borrelieninfektionen des Nervensystems. Klinik und Differentialdiagnose Schmerz 1: 107–113

Kohlhepp W, Mertes HG, Oschmann P, Rohrbach E (1987) Akute und chronische Erkrankungen bei zeckenvermitteler Borreliose. Nervenarzt 58: 557–563

Kristoferitsch W (1989) Neuropathien bei Lyme–Borreliose. Springer, Wien New York

Kristoferitsch W, Spiel G, Wessely P (1983) Zur Meningopolyneuritis (Garin–Bujadoux, Bannwarth). Klinik und Laborbefunde. Nervenarzt 54: 640–646

Kristoferitsch W, Stanek G, Kunz C (1986) Doppelinfektion mit Frühsommermeningoenzephalitis (FSME) – Virus und *Borrelia burgdorferi.* Dtsch Med Wochenschr 111: 861–864

Krüger H, Englert D, Pflughaupt KW (1981) Demonstration of oligoclonal immunoglobulin G in Guillain–Barré syndrome and lymphocytic meningoradiculitis by isoelectric focusing. J Neurol 226: 15–24

Krüger H, Reuss K, Pulz M, Rohrbach E, Pflughaupt KW, Martin R, Mertens HG (1989) Meningoradiculitis and encephalomyelitis due to *Borrelia burgdorferi:* a follow–up study of 72 patients over 27 years. J Neurol 236: 322–328

Krüger H, Kohlhepp W, König S (1990) Follow up of antibiotically treated and untreated neuroborreliosis. Acta Neurol Scand 82: 59–67

Kuntzer T, Bogousslavsky J, Miklossy J, Steck AJ, Janzer R, Regli F (1991) Borrelia Rhombencephalomyelopathy. Arch Neurol 48: 832–836

Maida E, Kristoferitsch W, Spiel G (1986) Liquorveränderungen bei Meningoradiculitis Garin–Bujadoux–Bannwarth. Nervenarzt 57: 149–152

Martin R, Ortlauf J, Sticht–Groh V, Bogdahn U, Goldmann SF, Mertens HG (1988) *Borrelia burgdorferi*–specific and autoreactive T–cell lines from cerebrospinal fluid in Lyme radiculomyelitis. Ann Neurol 24: 509–516

May EF, Jabbari B (1990) Stroke in neuroborreliosis. Stroke 21: 1232–1235

Meier C, Grahmann F, Engelhardt A, Dumas M (1989) Peripheral nerve disorders in Lyme Borreliosis. Nerve biopsy studies from eigth cases. Acta Neuropath 79: 271–278

Meier C, Reulen HJ, Huber P, Mummentahler M (1989) Meningoradiculoneuritis mimicking vertebral disk herniation. A "neurosurgical" complication of Lyme borreliosis. Acta Neurochir (Wien) 98: 42–46

Meyer–Rienecker HJ, Hitzschke B (1977) Lymphocytic meningoradiculitis (Bannwarth's syndrome). In: Vinken PJ, Bruyn GW(eds) pp 571–586 Handbook of clinical neurology vol 34, Amsterdam

Murray N, Kristoferitsch W, Stanek G, Steck AJ (1986) Specificity of CSF antibodies against components of *Borrelia burgdorferi* in patients with meningopolyneuritis Garin-Bujadoux-Bannwarth. J Neurol 233: 224–227

Neubert U, Krampitz HE, Engl H (1986) Microbiological findings in erythema (chronicum) migrans and related disorders. Zentralbl Bakteriol Hyg A 263: 237–252

Pachner AR, Steere AC (1985) The triad of neurologic manifestations of Lyme disease: meningitis, cranial neuritis, and radiculoneuritis. Neurology 35: 47–53

Pfister HW, Einhäupl KM (1988) Die neurologisch manifeste Lyme–Borreliose. In: Elger CE, Dengler R (eds) Jahrbuch der Neurologie. Biermann S 63–70

Pfister HW, Einhäupl KM, Preac–Mursic V, Wilske B (1985) Similarity of neurologic manifestations of *Borrelia burgdorferi* infections in America and Europe. Neurology 35: 1393–1394

Pfister HW, Einhäupl KM, Wilske B, Preac–Mursic V (1986) Bannwarth's syndrome and the enlarged neurological spectrum of arthropod borne borreliosis. Zentralbl Bakteriol Hyg A 263: 343–347

Pfister HW, Wilske B, Preac–Mursic V, Einhäupl KM (1989) Clinical and serological follow up of patients with Bannwarth's syndrome: comparison of patients with and without penicillin treatment. In: Zentralbl Bakteriol Suppl 18: 276–279

Pfister HW, Rieder G, Förderreuther S, Wilske B, Preac–Mursic V, Kapfhammer HP, Einhäupl KM (1991) Atypical neurologic manifestations of Lyme Borreliosis. Presented at the 17th International Congress of Chemotherapy 23 – 28 June 1991, Berlin (abstract)

Queisser H (1962) Beobachtungen über einige Fälle von Zeckenenzephalitis in Deutschland. Munch Med Wochenschr 104: 2288–2293

Radl B, Ladurner G, Lechner H, Stunzer D (1983) Klinik und Verlauf der Frühsommermeningoenzephalitis (FSME, TBE). Neuropsychiatr Clin 2: 131–135

Raucher HS, Kaufman DM, Goldfarb J, Jacobsen RI, Roseman B, Wolff RR (1985) Pseudotumor cerebri and Lyme disease: a new association. J Pediatr 6: 931–933

Razavi–Encha F, Fleury–Feith J, Gherardi R, Bernaudin JF (1987) Cytologic features of cerebral spinal fluid in Lyme disease. Acta Cytologica 31: 439–440

Reimers CD, Pongratz DE, Neubert U, Pilz A, Hübner G, Naegele M, Wilske B, Duray PH, DeKoning J (1989) Myositis caused by *Borrelia burgdorferi*. Report of 4 patients. J Neurol Sci 91: 215–226

Reisner H (1969) Die akute Polyradikulitis. Z Nervenheilkunde 27: 269–276

Reisner H (1981 Clinic and treatment of tick–borne encephalitis (TBE). Introduction. In: Kunz C (ed) Facultas, Vienna, pp 1–5 Tick–borne encephalitis. International symposium, Baden/Vienna 1979

Rohmer F, Collard M, Jesel M, Warter JM, Coquillat G, Class JJ (1974) Les meningoradiculitides: donnèes cliniques, electromyographiques et ètiologiques á propos de 36 observations. Limites nosologiques. Rev Neurol 130: 415–431

Schaltenbrand G (1962) Radiculomyelomeningitis nach Zeckenbiß. Munch Med Wochenschr 104: 829–834

Schaltenbrand G (1967) Durch Arthropoden übertragene Infektionen der Haut und des Nervensystems. Verh Dtsch Ges Inn Med 72: 975–1005

Schmidt R, Ackermann R (1985) Durch Zecken übertragene Meningo–Polyneuritis (Garin–Bujadoux, Bannwarth). Erythema chronicum migrans Krankheit des Nervensystems. Fortschr Neurol Psychiatr 53: 145–153

Schmidt R, Kabatzki J, Hartung S, Ackermann R (1985) Erythema–migrans–Borreliose in der Bundesrepublik Deutschland. Dtsch med Wochenschr 110: 1803–1807

Schmutzhard E, Stanek G, Pohl P (1985) Polyneuritis cranialis associated with *Borrelia burgdorferi*. J Neurol Neurosurg Psychiatr 48: 1182–1184

Schmutzhard E, Willeit J, Gerstenbrand F (1986) Meningopolyneuritis Bannwarth with focal nodular myositis. Klin Wochenschr 64: 1204–1208

Schmutzhard E, Pohl P, Stockhammer G, Kleedorfer B, Stanek G (1988) Unusual neurological manifestations of Lyme borreliosis. Ann NY Acad Sci 539: 495–496

Schoenen J, Sianard–Gainko J, Carpentier M, Reznik M (1989) Myositis during *Borrelia burgdorferi* infection (Lyme disease). J Neurol Neurosurg Psychiatry 52: 1002–1005

Schroeter V, Belz GG, Blenk H (1988) Paralysis of recurrent laryngeal nerve in Lyme disease. Lancet II: 1245

Sigal LH, Tatum AH (1988) Lyme disease patients' serum contains IgM antibodies to *Borrelia burgdorferi* that cross–react with neuronal antigens. Neurology 38: 1439–1442

Sköldenberg B, Stiernstedt G, Garde A, Kolmodin G, Carlström A, Nord CE (1983) Chronic meningitis caused by a penicillin–sensitive microorganism? Lancet II: 75–78
Sterman AB, Nelson S, Barclay P (1982) Demyelinating neuropathy accompanying Lyme disease. Neurology 32: 1302–1305
Stiernstedt G, Sköldenberg B, Vandvik B, Hederstedt B, Garde A, Kolmodin G, Jörböck H, Svenungsson B (1984) Chronic meningitis and Lyme disease in Sweden. Yale J Biol Med 57: 491–497
Stiernstedt GT, Granström M, Hederstedt B, Sköldenberg B (1985) Diagnosis of spirochetal meningitis by enzyme–linked immunosorbent assay and indirect immunofluorescence assay in serum and cerebrospinal fluid. J Clin Microbiol 21: 819–825
Stiernstedt GT, Sköldenberg B, Garde A, Kolmodin G, Jörbeck H, Svenungsson B, Carlström A (1986) Clinical manifestations of borrelia infections of the nervous system. Zentralbl Bakteriol Hyg A 263: 289–296
Stiernstedt G, Gustafsson R, Karlsson M, Svenungson B, Sköldenberg B (1988) Clinical manifestation and diagnosis of neuroborreliosis. Ann NY Acad Sci 539: 46–55
Suchanek G, Kristoferitsch W, Stanek G, Bernheimer H (1986) Anti-myelin antibodies in cerebrospinal fluid and serum of patients with meningopolyneuritis Garin-Bujadoux-Bannwarth and other neurological diseases. Zentralbl Bakteriol Hyg A 263: 160–168
Vallat JM, Leboutet MJ, Lubeau M, Dumas M, Hugon J, Corvisier N (1984) Tick bite neuropathy: an analysis of nerve biopsies from seven patients. Neurology 34 (Suppl 1): 180–181
Vallat JM, Hugon J, Lubeau M, Leboutet MJ, Dumas M, Desproges–Gotteron (1987) Tick–bite meningoradiculoneuritis: clinical, electrophysiologic, and histologic findings in 10 patients. Neurology 37: 749–753
Veenendaal–Hilbers JA, Perquin WVM, Hoogland PH, Doornbos L (1988) Basal meningovasculitis and occlusion of the basilar artery in two cases of *Borrelia burgdorferi* infection. Neuroradiology 38: 1317–1319
Weber K (1974) Erythema–chronicum–migrans–Meningitis – eine bakterielle Infektionskrankheit? Munch Med Wochenschr 116: 1993–1998
Weber K, Neubert U (1986) Clinical features of early erythema migrans disease and related disorders. Zentralbl Bakteriol Hyg (A) 263: 209–228
Weber T, Jürgens S, Lüer W (1987) Cerebrospinal fluid immunoglobulins and virus–specific antibodies in disorders affecting the facial nerve. J Neurol 234: 308–314
Wenig C (1975) Chronische lymphozytäre Meningitis mit polytoper Beteiligung des Nervensystems (nach Zeckenbiß?). Psychiatr Neurol Med Psychol (Leipz) 27: 592–600
Wolf G (1970) Über die chronische lymphozytäre Meningitis unter dem Bilde der Polyneuritis (Bannwarth). Fortsch Neurol Psychiatr 38: 221–234
Wulff C, Hansen K, Strange P, Trojyborg W (1983) Multiple mononeuritis and radiculitis with erythema, pain, elevated CSF protein and pleocytosis (Bannwarth's syndrome). J Neurol Neurosurg Psychiatry 46: 485–490

12 Joint Manifestations

P. HERZER

12.1 The Delayed Recognition of Lyme Arthritis

Though joint symptoms had been previously mentioned in case reports of patients in Europe with erythema chronicum migrans (Stadelmann 1934; Weber 1974) or lymphocytic meningitis (Bannwarth 1941; Schaltenbrand 1949; Erbslöh und Kohlmeyer 1968), the association of arthritis with erythema migrans and neurological disease was not recognized until Lyme arthritis was described in North America (Steere et al. 1976, 1977a, b). However, joint abnormalities in patients with acrodermatitis chronica atrophicans (ACA) had been repeatedly described in the European dermatological literature since 1922 (Jessner 1922; Jessner and Loewenstamm 1924; Hövelborn 1931; Benjamowitsch and Maschkilleisson 1933; Gans 1933; Gans and Landes 1952; Hopf and Klingmüller 1966; Müller 1969) and also in a case report from North America (Sweitzer and Laymon 1935); this even led to the term "*Akrodermatitis atrophicans arthropathica*" being proposed (Gans and Landes 1952). However, a causal relation of joint and bone abnormalities with ACA had been called in question (Hauser 1955). In the earlier rheumatological literature, there is only one short reference to five cases of ACA with polyarticular involvement of joints beneath affected skin (Schilling 1970). Nowadays, in view of the clinical spectrum of Lyme borreliosis, the association of ACA and Lyme arthritis is obvious. Furthermore, recent reports (Asbrink et al. 1986; Hovmark et al. 1986) have confirmed the earlier presumptions of specific joint and bone abnormalities under the affected skin in ACA.

The first case reports on the association of erythema migrans and arthritis in Europe in parallel with Lyme disease in North America were published in 1980 (Ackermann et al. 1980; Hewitt et al. 1980). Notwithstanding the following suggestions that Lyme arthritis might be underestimated in Europe (Herzer et al. 1983; Stiernstedt et al. 1985; Herzer and Wilske 1986; Hovmark et al. 1986), the conviction that arthritis is a rare manifestation of European Lyme borreliosis had become firmly established. Misconceptions concerning the rheumatological manifestations of the disease in Europe were mainly due to unjustified conclusions from seroepidemiological studies (Stanek et al. 1985; Muhlemann and Wright 1987) which did not take into consideration the relative frequency of diagnostic indications for serological testing. A diagnostic odyssey in our first patients with Lyme arthritis reflected the lack of awareness of this new rheumatological disease. Thus, there is good reason to assume that patients with Lyme borreliosis in Europe initially represented a bias of dermatological and neurological manifestations. In contrast, the definition of Lyme disease in North America resulted from the identification of patients with Lyme arthritis and/or erythema migrans only (Steere et al. 1977b). Comparisons of the disease in Europe and North America should take account of the different evolution of Lyme borreliosis on both sides of the Atlantic (Herzer 1988). Lastly, it should be recalled that antibiotic therapy of erythema migrans was widely practised in Europe since 1951 (Hollström 1951) which may have prevented endemics of arthritis as observed in Lyme.

A first comprehensive analysis of joint manifestations in European Lyme borreliosis has been previously delineated in a monograph (Herzer 1989).

12.2 Epidemiology

12.2.1 Incidence of Lyme arthritis

In accordance with the widespread occurrence of Lyme borreliosis in Europe, patients with Lyme arthritis have meanwhile been reported from many countries, e.g., Belgium (Bigaignon et al. 1987; Huaux et al. 1988), Britain (Macallan et al. 1987), Czechoslovakai (Valešová and Trnavský 1990), Denmark (Kryger et al. 1990), France (Illouz and Hewitt 1981; Mallecourt et al 1982; Charmot et al. 1982; Dougados et al. 1983), Germany (Ackermann et al. 1980; Herzer et al. 1983, 1986; Weber et al. 1983; Goebel et al. 1985; Herzer and Wilske 1986; Herzer 1989, 1991), Italy (Crovato et al. 1985), the Netherlands (Blaauw and van der Linden 1990), Sweden (Stiernstedt and Granström 1985; Hovmark et al. 1986), Switzerland (Gerster et al. 1981; Müller 1984; Satz et al. 1986; Schaad et al. 1986), and the Russia (Dekonenko 1988).

However, no reliable data are available on the true incidence of Lyme arthritis. Moreover, the relative frequency of joint manifestations in Lyme borreliosis cannot be assessed because of bias in patient selection from various medical specialities. Since antibiotic therapy is mandatory, there is no large body of data on the spontaneous course of the disease. Still, it has been reported that only one of 16 Swedish patients developed arthritis after spontaneous healing of erythema migrans (Asbrink and Olsson 1985). In contrast, of 55 North American patients with erythema migrans who were not treated with antibiotics, 10 (18%) developed arthralgias alone and 34 (62%) began to have arthritis (Steere et al. 1987). Moreover, our studies have shown that rheumatologists will probably more often see cases of Lyme arthritis without extraarticular features of the disease (Herzer 1989). Anyhow, by the time that erythema migrans and the neurological manifestations come to be treated more readily with anti-

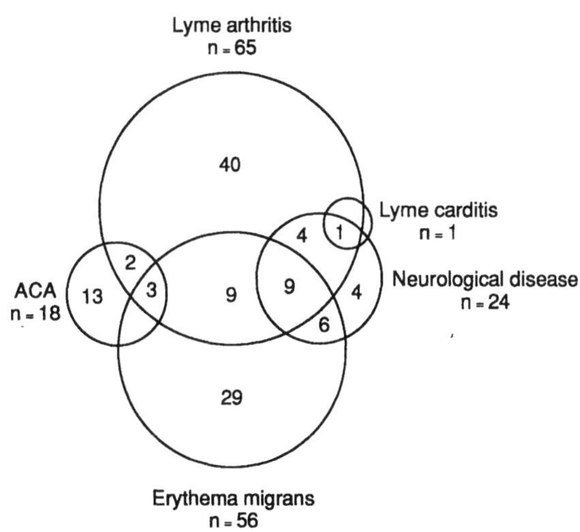

Fig. 1 Clinical manifestations in 120 consecutive patients with Lyme borreliosis (Herzer 1991). Lyme arthritis without a history of well–defined Lyme borreliosis or concurrent extrarticular disease manifestations was diagnosed by the detection of specific IgG antibodies (titers ≥ 98% percentile of 275 healthy controls), by the exclusion of differential diagnoses, and only where the clinical features corresponded to those seen in "definitive" Lyme arthritis with well–defined manifestations of Lyme borreliosis

biotics, cases of classical Lyme arthritis with preceding erythema migrans or neurological disorders will become increasingly rare.

An analysis of 120 consecutive patients with Lyme borreliosis seen at our general outpatient department revealed arthritis to be the most frequent disease manifestation (Fig. 1) (Herzer 1991). This specification merely reflects the spectrum of the disease in a medical clinic, in which rheumatology is a special discipline. Nevertheless, our approach to Lyme borreliosis allows us to draw a parallel with the initial description of Lyme disease in North America, in which only 13 of 51 patients were reported to have had erythema migrans before they developed arthritis (Steere et al. 1977a). Thus, our experience indicates that conjectures as to the rare occurrence of arthritic complications in European Lyme borreliosis compared to the disease in North America may be attributed to the earlier lack of awareness of Lyme arthritis in Europe. Meanwhile, in our experience it balances each other out that Lyme arthritis is still not recognized and that Lyme arthritis is overdiagnosed due to misinterpretation of serological findings.

12.2.2 Age and Sex Incidence

Lyme arthritis affects both children and adults. The first comprehensive study on Lyme borreliosis in children from Europe (Schaad et al. 1986) refers to patients aged between 1 and 15 years (median 9 years). However, the age of the patients with joint manifestations was not specified in this report. In 65 patients with Lyme arthritis

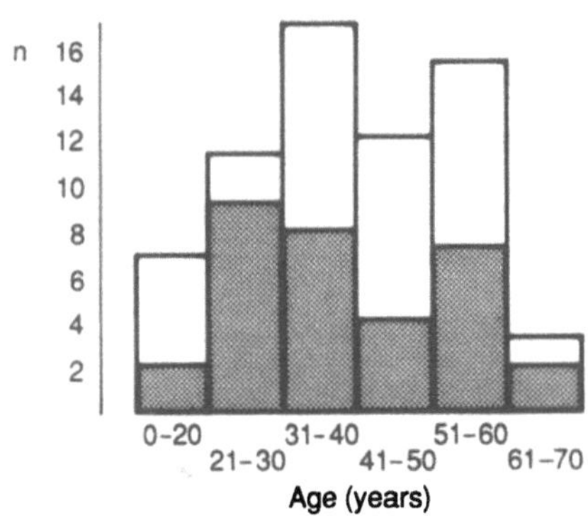

Fig. 2. Age at onset and sex incidence (*shaded areas*, women; *white areas*, men) in 65 patients with Lyme arthritis

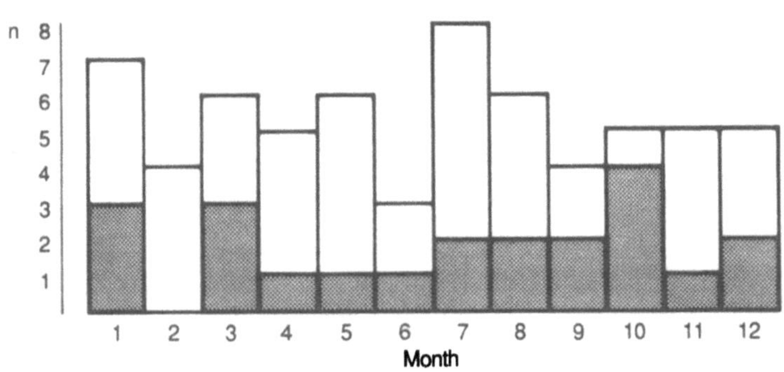

Fig. 3. Seasonal onset of Lyme arthritis (shaded areas, arthritis with erythema migrans and/or neurological disease; white areas, arthritis only) in 64 patients

(Herzer 1991), the age at onset ranged from 13 to 70 years (median 39 years). Susceptibility to Lyme arthritis does not seem to be influenced by sex–related factors (Fig.2).

12.2.3 Seasonal Onset

Since the latent period between infection and onset of Lyme arthritis is highly variable and mostly runs to several months, there is no seasonal peak in the occurrence of Lyme arthritis (Fig.3).

12.3 Clinical Description

Several distinct patterns of rheumatic manifestations can be seen throughout the different stages of Lyme borreliosis. Essentially, the spectrum of articular manifestations can be classified in three categories:
1. Arthralgias (musculoskelatal pain) without objective findings,
2. Arthritis (intermittent or chronic) with objective physical findings
of synovial thickening or joint effusion, and
3. Chronic joint and bone involvement under affected skin in ACA.

12.3.1 Arthralgias

Intermittent and migrating musculoskeletal pain affecting joints, entheses, tendons, muscles, bones, or the spine without objective clinical findings usually becomes apparent early in the course of the illness. Arthralgias have been reported to occur in as many as 48 % of patients with erythema migrans in North America (Steere et al. 1983a), but only in 22 % at the most of European patients (Weber and Neubert 1986). A median duration of 6 months has been reported for the joint symptoms in the European patients, most of whom have been treated with antibioties (Weber and Neubert 1986). In North American patients who had no antibiotic therapy for erythema migrans, the total duration of intermittent musculoskeletal pain ranged from 1 month to 6 years (mean, 3.1 years) (Steere et al. 1987).

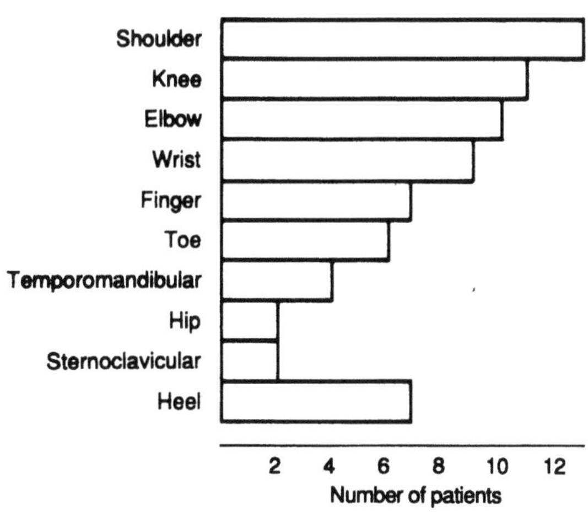

Fig. 4. Arthralgias preceding and/or coexisting with Lyme arthritis in 25 patients; frequency of joints (and heels) affected by pain without objective clinical findings

The succession or the coexistence of intermittent attacks of musculoskeletal pain and arthritis are particular indications of Lyme arthritis. In 25 of our 65 patients with Lyme arthritis, episodes of severe pain in joints or periarticular sites had either preceded arthritis for several weeks and months (*n*=8), had preceded and continued after the onset of arthritis (*n*=5), or had developed as late as arthritis (*n*=12) (Herzer 1991). Particular episodes lasted from some hours to several days and were separated by days to months of remission. Episodes of arthralgias sometimes alternated with attacks of arthritis. Large and small joints (Fig.4) were affected in an often migratory pattern, but commonly only one or two sites were affected at one time. In one patient who had not received antibiotic treatment, attacks of joint pain still occurred 9 years after remission of Lyme arthritis (Herzer 1989). In principle, this pattern of musculoskeletal involvement corresponds to that described in North America (Steere et al. 1987) and has also been observed by other European rheumatologists (Valešová and Trnavský 1990).

12.3.2 Arthritis

12.3.2.1 Onset of Arthritis

When patients from North America who did not receive antibiotic treatment for erythema migrans were followed up for a mean duration of 6 years (range, 3 – 8 years) arthritis occurred from 4 days to 2 years after disease onset (mean, 6 months) (Steere et al. 1987). In our patients, the latent period between tick bites or erythema migrans and the onset of arthritis ranged from 10 days to 16 months (mean, 2.5 months and 3 months, respectively); in this study, tick bites and skin lesions suggestive of erythema migrans were only taken into account when they did not precede the onset of arthritis by more than 2 years (Herzer 1991). On the example of patients in whom tick bite and erythema migrans had preceded arthritis by 4 and 12 years, respectively, it has been illustrated that it is difficult or even impossible to determine the maximum duration of the latent period between infection and onset of arthritis (Herzer 1989). Particularly in cases with long intervals between recallable tick bites or early disease manifestations and arthritis, the eventuality of an unnoticed (re)infection or the succession of unrelated incidents has to be considered.

It has been emphasized by various European authors (Herzer 1989; Kryger et al. 1990; Valešová and Trnavský 1990) that Lyme arthritis often begins at the same extremity that was affected by a tick bite or erythema migrans; for instance, this correlation has been observed in 15 of 18 patients in whom erythema migrans had preceded arthritis (Herzer 1991).

12.3.2.2 Pattern of Joint Involvement and Course of Arthritis

Brief attacks of mono- or oligoarticular arthritis, primarily in the knee (Fig. 5) have been emphasized as the characteristic feature of Lyme arthritis (Steere et al. 1977a, b). One to 12 episodes of arthritis during a period of 1 – 8 years have occurred in 28 of 55 patients with untreated erythema migrans; six of the 55 patients developed chronic arthritis defined as continuous arthritis for 1 year or more (Steere et al. 1987).

The clinical features in our patients with Lyme arthritis are summarized in Table 1. In the patients with intermittent arthritis, the episodes lasted from 1 day to 5 months and were separated by days or up to 4 years of complete remission. Individual patients experienced 12 attacks of arthritis at the most, and the maximum period of recurrent arthritis was 10 years. Ten patients had unremitting joint swelling for 1 year or more. In four of these patients, an intermittent arthritis had preceded chronic joint involvement. Mono- or oligoarticular arthritis migrated to as many as nine joints. One patient had symmetrical involvement of the wrists and the knees. In the other patients with

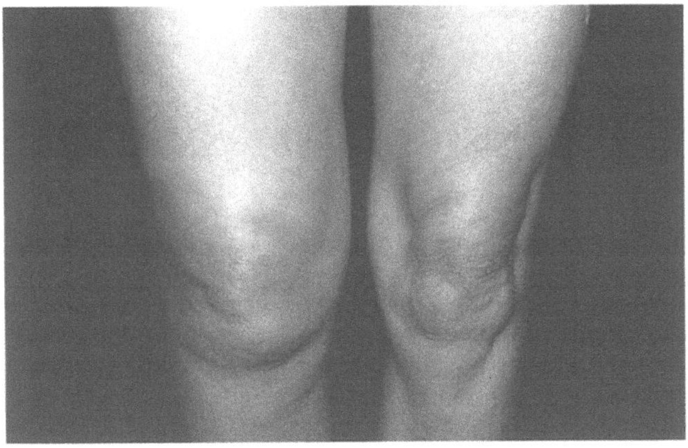

Fig. 5. Knee arthritis, the most common presentation of Lyme arthritis. Generally, swelling is out of proportion to pain

Table 1. Clinical features of Lyme arthritis in 65 patients (Herzer 1991)

		n
Course	Intermittent	55
	First intermittent, then chronic	4
	Unremitting (chronic)	6
Pattern of joint involvement	Monoarticular	39[a]
	Oligoarticular	21
	Polyarticular[b]	4
	Dactylitis (Sausage digits)	15
	Heel swelling[c]	6

[a] Onset of arthritis was monoarticular in 55 patients
[b] At least four joints affected simultaneously (without dactylitis beeing taken into account)
[c] One patient had heel swelling only

polyarthritis, large and small joints were affected in an asymmetrical pattern. None of the patients with polyarticular involvement developed chronic arthritis, and none had subcutaneous nodules. Special emphasis has been laid on the occurrence of dactylitis (sausage digits) (Fig. 6) and heel swelling, which we have indicated to be a feature of Lyme arthritis (Herzer and Wilske 1986; Herzer 1986). It should be also stressed that unilateral diffuse hand and finger swelling was a striking feature in all three of our patients, in whom the onset of arthritis could be related to the early stage of the disease, either because of the simultaneous existence of erythema migrans (Fig. 7), neurological disease, or respectively seroconversion (Herzer 1989). One patient had heel swelling only. The most common joint involved was the knee (Fig. 8). Affected knees were generally very swollen, only moderately warm, but not red. Pain was often out of proportion to the degree of swelling. Baker cysts were found in 12 of 24 patients investigated by ultrasound. Attacks in other joints were occasionally very painful and resembled gout, e.g., podagra (Herzer et al. 1986), or septic arthritis. Chronic arthritis affected the knee (n=7), the ankle, the hip, and finger joints (sausage digit).

In conclusion, our observations suggest that there are no fundamental differences in

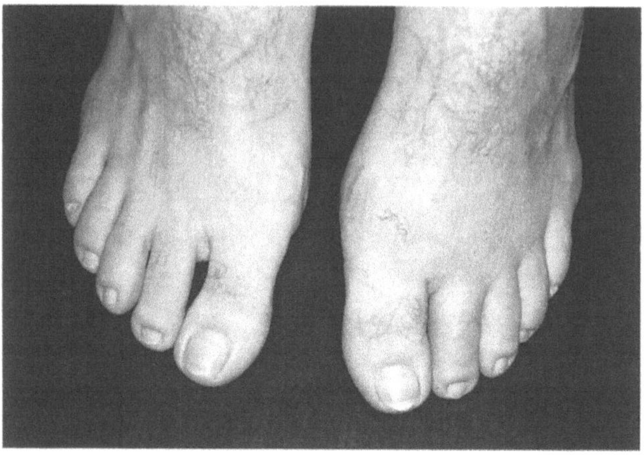

Fig. 6. Sausage toes in a patient with Lyme arthritis

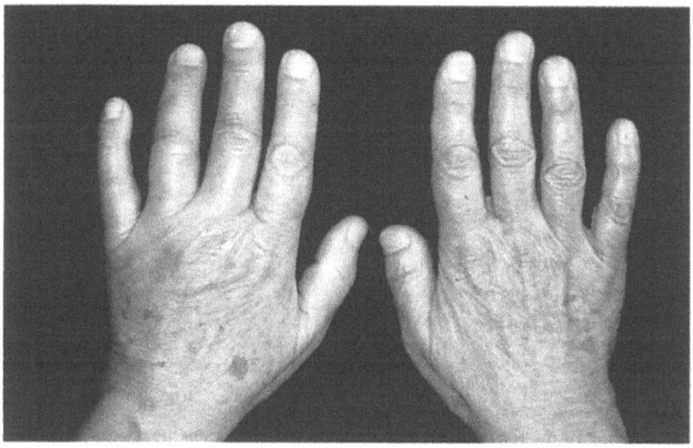

Fig. 7. Arthritis affecting the left hand of a patient who simultaneously had erythema migrans on the left shoulder

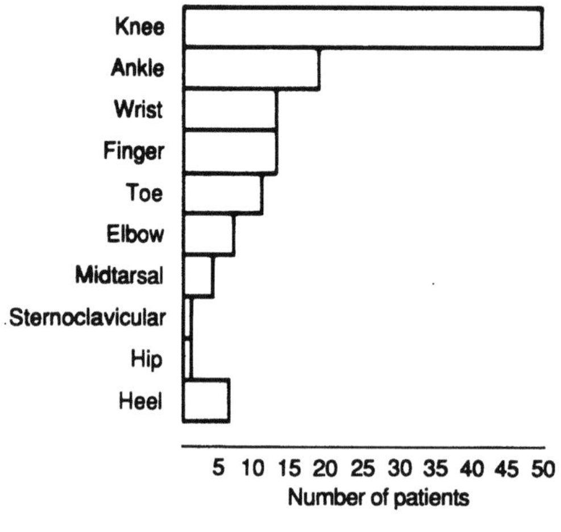

Fig. 8. Frequency of joints (and heels) involved in 65 patients with Lyme arthritis (Herzer 1991)

the clinical characteristics of Lyme arthritis in Europe and North America (Steere et al. 1987). In principle, most of the other European reports also correspond to these descriptions. However, it has been postulated by European authors that the spectrum of Lyme arthritis also includes Reiters' syndrome (Weyand and Goronzy 1989), and involvement of the sacroiliac joints (Weyand and Goronzy 1989; Valešová and Trnavský 1990). With regard to the often uncertain diagnostic criteria in individual patients, however, it should be mentioned that these features were not seen in a prospective monitoring of 55 patients with untreated erythema, 34 of whom developed arthritis (Steere et al. 1987). Meanwhile, we have also seen one patient with Reiter's syndrome and another single patient with sacroiliitis who had significant immunoglobulin G (IgG) antibody titers against *Borrelia burgdorferi*. However, we did not deem them as having Lyme arthritis since we never observed the respective criteria in patients with preceding extraarticular manifestations of Lyme borreliosis.

12.3.2.3 Constitutional Symptoms

Although constitutional symptoms mostly occur early in the disease (Steere et al. 1983a), they occasionally outlast the initial stages and may accompany arthritis (Steere et al. 1987). On the other hand, constitutional symptoms also may develop as late as arthritis (Herzer 1989). Altogether, nine of our 65 patients with Lyme arthritis also had fatigue, malaise, fever, or night sweats.

12.3.3 Chronic Joint and Bone Involvement Under Affected Skin in ACA

Arthralgias or typical Lyme arthritis may precede ACA or may coincide with ACA (Gans 1933; Weber et al. 1984; Asbrink et al. 1986; Hovmark et al. 1986; Satz et al. 1986; Herzer 1989). In addition, distinctive joint and bone abnormalities under the affected skin may develop in patients with long-standing ACA (Asbrink et al. 1986; Hovmark et al. 1986). These joint abnormalities are distinguished by finger and toe abnormalities due to subluxations or luxation of the joints without obvious erosive lesions and thus may be considered a Jaccoud-like arthropathy in Lyme borreliosis (Herzer 1989). In some patients, X-rays have revealed periosteal thickening under affected skin (see Chap. 14).

12.4 Radiological Findings

A wide range of radiological findings have been described in patients from North America with severe arthritic manifestations (Lawson and Steere 1985). Soft tissue swelling, evidence of effusion and edema of the infrapatellar fat pad were the most common findings. In some cases, tendons and ligaments were found to be thickened, calcified, or ossified. Calcifications of the articular cartilage and of the menisci were also associated with Lyme arthritis. Some patients had developed typical changes of inflammatory arthritis, including juxta-articular osteoporosis, cartilage loss, and cortical or marginal bone erosions. A few patients demonstrated a more degenerative pattern consisting of cartilage loss, subarticular sclerosis, and osteophytosis. Most abnormalities have been noted about the knees, and significant changes were also seen in hips and shoulders.

Apart from findings of soft tissue swelling and joint effusions, radiographic abnormalities were noted in nine of our 65 patients with Lyme arthritis (Herzer 1991). The most frequent finding was marked juxta-articular osteoporosis (Fig. 9) in the knee (n=6). In one patient with dactylitis, diffuse osteoporosis of a finger was noted; in this

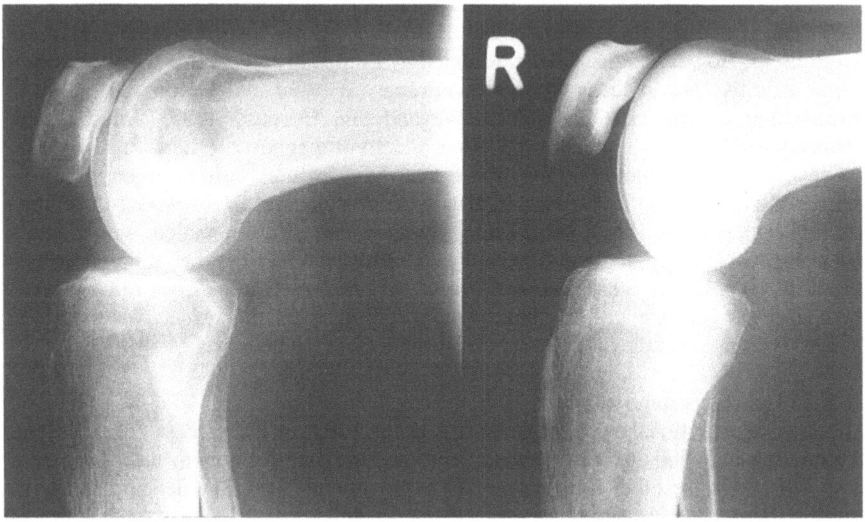

Fig. 9. Juxta-articular osteoporosis in chronic Lyme arthritis (Herzer 1989)

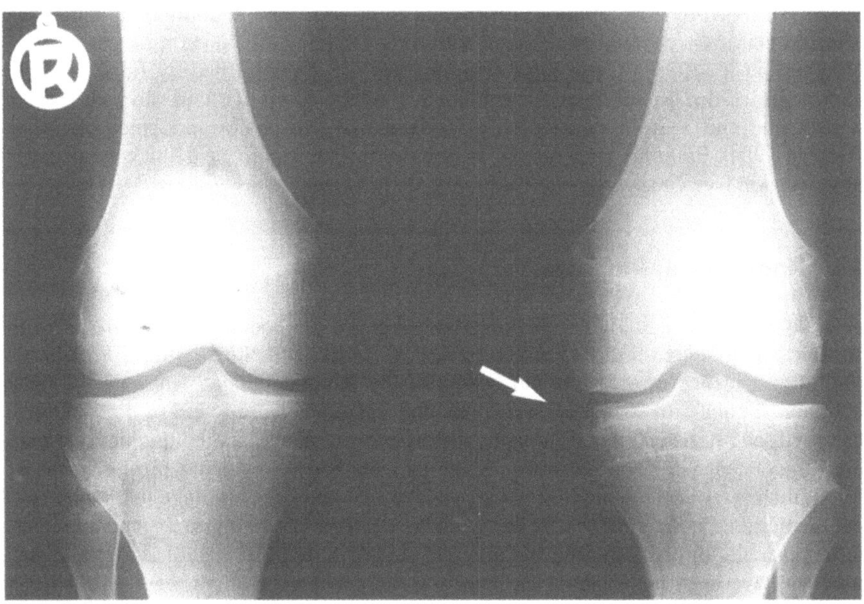

Fig. 10. Marginal erosion at the left tibia in a patient with chronic knee arthritis (Herzer et al. 1986)

patient, the inflammatory involvement of both the finger joints and the interarticular tissues was shown by scintigraphy (Herzer 1989). Two patients developed sharply localized erosions at the margin of the tibia (Fig. 10). Loss of the articular cartilage and subarticular cysts in the hip were seen in one patient. Two patients with heel swelling demonstrated calcaneal erosions at the insertion of the Achilles tendon or the plantar aponeurosis. In a patient with intermittent knee arthritis, calcific densities in the quadriceps tendon were seen (Herzer 1989); this patient also had marked calcific densities in her shoulders, which were affected by intermittent arthralgias. As already mentioned, radiological evidence of sacroiliitis has been reported in patients with presumed Lyme arthritis (Weyand and Goronzy 1989; Kryger et al. 1990; Valesova and Trnavsky 1990). However, it is still not clear whether Lyme arthritis indeed affects the sacroiliac joints.

12.5 Laboratory Findings

12.5.1 Serological Diagnosis

In Lyme arthritis, microscopic demonstration (Snydman et al. 1986) or cultivation (Schmidli et al. 1988) of *B. burgdorferi* from synovial fluid or synovium specimens are low yield procedures (Steere et al. 1984; Herzer 1989). Thus, serological testing is essentially the only choice for laboratory confirmation of the diagnosis.

Although increased IgM antibody titers against *B. burgdorferi* may be demonstrated in 24 % (Herzer 1989) to 41 % (Steere et al. 1983b) of the patients with Lyme arthritis, the significance of positve IgM titers is modified by the observation of false-positive IgM reactions in immunofluorescence (IFT) studies (Wilske et al. 1984; Herzer 1989). IgM antibodies against the 41–kDa protein of *B. burgdorferi* have been found in the sera of patients with various rheumatic diseases (Craft et al. 1986). Seroconversion will be only observed in the rare cases of early Lyme arthritis (Herzer 1989). Thus, serodiagnosis of Lyme arthritis essentially depends on the detection of specific IgG antibodies. When Lyme arthritis is present, titers of specific IgG antibodies generally show maximum levels, and serial determinations do not reveal significant changes of the titers for months and years, irrespective of the course of arthritis or antibiotic treatment (Herzer 1989). Therefore, it is not possible to differ between serological findings of diagnostic relevance and specific antibodies due to remote subclinical infections.

The potential and limits of serological testing for Lyme arthritis have been reported by us in detail (Herzer 1989; Herzer et al. 1989). In general, significant IgG antibody titers are confirmatory findings in patients with preceding early manifestations of Lyme borreliosis ("definitve" Lyme arthritis). However, often the question arises whether or not significant IgG antibody titers agains *B. burgdorferi* permit the diagnosis of Lyme arthritis in patients without evidence of extra–articular manifestations of Lyme borreliosis. When sera from 185 patients whose history did not suggest pathognomonic manifestations of Lyme borreliosis and who suffered from undifferentiated arthritis were studied by immunofluorescence test (IFT), by IFT after absorption with *Treponema phagedenis* (IFT–Abs), and by enzyme–linked immunosorbent assay (ELISA), significant IgG antibody titers were found in 35 patients. Although our assays were highly specific (>97%), the diagnosis of Lyme arthritis had to be questioned in nine of the 35 patients on account of potential differential diagnoses and in three patients because of an incomplete rheumatological check–up. These findings clearly demonstrate that even highly specific serological findings should not substitute for a careful clinical evaluation or for the exclusion of potential differential diagnoses.

Serological investigations of paired serum/synovial fluid samples did not reveal significant differences of specific antibody titers (Herzer 1989; Herzer et al. 1989). By comparing specific antibody titers with total IgG or albumin in synovial fluid and serum, we found indices of >1 in seven of eight patients investigated; this gave rise to the assumption of intraarticular antibody production. However, these indices were lower than those regarded significant for intrathecal production of antibodies against *B. burgdorferi* in patients with Bannwarth's syndrome (Stiernstedt et al. 1985; Wilske et al. 1986) and thus proved not to be helpful for diagnostic confirmation. However, the analysis of joint fluid and serum from a patient with Lyme arthritis by agarose isoelectricfocusing and immunoblot revealed intrasynovial production of oligoclonal anti-*B. burgdorferi* IgG antibodies (Cruz et al. 1990).

12.5.2 Nonspecific Laboratory Findings

Lyme arthritis is one of the rare inflammatory joint diseases in which routine laboratory parameters are often completely normal; Table 2 shows the findings in our patients. Only about half of the patients have elevated erythrocyte sedimentation rates above 20 mm/h (Steere et al. 1987; Herzer et al. 1986; Herzer 1989; Kryger et al. 1990; Valešová and Trnavský 1990). Some patients have white cell counts slightly above 10.000/µl and shifts to the left in the differential count (Steere et al. 1977b; Herzer et al. 1986). Low serum iron levels can be noted (Herzer et al. 1986). Elevated total serum IgM levels, circulating immune complexes, and cryoglobulins have been reported by several authors (Steere et al. 1977a, 1979c, 1987; Hardin et al. 1979; Herzer et al. 1986; 1989). Low titer rheumatoid factors can be seen in some patients (Kujala et al. 1987; Herzer 1989). The finding of antinuclear antibodies in patients with Lyme arthritis (Goebel et al. 1985; Valešová and Trnavský 1990) is of questionable significance.

12.5.3 Synovial Fluid Analysis

Joint fluid white cell counts in our patients ranged from 1500 to 35000 cells/mm^3 (Herzer 1989); a range of 500 to 110 000 cells/mm^3 (median 26 000) has been reported

Table 2. Percentage of abnormal nonspecific laboratory findings in Lyme arthritis (Herzer 1991)

Findings	%
Circulating immune complexes (Clq binding)	50
ESR > 20 mm/h[a]	43
Total IgM > 280 mg/dl[b]	38
Low serum iron	34
Cryoglobulins	28
Leukocytosis (10.000 to 13.800/µl)	20
Shift to the left (5 – 20 band forms)	12
Rheumatoid factors (low titer)	16

ESR, erythrocyte sedimentation rate; IgM, immunoglobulin M
[a] ESR range 2 – 65 mm/h, median 15 mm/h
[b] Total IgM range 74 – 960 mg/dl, median 243 mg/dl

by other authors (Steere et al. 1977, 1979c). Generally, polymorphonuclear cells predominate (average 70% – 80%). Rarely, eosinophilia has been found (Kay et al. 1988). Total protein commonly ranges from 3.5 to 5.6 (Steere et al. 1977a, 1987; Herzer 1989). Cryoglobulins (Steere et al. 1979c) and abnormal C1q binding consistent with antigen–antibody complexes (Hardin et al. 1979) are commonly present in synovial fluid.

12.6 Genetic Susceptibility?

At first, chronic Lyme arthritis was thought to be associated with the B cell alloantigen DR2 (Steere et al. 1979a). Thereupon, further studies of class I and class II HLA-antigens by the same North American authors showed that the frequency of this immunogenetical marker was also significantly increased in patients with Lyme arthritis that persisted only for 1 month to 1 year, as well as in patients with neurological complications of the disease (Steere et al. 1979b). However, HL–DR typing of patients with Bannwarth's syndrome in Austria did not reveal any significant HLA–DR association (Kristoferitsch and Mayr 1984), whereas in Austrian patients with ACA HLA–DR2 was found significantly more often than in controls (Kristoferitsch et al. 1986). A study from Germany of 183 persons with various stages of an infection with *B. burgdorferi* distinguished HLA–Cw3 or the phenotype combination HLA–A2 and –Cw3 as risk factors for the development of Lyme borreliosis (Pflüger et al. 1989). These authors even showed a trend for a negative association between HLA–DR2 and –DR3 with the progression of the disease. However, this study did not include cases of Lyme arthritis.

Determination of the HLA types of 80 consecutive patients with Lyme arthritis did not enable Steere et al. (1988) to reproduce their initial results. This time, they reported on a significant association of chronic Lyme arthritis with HLA–DR4 and moreover considered the importance of both the DR4 and DR3 loci in chronic Lyme arthritis. In the final analysis, chronic Lyme arthritis has been associated with HLA–DR4 and HLA-DR2 (Steere et al. 1990).

Identification of the HLA–A, –B,–C, and –DR specificities in 40 patients with Lyme arthritis and ten patients with ACA revealed that the frequencies of these HLA antigens did not significantly differ from those in controls (Herzer 1989). A reevaluation of our data with due regard to patients with chronic Lyme arthritis did not give evidence of an immunogenetic basis for the development of chronic Lyme arthritis, either (Herzer et al. 1990).

12.7 Diagnostic Approaches and Differential Diagnoses

A clinical history of well–defined Lyme borreliosis, in particular of erythema migrans and neurological disease manifestations, is the most reliable key to Lyme arthritis. However, in patients with long intervals between preceding extra–articular symptoms suggestive of Lyme borreliosis and arthritis, the significance of such a history may be questionable. The great range of presentations of Lyme arthritis can make recognition difficult. Moreover, Lyme arthritis has many features in common with several other rheumatic diseases. On account of these ambiguous clinical criteria, the improbability of being able to culture *B. burgdorferi* from joint fluid or synovium, and the pitfalls of the serological diagnosis, a careful exclusion of potential differential diagnoses (Table 3) is a essential prerequisite for the diagnosis of Lyme arthritis.

Table 3. Essential differential diagnoses of Lyme arthritis with regard to its particular clinical features

Clinical feature	Differential diagnosis
Acute monoarticular onset	Crystal–induced arthritis (gout, pseudogout) Septic arthritis Sarcoid arthritis
Migratory arthritis	Gonococcal arthritis Viral arthritis Rheumatic fever
Intermittent course	Intermittent hydarthrosis Palindromic rheumatism Whipple's disease
Pattern of joint involvement	Reactive arthritis Psoriatic arthritis Peripheral arthritis in ankylosing spondylitis Enteropathic arthritis HLA–B27–positive juvenile oligoarthritis Antinuclear antibody positive juvenile oligoarthritis Early rheumatoid arthritis Systemic lupus erythematosus
Arthralgias	Psychogenic rheumatism Fibromyalgia Osteoarthrosis

The acute presentation of Lyme arthritis can be mistaken for infective (septic) or crystal–induced arthritis. Particularly when attacks of Lyme arthritis affect the first metatarsophalangeal joint alone, it may be confused with gout; significant IgG antibodies against *B. burgdorferi* and hyperuricemia may be conflicting diagnostic criteria in a patient with acute monoarticular arthritis unless crystals are demonstrated in synovial fluid. Since erythema nodosum is not mandatory in acute sarcoid arthritis, which commonly affects the ankles, it may be wrongly diagnosed as Lyme arthritis in seropositive individuals.

Migratory arthritis in Lyme borreliosis is similar to that in rheumatic fever, disseminated gonococcal infection, and viral infections, Diffuse hand swelling, as in early Lyme arthritis, may also occur in viral infections, e.g., in parvovirus B19 infection.

With regard to the intermittent course, Lyme arthritis may be mistakenly diagnosed as intermittent hydarthrosis or palindromic rheumatism. Recurrent episodes of arthritis may precede pathognomonic symptoms of Whipple's disease.

The pattern of joint involvement in Lyme arthritis closely resembles that seen in seronegative spondarthropathies. Heel involvement and sausage digits, previously regarded as unique markers of the seronegative spondyloarthropathies, can also be seen in Lyme arthritis. Considering the problematic interpretation of serological findings, a definitive differentiation between Lyme arthritis and postenteric or postveneric reactive arthritis can be difficult. In patients who have psoriatic skin lesions and who are also positive for *B. burgdorferi*–specific antibodies, the definitive diagnosis may remain undecided. Though ocular involvement is infrequent in Lyme borreliosis, this aspect can also cause diagnostic difficulties. Conjunctivitis may be noted in early Lyme borreliosis (Steere et al. 1983a), but the succession of conjunctivitis and arthritis commonly indicates Reiter's syndrome. Uveitis has also been seen in Lyme borreliosis (Steere et al. 1985; Jacobs et al. 1986). Generally, the association of arthritis and uveitis is suggestive of HLA–B27–positive spondyloarthropathies and uveitis is a typical feature of antinuclear antibody–positive pauciarticular juvenile arthritis.

Chronic symmetrical small joint involvement, typical for rheumatoid arthritis, is an unusual presentation for Lyme arthritis. However, on account of the variable patterns of early rheumatoid arthritis, it can be difficult to separate Lyme arthritis from rheumatoid arthritis. It should also be mentioned that Lyme borreliosis and systemic lupus erythematosus have many clinical features in common.

Musculoskeletal pain in Lyme borreliosis may be mistaken for psychogenic rheumatism or fibromyalgia. Conversely, fibromyalgia may be wrongly diagnosed as Lyme borreliosis. In contrast to the distinctive intermittent and migratory pattern of musculoskeletal pain in Lyme borreliosis, fibromyalgia is characterized by more generalized, chronic pain and by symmetrical tender points.

In *conclusion*, diagnosing Lyme arthritis resembles a jigsaw puzzle in which clinical history, typical clinical presentation, laboratory findings, and the exclusion of other rheumatic entities are the elements to be assembled. The great variety of arthritides with similar clinical features or the fact that many clinical complexes or syndromes in rheumatology are poorly defined by clinical criteria only, e.g., intermittent hydarthrosis and palindromic rheumatism, easily explains why Lyme arthritis obviously existed for a long time unrecognized.

References

Ackermann R, Runne U, Klenk V, Dienst C (1980) Erythema chronicum migrans mit Arthritis. Dtsch Med Wochenschr 105:1779–1781

Asbrink E, Olsson I (1985) Clinical manifestations of erythema migrans Afzelius in 161 patients. Acta Derm Venereol (Stockh) 65:43–52

Asbrink E, Brehmer–Andersson E, Hovmark A (1986) Acrodermatitis chronica atrophicans – a spirochetosis. Clinical and histopathological picture based on 32 patients; course and relationship to erythema chronicum migrans Afzelius. Am J Dermatopathol 8:209–219

Bannwarth A (1941) Chronische lymphocytäre Meningitis, entzündliche Polyneuritis und "Rheumatismus". Ein Beitrag zum Problem "Allergie und Nervensystem" in zwei Teilen. Arch Psychiatr Nervenkr 113:284–376

Benjamowitsch E, Maschkilleisson LN (1933) Weitere Beiträge zur Frage über die Atrophie der Haut. III. Acrodermatitis chronica atrophicans und ihre Beziehungen zur Sclerodermia. Acta Derm Venereol (Stockh) 14:313–341

Bigaignon G, Goubau P, Desmyter J, Vandepitte J (1987) Lyme borreliosis in Belgium. Lancet I:557

Blaauw AAM, Nohlmans, van der Linden Sj (1990) Arthritis due to Lyme borreliosis in the Netherlands. IVth international conference on Lyme borreliosis, Stockholm, Sweden, 18–21, June, 1990, abstract W/TH–P77

Charmot G, Rodhain F, Perez C (1982) Un cas d'arthrite de Lyme observé en France. Presse Med 11:207–208

Craft JE, Fischer DK, Shimamota GT, Steere AC (1986) Antigens of *Borrelia burgdorferi* recognized during Lyme disease. Appearance of a new immunoglobulin M response and expansion of the immunoglobulin G response late in the illness. J Clin Invest 78:934–939

Crovato F, Nazarri G, Fumarola D, Rovetta G, Cimmino MA, Bianchi G (1985) Lyme disease in Italy: first reported case. Ann Rheum Dis 44:570–571

Cruz M, Hansen K, Faarvang KL, Link H (1990) Intrasynovial production of oligoclonal specific antibodies in Lyme arthritis. N Engl J Med 322:1889–1890

Dekonenko EP (1988) Clinical manifestation of tick–borne erythema in Russia. Ann NY Acad Sci 539:452–453

Dougados M, Kahan A, Vannier A, Amor B (1983) Arthrite de Lyme. Deux nouveaux cas de français. Rev Rhum Mal Osteoartic 50:299–302
Erbslöh F, Kohlmeyer K (1968) Über polytope Erkrankungen des Nervensystems bei lymphozytärer Meningitis. Fortschr Neurol Psychiatr 35:321–342
Gans O (1933) Akrodermatitis atrophicans und Sklerodermie. Zentralbl Haut Geschlechtskr 45:678–679
Gans O, Landes E (1952) Akrodermatitis atrophicans arthropathica. Hautarzt 3: 151–155
Gerster JC, Guggi S, Perroud H, Bovet R (1981) Lyme arthritis appearing outside the United States: a case report from Switzerland. Br Med J 283:951–952
Goebel KM, Krause A, Goebel HH (1985) Lyme–Krankheit mit transitorischer Autoimmunität. Inn Med 12:209–214
Hardin JA, Steere AC, Malawista SE (1979) Immune complexes and the evolution of Lyme arthritis. Dissemination and localisation of abnormal C1q binding activity. N Engl J Med 301:1358–1363
Hauser W (1955) Zur Kenntnis der Akrodermatitis chronica atrophicans. (Unter besonderer Berücksichtigung der Veränderungen an den hautnahen Lymphknoten, des Knochenmarks, der Serumeiweißverhältnisse sowie der Ätiologie und der Pathogenese). Arch Dermatol Syph (Berl) 199:350–393
Herzer P (1986) Die Klinik der Lyme–Arthritis: Unterschiedliche Befallsmuster und Krankheitsverläufe. Z Rheumatol 45:175
Herzer P (1988) Lyme arthritis in Europe: comparisons with reports from North America. Ann Rheum Dis 47:789–791
Herzer P (1989) Lyme–Borreliose. Epidemiologie, Ätiologie, Diagnostik, Klinik und Therapie. Steinkopff, Darmstadt
Herzer P (1991) Joint manifestations of Lyme borreliosis in Europe. Scand J Infect Dis – Suppl 77:55–63
Herzer P, Wilske B (1986) Lyme arthritis in Germany. Zentralbl Bakteriol Hyg (A) 263:268–274
Herzer P, Schattenkirchner M, Zöllner N (1983) Lyme–Arthritis – eine zu selten bedachte Diagnose ? Verh Dtsch Ges Inn Med 89:299–302
Herzer P, Wilske B, Preac–Mursic V, Schierz G, Schattenkirchner M, Zöllner N (1986) Lyme arthritis: clinical features, serological, and radiographic findings of cases in Germany. Klin Wochenschr 64:206–215
Herzer P, Wilske B, Schewe S, Preac–Mursic V, Schierz G (1989) Serodiagnostik der Lyme-Arthritis. Akt Rheumatol 14:125–129
Herzer P, Schewe S, Scholz S, Albert E (1990) HLA–Antigens in late Lyme borreliosis (Lyme arthritis and acrodermatitis chronica atrophicans). IV international conference on Lyme borreliosis, Stockholm, Sweden, 18–21, June, 1990, abstract M/TU–P54
Hewitt J, Illouz G, Giroin D, Dolivo M, Leibowitch M, Gutman L (1980) Erythema chronicum migrans suivi d 'une polyarthrite séronégative. Premier cas européen d'arthrite de Lyme? Journées de la société française de Dermatologie, Paris, p. 87
Hollström E (1951) Successful treatment of erythema chronicum migrans Afzelii. Acta Derm Venereol (Stockh) 31:235–243
Hopf HC, Klingmüller G (1966) Acrodermatitis chronica atrophicans mit Gelenkbeteiligung und neurologischen Ausfällen. Nervenarzt 36:364–366
Hövelborn C (1931) Gelenkveränderungen bei Acrodermatitis chronica atrophicans. Arch Dermatol Syph (Berl) 164:349–356
Hovmark A, Asbrink E, Olsson I (1986) Joint and bone involvement in Swedish patients with *Ixodes ricinus*–borne borrelia infection. Zentralbl Bakteriol Hyg (A) 263:275–284
Huaux JP, Bigaignon G, Stadtsbaeder S, Zangerlé PF, Nagant de Deuxchaisnes Ç (1988) Pattern of Lyme arthritis in Europe: report of 14 cases. Ann Rheum Dis 47:164–165
Illouz G, Hewitt J (1981) A propos de l'arthrite de Lyme. Polyarthrite inflammatoire après un érythème annulaire migrant. Rev Rhum Mal Osteoartic 48:813–815
Jacobs JC, Stevens M, Duray PH (1986) Lyme disease simulating septic arthritis. JAMA 256:1138–1139
Jessner M (1922) Weiterer Beitrag zur Kenntnis der Acrodermatitis chronica atrophicans. Arch Dermatol Syph (Berl) 139:294–305

Jessner M, Loewenstamm A (1924) Bericht über 66 Fälle von Acrodermatitis chronica atrophicans. Dermatol Wochenschr 79:1171–1177

Kay J, Eichenfield AH, Athreya BH, Doughty RA, Schumacher HR (1988) Synovial fluid eosinophilia in Lyme disease. Arthritis Rheum 31:1384–1389

Kristoferitsch W, Mayr WR (1984) HLA–DR in meningopolyneuritis of Garin-Bujadoux-Bannwarth: contrast to Lyme desease. J Neurol 231:271–272

Kristoferitsch W, Mayr WR, Partsch H, Neumann R, Stanek G (1986) HLA–DR in Lyme borreliosis. Lancet II:278

Kryger P, Hansen K, Vinterberg H, Pedersen FK (1990) Lyme borreliosis among Danish patients with arthritis. Scand J Rheumatol 19:77–81

Kujala GA, Steere AC, Davis IV JS (1987) IgM rheumatoid factor in Lyme disease: correlation with disease activity, total serum IgM, and IgM antibody to *Borrelia burgdorferi*. J Rheumatol 14:772–776

Lawson JP, Steere AC (1985) Lyme arthritis: radiologic findings. Radiology 154:37–43

Macallan DC, Hughes CA, Bradlow A (1987) Lyme arthritis in southern England. Br Med J 294: 1062–1063

Mallecourt J, Landureau M, Wirth AM (1982) La maladie de Lyme. Un cas clinique observé en Eure–et–Loire. Presse Med 11:39–41

Muhlemann MF, Wright DJM (1987) Emerging pattern of Lyme disease in the United Kingdom and Irish Republic. Lancet I:260–262

Müller H (1969) Akrodermatitis atrophicans arthropathica mit Pseudosklerodermie, Sklerodermia circumscripta und muskulärer Beteiligung. Z Haut Geschlechtskr 44:1–12

Müller W (1984) Die Lyme–Arthritis (Erythema–migrans–Arthritis). Schweiz Med Wochenschr 114:265–269

Pflüger K–H, Reimers CD, Neubert U, Meisel C, Trapp B, Leititis J, Völker B, Münchhoff P, Litzenberger J, Holthausen H, Pongratz E (1989) Lyme borreliosis and possible association with HLA–antigens. Tissue Antigens 33:375–381

Satz N, Ott A, Zogg F, Knoblauch M (1986), Die Erythema–migrans–Krankheit. Das klinische Spektrum der Infektion mit *Borrelia burgdorferi*. Schweiz Med Wochenschr 116:763–769

Schaad UB, Flüeler U, Schaub H, Suter H, Vischer D, Caflisch U, Tschumi A, Wick H, Vest M, Durrer D (1986) Durch *Ixodes–ricinus*–Spiochäten (*Borrelia burgdorferi*) verursachte Krankheitsbilder (Lyme–Krankheit) bei pädiatrischen Patienten in der Schweiz. Schweiz Med Wochenschr 116:1426–1430

Schaltenbrand G (1949) Chronische aseptische Meningitis. Nervenarzt 20:433–442

Schilling F (1970) Die symptomatischen Arthritiden. Heilkunst 83:1–6

Schmidli J, Hunziker T, Moesli P, Schaad UB (1988) Cultivation of *Borrelia burgdorferi* from joint fluid three months after treatment of facial palsy due to Lyme borreliosis. J Infect Dis 158:905–906

Snydman DR, Schenkein DP, Berardi VP, Lastavica CC, Pariser KM (1986) *Borrelia burgdorferi* in joint fluid in chronic Lyme arthritis. Ann Intern Med 104:798–800

Stadelmann R (1934) Ein Beitrag zum Krankheitsbild des Erythema chronicum migrans Lipschütz. Inauguraldissertation, University of Marburg

Stanek G, Wewalka G, Groh V, Neumann R, Kristoferitsch W (1985) Differences between Lyme disease and European arthropod–borne *Borrelia* infections. Lancet I:401

Steere AC, Malawista SE, Snydman DR, Andiman WA (1976) A cluster of arthritis in children and adults in Lyme, Connecticut. Arthritis Rheum 19:824

Steere AC, Malawista SE, Snydman DR, Shope RE, Andiman WA, Ross MR, Steele FM (1977a) Lyme arthritis. An epidemic of oligoarticular arthritis in children and adults in three Connecticut communities. Arthritis Rheum 20:7–17

Steere AC, Malawista SE, Hardin JA, Ruddy S, Askenase PW, Andiman WA (1977b) Erythema chronicum migrans and Lyme arthritis. The enlarging clinical spectrum. Ann Intern Med 86:685–698

Steere AC, Gibofsky A, Pattaroyo ME, Winchester FJ, Hardin JA, Malawista SE (1979a) Chronic Lyme arthritis. Clinical and immunogenetic differentiation from rheumatoid arthritis. Ann Intern Med 90:896–901

Steere AC, Gibofsky A, Hardin JA, Winchester RJ, Malawista SE (1979b) Lyme arthritis: immunologic and immunogenetic markers. Arthritis Rheum 22:622–623

Steere AC, Hardin JA, Ruddy S, Mummaw JG, Malawista SE (1979c) Lyme arthritis. Correlation of serum and cryoglobulin IgM wit activity, and serum IgG with remission. Arthritis Rheum 22:471–483
Steere AC, Bartenhagen NH, Craft JE, Hutchinson GJ, Newman JH, Rahn DW, Sigal LH, Spieler PN, Stenn KS, Malawista SE (1983a) The early clinical manifestations of Lyme disease. Ann Intern Med 99:76–82
Steere AC, Grodzicki RL, Kornblatt AN, Craft JE, Barbour AG, Burgdorfer W, Schmid GP, Johnson E, Malawista SE (1983b) The spirochetal etiology of Lyme disease. N Engl J Med 308:733–740
Steere AC, Grodzicki RL, Craft JE, Shrestha M, Kornblatt AN, Malawista SE (1984) Recovery of Lyme disease spirochetes from patients. Yale J Biol Med 57:107–110
Steere AC, Duray PH, Kauffmann DJH, Wormser GP (1985) Unilateral blindness caused by infection with the Lyme disease spirochete, *Borrelia burgdorferi*. Ann Intern Med 103:382–384
Steer AC, Schoen RT, Taylor E (1987) The clinical evolution of Lyme arthritis. Ann Intern Med 107:725–731
Steere AC, Feld J, Winchester R (1988) Association of chronic Lyme arthritis with increased frequencies of DR4 and DR3. Arthritis Rheum 31 (Suppl):S98
Steere AC, Dwyer E, Winchester R (1990) Association of chronic Lyme arthritis with HLA-DR4 and HLA–DR2. N Engl J Med 323:219–223
Stiernstedt G, Granström M (1985) *Ixodes ricinus* spirochete infection as the cause of potinfectious arthritis in Sweden. Scand J Rheumatol 14:336–342
Stiernstedt G, Granström M, Hederstedt B, Sköldenberg B (1985) Diagnosis of spirochetal meningitis by enzyme–linked immunosorbent assay and indirect immunofluorescence assay in serum and cerebrospinal fluid. J Clin Microbiol 21:819–825
Sweitzer SE, Laymon CW (1935) Acrodermatitis chronica atrophicans. Arch Derm Syph (Chicago) 35:196–212
Valešová M, Trnavský K (1990) Joint manifestations of Lyme borreliosis in Czechoslovakian patients. Z Rheumatol 49:192–196
Weber K (1974) Erythema–chronicum–migrans–Meningitis – eine bakterielle Infektionskrankheit ? Munch Med Wochenschr 116:1993–1998
Weber K, Neubert U (1986) Clinical features of early erythema migrans disease and related disorders. Zentralbl Bakteriol Hyg (A) 263:209–228
Weber K, Puzik A, Becker T (1983) Erythema–migrans–Krankheit. Beitrag zur Klinik und Beziehung zur Lyme–Krankheit. Dtsch Med Wochenschr 108:1182–1190
Weber K, Schierz G, Wilske B, Preac–Mursic V (1984) Zur Klinik und Ätiologie der Acrodermatitis chronica atrophicans. Hautarzt 35:571–577
Weyand CM, Goronzy JJ (1989) Immune responses to *Borrelia burgdorferi* in patients with reactive arthritis. Arthritis Rheum 32:1057–1064
Wilske B, Schierz G, Preac–Mursic V, Weber K, Pfister HW, Einhäupl K (1984) Serological diagnosis of erythema migrans disease and related disorders. Infection 12:331–337
Wilske B, Schierz G, Preac–Mursic V, von Busch K, Kühbeck R, Pfister HW, Einhäupl K (1986) Intrathecal production of specific antibodies against *Borrelia burgdorferi* in patients with lymphocytic meningoradiculitis (Bannwarth's syndrome). J Infect Dis 153:304–314

13 Myalgia, Myositis, and Fasciitis

C. D. REIMERS and E. SCHMUTZHARD

13.1 Clinical Findings

13.1.1 Definitions

Myalgia is a pain localized in the muscle bulk.
The hallmark of a *myositis* is the presence of inflammatory infiltrates in the muscle. From the histopathological point of view, inflammatory myopathies can be divided into three major groups. A purely interstitial myositis is characterized by predominantly lymphohistiocytic infiltrates in the interstitium without any or with only slight nonspecific changes in muscle fibres. In a focal nodular myositis, localized, often perivascular infiltrates with fiber degenerations in their neighborhood are found. A polymyositis is defined by a diffuse infiltration of the interstitium combined with fiber necroses.

A *fasciitis* is characterized by inflammatory infiltrates in the muscle fascia. Typical features of one particular kind of fasciitis, eosinophilic fasciitis (Shulman syndrome; Shulman 1974), are pain, stiffness, swelling, induration of the extremities, and blood eosinophilia. In about half of the patients, eosinophils can be demonstrated in the fascia. Symptoms are symmetrical in 90% of the patients, and, in the majority immunoglobulin G (IgG) levels and erythrocyte sedimentation rate are markedly elevated.

13.1.2 Age

All of the European patients suffering from *Borrelia burgdorferi* myositis or fasciitis seen up to now were adults. Their median age was 57 years (range 22 – 70 years).

13.1.3. Sex Ratio

Three men and five women suffering from *B. burgdorferi* myositis as well as one woman and two men with a fasciitis due to a *B. burgdorferi* infection have been observed in Europe up to now.

13.1.4 Clinical Description of Manifestations

13.1.4.1 Myalgias

Myalgias are frequent complaints in stage I and II Lyme borreliosis in Europe as well as in the United States. Steere et al. (1983,1986) observed myalgias in 43% of 314 patients with erythema migrans. Some of these patients reported generalized aches and stiffness in muscles and joints. Others complained about severe cramping pain, particularly in the calves, thighs, and lower and upper back. The pain was usually migratory and sometimes lasted only hours in a given location. During examination, a few patients had localized muscle tenderness to palpation. There was no muscle weakness.

Usually signs and symptoms appeared intermittently and usually disappeared within several weeks. Weber and Neubert (1986) reported myalgias in 18 of their 104 German patients with erythema migrans. The pain was generalized in three of them and localized in the muscles adjacent to the skin lesion or migratory in the others. Sigal et al. (1990) reported fibromyalgic complaints in the course of Lyme borreliosis. They assumed that this pain might be caused by disturbed sleep due to encephalopathy. Antibiotics were effective only if there was coexistent active Lyme borreliosis.

13.1.4.2 Myositis

The first myopathy caused by *B. burgdorferi* proven by electrophysiological and histological examination has been described by Schmutzhard et al. (1986). These authors reported on a woman suffering from meningoradiculoneuritis (Bannwarth's syndrome) accompanied by a myositis. From 1987 to 1989, Atlas et al. (1988) and Duray and Steere (1988) in the United States of America as well as Midgard and Hofstad (1987), Reimers et al. (1988,1989), Detmar et al. (1989), and Schoenen et al. (1989) in Europe presented further patients with myositis wich was obviously caused by a *B. burgdorferi* infection. Up to now, Atlas et al. (1988) and Reimers et al. (1989) have succeeded in demonstrating a few spirochetes in a muscle specimen by silver staining.

In 1952, Gans and Landes described intramuscular lymphoplasmocellular infiltrates in a patient with acrodermatitis chronica atrophicans. In 1968, Müller reported on

Table 1. Clinical features in *Borrelia burgdorferi* myositis (European cases)

Reference	Cutaneous manifestations	Neurological manifestations
Detmar et al. (1989)	acrodermatitis chronica atrophicans	feeling of weakness of the legs; results of a neurological examination not reported
Midgard and Hofstad (1987)	"exanthema"	meningitis, polyradiculitis with quadriceps atrophy
Reimers et al. (1988, 1989) case 1	acrodermatitis chronica atrophicans	weakness of proximal leg muscles; mild neuropathy
Reimers et al. (1989) case 2	acrodermatitis chronica atrophicans	creatine kinase level elevated; pathological electromyogram
CD Reimers (unpublished data)	possibly erythema	lymphocytic meningo-radiculoneuritis
CD Reimers (unpublished data)	—	weakness of proximal arm muscles and tibialis anterior
Schmutzhard et al. (1986)	erythema	lymphocytic meningo-radiculoneuritis
Schoenen et al. (1989)	erythema	headache, myalgias, weakness of neck flexor, and bulbar and proximal limb muscles

another patient with acrodermatitis chronica atrophicans, pseudoscleroderma, localized scleroderma, and polyarthritis accompanied by an interstitial myositis. These inflammatory myopathies were probably also caused by *B. burgdorferi*.

We have observed five patients, four women and one man, 52 – 66 years old, with a histopathologically proven *B. burgdorferi* myositis. The clinical features of all patients suffering from histologically proven *B. burgdorferi* myositis are summarized in Table 1. Patients 1 – 4 and patient 6 were in the chronic stage of the disease, the remaining patients became ill within a few weeks or months after infection. In both patients with lymphocytic meningoradiculoneuritis, a myositis was considered because of an elevated serum creatine kinase (CK) activity. In patient 2, described by Reimers et al. (1989), magnetic resonance imaging studies of the calves gave further hints of muscular involvement. In another male patient (Schmutzhard et al., unpublished data) suffering from acrodermatitis chronica atrophicans of the left arm, elevated CK levels and a myopathic electromyographic pattern in the biceps and triceps brachii muscle point to an accompanying myopathy. A muscle biopsy was not taken.

Two of the three patients with *B. burgdorferi* myositis seen in the United States of America (Atlas et al. 1988; Reimers et al. 1989) were characterized by muscle weakness, and two of them complained about muscular pain. One patient reported, that he had had a rash for about 14 days. The three American patients suffered from an acute illness, and their general condition was impaired.

13.1.4.3 Fasciitis

Grahmann et al. (1989) presented a patient with diffuse fasciitis, interstitial myositis in the superficial layers, and eosinophilia followed by an acrodermatitis chronica atrophicans after 3 years. Stanek et al. (1987) described a second patient with Shulman's syndrome following three tick bites 4 months previously. The man suffered from progressive swelling and induration of the skin resulting in flexion contractures. Ferradini et al. (1987) doubt that Stanek's patient was really a patient with Shulman's syndrome as eosinophilia in the blood and fascia and IgG hypergammaglobulinemia were absent. Sepp et al. (1988) saw a patient with Shulman's syndrome complicated by a carpal tunnel syndrome in a patient with B. burgdorferi infection. The patient presented morphea–like lesions and diffuse swelling and induration of both forearms. Duray and Steere (1988) and Duray (1989) observed further patients with Lyme-associated Shulman's syndrome in the United States. Reimers et al. (1989) reported on a patient with noneosinophilic fasciitis due to a *B. burgdorferi* infection (see Sect. 13.2.2.2).

13.2 Special Procedures

13.2.1 Electrophysiological Studies

In four of seven European patients sensory and/or motor nerve conduction velocities were moderately reduced, indicating neuropathy (patients 3,4,5 and 7; Table 1).

Electromyography revealed spontaneous activity (fibrillations and positive sharp waves) in all but two patients (patients 2 and 6; Table 1). At voluntary innervation, two patients showed a myopathic pattern (polyphasic motor units of short duration and low amplitudes), three exhibited neuropathic changes (long–duration, high–amplitude motor units), and two a mixture of both patterns. In patient 2 of Reimers et al. (1989) there was an interference pattern of normal motor units.

13.2.2 Myopathology

13.2.2.1 Myositis

All but one (patient 8; Table 1) of the muscle specimens contained inflammatory infiltrations consisting of mononuclear cells (lymphocytes, plasma cells, and macrophages; Fig. 1) which were localized predominantly around small vessels (Fig. 2). In addition, fiber degenerations were seen in five patients (patients 2 – 6; Table 1) which were consistent with the myopathological diagnosis of focal nodular myositis (Fig. 2) (Pongratz and Pilz 1988). Detmar's, Schmutzhard's, Atlas's, and two of Duray's muscle biopsy samples obviously showed only slight fiber abnormalities such as fiber atrophies, pathological variations of fiber size, and centrally located nuclei. These muscle preparations fulfilled the histopathological criteria of a purely interstitial myositis (Pongratz and Pilz 1988) or, according to Adams et al. (1967), of a "lymphocytosis of muscle", although elevated serum CK in three of these four patients might be a sign for fiber damage elsewhere. Schoenen et al. (1989) found a necrotizing myopathy without inflammatory reaction.

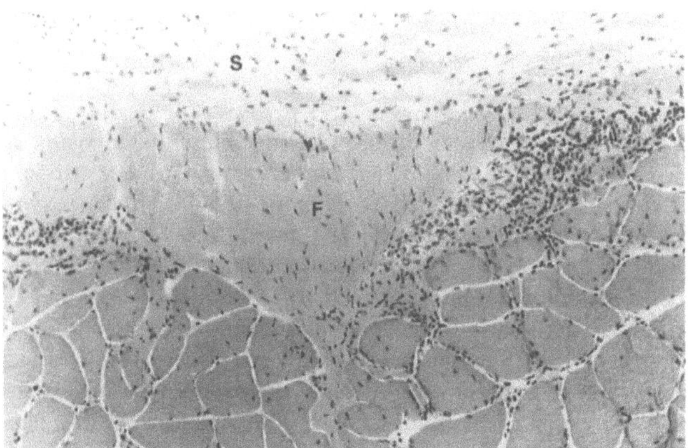

Fig. 1. Lymphohistiocytic infiltration of fascia and muscle in a 61–year–old man with *Borrelia burgdorferi* myositis and fasciitis (S, subcutaneous fat; F, fascia; hematoxylin and eosin, original magnification, x 100)

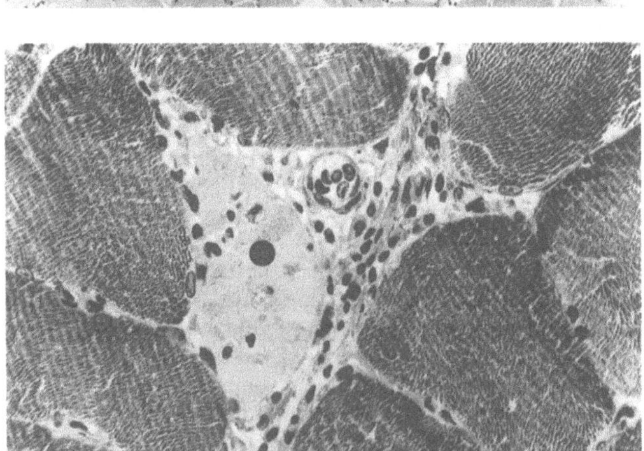

Fig. 2. Muscle fiber degeneration (*black dot*) adjacent to a small perivascular mononuclear infiltrate in a 66–year–old woman with *Borrelia burgdorferi* myositis, hematoxylin and eosin, original magnification, x 400.

Immunohistology performed in two of our specimens (from patients 3 and 4; Table 1) showed that infiltrates consisted of many T4 helper cells, many of them expressing human leukocyte antigen (HLA)–DR antigens and few interleukin 2 receptors. Only a small number of cytotoxic T8 lymphocytes was visible. None of them invaded muscle fibers. The high percentages of B and T4 lymphocytes, the low number of cytotoxic T cells, and the absence of actual muscle fiber invasion by lymphocytes in inflammatory infiltrations suggested that, similar to dermatomyositis, *B. burgdorferi* myositis is predominantly due to a local humeral response (Engel and Arahata 1986). In *B. burgdorferi* myositis autoimmunity initiated by the spirochetes themselves or by a spirochetal antigen may play a role. Electron–microscopic examination revealed extracellular lysosomes in some muscle fibers in one of our three specimens (from patient 3; Table 1). This seems to be a unique finding (Ghadially 1985).

Silver stainings depicted spirochetes in three of our four patients and in one of the patients of Detmar and Maciejewski (Chap. 20). The patient in whom silver stainings failed to show spirochetes had been treated with high–dose ceftriaxone before biopsy. An attempt to isolate *B. burgdorferi* in four of our muscle specimens was not successful.

13.2.2.2 Fasciitis

An en bloc biopsy of skin, fascia, and muscle in patient 2 of Reimers et al. (1989) also revealed mononuclear infiltrates in the fascia (Fig. 1). There were no eosinophils and only a slight blood eosinophilia, which might have been due to the patient's allergic diathesis. Thus, the diagnosis of eosinophilic fasciitis (Shulman's syndrome) could not be substantiated. Histopathological examination showed thickening and fibrosis of the fascia, and fascial infiltrates consisting of lymphocytes, plasma cells and histiocytes, but only a few eosinophils were present.

In Grahmann's (1989) patient with fasciitis, inflammatory infiltrates consisting mainly of lymphocytes and eosinophils in the dermis and subcutis were seen. The fascia and endomysial tissues were thickened and showed collagenous hypertrophy and perivascular infiltrates of lymphocytes, histiocytes, and eosinophils. The histopathology of Sepp's (1988) patient was suggestive of morphea; an inflammatory infiltrate with many mononuclear cells and eosinophils could be seen in the fascia (Figs. 3, 4). Stanek's (1987) patient with fasciitis presented thickening and fibrosis of the fascia and fascial infiltrates composed of lymphocytes, plasma cells, and histiocytes, but only a few eosinophils.

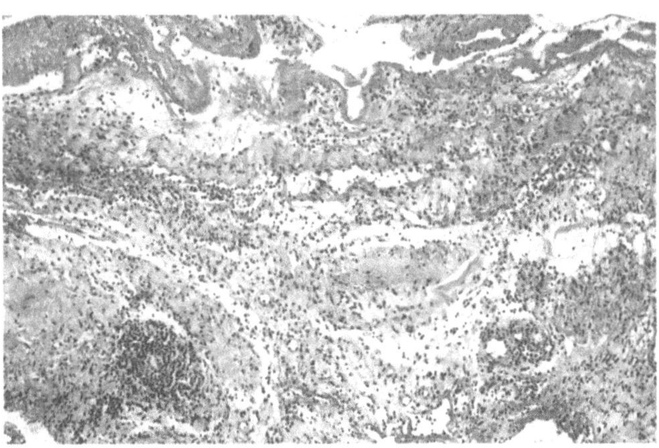

Fig. 3. Diffuse inflammatory infiltration in the fascia of a 59–year–old man with a Shulman–like syndrome due to *Borrelia burgdorferi* infection (according to Sepp et al. 1988), hematoxylin and eosin, original magnification, x 150.

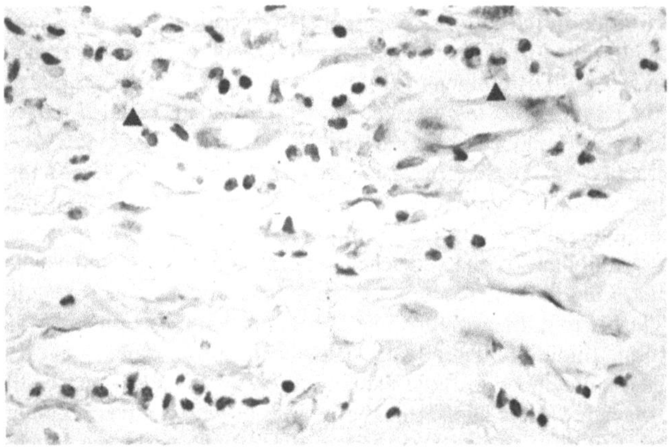

Fig. 4. The same patient as in Fig. 3: Several eosinophils (*black triangles*) can be seen in the fascia, Giemsa; original magnification, x 400.

13.3 Special Laboratory Examinations

13.3.1 Nonspecific Laboratory Findings

The patient of Schoenen et al. (1989) had CK levels of up to 8000 U/l. The remaining patients had normal or slightly elevated CK levels (up to three times the upper limit of the normal range). Besides erythrocyte sedimentation rates of 40 mm/h and 98 mm/h in two patients, there were no further abnormalities in routine laboratory investigations.

13.3.2 Serology

All of the patients presented in Table 1 had elevated IgG (IFT IgG 1:128 to 1:20480; normal below 1:64) and/or IgM antibody titers (IFT IgM 1:64 to 1:512; normal below 1:32) against *B. burgdorferi*. The antibody titers in the cerebrospinal fluid were increased in the patient of Midgard and Hofstad (1987), but negative in five other patients. Only patient 2 of Reimers et al. (1989) showed a slight disturbance of the blood–brain barrier. In the four patients with fasciitis, IgG antibody titers against *B. burgdorferi* were elevated, whereas the IgM antibody titers were normal.

13.4 Diagnosis

In the case of acute muscular pain combined with other signs or symptoms of early Lyme borreliosis, borrelial muscular involvement may be easily diagnosed. On the other hand, a *B. burgdorferi* infection should be considered if acute myalgias cannot be explained otherwise. Because of the high prevalence of asymptomatic *B. burgdorferi* infections in many regions (Neubert et al. 1988; Schmutzhard et al. 1988), elevated IgG antibody titers alone do not, however, prove *B. burgdorferi* etiology. In case of doubt, antibiotic treatment is indicated. Up to now, according to the literature as well as to our own experiences, there is no evidence that chronic myalgias may be caused by *B. burgdorferi* infection.

A *B. burgdorferi* myositis should be considered in patients with muscle weakness, muscular pains, elevated serum CK, or myopathic patterns in electromyography studies, and when there are clinical or serological signs of *B. burgdorferi* infection. In addition to electromyography, myosonography or magnetic resonance imaging of muscles may be helpful in searching for the optimal biopsy site (Reimers et al. 1989); computerized tomography seems to be less sensitive (Atlas et al. 1988; Reimers et al. 1989). The clinical diagnosis should be proven by myopathology.

13.5 Differential Diagnosis

B. burgdorferi – induced myalgias and myositis have to be differentiated from poly– or dermatomyositis, myositis in collagenoses or malignancies, meta–infective interstitial myositis, myositis caused by other pathogens (e.g., Coxsackie viruses), polymyalgia rheumatica, metabolic abnormalities (e.g., Mc Ardle's syndrome), toxic myopathies (e.g., alcohol), or so–called fibrositis.

In cases of fasciitis, systemic scleroderma, localized scleroderma, and other types of chronic fascial inflammations, e.g., in juvenile dermatomyositis, mixes connective tissue disease, and systemic lupus erythematosus (Simon et al. 1982), have to be considered.

Staining of spirochetes or cultivation of *B. burgdorferi* from muscles seems to be very difficult, but may be successful and indicate the spirochetal etiology of a myositis.

13.6 Summary

Up to now, there have only been a few reports of myositis or fasciitis due to *B. burgdorferi* infection. The spectrum of clinical features is very heterogeneous. It includes acute and chronic cases of muscle pain and weakness. From the histopathological point of view, *B. burgdorferi* myositis is an interstitial or focal nodular myositis. Borrelial fasciitis appears as a Shulman–like syndrome.

Acknowledgements

We are grateful to Dr. rer. nat. Vera Preac–Mursic, Max von Pettenkofer Institute, University of Munich, for attempts to cultivate *B. burgdorferi* from muscle biopsies and J. de Koning, Laboratorium voor de Volksgezondheid in Friesland, Leeuwarden, The Netherlands, for silver stainings of muscle specimens.

References

Adams RD, Denny–Brown D, Pearson CM (1967) Diseases of muscle. A study in pathology, 2nd edn. Harper and Row, New York, pp 414 – 457

Atlas E, Novak StN, Duray PH, Steere AC (1988) Lyme myositis: muscle invasion by *Borrelia burgdorferi*. Ann Intern Med 109:245 – 246

Detmar U, Maciejewski W, Link C, Breit R, Sigl H, Robl H, Preac–Mursic V (1989) Ungewöhnliche Erscheinungsformen der Lyme–Borreliose. Ein Beitrag zum klinischen Spektrum dieser Krankheitsgruppe. Hautarzt 40:423 – 429

Duray PH (1989) Visceral histopathology in Lyme borreliosis: new obserbations. Zentralbl Bakteriol (Suppl) 18:116 – 125

Duray PH, Steere AC (1988) Clinical pathologic correlations of Lyme disease by stage. Ann NY Acad Sci 539:65 – 79

Engel AG, Arahata K (1986) Mononuclear cells in myopathies: quantitation of functionally distinct subsets, recognition of antigen–specific cell–mediated cytotoxicity in some diseases, and

implications for the pathogenesis of the different inflammatory myopathies. Hum Pathol 17: 704 – 721
Ferradini L, Bagot M, Roujeau JC, Touraine R (1987) *Borrelia burgdorferi* and Shulman Syndrome. Lancet II:624
Gans O, Landes E (1952) Akrodermatitis atrophicans arthropathica. Hautarzt 3:151 – 155
Ghadially FN (1985) Ultrastructural pathology of the cell and matrix, 2nd edn. Butterworth, London, pp 786 – 796
Grahmann F, Schmidli J, Meier C (1989) Shulman syndrome as precursor of acrodermatitis chronica atrophicans Herxheimer. Another manifestation of Lyme borreliosis. Zentralbl Bakteriol (Suppl) 18:172 – 177
Midgard R, Hofstad H (1987) Unusual manifestation of nervous system *Borrelia burgdorferi* infection. Arch Neurol 44:781 – 783
Müller H (1969) Akrodermatitis atrophicans arthropathica mit Pseudosklerodermie, Sklerodermia circumscripta und muskulärer Beteiligung. Z Haut–GeschlKr 44:1 – 12
Neubert U, Münchhoff P, Völker B, Reimers CD, Pflüger KH (1988) *Borrelia burgdorferi* infections in Bavarian forest workers. A follow–up study. Ann NY Acad Sci 539:476 – 479
Pongratz DE, Pilz A (1988) Histologische und immunhistologische Befunde bei entzündlichen Muskelerkrankungen. Aktuel Rheumatol 13:229 – 233
Reimers CD, Müller W, Neubert U, Knutzner H, Pongratz DE (1988) Noduläre Herdmyositis: behandelbare Spätkomplikation bei *Borrelia burgdorferi*-Infektion. In: Mortier W, Pothmann R, Kunze K (eds) Aktuelle Aspekte neuromuskulärer Erkrankungen. Therapie, Früherkennung, Genetik, Mitochondriopathien. Thieme, Stuttgart, pp 64 – 67
Reimers CD, Pongratz DE, Neubert U, Pilz A, Hübner G, Naegele M, Wilske B, Duray PH, de Koning J (1989) Myositis caused by *Borrelia burgdorferi*. Report of 4 cases. J Neurol Sci 91:215 – 226
Schmutzhard E, Willeit J, Gerstenbrand F (1986) Meningopolyneuritis Bannwarth with focal nodular myositis. A new aspect in Lyme borreliosis. Klin Wochenschr 64:1204 – 1208
Schmutzhard E, Pohl P, Stanek G (1988) *Borrelia burgdorferi* antibodies in patients with relapsing/remitting form and chronic progressive form of multiple sclerosis. J Neurol Neurosurg Psychiatry 51:1215 – 1218
Schoenen J, Sianard–Gainko J, Carpentier M, Reznik M (1989) Myositis during *Borrelia burgdorferi* infection (Lyme disease). J Neurol Neurosurg Psychiatry 52:1002 – 1005
Sepp N, Schmutzhard E, Fritsch P (1988) Shulman syndrome associated with *Borrelia burgdorferi* and complicated by carpal tunnel syndrome. J Am Acad Dermatol 18:1361 – 1362
Shulman LE (1974) Diffuse fasciitis with hypergammaglobulinemia and eosinophilia: a new syndrome ? J Rheuma 1:82
Sigal LH, Johnson RW (1990) Fibromyalgia: mimic of Lyme disease (LD), part of the spectrum of musculoskeletal complaints of LD. Poster at the IV International Conference on Lyme Borreliosis, Stockholm, Sweden
Simon DB, Ringel StP, Sufit RL (1982) Clinical spectrum of fascial inflammation. Muscle Nerve 5:525 – 537
Stanek G, Konrad K, Jung M, Ehringer H (1987) Shulman Syndrome, a scleroderma subtype caused by *Borrelia burgdorferi*. Lancet I:1490
Steere AC, Bartenhagen NH, Craft JE, Hutchinson GJ, Newman JH, Rahn DW, Sigal LH, Spieler PN, Stenn KS, Malawista SE (1983) The early clinical manifestations of Lyme disease. Ann Intern Med 99:76 – 82
Steere AC, Bartenhagen NH, Craft JE, Hutchinson GJ, Newman JH, Pachner AR, Rahn DW, Sigal LH, Taylor E, Malawista SE (1986) Clinical manifestations of Lyme disease. Zentrabl Bakteriol Hyg (A) 263:201 – 205
Weber K, Neubert U (1986) Clinical features of early erythema migrans disease and related disorders. Zentralbl Bakteriol Hyg (A) 263:209 – 228

14 Acrodermatitis Chronica Atrophicans

E. ASBRINK, A. HOVMARK, and K. WEBER

14.1 Introduction

Acrodermatitis chronica atrophicans (ACA) has hitherto mainly been described from northern, central and eastern parts of Europe. In recent years, there have been a few American case reports on ACA in connection with Lyme disease (Lavoie et al. 1986; Kaufman et al. 1989). Previously the majority of cases described from the United States referred to immigrants from Europe (Montgomery and Sullivan 1945).

Besides ACA, many other names have been given to this condition, a fact which reflects the wide clinical spectrum. As it can be difficult to diagnose ACA on clinical and histopathologic grounds alone, the older literature, i.e., that written before the etiology was revealed and specific serologic methods became available, must be read with caution. With time the clinical presentations have also changed. Previously most patients consulted a dermatologist when the disease was in an advanced atrophic phase, which is also reflected by the many descriptions of ACA in its atrophic form in the older dermatologic literature. Since the introduction of penicillin, and especially now that antibiotics are widely used, patients with advanced atrophy are more seldom seen. Subcurative doses of antibiotics given for other infections may modify the clinical picture and course of ACA. Due to the development of specific serologic tests the clinical spectrum has also been widened, and early cases are nowadays more often diagnosed.

14.2 Definition

ACA is a late (tertiary) skin manifestation of Lyme borreliosis. It usually begins in the later decades of life and predominates in women. The predilection sites are the acral parts of the limbs, usually on their extensor surfaces. The course is chronic and usually progressive. It begins with a characteristic bluish–red discoloration, often with a doughy infiltration, progresses to atrophy of the skin, and is sometimes complicated by sclerodermic changes. Extracutaneous involvement of the joints and the peripheral nervous system is not uncommon.

14.3 Course Preceding Acrodermatitis Chronica Atrophicans

In the 1950s it was noted that ACA was more common in rural populations in Germany than in urban ones, and that the geographical distribution was similar to the distribution of *Ixodes ricinus* (Hauser 1955a,b). Hauser also claimed that almost all patients with ACA had a history of tick bites. In a recent study, about one–third of 50 patients with ACA had never noticed a tick bite, another third reported having had at least one tick bite during their lifetime, and the rest belonged to the group who had one or several tick bites every summer (Asbrink et al. 1986b). In another study, five of nine patients with ACA had a history of tick bites (Weber et al. 1984b). As the time that elapses from the occurrence of a causative tick bite to the development of ACA lesions may be several years, it is not surprising that the connection with a tick bite is seldom suspected or proposed by the patients.

Analogous to syphilis, the natural history of Lyme borreliosis can be divided into three stages, where stages I and II may be classified as early and stage III as late borreliosis (Asbrink and Hovmark 1988). In the individual patient, manifestations from any or all of the three stages may occur, and sometimes the infection will not become manifest until ACA arises. It is likely that an untreated borrelial infection will in the majority of cases resolve spontaneously after the primary or secondary stage. In some cases, there may be a latent stage before chronic manifestations appear, and in others overlapping between stages may occur. As evidenced by the cultivation of spirochetes from an ACA lesion of more than 10 years duration (Asbrink and Hovmark 1985), the borreliae may survive in humans for a considerable length of time.

In the older literature, there are only single case reports of erythema migrans preceding ACA (Hauser 1955a) or of coexisting erythema migrans and ACA (Hauser 1955a). Recent studies of 50 Swedish patients with ACA have revealed that about one–fifth have a history consistent with a preceding spontaneously healing erythema migrans, usually on the extremity where the ACA lesions developed 6 month – 8 years later (Asbrink et al. 1984, 1986b; Asbrink 1985). A preceding erythema migrans was also described in two of 21 German patients with ACA (Weber and Neubert 1986). Some of the patients with ACA also have a history of preceding neurologic and/or rheumatologic complaints. In certain cases there is a characteristic history of an untreated Bannwarth's syndrome with radicular pains and cranial nerve palsy such as facial palsy, but in others the clinical presentations are more difficult to evaluate in retrospect. An attempt to estimate the frequency of preceding untreated secondary stage manifestation in patients with ACA is, therefore, hazardous.

The ACA lesions usually develop slowly. At the time of diagnosis the cutaneous manifestations have not seldom been present for many years and sometimes for decades. When estimating the duration of the infection, it should be remembered that it may have started several years prior to the development of the ACA lesions. As a rule, however, the precise onset cannot be ascertained solely on the basic of a preceding tick bite, but can be established with near certainty if there is a history of erythema migrans on the extremity where ACA later developed.

14.4 Clinical Findings

14.4.1 Sex and Age

A female predominance of between 65% and 80% is generally found in studies of ACA patients (Göhr 1935; Montgomery and Sullivan 1945; Thyresson 1949; Hauser 1955b; Hopf 1966; Asbrink et al. 1986b; Weber et al. 1988). The usual age of involvement is 40 to 70 years. There are single case reports in children. In 168 consecutive patients with ACA, 67% were women in the age range 13 – 89 years (median 64 years) (unpublished data).

14.4.2 Initial Characteristic Features

ACA starts to develop on the extremities, and the extensor surfaces are mainly affected. The most common site is the lower leg, with initial involvement of a foot or of the knee region. More patchy or diffuse erythematous lesions spread over the leg may also appear (Fig. 1, see color plate II between pp 118–119). The initial site of involvement on the upper extremity is usually the olecranon area or the dorsal aspect of the hand (Fig. 2, see color plate II between pp 118–119).

Despite a duration of a number of years, the disease my affect only one extremity. However, with time symmetric involvement or involvement of the homolateral or contralateral limbs may also develop. The soles of the feet, the palms of the hands and the toes and the fingers may be involved. In some patients the cutaneous manifestations are confined to a heel. A swollen, sometimes discolored and painful heel is typical of ACA (Fig. 3), but can be missed if the patient is not turned around and examined from the back (Asbrink and Hovmark 1988). Although the limbs, sometimes in combination with the buttocks, are the most commonly involved parts of the body, lesions on the trunk, which may be extensive, and on the face may also occur in rare cases. The first dermatologic signs consist of livid red skin changes, either diffuse or localized, and the areas involved may appear edematous with a pasty consistency. Sometimes the erythema may be inconspicuous, and swelling may dominate the clinical picture and suggest angioneurotic edema or lymphedema, or, in advanced cases, it may resemble elephantiasis. A typical sign is that one of the feet gradually increases in size and that the need for larger and larger shoes arises. Initially the erythema and swelling may also vary in intensity. A concomitant annular migrating erythema, as in erythema migrans, may sometimes appear in the periphery of the typical ACA lesions.

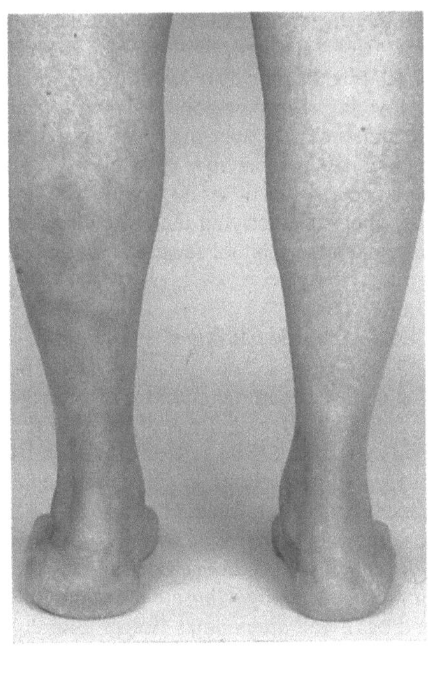

Fig. 3. Acrodermatitis chronica atrophicans; slight erythema on one of the lower legs and characteristic swelling of the foot

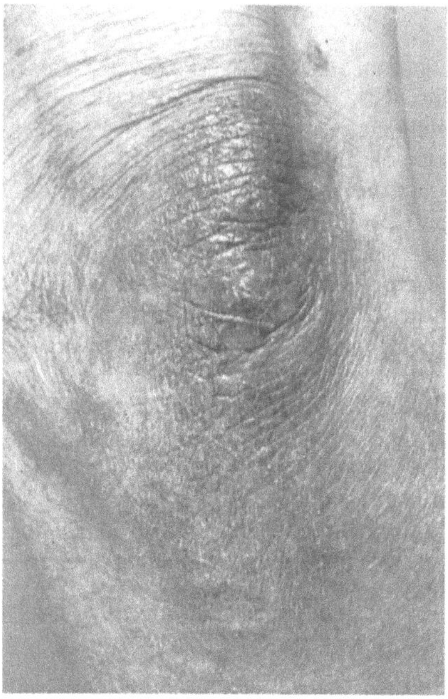

Fig. 4. Acrodermatitis chronica atrophicans; bluish–red inflammatory lesions on the olecranon area and fibrous nodules

14.4.3 Fibrous Thickening

Fibrous thickening of the skin may develop and manifest itself as indurated bands, especially in the ulnar or tibial region, or as nodules near joints. The so–called ulnar bands, which may extend from the elbow along the ulna towards the wrist, are more or less sharply defined, light or intense bluish–red, and of a firm or doughy consistency. Similar bands may also develop in the region of the knee joint but are more uncommon at this site. Fibrotic nodules may appear in the cutaneous and subcutaneous tissue, characteristically over the olecranon area (Fig. 4) but also occasionally in other areas such as the patellar region or close to joints on the dorsa of the hands or feet. They may occur solitarily or in groups. The nodules may vary in diameter from 0.5 to 2 – 3 cm. The skin overlying them may be skin–colored, yellowish, or bluish–red. They are firm and generally not sensitive to touch and usually fixed to the underlying tissue.

14.4.4 Atrophic Changes

The course is chronic and as a rule spontaneous healing does not occur. After years or decades the affected areas of the skin gradually become atrophic and develop a wrinkled appearance (Fig. 5, see color plate III between pp 118–119; Fig. 6).

Hyperpigmentation as well as depigmentation may develop. The network of blood vessels appears prominent because of the translucence of the thin epidermis. Besides diffuse atrophy, macular atrophy may occur. The term "atrophia maculosa cutis" (Oppenheim 1906) or anetodermia has been used for this uncommon, disseminated form of macular atrophy. Chronic ulceration of the atrophic skin may develop. Squamous cell and basal cell carcinomas (Lagerholm et al. 1970), malignant lymphoma (Goos 1971), and also other tumors have been reported to occur in skin areas affected by ACA.

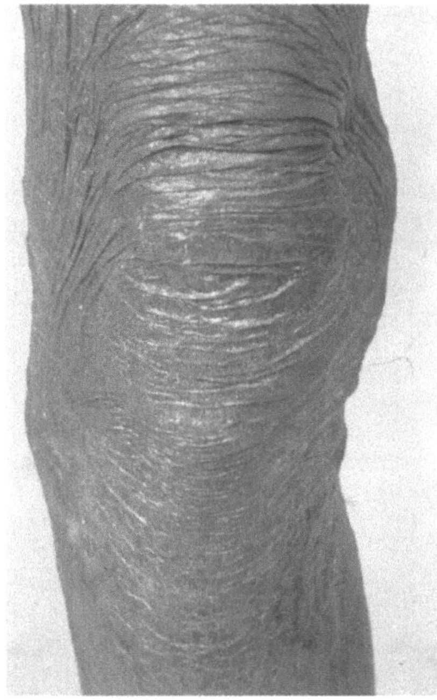

Fig. 6. Acrodermatitis chronica atrophicans; pronounced atrophy in the knee region

14.4.5 Sclerotic Changes

ACA is sometimes described as resembling morphea. Areas with sclerotic skin lesions both clinically and histopathologically indistinguishable from localized scleroderma (morphea) or lichen sclerosus et atrophicus may develop in patients with ACA (Asbrink et al. 1985). There may be merging of regions of typical inflammatory ACA with those of fibrosis. In the older literature, there are many reports of scleroderma–like lesions coexisting with ACA. Göhr (1935) found sclerodermatous changes in 13% of his 431 patients with ACA. Donnermann and Heite (1959), who found the same frequency in an analysis of 840 cases reported in the literature, were of the opinion that it could be explained by accidental coexistence. In a recent work, about 10% of patients with typical inflammatory ACA revealed sclerotic lesions as well (Asbrink and Hovmark 1987) (Fig. 7). The term "pseudoscleroderma" has been proposed for the sclerotic skin lesions that may occur in ACA (Jablonska 1975).

14.4.6 Benign Lymphocytic Skin Infiltrates

Benign lymphocytic skin infiltrates consistent with borrelial lymphocytomas have also been described in patients with ACA (Gottron 1938; Bäfverstedt 1943).

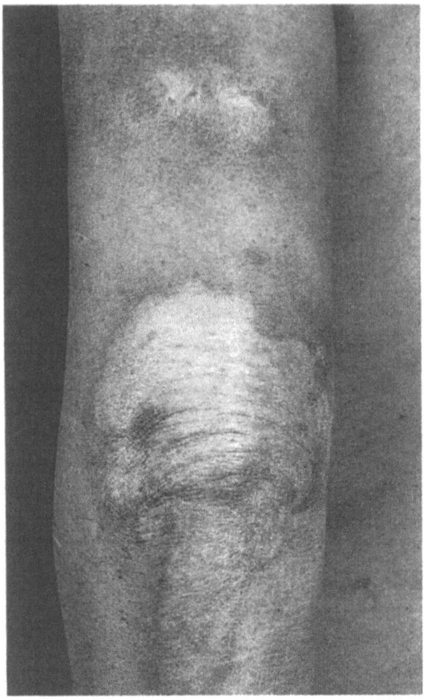

Fig. 7. Acrodermatitis chronica atrophicans concomitant with sclerotic lesions on one arm

14.4.7 Other Associated Signs and Symptoms

Enlarged regional lymph nodes are common, and sometimes a more widespread lymphadenopathy may develop (Hauser 1955b; Weber et al. 1984b). Some patients complain of nonspecific symptoms such as headache and migrating myalgias or arthralgias. Occasionlly involuntary loss of weight may occur, as has also been described in the second stage of this borrelial infection (Asbrink et al. 1985).

14.5 Neurologic Manifestation

14.5.1 Peripheral Neuropathy

In systematic studies in the 1960s and 1970s, an association between ACA and peripheral neuropathy has been established (Hopf 1966, 1975; Kaiser 1972). In these reports, which included both treated and untreated patients, nearly half of the patients with ACA showed signs of a predominantly sensory polyneuropathy often most pronounced in the limbs with cutaneous involvement.

Fairly frequent complaints by ACA patients are hyperesthesia/dysesthesia, muscle cramps, weakness in the muscles, and/or sensations of heaviness, mainly of the affected limb. Some patients complain of intermittent lightning pains in the extremities. According to recent data (Kristoferitsch et al. 1988; Kindstrand et al. 1988), the majority of untreated patients with ACA have some kind of neuropathy as diagnosed by clinical and/or electroneurographic examinations. The findings are very variable, but the neuropathy is mostly mild or moderate. Sensory and motor mono- or polyneuropathy or patchy dysesthesia at the site of the cutaneous lesions may develop.

14.5.2 Involvement of the Central Nervous System

Ther are also hints that CNS disturbances may occur in patients with ACA. In a case report from 1952, Gans and Landes reported on mental disturbances in a patient with ACA. Pathologic auditory brain–stem responses that were restored to normal or improved after intravenous antibiotic treatment have been found in patients with ACA (Sandström et al. 1989). Profound, sometimes incapacitating fatigue and personality changes may at times accompany ACA. However, cerebrospinal fluid (CSF) abnormalities such as pleocytosis, an elevated CSF protein level or intrathecal immunoglobulin (Ig) production with raised IgG index and oligoclonal bands are seldom found in patients with ACA (Hopf 1966, own unpublished data).

14.6 Joint Manifestations

14.6.1 Earlier Reports

In the older literature, there are several reports of joint and bone manifestations in patients with ACA (Jessner and Loewenstamm 1924; Hövelborn 1931; Müller 1969). However, a connection between ACA and joint abnormalities was neither generally recognized nor accepted. The findings of joint changes were attributed by some to a coincidental occurrence of arthrosis deformans in the elderly (Hauser 1955b). However, several authors have called attention to the fact that the skin lesions have a predilection for sites overlying joints such as the olecranon, the knee, and the small joints of the hands and feet. At the beginning of the twentieth century the term "arthrodermatitis chronica" and later the designation "acrodermatitis atrophicans arthropathica" were proposed (Gans and Landes 1952). There have also been several other old case reports of patients with ACA and joint changes, with subluxations/luxations of the small joints in the hands or feet (Hövelborn 1931; Müller 1969) sometimes complicated by bone atrophy (Raschke 1958).

14.6.2 Recent Findings

In a systematic investigations of 50 patients with ACA, radiographic examination revealed subluxations and/or luxations of small joints of the hands or feet in 11 patients (Hovmark et al. 1986) (see also Fig. 8 on color plate III between pp 118–119). Forefoot involvement with clawed toes had not seldom been diagnosed as hammer toes without the underlying cause being revealed. In some of the patients foot plates/inlays or physiotherapy had been prescribed, and in others surgical treatment for correction had been performed (Asbrink 1987). Periosteal thickening of bones was observed in four of the 50 patients. This finding is similar to the bone involvement – the so–called dactylitis syphilitica – that may occur during the late phase of syphilis. The affected joints or bones were usually located underneath the skin lesions, but sometimes other joints were also involved.

In the patients with the above–mentioned joint and/or bone manifestations enzyme–linked immunosorbent assay (ELISA) revealed a higher mean serum immunoglobulin (Ig) antibody titer against *B. burgdorferi* than in those without skeletal manifestations, and these patients had usually had their disease for a longer period than the patients with skin lesions alone. In 17 (20%) of 86 patients with ACA investigated episodic attacks of joint effusions of a knee were found to have preceded or to occur simultaneously with the ACA lesions (Asbrink 1987). A characteristic sign described by several authors is hyperpathia arising from impacts against bones underlying ACA lesions (Götz 1955; Weber et al. 1984b; Asbrink et al. 1986b). Periarticular manifestations such as knee or olecranon bursitis, epicondylitis, retro– and subcalcaneal bursitis, and Achilles tendinitis on the same extremity as the cutaneous involvement have been found subsequent to an untreated erythema migrans and/or preceding ACA (Herzer et al. 1986; Asbrink et al. 1986b; Asbrink 1987).

14.7 Histopathology

In the early phase of ACA there are usually no epidermal changes. In the dermis, there is an inflammatory reaction with varying degrees of edema. Later the epidermis gets thin, with flattening or absence of the papillae, and in some cases a variable hyperkeratosis may be observed. An often dense, patchy, and/or interstitial lymphocytic infiltrate with a moderate or rich admixture of plasma cells may involve the whole or part of the dermis and sometimes also the subcutaneous fat. In time there will be degeneration of the collagen and a reduction in elastic tissue. The veins of the upper and middle dermis become dilated to varying degrees, and the capillaries may display endothelial proliferation. Scattered vacuoles or groups of vacuoles on different levels of the dermis, often very close to the epidermis, have been described by different authors (Benjamowitsch and Maschkilleisson 1933; Montgomery and Sullivan 1945). They have previously been considered to be due to fatty degeneration of collagen fibers. In a recent study, however, we found reason to question whether the vacuoles really are fat cells and not an expression of lymphedema (Asbrink et al. 1986a). In the fibrotic nodules, there is a varying degree of infiltration of lymphocytes and plasma cells, especially in the periphery, and in the central parts there is hyalinization and degeneration of the collagen sometimes in an onion–skin–like pattern (Hardmeier 1968).

As the disease progresses, the histologic changes become striking. Epidermal atrophy, dilated dermal vessels often surrounded by lymphocytes and plasma cells in varying numbers, and degeneration of elastin and collagen are almost pathognomonic of ACA. After many years advanced atrophy of the dermis, including atrophy of hair follicles and of sweat and sebaceous glands, follows the inflammatory changes. Later the inflammatory cells decrease and almost disappear, and a pronounced thinning of the entire dermis and also the subcutis develops.

Immunohistologic staining of ACA lesions indicates a chronic hypersensitivity immune response to foreign antigens. The infiltrating T cells are dominated by "helper/inducer" cells and many lymphocytes express human leukocyte antigens HLA–DR and HLA-DQ antigens. In ACA, not only HLA–DR but also HLA–DQ antigens have also been found on the keratinocytes. After antibiotic treatment these antigens disappeared (Tjernlund et al. 1986).

14.8 Laboratory Findings

14.8.1 Serology

In serologic studies, elevated IgG antibody titers of antibodies to *B. burgdorferi* have been found almost invariably in patients with ACA (Ackermann et al. 1984; Weber et al. 1984a,b; Asbrink et al. 1984, 1985). In the currently used serologic tests, i.e., the indirect immunofluorescence assay and the ELISA with whole cells or whole–cell sonic extracts as antigen, most patients show very high IgG titers, but about 15% have borderline titers, defined as the highest titers found in 5% of healthy individuals. However, IgM reactivity has been shown to be due to IgM rheumatoid factor activity (Wilske et al. 1984; Hansen and Asbrink 1989).

14.8.2 Nonspecific Laboratory Findings

It has long been known that patients with ACA may have an increased erythrocyte sedimentation rate and elevated serum immunoglobulin levels. (Thyresson 1949; Hauser 1955B; Asbrink et al. 1986b; Weber and Neubert 1986). In two studies, elevated serum IgM levels were observed in 26% of the patients, and in one of the studies, circulating immune complexes were found in 23%. A reactive hyperplasia with a plasma cell infiltrate in the bone marrow of patients with ACA has been described by Hauser (1955).

14.9 Diagnosis

The diagnosis of ACA is based primarily on clinical and serologic criteria and histologic examination serves a confirmatory purpose. A bluish–red discoloration, often with swelling and/or atrophy of the skin, on an acral part of an extremity, together with elevated serum IgG antibody titer to *B. burgdorferi*, is consistent with the diagnosis of ACA. The histopathologic picture can help to establish the diagnosis or rule out other diagnoses; however, in the early phase, the histopathologic changes may be slight. The diagnosis may be further confirmed by cultivation of *B. burgdorferi* (Asbrink et al. 1984; Preac–Mursic et al. 1985; Stanek 1985; Neubert et al. 1986) or by morphologic identification of spirochetes in histologic sections (Fritz and Lagerholm 1983; Neubert et al. 1986).

ACA may be the first sign of a borrelial infection, but many patients have histories consistent with preceding primary and/or secondary stage manifestations. There are variations, however, in the time of appearance of the manifestations and in their severity. A history of an untreated erythema migrans, the hallmark of Lyme borreliosis, on the same extremity where livid, edematous skin lesions later develop is suggestive of ACA. Patients very seldom recognize tick bites as relevant or important in the evolution of ACA.

14.10 Differential Diagnosis

The early ACA manifestations are often clinically overlooked or misdiagnosed. To consider ACA and include it in the differential diagnosis is one–half of the diagnosis. It is not uncommon for patients with ACA to have visited a considerable number of physicians before the correct diagnosis is made. A bluish–red or swollen leg, often found in elderly women, is frequently felt to represent circulatory problems. In such patients a history of circulatory investigation is the rule rather than the exception. In others the skin lesions are not seldom mistaken for trophic skin changes, sequelae to frostbite, or connective tissue disease, or attributed to aging. Sclerotic lesions in patients with ACA are often diagnosed as morphea or lichen sclerosus et atrophicus.

Because of their location in the vicinity of joints, fibrotic nodules are sometimes confused with rheumatic nodules or gouty tophi and, if they have a yellowish color, with xanthoma.

In some patients with ACA joint deformities, enlarged lymph nodes or pain may be the predominant sign or symptom, and the inflammatory lesions of the skin may be neglected or overlooked by the patient as well as by the doctor, as they may be discreet. Biopsies of enlarged regional, often inguinal, lymph nodes are sometimes performed in patients with ACA. Examinations for malignancy may also be initiated by reason of the lymphadenopathy, especially if this appears in combination with weight loss and profound fatigue (Asbrink 1985).

Acrocyanosis with a blue or reddish discoloration of the hands or feet must be differentiated from ACA. Other vascular disturbances such as livedo reticularis may sometimes also be mistaken for ACA. The histopathologic picture without an inflammatory infiltrate with plasma cells and negative or low antiborrelial antibody titers will help to exclude ACA.

References

Ackermann R, Boisten HP, Kabatzki J, Runne U, Krüger K, Herrmann WP (1984) Serumantikörper gegen *Ixodes–ricinus* Spirochäte bei Acrodermatitis chronica atrophicans (Herxheimer). Dtsch Med Wochenschr 109:6–10

Asbrink E (1985) Erythema chronicum migrans Afzelius and acrodermatitis chronica atrophicans. Early and late manifestations of *Ixodes ricinus*–borne *Borrelia* spirochetes. Acta Derm Venereol (Suppl) (Stockh) 118:1–63

Asbrink E (1987) Erythema migrans and acrodermatitis chronica atrophicans: association with joint manifestations. In: Andrianakos A, Kappou I, Mavrikakis M, Moutsopoulos H (eds) Eurorheumatology. Tagas, Athens, pp 246–248

Asbrink E, Hovmark A, (1985) Successful cultivation of spirochetes from skin lesions of patients with erythema chronicum migran Afzelius and acrodermatitis chronica atrophicans. Acta Pathol Microbiol Immunol Scand (B) 93:161–163

Asbrink E, Hovmark A (1987) Cutaneous manifestations in *Ixodes*–borne *Borrelia* spirochetosis. Int J Dermatol 26:215–223

Asbrink E, Hovmark A (1988) Early and late cutaneous manifestations in *Ixodes*–borne borreliosis. (Erythema migrans borreliosis, Lyme borreliosis) Ann NY Acad Sci 539:4–15

Asbrink E, Hovmark A, Hederstedt B (1984) The spirochetal etiology of acrodermatitis chronica atrophicans Herxheimer. Acta Derm Venereol (Stockh) 64:506–512

Asbrink E, Hovmark A, Hederstedt B (1985) Serologic studies of erythema chronicum migrans Afzelius and acrodermatitis chronica atrophicans with indirect immunofluorescence and enzyme–linked immunosorbent assays. Acta Derm Venereol (Stockh) 65:509–514

Asbrink E, Brehmer–Andersson E, Hovmark A (1986a) Acrodermatitis chronica atrophicans – a spirochetosis. Clinical and histopathological picture based on 32 patients. Am J Dermatopathol 8:209–219

Asbrink E, Hovmark A, Olsson I (1986b) Clinical manifestations of acrodermatitis chronica atrophicans in 50 Swedish patients. Zentrabl Bakteriol Hyg (A) 263:253–261

Bäfverstedt B (1943) Über Lymphadenosis benigna cutis. Eine klinische und pathologisch–anatomische Studie. Norstedt, Stockholm

Benjamowitsch E, Maschkilleisson LN (1933) Weitere Beiträge zur Frage über die Atrophie der Haut. Acrodermatitis chronica atrophicans und ihre Beziehung zur Sclerodermia. Acta Derm 14:313–341

Donnermann Ch, Heite HJ (1959) Beitrag zur Symptomatologie der Akrodermatitis chronica atrophicans (Pick–Herxheimer). Arch Klin Exp Dermatol 208:516–527

Fritz A, Lagerholm B (1983) Acrodermatitis chronica atrophicans, Erythema chronicum migrans and Lymphadenosis benigna cutis – spirochetal diseases? Acta Derm Venereol (Stockh) 63:432–436

Gans O, Landes E (1952) Acrodermatitis atrophicans arthropathica. Hautarzt 3:151–155

Göhr A (1935) Zur Klinik der Acrodermatitis chronica atrophicans. Dissertation, Berlin

Goos M (1971) Acrodermatitis chronica atrophicans and malignant lymphoma. Acta Derm Venereol (Stockh) 51:457–459

Gottron H (1938) Lymphadenosis cutis circumscripta im Bereich der Mamille bei gleichzeitiger Acrodermatitis chronica atrophicans der Extremitäten. Zentralbl Hautkr 59:633

Götz H (1955) Die Acrodermatitis chronica atrophicans als Infektionskrankheit. Hautarzt 6: 249–252

Hansen K, Asbrink E (1989) The *Borrelia burgdorferi* flagellum ELISA improves serodiagnosis in erythema migrans and acrodermatitis chronica atrophicans. J Clin Microbiol 27:545–551

Hardmeier T (1968) Zur Histopathologie der fibroiden Knoten bei Akrodermatitis chronica atrophicans. Arch Klin Exp Derm 232:373–389

Hauser W (1955a) Zur Klinik, Ätiologie und Pathogenese der Acrodermatitis chronica atrophicans. Hautarzt 6:77–80

Hauser W (1955b) Zur Kenntnis der Acrodermatitis chronica atrophicans. Arch Dermatol 199:350–393

Herzer P, Wilske B, Preac–Mursic V, Schierz G, Schattenkirchner M, Zöllner N (1986) Lyme arthritis:clinical features, serological, and radiographic findings of cases in Germany. Klin Wochenschr 64:206–215

Hopf HC (1966) Acrodermatitis chronica atrophicans (Herxheimer) und Nervensystem. Springer, Berlin, Heidelberg, New York (Monographien aus dem Gesamtgebiete der Neurologie und Psychiatrie, vol. 114)

Hopf HC (1975) Peripheral neuropathy in acrodermatitis chronica atrophicans (Herxheimer). J. Neurol Neurosurg Psychiat 38:452–458

Hovmark A, Asbrink E, Olsson I (1986) Joint and bone involvement in Swedish patients with *Ixodes ricinus*–borne spirochetal infection. Zentralbl Bakteriol Hyg (A) 263:275–284

Hövelborn CI (1931) Gelenkveränderungen bei Acrodermatitis chronica atrophicans. Arch Dermatol 164:349–356

Jablonska S (1975) Acrodermatitis atrophicans and its sclerodermiform variety; relation to scleroderma. Scleroderma and pseudoscleroderma. PZWL, Warsaw

Jessner M, Loewenstamm A (1924) Bericht über 66 Fälle von Acrodermatitis chronica atrophicans. Dermatol Wochenschr 79:1169–1177

Kaiser M (1972) Neurologische Komplikationen bei Akrodermatitis Chronica Atrophicans (Herxheimer) und ihre Beeinflussung durch die Penicillintherapie. Dissertation, University of Göttingen

Kaufman LD, Gruber BL, Phillips ME, Benach JL (1989) Late cutaneous Lyme disease: acrodermatits chronica atrophicans. Am J Med 86:828–830

Kindstrand E, Asbrink E, Nilsson BY, Hovmark A (1988) Neuropati vid kronisk Borrelia–infektion. Läkaresällskapets Riksstämma, Stockholm.

Kristoferitsch W, Sluga E, Graf M, Partsch H, Neumann R, Stanek G, Budka H (1988) Neuropathy associated with acrodermatitis chronica atrophicans: clinical and morphological features. Ann NY Acad Sci 539:35–45

Lagerholm B, Molin L, Gip L (1970) Basal cell carcinoma in association with acrodermatitis chronica atrophicans Herxheimer. Acta Derm Venereol (Stockh) 50:218–220

Lavoie PE, Wilson AJ, Tuffanelli DL (1986) Acrodermatitis chronica atrophicans with antecedent Lyme disease in California. Zentralbl Bakteriol Hyg (A) 263:262–265

Montgomery H, Sullivan R (1945) Acrodermatitis chronica atrophicans. Arch Dermatol Syph 51:32–45

Müller H (1969) Akrodermatitis atrophicans arthropathica mit Pseudosklerodermie, Skleroderma circumscripta und muskulärer Beteiligung. ZHaut Geschlechtskr 44:1–12

Neubert U, Krampitz HE, Engl H (1986) Microbiological findings in erythema (chronicum) migrans and related disorders. Zentralbl Bakteriol Hyg A 263:237–252

Oppenheim M (1906) Zur Kenntnis der Atrophia maculosa cutis. Arch Dermatol 81:127–146, 291–312

Preac–Mursic V, Wilske B, Herzer P, Schierz G, Bauer M (1985) Acrodermatitis chronica atrophicans – eine Borreliose! Hautarzt 36:691–693

Raschke G (1958) Über den Zusammenhang der Akrodermatitis chronica atrophicans Pick–Herxheimer und der Akroosteolyse. Dermatol Wochenschrift 9:218–221

Sandström M, Bredberg G, Asbrink E, Hovmark A, Holmkvist C (1989) Brainstem response audiometry in chronic *Ixodes*–borne borreliosis. Scand Audiol 18:205–210

Stanek G (1985) Lyme disease and related disorders. Microbiol Sci 2:231–234

Tjernlund U, Scheynius A, Asbrink E, Hovmark A (1986) Expression of HLA–DQ antigens on keratinocytes in *Borrelia* spirochete–induced skin lesions. Scand J immunol 23:383–388

Thyresson N (1949) The penicillin treatment of acrodermatitis chronica atrophicans. Acta Derm Venereol (Stockh) 29:572–621

Weber K, Neubert U (1986) Clinical features of early erythema migrans disease and related disorders. Zentralbl Bakteriol Hyg (A) 263:209–228

Weber K, Schierz G, Wilske B, Preac–Mursic V (1984a) European erythema migrans disease and related disorders. Yale J Biol Med 57:463–471

Weber K, Schierz G, Wilske B, Preac–Mursic V (1984b) Zur Klinik und Ätiologie der Acrodermatitis chronica atrophicans. Hautarzt 35:571–577

Weber K, Preac–Mursic V, Neubert U, Thurmayr R, Herzer P, Wilske B, Schierz G, Marget W (1988) Antibiotic therapy of early European Lyme borreliosis and acrodermatitis chronica atrophicans. Ann NY Acad Sci 539:324–345

Wilske B, Schierz G, Preac–Mursic V, Weber K, Pfister H–W, Einhäupl K (1984) Serological diagnosis of erythema migrans disease and related disorders. Infection 12:331–337

15 Chronic Central Nervous System Involvement

R. MARTIN, W. KOHLHEPP, and H. G. MERTENS

15.1 Introduction

Although the meninges and nerve roots are more commonly involved during Lyme borreliosis, especially in Europe, numerous reports from Europe and the United States have documented the occurrence of acute and chronic manifestations in the brain or spinal cord (Ackermann et al. 1985; Klenk et al. 1985; Pfister et al. 1985; Reik et al. 1986; Kohlhepp et al. 1987; Behringer and Wirbatz 1987; Weder et al. 1987; Wokke et al. 1987a; Kohler et al. 1988; Halperin et al. 1989). The following chapter describes the clinical features and necessary investigations which help in the diagnosis of chronic central nervous system (CNS) involvement during Lyme borreliosis.

15.2 Clinical Findings

15.2.1 Definition of Manifestation

This chapter will examine cases of inflammation of the brain and spinal cord. The meninges, cranial, and peripheral nerve roots are almost always affected during the infection and will also be discussed here to a limited extent.

The term CNS Lyme borreliosis includes all forms of encephalitis, encephalomyelitis, meningoencephalitis, and radiculomyelitis. The definition of what constitutes chronic disease with respect to infection with *Borrelia burgdorferi* is difficult, and there is no general agreement in terms of duration so far. Ackermann et al. (1988) and Kohlhepp et al. (1987) have defined chronic encephalomyelitis as lasting either longer than 6 (Ackermann) or 12 months (Kohlhepp). Steere (1989) mentions the chronic forms as manifestations of late or stage 3 Lyme disease, which occur with or without previous disease stages 1 and 2, but gives no definite requirements for disease duration. Several problems hamper the exact determination of disease duration: Up to 70% of the patients remember neither a tick bite nor an erythema migrans (Pfister et al. 1984; Stiernstedt et al. 1984; Kohlhepp et al. 1987). Out of 44 patients with progressive Borrelia encephalomyelitis reported by Ackermann et al. (1988) only five mentioned a previous meningoradiculoneuritis, and none remembered an erythema migrans. Initial symptoms may be subtle and are often misleading such as fatigue, migrating pain, deficits in concentration and memory, or other psychoorganic abnormalities. The onset of complaints is therefore difficult to assess. The infection may be latent with no or minimal clinical signs and normal cerebrospinal fluid (CSF) findings (Pfister et al. 1989). It is not known how long *B. burgdorferi* can stay in the latent phase, whether Lyme borreliosis always becomes symptomatic, or whether tissue damage can be demonstrated for example by magnetic resonance imaging (MRI) in asymptomatic individuals. In addition, the neurologic deficits and pathological CSF findings are often not completely reversible, even after repeated antibiotic treatment (Kohlhepp et al. 1987; Ackermann et al. 1988). In these cases, it is unclear whether the patient still harbors the bacterium or has symptoms related to residual effects of the infection. Finally, several reports describe autoimmune phenomena induced by *B. burgdorferi* (Suchanek et al. 1986; Martin et al. 1988b; Aberer et al. 1989; Schlüsener et al. 1989). It is unknown whether these immune responses to brain and peripheral nervous system antigens require the continuing presence of borreliae or if they, once initiated, lead itself to chronic CNS damage. It is also uncertain whether an association between certain human leukocyte antigens (HLA) (DR2) and chronic disease courses documented in patients with Lyme arthritis (Steere et al. 1979) also exists for nervous system involvements. So far, Wokke et al. (1988) observed that individuals with HLA–DR2 suffering from meningoradiculoneuritis secreted higher antibody titers against *B. burgdorferi* and might, therefore, be less prone to severe disease courses, provided that these antibodies are protective. This finding was further supported by a recent study by Pflüger et al. (1988) which showed increased frequencies of HLA–A2 and Cw3 in patients with *B. burgdorferi* infection and steadily decreasing frequencies of HLA–DR2 with more severe disease courses. Both studies are, however, preliminary, and further investigations are necessary, especially with respect to chronic CNS manifestations.

From the above–mentioned reasons, it is obvious that a clear–cut determination of disease duration is often impossible. We, therefore, prefer to use the term chronic CNS Lyme borreliosis for any *B. burgdorferi* infection of the brain and/or spinal cord with and without meningeal, cranial, or peripheral nerve root involvement which lasts longer than 6 months beginning from the first symptoms. In addition, chronic disease should usually show signs of progression, since residual symptoms of a meningoradiculoneuritis may also be seen for several months, but clearly tend to improve during this period.

15.2.2 Age and Sex Distribution

The age at onset of chronic CNS borreliosis ranges from 6 – 81 years (Kohlhepp et al. 1987; Ackermann et al. 1988; Halperin et al. 1989; Pachner et al. 1989) with slight peaks in the third and sixth decade. No major differences in age distribution have been observed between the United States and Europe. Both sexes are equally affected. A seasonal distribution is neither described for chronic manifestations of the CNS nor for those of the skin or joints. The time between first symptoms and medical attention ranges from several weeks up to 15 years (Bensch et al. 1987), although this time span gives no information about the onset of the infection.

As a mean disease duration of chronic Borrelia encephalomyelitis, Ackermann et al. (1988) mention 2,8 years, with a range from 7 month up to 12 years.

15.2.3 Clinical Description of Manifestations

The multitude of different nervous system involvements to some extent parallel those seen in neurosyphilis (Burke and Schaberg 1985). So far, it seems that chronic and progressive CNS manifestations occur less frequently during Lyme borreliosis than in neurosyphilis (Rowland 1989). In a follow–up study by Krüger et al. (1989) mild sequelae, but no progressive disease occurred in 23 out of 55 patients who were examined 5–27 years after Lyme meningoradiculoneuritis or encephalomyelitis and who had received no antibiotic treatment during or after the disease. Severe progressive disease courses have, however, been described by other authors (Vandvik et al. 1984; Ackermann et al. 1985, 1989; Klenk et al. 1985; Reik et al. 1985; Hansen et al. 1986; Hänny and Häuselmann 1987; Kohlhepp et al. 1987; Kohler et al. 1988; Weder et al. 1987). The following chapter will describe the general symptoms as well as the clinical characteristics of chronic CNS manifestations.

Nonspecific and vague complaints of the patients frequently precede the onset of myelitic or encephalitic symptoms for several weeks to several years. The prodromal symptoms may resemble those found in psychosomatic disorders, and, therefore, patients sometimes see several physicians before the diagnosis is established.

The clinical manifestations of chronic *B. burgdorferi*–induced CNS disease are shown in Table 1. As has long been known for neurosyphilis (Burke and Schaberg 1985), any part of the CNS from the cerebral vessels and meninges, cranial and spinal roots to the parenchyma of the brain and spinal cord can be involved during the inflammatory process associated with the *B. burgdorferi* infection. Chronic CNS involvements during Lyme borreliosis account for up to 4,7% of all neurologic manifestations of Lyme disease, when 12 months are taken as a minimal disease duration (Mertens et al. 1989), although Halperin et al. (1989) found a mild form of chronic encephalopathy to be particularly common among his patients (41 out of 85). Since CSF investigation in these patients often showed minor or no signs of inflammation, they cannot be compared with those mentioned by Mertens et al. (1989).

The often insidious onset of the disease may include headaches and meningism. Although pain is more characteristic for a meningoradiculoneuritis, patients with myelitis or encephalitis sometimes also suffer from pain because of additional radicular involvement. The long–lasting and remittent pain may be so intense that patients lose weight and have difficulties in carrying on in their professions. A considerable number of our patients complain of pain, especially during the night, which sometimes leads to severe sleeping difficulties.

Other complaints include influenza–like symptoms with myalgia, arthralgia without joint swelling, fever, fatigue, general malaise, weight loss, and, rarely, nausea, vomiting (Kohlhepp et al. 1987; Pachner et al. 1989), or transient weakness of an extremity (Kohler et al. 1988). Vegetative stigmatization with tachycardia, profuse sweating, and abnormal anxiety reactions may also occur (Kohler et al. 1988).

Cranial nerve abnormalities are present in 50% of the patients (Schaltenbrand 1962; Kohlhepp et al. 1987; Weder et al. 1987; Ackermann et al. 1988; Mertens et al. 1989). A mono– or diplegia of the facial nerve (VII) accounts for the largest part of cranial nerve deficits, although it is not as frequently observed here as it is during acute diseases. It is usually caused by a neuritis of the peripheral portion of the nerve and presents as peripheral paresis, although a central hemiparesis or nuclear paresis of sin-

Table 1. Major manifestations of chronic central nervous system Lyme borreliosis

Disorder	*Clinical findings/symptoms*
Encephalitis	Spasticity; urinary incontinence; paresis; cranial nerve deficits; ataxia; extrapyramidal disorders; headaches and meningism; neuropsychiatric disorders
Radiculomyelitis/transverse myelitis	Paresis; sensory level: spasticity; radicular pain; urinary incontinence
Stroke–like vasculitic disorder and cranial nerve deficits	Transitoric or permanent hemi- or monoparesis

gle cranial nerves may be superimposed (Kohler et al. 1988). Papilledema, impaired vision due to optic neuritis, bitemporal hemianopia, strabismus, facial numbness, hypacusis and vertigo, dysarthria, and difficulties in swallowing have been documented in a number of patients and indicate that any cranial nerve can be involved (Behringer and Wirbatz 1987; Kohlhepp et al. 1987; Weder et al. 1987; Kohler et al. 1988; Halperin et al. 1989). Whether these deficits are due to a vasculitis of neural vessels or to demyelinations or axonal destructions caused by the spirochete is not yet clear.

The clinical disorders are further characterized by different forms of spastic paresis with brisk tendon reflexes, positive Babinski sign, spasticity, and accompanying urinary incontinence and, rarely, myoclonic jerks (Ackermann et al. 1985; Klenk et al. 1985; Bensch et al. 1987; Kohlhepp et al. 1987; Weder et al. 1987; Kohler et al. 1988). Some patients develop acute hemiparesis and aphasia closely resembling stroke (Midgard and Hofstad 1987; Weder et al. 1987; Uldry et al. 1987; Kohler et al. 1988; Veenendaal–Hilbers et al. 1988; May and Jabbari 1990). From the parenchymal brain defects observed by computed tomography (CT) or MRI and from partial vessel occlusions detected by Doppler sonography and arteriography (Midgard and Hofstad 1987; Kohler et al. 1988; Uldry et al. 1987; Veenendaal–Hilbers et al. 1988; May and Jabbari 1990), it was concluded that these lesions are due to a vasculitis caused by *B. burgdorferi*. Short–term one–sided paresis mimicking transient ischemic attacks has also been reported (Kohler et al. 1988). Diagnostic difficulties may result when the spasticity is accompanied by peripheral nerve deficits and muscle weakness (Wokke et al. 1987b). Pareses, spasticity, and micturition disturbances have been observed for up to years after meningoradiculoneuritis or encephalomyelitis. Sensory deficits other than paresthesia and hyperesthesia are not common for central neuroborreliosis (Kohler et al. 1988), although a clear–cut level for pin, touch, temperature, and vibratory sensation was found in cases with transverse myelitis (Klenk et al. 1985; Rousseau et al. 1986; Kohlhepp et al. 1987).

Motor and sensory systems are not the only ones affected by chronic CNS Lyme borreliosis. The frequent occurrence of ataxia of legs and gait (Kohlhepp et al. 1987; Weder et al. 1987; Kollikowski et al. 1988; Mertens et al. 1989) in up to 60% of patients, as well as dysdiadochokinesia, are suggestive of cerebellar damage. Recently, Kohlhepp et al. (1987,1989) have reported five patients suffering from Lyme borreliosis who presented initially with signs of Parkinson syndrome including tremor, rigor, bradykinesia, and vegetative signs or, rarely, with choreatic movements. Besides antibiotic treatment, L–dopa administration was required, but could later gradually be decreased or completely withdrawn. This syndrome was improved in all patients by antibiotic treatment, although mild symptoms may remain.

Like other encephalitic diseases, chronic CNS Lyme borreliosis often presents with neuropsychiatric symptoms. Deficits in concentration (Wokke et al. 1987a), fatigue, and depressiveness may long precede the onset of dementia (Bensch et al. 1987) Symptoms of dementia included impairment of cognition, memory, orientation (Bensch et al. 1987; Wokke et al. 1987a; Kollikowski et al. 1988; Halperin et al. 1989; Pachner et al. 1989), and also psychiatric disorders (Kohlhepp et al. 1987; Halperin et al. 1989; Pachner et al. 1989) such as recurrent depression or schizoaffective disorder. In one patient who was suffering from a mild organic brain syndrome of recent onset, Alzheimer's disease was diagnosed before the patient died (Macdonald and Miranda 1987). At autopsy, *B. burgdorferi* was demonstrated in and cultivated from brain tissue. It is, however, too early to conclude from this single case that *B. burgdorferi* is able to induce a chronic dementing process, since Papolla et al. (1989) were not able to cultivate *B. burgdorferi* from the brain of patients who had died from Alzheimer's disease or detect the bacteria by immunofluorescence or transmission electron microscopy. Other signs of encephalitis such as seizures or altered consciousness, and focal brain stem symptoms are rarely observed (Bensch et al. 1987; Halperin et al. 1989; Midgard and Hofstad 1987; Pachner et al. 1989; Reik et al. 1985; Weder et al. 1987).

15.3 Special Procedures

Although none of the investigations discussed below is specific for Lyme neuroborreliosis, they are important for establishing the extent of tissue damage and involvement of different systems of the CNS.

15.3.1 Imaging Techniques

As soon as an encephalitic or myelitic disease is suspected either a CT and/or MRI should be performed. Low–density lesions (CT) have been observed and are most often located in the periventricular white matter, the basal ganglia, and the thalamus, and less frequently in the grey matter (Reik et al. 1985; Wokke et al. 1987a; Kohlhepp et al. 1987, 1989; Kohler et al. 1988; Pachner et al. 1989). Contrast enhancement and mass effects indicating either a local disruption of the blood–brain barrier, intense inflammatory infiltrates, or vasculitic changes have been reported (Reik et al. 1985; Weder et al. 1987) as well as a multi–infarct–like picture. If the disease has been of long duration, asymmetrical enlargement of the lateral ventricles, general cerebral atrophy, parenchymal calcifications, and communicating hydrocephalus have been seen by CT (Bensch et al. 1987; Kohler et al. 1988).

MRI is more sensitve in detecting such foci. Multiple lesions of high signal intensity in the periventricular white matter are reminiscent of the demyelinated areas occurring in multiple sclerosis (Fig. 1) (Kohler et al. 1988). The foci are better visible

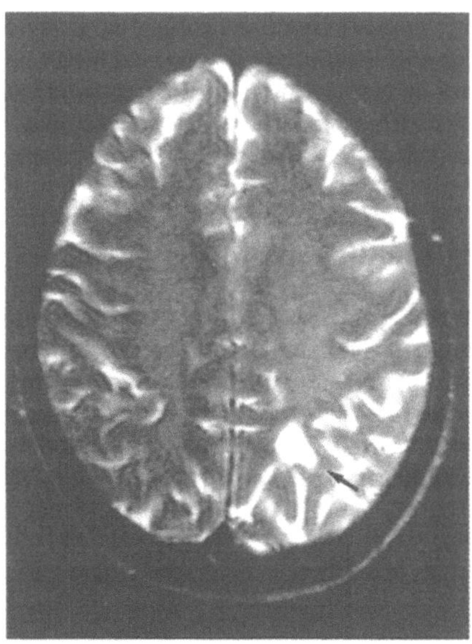

Fig. 1. Axial T_2–weighted MRI scan of the brain showing a hyperintense white matter lesion in the left occipital lobe (*arrow*). The female patient had been suffering from chronic Lyme encephalomyelitis for 8 months at the time of investigation. (Courtesy of Prof. M. Nadjmi, Department of Neuroradiology, University of Würzburg, Germany)

in T2–weighted than in proton–density–weighted scans. At follow–up they may have disappeared in treated patients (Halperin et al. 1989). In cases of long duration, severe atrophy of brain parenchyma, pons, and medulla oblongata have also been reported (Bensch et al. 1987).

Arteriography and myelography are rarely performed, but may be appropriate when either a stroke–like course is seen or lumbar arachnitis is suspected. Findings in both investigations have been normal in most of the few patients in whom they were performed (Reik et al. 1985; Wokke et al. 1987a; Kohler et al. 1988), but multiple stenoses of small and large intracerebral arteries and poststenotic dilatations indicating arteritis could also be documented by arteriography (Midgard and Hofstad 1987).

15.3.2 Electrophysiological Investigations

Although it is applied infrequently, electroencephalography (EEG) may be useful in the early detection of an encephalitic involvement in CNS Lyme borreliosis. Diffuse dysrythmia and circumscribed or diffuse slowing with interspersed sharp waves during Lyme encephalomyelitis and severe mental disturbance have been seen (Ackermann et al. 1985; Reik et al. 1985; Kollikowski et al. 1988; Pachner et al. 1989). Although anticonvulsant therapy resulted in an improvement in these patients, the EEG remained abnormal.

Somatosensory, visual, or acoustic evoked potentials often show prolonged latencies consistent with a demyelinating process (Ackermann et al. 1985; Kohler et al. 1988; Halperin et al. 1989; Krüger et al. 1989), although the frequency of these abnormalities for chronic CNS manifestations cannot be presented since only a few patients have been reported. Determinations of nerve conduction velocities, sensory and motor potential as well as F responses and H reflexes are only required in the diagnosis of peripheral neuropathy and will be discussed in detail elsewhere in this volume (see Chap. 16).

Electromyographic (EMG) investigation of affected muscles showed signs of denervation–like fibrillation potentials and positive sharp waves (Wokke et al. 1987b). In a previously described patient suffering from a subacute myelitis presenting as a partial stiff–man syndrome, normal motor units revealed tonic firing and a reflex myoclonus including long–lasting decrescendo activity, but no signs of denervation (Martin et al. 1990). Since the cardiac muscle may be involved during Lyme borreliosis, an electrocardiogram should always be performed.

15.3.3 Neuropsychological Testing

When neuropsychological tests are performed, they help to quantify and follow impairments in cognition, concentration, and memory, especially when these are subtle. These findings have so far been reported more frequently in patients with mild encephalitis and encephalopathy by American authors. Halperin et al. (1989) prefer the bedside Mini Mental State to the sophisticated and time–consuming battery of tests that they had used previously.

15.3.4 Histological Investigation and Cultivation of Spirochetes from CSF or Tissue Specimen

Proof or hints that a CNS disorder is caused by *B. burgdorferi* are provided by the demonstration of this bacterium in affected tissues or by its isolation from the brain or CSF. Both procedures cannot, of course, be generally recommended because of the high risk involved in brain biopsy in this treatable disease and because of the difficulties in growing the borreliae from the CSF (Barbour 1988). So far, no successful isolation of *B. burgdorferi* from brain tissue during or after chronic CNS Lyme borreliosis has been described. There are a few reports showing the agent in biopsy or autopsy specimen. In one patient with suspected herpes encephalitis, borreliae could be demonstrated in brain biopsy tissue (Pachner et al. 1989). Spirochetes which were morphologically compatible with *B. burgdorferi* were detected in the temporal lobe by Dieterle's silver stain; foci of microglial cells but without perivascular cuffing were also seen. In a newborn whose mother had suffered from Lyme borreliosis early during pregnancy and which died 23 h after birth, *B. burgdorferi* could be demonstrated by silver staining and immunocytochemistry in the brain and liver (Weber et al. 1988).

The successful isolation of *B. burgdorferi* is especially helpful in cases of latent Lyme neuroborreliosis where the CSF does not show inflammatory signs (Pfister et al. 1989).

15.4 Special Laboratory Examinations

15.4.1 Cerebrospinal Fluid

After the patient's history of a tick bite or/and erythema migrans and the serology, examination of the CSF is the most important diagnostic step. A spinal tap should be performed as early as possible when CNS involvement is suspected. Although CSF abnormalities resemble those found in other inflammatory or autoimmune diseases of the nervous system such as viral encephalitis, tuberculous meningitis, or multiple sclerosis, the composite of CSF findings found with *B. burgdorferi* infection is so characteristic that it often allows the diagnosis. There are no major differences between the CSF findings in meningoradiculoneuritis and chronic encephalo– or radiculomyelitis (Wilske et al. 1986; Ackermann et al. 1988; Martin et al. 1988a; Mertens et al. 1989). A proper analysis of the CSF should always include the determination of the following characteristics:

1. Lymphomonocytic pleocytosis (66 – 670 cells/ul; mean 230 cells/μl) (Mertens et al. 1989) mainly composed of CD4 T lymphocytes, some plasma cells, monocytes, and granulocytes.
2. Elevated total protein (mean 180mg/dl) (Mertens et al. 1989).
3. Blood–brain barrier dysfunction at the beginning, later gradually disappearing.
4. Markedly elevated immunoglobulins of classes IgA, IgG and IgM. IgG and IgA are intrathecally secreted for more than 1,5 years after antibiotic treatment. The intrathecal synthesis can be calculated according to Delpech and Lichtblau (1972) or Reiber (1980).
5. Oligoclonal, *B. burgdorferi*–specific IgG bands can be detected in the CSF by immunoblotting.
6. Positive serology for *B. burgdorferi*–specific IgG or IgM or both using the immunofluorescence assay (IFA) or the enzyme–linked immunosorbent assay (ELISA).

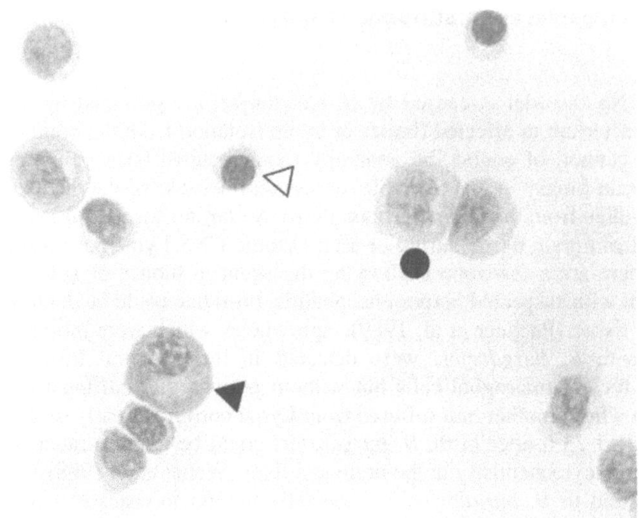

Fig. 2. Lymphomonocytic pleocytosis during chronic CNS Lyme borreliosis (Pappenheim stain). Lymphocytes are prominent (*open triangle*), but monocytes (*closed circle*) and plasma cells (*closed triangle*) are also seen. (Courtesy of Dr. K.W. Pflughaupt, Clinical Laboratory, Department of Neurology, University of Würzburg, Germany)

The intrathecal synthesis of specific IgG can be assessed by adjusting paired CSF/blood samples to the same IgG content for ELISA and calculation quotients as mentioned under point 4. (For more details, see Chapt. 21).

Outstanding features are a lymphomonocytic pleocytosis, reaching rarely up to 1000 cells/µl which are mainly composed of CD4+ T cells (Fig. 2) (Martin et al. 1988b), an elevated total protein and albumin, raised levels of immunoglobulin IgA, IgG, and IgM (Wilske et al. 1986; Kohlhepp et al. 1987; Martin et al. 1988a) due to increased numbers of Ig–secreting B cells (Henriksson et al. 1986). Rarely, granulocytes are prominent early in the course of the disease. Specific CSF IgM is less pronounced in chronic courses than in acute ones (Kohlhepp et al. 1987), IgG, however, is persistently secreted within the CSF by a restricted number of B cells (Murray et al. 1986; Martin et al. 1988a). The intrathecal synthesis of *B. burgdorferi*–specific IgG, however, has not only been observed in patients suffering from chronic Lyme radiculomyelitis (Martin et al. 1988a), but also some years after acute and untreated meningoradiculoneuritis (Krüger et al. 1989). Other CSF parameters such as glucose and lactate are noncharacteristic and usually found within the normal range. The serologic findings will be mentioned below and in detail in Chapt. 21.

During follow–up, the initial blood–brain–barrier disturbance indicated by excessively elevated protein values improves. Later, the most characteristic parameters are intrathecal IgG synthesis and oligoclonal *B. burgdorferi*–specific IgG bands, which can be demonstrated by isoelectric focusing and subsequent blotting onto antigen-coated filters, i.e., immunoblotting (Murray et al. 1986; Martin et al. 1988a; Krüger et al. 1989). Ongoing disease activity and therapeutic effectiveness is judged by the degree of pleocytosis. Normalization of the cell counts can generally be expected within 6 months following successful treatment. If more than 10 cells/µl are found at this point, the patient should be treated again. The continuous presence of cells may either be due to a persistent infection or to autoimmune reactions induced by *B. burgdorferi* (Martin et al. 1988b).

15.4.2 Polymerase Chain Reaction

Polymerase chain reaction is a powerful technique which allows the enzymatic amplification of only a few copies of nucleic acids. It is already widely applied for the detection of a great variety of microorganisms in tissue or body fluids. Rosa and Schwan (1989) recently described it for the investigation of *B. burgdorferi* deoxyribonucleic acid (DNA). Its extreme sensitivity will be especially helpful in studying whether *B. burgdorferi* can persist in the nervous system or CSF. However, its sensitivity may also be a cause of false–positive results if reaction conditions are not optimized. The detection of *B. burgdorferi* DNA in the CSF of patients with suspected chronic CNS manifestations who do not present with clear–cut serologic findings may be of comparable value for the diagnosis as a positive CSF culture.

15.4.3 Investigation of Antigen–Specific T Cells

Since both false–positive and false–negative serologies are seen in a considerable number of patients, additional evidence for a *B. burgdorferi*–caused disease can be gained by testing the patient's peripheral blood lymphocytes (PBL) or CSF mononuclear cells. For this purpose, these cells are incubated with *B. burgdorferi* and other

Table 2. Serologic findings and results of proliferative testings of PBL derived from seven patients suffering from subacute or chronic lymphomeningoradiculitis and radiculomyelitis

		Patient						
		1	2	3	4	5	6	7
Serum	IFA[a]	1:256	1:64b	1:256	1:1024	1:2048	1:64[b]	1:512
	ELISA IgG[c,d]	1:100+	1:100–	1:800	1:800	1:100+	1:100+	nd
	ELISA IgM[c,d]	1:100–	nd	nd	1:100–	1:100–	1:100–	nd
CSF	IFA[e]	1:64	1:8	1:64	1:256	1:256	1:32	1:128
	ELISA IgG[d,f]	1:80	1:10–	1:80	1:80	1:80	1:10+	nd
	ELISA IgM[d,f]	1:10+	nd	nd	1:10+	nd	1:10–	nd
Proliferation of PBL[g]								
	B. burgdorferi	3,9	10,0	26,9	2,4	3,1	6,1	1,6
	Tetanustoxoid	<1	3,9	53,8	14,2	5,1	27,3	28,8
	Leptospiral antigen	<1	<1	n.d.	1,2	2,0	1,3	<1
	Myelin basic protein	<1	1,2	1,2	<1	3,5	2,0	1,4

IFA, immunofluorescence assay; ELISA, enzyme–linked immunosorbent assay; CSF, cerebrospinalfluid, nd, not done; PBL, peripheral blood lymphocytes
[a] Positive titers are ≥1:64.
[b] Patients 2 and 6 had been pretreated with antibiotics.
[c] Positive titers are ≥1:100.
[d] Symbols +/– indicate the evaluation of the titer in each particular assay. It was scored positive if it exceeded 3 standard deviations above the mean of 8 negative controls included in each assay.
[e] Positive titers are ≥1:8.
[f] Positive titers are ≥1:10
[g] Stimulation indices are given as results of the proliferative assays (^{3}H–thymidine incorporation with antigen/incorporation without antigen). Indices >3 are scored positive.

antigens before tritiated thymidine is incorporated as a marker to determine the antigen–specific proliferative response (Pachner et al. 1985; Martin et al. 1988b) (Table 2). This assay provides an additional means of determining whether the patient had already been exposed to *B. burgdorferi*. In most of the patients, positive proliferative responses of either PBL or CSF lymphocytes are seen. The assay is of special value in patients with suspected Lyme disease and negative serology (Table 2, patient 2). The positive proliferative response indicates that the patient has encountered the bacterium. The response stays positive over long periods of time, even when the serologic response has become negative. Because of its technical difficulties and because it is far less standardized than serologic assays, the proliferative testing of PBL cannot, however, be recommended as a screening procedure. T cell lines established from the CSF during chronic Lyme radiculomyelitis responded to *B. burgdorferi* but also to autoantigens such as myelin basic protein (MBP), galactocerebrosides, and peripheral myelin and cardiolipin (Martin et al. 1988b). This indicated that not only humoral autoimmunity, but also self–reactive T cells can be induced by the bacterium. Whether they contribute to the demyelination seen during CNS Lyme borreliosis or rather represent cells which had been stimulated by released myelin and do not mediate damage themselves is still unclear.

15.5 Laboratory Changes

15.5.1 Serology

Since the isolation of *B. burgdorferi* is difficult, and histological demonstration of this agent in affected tissues often not feasible, serologic testing is still the most important laboratory parameter. *B. burgdorferi*–specific antibodies are detected by either IFA or ELISA. The validity of these assays will be discussed in detail in Chap. 21. In patients with suspected CNS Lyme borreliosis who are weakly seropositive or negative, it is recommended to try isolation of the borreliae from the CSF, use the immunoblot or Western blot techniques or test the T cells for antigen–specific proliferation. This also holds true for patients presenting with a clinical syndrome consistent with CNS Lyme borreliosis and positive serum titers, but negative CSF antibody titers. These patients should not be regarded as suffering from CNS Lyme borreliosis unless at least one of the assays mentioned above clearly indicates that *B. burgdorferi* caused the infection. After successful treatment, the titers may or may not continuously drop over the next few months. It remains to be determined whether significantly elevated titers which are seen in a number of healthy individuals (Sticht–Groh et al. 1988) and also for a long time after meningoradiculoneuritis (Krüger et al. 1989) indicate latent or persistent infection and require treatment.

15.6 Diagnosis

Since CNS Lyme borreliosis includes such a variety of different signs and symptoms, the diagnosis poses more difficulties than that of typical erythema migrans, meningoradiculoneuritis or acrodermatitis chronica atrophicans. Therefore, it usually requires several steps beginning with the patient's history, a clinical examination, and certain laboratory tests. With regard to the patient's history, special attention should be given to identifying a previous tick bite or erythema migrans, long–lasting pain, fatigue, and deficits in concentration and memory. The neurologic signs and physical examination have been mentioned in detail above.

Appropriate investigations to exclude other diseases and document the extent of tissue destruction are necessary. This is not only important for assessing the severity of the infection, but also for follow–up including the determination of therapeutic effectiveness. Of all laboratory parameters, serology and a thorough examination of the CSF provide the best information about the cause of the disease. If there is still doubt as to whether a patient suffers from CNS Lyme borreliosis, the investigation can be continued with an immunoblot or Western blot from CSF and serum, or with tests of T cell proliferation. In the future, the polymerase chain reaction applied to the CSF will provide additional information. Especially in chronic cases, IgM antibodies against *B. burgdorferi* may be absent, and an increased IgG titer is characteristic. However, an elevation of the IgG antibody titer is not always observed. A causal relationship between the CNS disorder and *B. burgdorferi* can be made, if an intrathecal *B. burgdorferi*–specific antibody response has been documented. Finally, the usually excellent response to antibiotic treatment is of further diagnostic value, although patients do not completely recover if brain or spinal cord damage is already advanced.

Table 3. Differential diagnosis of chronic CNS Lyme borreliosis

Differential diagnosis	*Distinguishing findings or investigations*
Neurosyphilis	Serology, joint destructions. Reflexes are often absent, the pupils are more often affected
Viral encephalitis	Serology, CSF usually with less pronounced protein values and intrathecal IgG synthesis. Seizures are more frequently seen in herpes simplex encephalitis; unresponsive to antibiotic treatment
Chronic fatigue syndrome	CSF; serology; unresponsive to antibiotic treatment
Major depressive disorder, psychosomatic disorders	CSF; serology; unresponsive to antibiotic treatment; history
Sarcoidosis	Lymph node involvement; chest X–ray; angiotensin converting enzyme; serology
Meningoencephalitis due to *Mycobacterium tuberculosis*	Chest X–ray; cultures; low CSF glucose; demonstration of mycobacteria in the CSF, tuberculin skin test
Fungal meningoencephalitis *(Cryptococcus neoformans, Aspergillus fumigatus)*	Demonstration of fungi of fungal antigens in the CSF; serology. Culture of fungi from the CSF
Lumbar arachnitis	CSF without pleocytosis; serology; myelography
Carcinomatous meningitis	Tumor cells in the CSF; less pronounced CSF pleocytosis; history; serology
Multiple sclerosis	Serology; CSF protein and immunoglobulins less elevated pleocytosis lower; evoked potentials more often abnormal; history
Alzheimer's disease and secondary dementia	Serology; CSF normal in Alzheimer's disease; clinical investigation does usually not disclose motor or sensory involvement
Brain tumors	CSF; serology; MRI; CT; intracranial pressure; biopsy

CSF, cerebrospinal fluid; IgG, immunoglobulin G; MRI, magnetic resonance imaging; CT, computed tomography

15.7 Differential Diagnosis

It can easily be concluded from the complex picture of CNS manifestations during Lyme borreliosis that a number of other disorders have to be considered in the differential diagnosis. An overview of the most important disorders and the investigations which help to exclude them are presented in Table 3.

15.8 Summary

Although the etiology of Lyme borreliosis was discovered less than a decade ago, it has been known for much longer that meningoradiculoneuritis or even encephalitis may evolve after a tick bite. Compared to the overall incidence of neurologic manifestations during Lyme borreliosis, cases of chronic CNS borreliosis only account for up to 4,7%. It still needs to be clarified whether genetic factors of the patient or characteristics of different *B. burgdorferi* strains are responsible for the occurrence of chronic courses rather than the often favorable outcome. Because of the sometimes long time interval between the beginning of the infection, the often subtle symptons, and the wide span of CNS syndromes, the diagnosis may pose major difficulties. It is based on the case history, clinical examination, serology (including the demonstration of intrathecally produced specific antibodies), CSF findings, and, finally, the response to antibiotic treatment. In the past few years, the disease has gained widespread public attention and, therefore, overdiagnosis was anticipated. Nevertheless, it cannot be emphasized too often that a physician and especially a neurologist should consider borreliosis in cases of chronic CNS disorders since it can be well treated.

Acknowledgements. We thank H.F. McFarland M.D., Neuroimmunology Branch, National Institute of Neurological Disorders and Stroke, National Institutes of Health, for helpful discussion and for critically reviewing the manuscript.

References

Aberer E, Brunner C, Suchanek G, Klade H, Barbour A, Stanek G, Lassmann H (1989) Molecular mimicry and Lyme borreliosis: a shared antigenic determinant between *Borrelia burgdorferi* and human tissue. Ann Neurol 26:732–737

Ackermann R, Gollmer E, Rehse-Küpper B (1985) Progressive Borrelien-Enzephalomyelitis. Dtsch Med Wochenschr 110:1039–1042

Ackermann R, Rehse-Küpper B, Gollmer E, Schmidt R (1988) Chronic neurologic manifestations of erythema migrans borreliosis. Ann NY Acad Sci 539:16–23

Barbour AG (1988) Laboratory aspects of Lyme borreliosis. Clin Microbiol Rev 1:399–414

Behringer A, Wirbatz A (1987) Chronische Borrelien-Erkrankung des zentralen Nervensystems. Nervenarzt 58:564–567

Bensch J, Olcen P, Hagberg L (1987) Destructive chronic Borrelia meningoencephalitis in a child untreated for 15 years. Scand J Infect Dis 19:697–700

Burke JM, Schaberg DR (1985) Neurosyphilis in the antibiotic era. Neurology 35:1368–1371

Delpech B, Lichtblau E (1972) Étude quantitative des immunoglobulines G et de l'albumine du liquide céphalorachidien. Clin Chim Acta 37:15–23

Hänny PE, Häuselmann HJ (1987) Die Lyme-Krankheit aus der Sicht des Neurologen. Schweiz Med Wochenschr 117:901–915

Halperin JJ, Luft BJ, Anand AK et al. (1989) Lyme borreliosis: central nervous system manifestations. Neurology 39:753–759

Hansen K, Rechnitzer C, Pedersen NS, Arpi M, Jessen O (1986) *Borrelia* meningitis in Denmark. Zentralbl Bakteriol Hyg (A) 263:348–350
Henriksson A, Link H, Cruz M, Stiernstedt G (1986) Immunoglobulin abnormalities in cerebrospinal fluid and blood over the course of lymphocytic meningoradiculitis (Bannwarth's syndrome). Ann Neurol 20:337–345
Klenk W, Heitmann R, Ackermann R (1985) Rezidivierende Erythema-Chronicum-Migrans-Krankheit des Nervensystems: Querschnittsmyelitis als Rückfall einer Meningopolyneuritis Garin–Bujadoux–Bannwarth. Akt Neurol 12:20–23
Kohler J, Kern U, Kasper J, Rhese–Küpper B, Thoden U (1988) Chronic central nervous system in Lyme borreliosis. Neurology 38:863–867
Kohlhepp W, Mertens HG, Oschmann P, Rohrbach E (1987) Akute und chronische Erkrankungen bei zeckenvermittelter Borreliose. Nervenarzt 58:557–563
Kohlhepp W, Kuhn W, Krüger H (1989) Extrapyramidal features in central Lyme borreliosis. Eur Neurol 29:150–155
Kollikowski HH, Schwendemann G, Schulz M, Wilhelm H, Lehmann HJ (1988) Chronic *Borrelia* encephalomyeloradiculitis with severe mental disturbance: immunosuppressive versus antibiotic therapy. J Neurol 235:140–142
Krüger H, Reuss K, Pulz M, Rohrbach E, Pflughaupt KW, Martin R, Mertens HG (1989) Meningoradiculitis and encephalomyelitis due to *Borrelia burgdorferi*: a follow–up study of 72 patients over 27 years. J Neurol 236:322–328
Macdonald AB, Miranda JM (1987) Concurrent neocortical borreliosis and Alzheimer's disease. Hum Pathol 18:749–761
Martin R, Martens U, Sticht–Groh V, Dörries R, Krüger H (1988a) Persistent intrathecal secretion of oligoclonal, *Borrelia burgdorferi*–specific IgG in chronic meningoradiculitis. J Neurol 235:229–233
Martin R, Ortlauf J, Sticht–Groh V, Bogdahn U, Goldmann SF, Mertens HG (1988b) *Borrelia burgdorferi*–specific and autoreactive T cell lines from cerebrospinal fluid in Lyme radiculomyelitis. Ann Neurol 24:509–516
Martin R, Meinck HM, Schulte–Mattler W, Ricker K, Mertens HG (1990) *Borrelia burgdorferi* myelitis presenting as a partial stiff–man syndrome. J Neurol 237:51–54
May EF, Jabbari B (1990) Stroke in neuroborreliosis. Stroke 21:1232–1235
Mertens HG, Kohlhepp W, Martin R (1989) Die Borrelienenzephalitis und -myelitis. In Fischer PA, Baas H, Enzensberger W (eds) Gerontoneurologie, Enzephalitiden, Neurogenetik. Springer, Berlin Heidelberg New York, pp 396–407 (Verhandlungen der Deutschen Gesellschaft für Neurologie, vol. 5)
Midgard R, Hofstad H (1987) Unusual manifestations of nervous system *Borrelia burgdorferi* infection. Arch Neurol 44:781–783
Murray N, Kristoferitsch W, Stanek g, Steck AJ (1986) Specificity of CSF antibodies against components of *Borrelia burgdorferi* in patients with meningopolyneuritis Garin-Bujadoux-Bannwarth. J Neurol 233:224–227
Pachner AR (1989) Lyme disease. TINS 12:177–181
Pachner AR, Steere AC, Sigal LH, Johnson CJ (1985) Antigen-specific proliferation of CSF lymphocytes in Lyme disease. Neurology 35:1642–1644
Pachner AR, Duray P, Steere AC (1989) Central nervous system manifestations of Lyme disease. Arch Neurol 46:790–795
Papolla MA, Omar R, Saran B, Andorn A, Suarez M, Pavia C, Weinstein A, Shank D, Davis K, Burgdorfer W (1989) Concurrent neuroborreliosis and Alzheimer's disease: analysis of evidence. Hum Pathol 20:753–757
Pfister HW, Einhäupl K, Preac–Mursic V, Wilske B, Schierz G (1984) The spirochetal etiology of lymphocytic meningoradiculitis of Bannwarth (Bannwarth's syndrome). J Neurol 231: 141–144
Pfister HW, Einhäupl K, Preac–Mursic V, Wilske B (1985) Lyme disease: commentary from Europe. Similarity of neurologic manifestations of *Borrelia burgdorferi* infections in America and Europe. Neurology 35:1393–1394
Pfister HW, Preac–Mursic V, Wilske B, Einhäupl KM, Weinberger K (1989) Latent Lyme neuroborreliosis: presence of *Borrelia burgdorferi* in the cerebrospinal fluid without concurrent inflammatory signs. Neurology 39:1118–1120

Pflüger KH, Reimers CD, Neubert U et al. (1988) Erythema migrans borreliosis: an HLA-associated disease ? Ann NY Acad Sci 539:414–416

Reiber H (1980) The discrimination between different blood–CSF barrier dysfunctions and inflammatory reactions of the CNS by a recent evaluation graph for the protein profile of cerebrospinal fluid. J Neurol 224:89–99

Reik L, Smith L, Khan A, Nelson W (1985) Demyelinating encephalopathy in Lyme disease. Neurology 35:267–269

Reik L, Burgdorfer W, Donaldson JO (1986) Neurologic abnormalities in Lyme disease without erythema chronicum migrans. Am J Med 81:73–78

Rosa PA, Schwan TG (1989) A specific and sensitive assay for the Lyme disease spirochete *Borrelia burgdorferi* using the polymerase chain reaction. J Infect Dis 160:1018–1029

Rowland LP (ed) (1989) Textbook of neurology, 8th edn. Lea and Febiger (Philadelphia), pp 152–161

Rousseau JJ, Lust C, Zangerle PF, Bigaignon G (1986) Acute transverse myelitis as presenting neurological feature of Lyme disease. Lancet II: 1222–1223

Schaltenbrand G (1962) Radikulomyelomeningitis nach Zeckenbiss. Munch Med Wochenschr 18:829–834

Schlüsener HJ, Martin R, Sticht–Groh V (1989) Autoimmunity in Lyme disease: molecular cloning of antigens recognized by antibodies in the cerebrospinal fluid. Autoimmunity 2:323–330

Steere AC (1989) Lyme disease. N Engl J Med 321:586–596

Steere AC, Hardin JA, Ruddy S, Mummaw JG, Malawista SE (1979) Lyme arthritis: correlation of serum and cryoglobulin IgM with activity, and serum IgG with remission. Arthritis Rheum 22:471–483

Sticht–Groh V, Martin R, Schmidt–Wolf I (1988) Antibody titer determinations against *Borrelia burgdorferi* in blood donors and in two different groups of patients. Ann NY Acad Sci 539:497–499

Stiernstedt G, Sköldenberg B, Vandvik B et al. (1984) Chronic meningitis and Lyme disease in Sweden. Yale J Biol Med 57:491–497

Suchanek G, Kristoferitsch W, Stanek G, Bernheimer H (1986) Anti–myelin antibodies in cerebrospinal fluid and serum of patients with meningopolyneuritis Garin–Bujadoux–Bannwarth and other neurological diseases. Zentralbl Bakteriol Hyg (A) 263:160–168

Uldry PA, Regli F, Bogousslavsky J (1987) Cerebral angiopathy and recurrent strokes following *Borrelia burgdorferi* infection. J Neurol Neurosurg Psychiatry 50:1703–1704

Vandvik B, Sköldenberg B, Stiernstedt G (1984) Tick–borne spirochetal meningitis, meningoradiculitis and meningoencephalitis. A report on 15 cases identified by demonstration of intrathecal spirochete–specific IgG antibody responses. Acta Neurol Scand 70:379–380

Veenendaal–Hilbers JA, Perquin WVM, Hoogland PH, Doornbos L (1988) Basal meningovasculitis and occlusion of the basilar artery in two cases of *Borrelia burgdorferi* infection. Neurology 38:1317–1319

Weber K, Bratzke H–J, Neubert U, Wilske B, Duray PH (1988) *Borrelia burgdorferi* in a newborn despite oral penicillin for Lyme borreliosis during pregnancy. Pediatr Infect Dis J 7:286–289

Weder B, Wiedersheim P, Matter L, Steck A, Otto F (1987) Chronic progressive neurological involvement in *Borrelia burgdorferi* infection. J Neurol 234:40–43

Wilske B, Schierz G, Preac–Mursic V et al. (1986) Intrathecal production of specific antibodies against *Borrelia burgdorferi* in patients with lymphocytic meningoradiculitis (Bannwarth's syndrome). J Infect Dis 153:304–313

Wokke JHJ, van Gijn J, Elderson A, Stanek G (1987a) Chronic forms of *Borrelia burgdorferi* infection of the nervous system. Neurology 37:1031–1034

Wokke JHJ, de Koning J, Stanek G, Jennekens FGI (1987b) Chronic muscle weakness by *Borrelia burgdorferi* meningoradiculitis. Ann Neurol 22:389–392

Wokke JHJ, van Doorn PA, Brand A, Schreuder GMT, Vermeulen M (1988) Association of HLA–DR antigen with serum IgG antibodies against *Borrelia burgdorferi* in Bannwarth's syndrome. J Neurol 235:415–417

16 Chronic Peripheral Neuropathy

W. KRISTOFERITSCH

16.1 Introduction

Peripheral neuropathy is one of the main manifestations of early disseminated Lyme borreliosis. It is well known as lymphocytic meningoradiculoneuritis (Bannwarth's syndrome) and has a self–limiting course of less than 6 months in most cases (see Chap. 11). However, cases of chronic peripheral neuropathy in late or chronic Lyme borreliosis are less well known. They may occur together with other manifestations of late Lyme borreliosis such as acrodermatitis chronica atrophicans (ACA) (Hopf 1966, 1975; Kaiser 1972; Kristoferitsch et al. 1988; Kristoferitsch 1989) and chronic arthritis (Halperin et al. 1987, 1988, 1990), or they may on rare occasions represent a non–self–limiting form of Bannwarth's syndrome (Weder et al. 1987; Wokke et al. 1987 a,b; Martin et al. 1988). There is some evidence that chronic peripheral neuropathy without meningeal inflammation and without other manifestations of late Lyme borreliosis also exists (Klöter et al. 1986; Meier et al. 1989).

16.2 ACA–Associated Neuropathy

The first systematic studies on chronic neuropathy were performed in patients with ACA by H.C. Hopf (1966) and M. Kaiser (1972), but signs or symptoms associated with peripheral neuropathy had already been mentioned in the first reports on ACA at the turn of the century (quoted by Hopf 1966). Hopf's clinical results could be confirmed in a recent study, which also included nerve biopsy investigations (Kristoferitsch et al. 1988; Kristoferitsch 1989).

Peripheral neuropathy occurs frequently in ACA; 40% – 63% of unselected patients with ACA suffer from symptoms corresponding to peripheral neuropathy (Hopf 1966, 1975; Kaiser 1972; Asbrink 1985; Kristoferitsch et al. 1988). Clinical signs of peripheral neuropathy have been found in about 40% of patients with ACA (Hopf 1966, 1978; Kaiser 1972; Kristoferitsch et al. 1988), so the term ACA–associated neuropathy has been used (Ackermann et al. 1988; Kristoferitsch et al. 1988).

16.2.1 Clinical Picture

As with ACA (Jessner and Loewenstamm 1924; Jordan 1930; Hopf 1966; Asbrink 1985), ACA–associated neuropathy is mainly observed in elderly patients and there is a predominance of female patients (Hopf 1966; Kaiser 1972; Kristoferitsch et al. 1988). In most cases, clinical symptoms and signs are mild, but chronic. Sensory deficits occur more frequently than loss of tendon reflexes or pareses (Hopf 1966, 1975; Kaiser 1972; Kristoferitsch et al. 1988) (Tables 1,2).

Table 1 Symptoms associated with peripheral neuropathy in 55 patients with acrodermatitis chronica atrophicans. (Modified from Kristoferitsch 1989)

Symptoms	*n*	%
Pain	16	29.1
Paresthesia	13	23.6
Weakness	4	7.3
Muscle cramps	2	3.6
Total reported symptoms	25	45.5

Table 2 Neurological signs in 22 patients with ACA–associated neuropathy (Modified from Kristoferitsch 1989)

Signs	*n*	%
Paresis	4	18.2
Tendon reflexes <	11	50.0
Sensory deficit	21	95.5
Patch–like	8	
Asymmetric	5	
Distal symmetric	8	

< , Diminution or loss

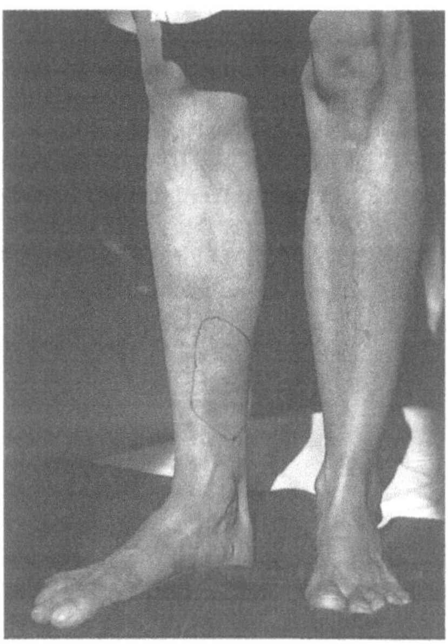

Fig.1. "Cutaneous" peripheral neuropathy in a patient with infiltrative acrodermatitis chronica atrophicans on the right lower extremity. *Circled area* indicates hypesthetic zone (from Kristoferitsch et al. 1988)

The distribution of sensory deficits may be patch–like and located within the region involved by ACA (Fig. 1), or of a distal asymmetric or distal symmetric type and may also involve extremities without ACA (Hopf 1966, 1975). Patch–like distributed hypesthesia or "peripheral neuropathy of the local type" (Hopf 1975) resembles cutaneous peripheral neuropathy in systemic vasculitis (Moore and Fauci 1981; Kristoferitsch 1989). Sensory deficits in radicular distribution or plexopathy are found exceptionally (Hopf 1966; Kristoferitsch et al. 1988). In patients with distal asymmetric or symmetric peripheral neuropathy, all sensory qualities are usually affected. Dissociated hypesthesia has been found rarely in some studies (Hopf 1966; Kristoferitsch 1989), but not in others (Kaiser 1972).

Loss or diminution of tendon reflexes or pareses are usually located on lower extremities and distributed distally. Pareses have exceptionally been observed in proximal distribution (cited in Hopf 1975; Kristoferitsch 1989) and have been attributed to a preceding Bannwarth's syndrome (Kristoferitsch 1989). Cranial nerve paresis has been found preceding ACA. This has also been interpreted as Bannwarth's syndrome prior to ACA (Asbrink 1985; Kristoferitsch 1989). On rare occasions, cranial nerve affections were seen simultaneously with ACA (Asbrink 1985; Kristoferitsch 1989). The connection between ACA–associated neuropathy and cranial nerve affection has not been studied systematically.

The clinical spectrum of ACA–associated neuropathy includes cutaneous peripheral neuropathy and asymmetric or symmetric, sensory or sensorimotor, distal polyneuropathy. Asymmetric and symmetric polyneuropathy seem to represent overlapping mononeuritis multiplex (Fig. 2). This pattern has been described in several infectious disorders (Dyck et al. 1984; Neundörfer et al. 1987), but also in vasculitis (Moore and Fauci 1981).

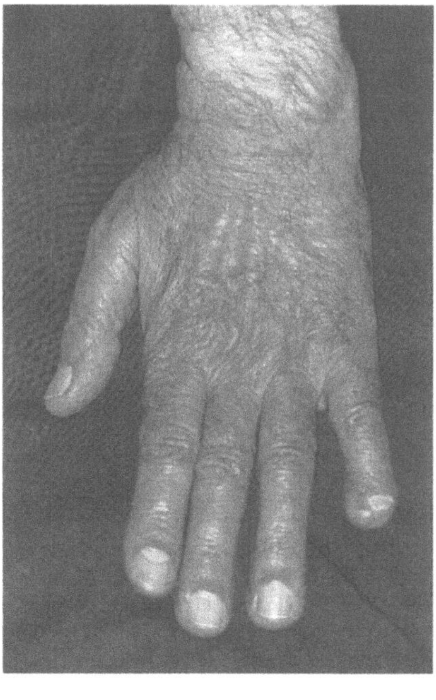

Fig.2 Neuritis of the left ulnar nerve in a patient with acrodermatitis chronica atrophicans of the left hand and mononeuritis multiplex. Note atrophy of the interosseous muscles

16.2.2 Peripheral Neuropathy and Dermatological Alterations

Although ACA–associated neuropathy even occurs on extremities without cutaneous involvement, it is more prominent in the limbs affected with ACA. No patients with ACA have been observed to present signs of peripheral nerve involvement exclusively on extremities without ACA (Hopf 1966, 1975; Kaiser 1972; Kristoferitsch 1989). No significant correlation was found between the duration of ACA, number of ACA-involved extremities, or stage of ACA on the one hand and manifestation of peripheral neuropathy on the other. Signs and symptoms of peripheral neuropathy were found in several patients, even when cutaneous alterations had disappeared after therapy (Kaiser 1972; Kristoferitsch et al. 1988; Weber et al. 1988; Kirstoferitsch 1989). Local factors in the region of ACA seem to influence the beginning, but not the further course of peripheral neuropathy.

16.2.3 Electroneurographical Testing

Electroneurographical (ENG) investigations in 55 patients with ACA have shown multiple nerve involvement (Kristoferitsch 1989). This corresponds to the clinical pattern

Table 3. Electroneurographical findings (peroneal and sural nerves) in 18 patients with ACA–associated peripheral neuropathy[a]

Patient	Peroneal nerve mNCV	SD	dL	CAP	F–L	SD	Sural nerve sNVC	SD	NAP
1	43	*	3.9	9.0	**49**	2.3	68	*	7.5
2	**34**	3.4	4.4	10.0	42	*	55	*	**1.5**
3r	**40**	2.4	4.4	**0.5**	**52**	3.6	45	*	13.6
3l	47	*	4.0	**2.0**	38	*	51	*	20.0
4r	**37**	2.8	**5.4**	3.7	**49**	2.6	40	*	3.0
4l	49	*	**5.5**	10.0	46	*	42	*	7.8
5	43	*	4.9	**0.5**	**57**	5.5	**36**	2.5	**1.3**
6	41	*	4.1	5.0	**48**	2.3	59	*	25.0
7	51	*	3.8	4.0	**48**	2.1	42	*	8.0
8	56	*	4.8	**1.2**	39	*	59	*	9.2
9	43	*	4.2	3.5	44	*	**29**	3.9	3.0
10	48	*	3.7	9.2	33	*	**32**	3.2	10.0
11r	47	*	**5.7**	3.7	48	*	**34**	2.8	**1.0**
11l	48	*	**5.2**	9.5	42	*	42	*	**2.0**
12	47	*	**5.1**	**2.4**	**56**	5.1	53	*	11.0
13	43	*	**7.2**	**0.7**	**60**	6.5	**33**	3.2	3.6
14r	48	*	4.5	**1.4**	47	*	61	*	**1.9**
14l	47	*	4.1	**1.1**	44	*	54	*	3.3
15	48	*	3.6	**2.0**	41	*	62	*	12.0
16	42	*	**5.3**	**1.0**	**61**	6.3	**24**	4.6	3.0
17	49	*	**6.6**	**0.8**	**49**	2.3	41	*	5.0
18	48	*	5.0	**2.3**	44	*	71	*	**1.0**

mNCV, motor nerve conduction velocity (m/s); SD, standard deviation; *, SD < 2.0; dL, distal latency (ms); CAP, compound action potential (mV); F–L, F wave latency (distal stimulation) (ms); sNCV, sensory nerve conduction velocity (m/s); NAP, nerve action potential (amplitude, μV)

[a] pathologic values are in bold face

of mononeuritis multiplex or overlapping mononeuritis multiplex. In the majority of patients, ENG alterations were compatible with axonal degeneration. In none of the peroneal nerves examined did the ENG results fulfil criteria for primary demyelination (Mamoli and Brunner 1984) (Table 3). Increased F wave latencies were found frequently. It is not clear if these findings represent nerve root involvement or if they are due to the higher sensitivity of F wave measurement in detecting subtle impairment of nerve conduction velocity (Conrad et al. 1975). In another study, electromyography and ENG findings were interpreted as "corresponding to a disorder affecting predominantly the distal part of peripheral nerves" (Hopf 1966). Up to 13% of patients with ACA show abnormal ENG alterations without clinical signs of peripheral neuropathy (Kristoferitsch 1989).

16.2.4 Morphology

Sural nerve biopsies from patients with ACA–associated neuropathy showed a perivascular infiltration with lymphocytes and plasma cells distributed around small and medium sized epineurial vessels and in the epineurial soft tissue (Kristoferitsch et al. 1988; Meier et al. 1989) (see also Chap.6). The infiltration was predominantly perivascular, including the external layers of the vessels. Occasionally vascular obliterations have been observed. Polymorphonuclear infiltration, necrosis or fibrinoid degeneration of the vessel wall, which are typical in necrotizing vasculitis (Fauci et al. 1978), were not seen. Inflammatory infiltration of endoneurial vessel was not present. Mononuclear cells could also be sporadically detected in the endoneurium of clinically severely affected patients. Semi–thin sections revealed disseminated loss of myelinated fibers. Teased–fiber preparations and electron–microscopical investigations showed that the fiber loss was caused primarily by axonal degeneration (Kristoferitsch et al. 1988; Meier et al. 1989). No staining for complement factor C3 and for fibrinogen was seen in immunofluorescence microscopy. No borrelial structures were found by silver staining, staining with immunoperoxidase–conjugated specific antibodies, or by electron microscopy.

There is a striking similarity between the nerve biopsy findings in ACA–associated neuropathy and findings reported in Bannwarth's syndrome (Camponovo and Meier 1986; Vallat et al. 1987; Engelhardt et al. 1989; Meier et al. 1989; Meurers et al. 1990). In the latter, perivascular cuffing of endoneurial vessels and degenerating perineurial cells could be detected additionally, thus indicating a more pronounced disorder. This has its counterpart in the clinically more acute and severe course of Bannwarth's syndrome.

16.2.5 Laboratory Findings

As with other laboratory findings, antibody titers against *Borrelia burgdorferi* did not differ in patients with ACA but without peripheral neuropathy and in those with ACA and peripheral neuropathy (Kristoferitsch 1989). Cerebrospinal fluid (CSF) was found normal in most patients with ACA (Hauser 1955) or with ACA–associated peripheral neuropathy. Pathologic CSF alterations were found only in 5 of 32 patients by Hopf (1966) and in 1 of 5 patients by Kristoferitsch (1989) and had been attributed to other concomitant neurological disorders.

16.2.6 Pathogenetic Aspects

The clinical pattern and ENG findings in ACA–associated neuropathy correspond to mononeuritis multiplex. Sural nerve biopsies show perivascular lymphocytic infiltration associated with a loss of myelinated fibers due to axonal degeneration. This has been accepted to be suggestive of vasculitic peripheral neuropathy (Wees et al. 1981). Involvement of different nerve fascicles and the relation between the intensity of vascular infiltration and the amount of nerve fiber loss point to an ischemic origin, yet the cause of the inflammatory infiltration remains unknown. It has yet to be proven whether ACA–associated neuropathy is of angiopathic–ischemic origin, or due to an immune reaction to borrelial antigens, or caused by both. Direct invasion of nerves with *B. burgdorferi* has not been detected so far.

The histomorphological alterations found were almost identical with those found in Bannwarth's syndrome (Camponovo and Meier 1986; Vallat et al. 1987; Engelhardt et al. 1989; Meier et al. 1989; Meurers et al. 1990) and similar to those seen in American patients with late manifestations of Lyme borreliosis (Halperin et al. 1987, 1988). It can be assumed that a common pathologic mechanism underlies the peripheral nerve involvement in these clinically different disorders.

16.2.7 Diagnosis

As soon as the diagnosis of ACA has been established, ACA–associated peripheral neuropathy will easily be recognized. It may then become apparent that clinical symptoms such as pain, paresthesia, hyperesthesia, weakness, or muscle cramps are not due to dermatological alterations, but are neurological symptoms. Further neurological examinations may confirm the diagnosis. In patients with ACA who present additional causes for peripheral neuropathy, nerve biopsy may be helpful in establishing the correct diagnosis.

16.2.8 Therapy

Therapeutic considerations in chronic peripheral neuropathy are described in Chap. 24.

16.3 Chronic Peripheral Neuropathy in Late Lyme Borreliosis Without ACA

Peripheral nerve involvement has been observed in morphea and lichen sclerosus et atrophicus, disorders of possible borrelial etiology (Hopf 1966; Aberer et al. 1988). Chronic non–self–limiting cases of radiculitis or radiculoneuritis have been reported on rare occasions (Weder et al. 1987; Wokke et al. 1987a; Martin et al. 1988). These show inflammatory CSF alterations, but lack the migratory pain which is characteristic for Bannwarth's syndrome. Camponovo and Meier have reported on sural nerve biopsies of a patient with neuritis, borderline CSF pleocytosis, and normal CSF proteins who was examined 2 years after the onset of neurological symptoms, erythema migrans and tick bite. In this patient peripheral neuropathy without radiculitis or meningeal inflammation may have followed Bannwarth's syndrome. In two other patients with chronic peripheral neuropathy but without signs of nerve root or meningeal involvement Lyme borreliosis was diagnosed exlusively on the basis of serodiagnostic findings (Klöter et al. 1986).

In 40% – 50% of American patients with evidence of late disseminated Lyme disease, i.e., cardiac, rheumatologic, or neurological abnormalities, neurophysiological alterations consistent with peripheral neuropathy were found. The ENG abnormalities were multifocal in nature, such as might be seen in mononeuritis multiplex and in ACA–associated neuropathy. As in patients with ACA–associated neuropathy, prolonged F wave latencies were found frequently. Although no clinical signs were evident in many of these patients, most of them had symptoms of intermittent limb paresthesias. A few patients only had radicular–type pain as in Bannwarth's syndrome. Nerve biopsy findings suggestive of a mild axonal process with secondary demyelination were described as qualitatively similar to those in Bannwarth's syndrome and ACA–associated peripheral neuropathy (Halperin et al. 1987, 1988, 1990).

16.4. Conclusion

Chronic peripheral neuropathy is a common manifestation of late Lyme borreliosis. It occurs in 40% – 63% of patients with ACA, which is a late cutaneous manifestation of Lyme borreliosis with European prevalence. It has been also observed in 40% – 50% of American patients with other manifestations of late Lyme borreliosis. The disorder may be easily overlooked because of the mild and subtle clinical symptoms which are – as in ACA–asssociated neuropathy – frequently regarded as non–neurological manifestations of Lyme borreliosis. Results of ENG and nerve biopsy studies seem to be qualitatively similar in ACA–associated neuropathy and in peripheral neuropathy of late Lyme borreliosis without ACA.

Lymphocytic vasculitis of the epineurial vessels with loss of myelinated fibers due to axonal degeneration are histomorphologically characteristic of ACA–associated neuropathy. In cases where other concomitant nonneurological manifestations of late Lyme borreliosis are lacking, serodiagnosis and nerve biopsy may be necessary to establish the correct diagnosis. There are no reports about the overall frequency of chronic peripheral neuropathy caused by Lyme borreliosis. Even if this disorder does not take a severe course, it is still of medical interest because it belongs to the few forms of chronic peripheral neuropathy which can be treated successfully.

References

Aberer E, Kollegger H, Kristoferitsch W, Stanek G (1988) Neuroborreliosis in morphea and lichen sclerosus et atrophicus. J Am Acad Dermatol 19:820–825

Ackermann R, Rehse–Küpper B, Gollmer E, Schmidt R (1988) Chronic neurologic manifestations of erythema migrans borreliosis. Ann NY Acad Sci 539:16–23

Asbrink E (1985) Erythema chronicum migrans Afzelius and acrodermatitis chronica atrophicans. Early and late manifestations of *Ixodes ricinus*–borne *Borrelia* spirochetes. Acta Derm Venereol (Suppl) (Stockh) 118:1–63

Asbrink E, Hovmark A (1988) Early and late cutaneous manifestations in *Ixodes*–borne borreliosis (Erythema migrans borreliosis, Lyme borreliosis). Ann NY Acad Sci 539:16–23

Camponovo F, Meier C (1986) Neuropathy of vasculitic origin in a case of Garin-Bujadoux-Bannwarth syndrome with positive borrelia antibody response. J Neurol 233:69–72

Conrad B, Aschoff JC, Fischler M (1975) Der diagnostische Wert der F–Wellen–Latenz. J Neurol 210:151–159

Dyck PJ, Thomas PK, Lambert EH, Bunge R (1984) Peripheral neuropathy, vol 2. Saunders, Philadelphia

Engelhardt A, Grahmann F, Neundörfer B (1989) Vasculitic neuropathy in a case of Bannwarth's syndrome. Zentralbl Bakteriol (Suppl) 18:136–140

Fauci AS, Haynes BF, Katz P (1978) The spectrum of vasculitis. Clinical, pathologic, immunologic, and therapeutic considerations. Ann Intern Med 89 (1):660–676

Halperin JJ, Little BW, Coyle PK, Dattwyler RJ (1987) Lyme disease: cause of a treatable peripheral neuropathy. Neurology 37:1700–1706

Halperin JJ, Pass HL, Anand AK, Luft BJ, Volkman DJ, Dattwyler RJ (1988) Nervous system abnormalities in Lyme disease. Ann NY Acad Sci 539:24–34

Halperin J, Luft BJ, Volkmann DJ, Dattwyler RJ (1990) Lyme neuroborreliosis: peripheral nervous system manifestations. Brain 113:1207–1221

Hauser W (1955) Zur Kenntnis der Akrodermatitis chronica atrophicans. (Unter besonderer Berücksichtigung der Veränderungen an den hautnahen Lymphknoten, des Knochenmarks, der Serumeiweißverhältnisse sowie der Ätiologie und der Pathogenese.) Arch Dermatol 199:350–393

Hopf HC (1966) Acrodermatitis chronica atrophicans (Herxheimer) und Nervensystem. Springer, Berlin Heidelberg New York (Monographien aus dem Gesamtgebiete der Neurologie und Psychiatrie, vol 114)

Hopf HC (1975) Peripheral neuropathy in acrodermatitis chronica atrophicans (Herxheimer). J Neurol Neurosurg Psychiatry 38:452–458

Jessner M, Loewenstamm A (1924) Bericht über 66 Fälle von Acrodermatitis chronica atrophicans. Dermatol Wochenschr 79:1169–1177

Jordan A (1930) Über die Ätiologie der idiopathischen progressiven Hautatrophie. Zentralbl Hautkr 34:312

Kaiser M (1972) Neurologische Komplikationen bei Akrodermatitis chronica atrophicans (Herxheimer) und ihre Beeinflussung durch die Penicillintherapie. Thesis, University of Göttingen

Klöter I, Adam T, Schabet M, Wiethölter H, Pfeiffer J (1986) *Borrelia*–induced meningoradiculitis – two different forms of the disease. Eur Neurol 25:262–268

Kristoferitsch W (1989) Neuropathien bei Lyme–Borreliose. Springer, Wien New York

Kristoferitsch W, Sluga E, Graf M, Partsch H, Neumann R, Stanek G, Budka H (1988) Neuropathy associated with acrodermatitis chronica atrophicans: clinical and morphological features. Ann NY Acad Sci 539:34–45

Mamoli B, Brunner G (1984) Elektrophysiologische Untersuchungen bei Neuropathien. Wien Klin Wochenschr 96 (Suppl 147):9–14

Meier C, Grahmann F, Engelhardt A, Dumas M (1989) Peripheral nerve disorders in Lyme–Borreliosis. Nerve biopsy studies from eigth cases. Acta Neuropathol (Berl) 79: 271–278

Martin R, Martens U, Sticht–Groh V, Dörries R, Krüger H (1988) Persistant intrathecal secretion of oligoclonal *Borrelia burgdorferi*–specific IgG in chronic meningoradiculomyelitis. J Neurol 235:229–233

Meurers B, Kohlhepp W, Gold R, Rohrbach E, Mertens HG (1990) Histopathological findings in the central and peripheral nervous system in neuroborreliosis. A report of three cases. J Neurol 237:113–116

Moore PM, Fauci AS (1981) Neurologic manifestation of systemic vasculitis: a retrospective and prospective study of the clinicopathologic features and responses to therapy in 25 patients. Am J Med 71:517–524

Neundörfer B, Schröder JM, Claus D (1987) Polyneuritiden und Polyneuropathien. In: Neundörfer B, Schimrigk K, Soyka D (eds) Praktische Neurologie, vol 2. VCH, Weinheim

Vallat JM, Hugon J, Lubeau M, Lebouted MJ, Dumas M, Desproges–Gotteron R (1987) Tick–bite meningoradiculoneuritis: clinical, electrophysiologic and histologic findings in 10 cases. Neurology 37:749–753

Weber K, Preac–Mursic V, Neubert U, Thurmayr R, Herzer P, Wilske B, Schierz B, Marget W (1988) Antibiotic therapy of early European Lyme borreliosis and acrodermatitis chronica atrophicans. Ann NY Acad Sci 539:324–345

Weder B, Wiedersheim P, Matter L, Steck A, Otto F (1987) Chronic progressive neurological involvement in *Borrelia burgdorferi* infection. J Neurol 234:40–43

Wees SJ, Sunwoo IN, Oh SJ (1981) Sural nerve biopsy in systemic necrotizing vasculitis. Am J Med 71:525–532
Wokke JHJ, de Koning J, Stanek G, Jennekens FGI (1987a) Chronic muscle weakness caused by *Borrelia burgdorferi* meningoradiculitis. Ann Neurol 22:389–392
Wokke JHJ, van Gijn J, Elderson A, Stanek G (1987b) Chronic forms of *Borrelia burgdorferi* infection of the nervous system. Neurology 37:1031–1034

17 Lyme Borreliosis in Childhood and Pregnancy

H.-J. CHRISTEN and F. HANEFELD

17.1 Introduction

Hitherto published reports on Lyme borreliosis in childhood deal mainly with single case observations. Only a few more comprehensive studies on pediatric patients have been published in Europe (Schaad et al. 1986; Christen et al. 1988; Naglo and Wide 1989; Huppertz 1990) and the USA (Steere et al. 1977; Williams et al. 1990). In addi-

Table 1. Prospective multicenter study on Lyme borreliosis in childhood with special emphasis on Lyme neuroborreliosis. Disorders found in the children investigated since 1986 (Christen and Hanefeld, unpublished)[a]

Disorders	Total (*n*)	Lyme borreliosis (*n*)
Neurologic disorders	*547*	*169*
Aseptic meningitis	283	46
Acute facial palsy	218	93
Meningoencephalitis	7	6
Recurrent headache	7	7
Bannwarth's syndrome	6	6
Guillain-Barré syndrome	5	3
Abducens nerve palsy	4	2
Oculomotor nerve palsy	1	1
Acute ataxia	2	1
Pseudotumor cerebri	2	1
Multiple sclerosis	9	–
Other neurologic symptoms	3	3
Dermatologic disorders	*30*	*30*
Erythema migrans	24	24
Borrelial lymphocytoma	5	5
Morphea	1	1
Acute arthritis	*15*	*9*
Gonarthritis	10	7
Oligoarthritis	2	2
Polyarthritis	3	–
Total	*592*	*208*

[a] The serological investigations were performed by Dr. med. H. Eiffert, Dr. vet. med. H. Lotter, Dr. med. M. Scriba and Prof. Dr. med. R. Thomssen (Department of Medical Microbiology, University of Göttingen).

tion to this material, our own results from a prospective, hospital–based, multicenter study are presented in this review. The methods of this study have been described in detail elsewhere (Christen et al. 1988 and 1990). Lyme borreliosis was diagnosed on the basis of the detection of specific antibodies against *Borrelia burgdorferi* using an immunoglobulin M (IgM) capture assay developed by Wassmann et al. (1989) and an IgG enzyme–linked immunosorbent assay (ELISA) according to Eiffert et al. (1990). During the first 3 years of this study, 592 children were examined. The distribution of the disorders investigated and the relative frequency of Lyme borreliosis are presented in Table 1. Based on the literature and our own results, the following description of Lyme borreliosis in childhood stresses the differences between children and adults.

17.2 Dermatological Manifestations

Erythema migrans has frequently been reported in children. As early as 1915, 2 years after Lipschütz (1913) had introduced the term "erythema chronicum migrans", Kerl reported on a 6-year-old girl presenting with this dermatological disorder. In 1920, Strandberg described "an unusual form of migratory erythema caused by tick bites", suggesting its etiology already at this time. He was one of the first to describe the combination of erythema migrans and borrelial lymphocytoma, which he observed in a 4-year-old girl. The second case of erythema migrans described by Lipschütz (1923) was in a 16-year-old girl who presented with a multilocular erythema migrans on the leg. In larger series about erythema migrans, children with this symptom accounted for about 10% of the patients (Hollström 1958; Sedlácek 1960).

Weber et al. (1983) and Åsbrink et al. (1986) have observed erythema migrans in 1-year-old children. Weber et al. (1986) reported on a reinfection with erythema migrans in a 6-year-old boy who had had primary erythema migrans at the age of 1 year. In the recent literature, there is only one pediatric report available describing details of erythema migrans in 13 patients (Schaad et al. 1986). Predominantly the head and neck region was affected in these children (8 out of 13 patients). We investigated 24 children with an erythema migrans and a history of a tick bite at the site of the erythema migrans. In three-fourths of the patients, the head and neck region was affected. Only 11 out of 24 children revealed elevated IgM antibody titers, and all of them had normal IgG antibody titers (Christen and Hanefeld, unpublished).

Borrelial lymphocytoma, a term introduced by Weber et al. (1985), is in many respects identical with the lymphadenosis benigna cutis, described by Bäfverstedt in 1943. In Bäfverstedt's original report, 26 out of 133 patients were less than 20 years old; nearly all of them presented with solitary lymphocytoma, mainly located in the face. Single case reports of lymphocytoma in children have been published since 1920 (Strandberg 1920; Bloedhorn 1923). In 1957, Paschoud succeeded in the first transmission of lymphocytoma from human to human which involved a 12-year-old boy. Höfer (1956) presented an evaluation of the literature on lymphocytoma; 12 out of 102 reported patients were less than 20 years old. In Mach and Wilgram's series (1966) on 115 patients with lymphocytoma, 28 pediatric patients were included.

The most common sites of borrelial lymphocytoma are the ears, the nipple, and the scrotum. This is in agreement with the reports of Weber and Neubert (1986) and Asbrink et al. (1989). Five out of Weber and Neubert's seven patients (1986) were children presenting with borrelial lymphocytoma exclusively on the ear. In the series of Asbrink et al. (1989), 13 out of 22 patients with borrelial lymphocytoma were below 13 years of age, 11 of whom presented with the lesion on the ear. In our series, five children presented with this sign located on the ears. Only two of them had increased IgM antibody titers, none of them had elevated IgG antibody titers. However, six out of seven of Weber and Neubert's patients (1986) and 14 of 19 patients of Asbrink et al. (1989) showed positive serum IgG antibody titers against *B. burgdorferi*.

Acrodermatitis chronica atrophicans (ACA) is a rare manifestation of Lyme borreliosis in childhood. In the classic report by Herxheimer and Hartmann (1902), a 15-year-old girl with this disease is described in detail. Reports on further pediatric patients have been published since 1896 (Nikulin 1896; Neumann 1899; Kerl 1920; Schiller 1921; Gottron 1939). In larger series, children accounted for about 10% – 20% of the patients with this disorder (Pirilä 1951; Kafka 1953; Hauser 1955). Göhr (1935) reported on only one pediatric patient with ACA in a series of 431 patients. Donnermann and Heite (1959) presented an evaluation of the literature on ACA in a total of 621 patients, 10% of whom were younger than 16 years of age. The youngest patient in the series of Hauser (1955) on 137 patients with ACA was 5 years old. In the recent literature, there is only one report on two well-documented patients with ACA in children (Nadal et al. 1988). In our series, we have not yet observed this disorder in a child.

17.3 Arthritis

Lyme arthritis usually consists of brief, self-limited attacks with joint effusion most commonly involving the knee (Steere et al. 1977; Culp et al. 1987). Besides this, chronic arthritis may sometimes occur (Steere et al. 1987). Other large joints such as the shoulder or the elbow joint may also be affected. Without a history of a preceding tick bite or an erythema migrans, the diagnosis of Lyme arthritis may be confused with bacterial septic arthritis, transient postinfectious arthritis, or pauciarticular juvenile rheumatoid arthritis (Schaad et al. 1986). Nonspecific laboratory findings are not helpful in the differential diagnosis. The diagnosis of Lyme arthritis should be based on a positive serological test in serum and, if available, in joint effusion and on exclusion of other forms of arthritis. False-positive serological results must be considered.

In our series, Lyme arthritis was diagnosed in nine children, seven of whom suffered from gonarthritis (Christen and Hanefeld, unpublished). Only three of these nine children presented with a history of a tick bite and/or erythema migrans. In the other six cases, antibody findings gave only presumptive evidence for infection with *B. burgdorferi*. All children revealed elevated IgM antibody titers, and four children elevated IgG antibody titers. In two children with arthritis, the diagnosis could additionally be confirmed by an increased IgM antibody titer in the synovial fluid.

17.4 Cardiac Abnormalities

Cardiac disorders represent additional but rare manifestations of Lyme borreliosis (see Chap.10) In pediatrics, two well-documented case reports are available. One child suffered from myocarditis due to Lyme borreliosis, which could be detected only by gallium scan while electrocardiography (ECG) and echocardiography findings were normal (Jacobs et al. 1984). The other child presented with conduction abnormalities after a preceding multilocular erythema migrans (Kishaba et al. 1988). Fifty of our children with Lyme neuroborreliosis had normal ECG findings (Christen and Hanefeld, unpublished).

17.5 Neurologic Disorders

17.5.1 Historical Review

The first pediatric patient with meningitis following an erythema migrans was described by Schirduan (1950). The distinct relationship between a tick bite, an erythema migrans in the same region, and a subsequent neurologic syndrome consisiting of optic neuritis and spastic signs in the presence of inflammatory cerebrospinal fluid (CSF) changes was observed by Schaltenbrand in a 9-year-old girl in 1953, but not published until 1966. In the classic report on lymphocytic meningoradiculitis by Bannwarth (1941), 4 out of 15 patients were children presenting with acute peripheral facial palsy associated with lymphocytic pleocytosis in the CSF. Bannwarth pointed out that the radicular pain characteristic in adults is missing in children, in whom facial palsy is the leading symptom, often presenting bilaterally. These observations were confirmed by Pette (1942), Zellweger (1946), Schaltenbrand (1949), and Keuth and Mennicken (1966).

17.5.2 Spectrum of Neurologic Disorders

17.5.2.1 Peripheral Facial Palsy

In the recent literature, acute peripheral facial palsy has been reported most frequently as a neurologic manifestation of Lyme borreliosis in children in Europe (Caflisch et al. 1984; Schaad et al. 1986; Jörbeck et al. 1987; Görke et al. 1989; Hanefeld et al. 1989; Naglo and Wide 1989) and in the USA (Williams et al. 1990). According to our observations (Christen and Hanefeld, unpublished), more than half of the children with Lyme neuroborreliosis presented with this manifestation (Fig. 1). Lyme borreliosis has been found to be the main cause of acute peripheral facial palsy in children (Christen et al. 1990). In 25 out of 72 consecutive children, Lyme borreliosis was diagnosed on the basis of detection of specific IgM antibodies in serum and CSF (Christen and Hanefeld, unpublished). In a comparable study in adults, Olsson et al. (1988) diagnosed Lyme borreliosis in 16 out of 82 patients with facial palsy.

According to our observations, the differences between the children suffering from Lyme borreliosis and those with facial palsy of unclear etiology were as follows (Christen et al. 1990; Christen and Hanefeld, unpublished): bilateral facial palsy occurred only in children with Lyme borreliosis (7 out of 93 cases); in contrast to cases

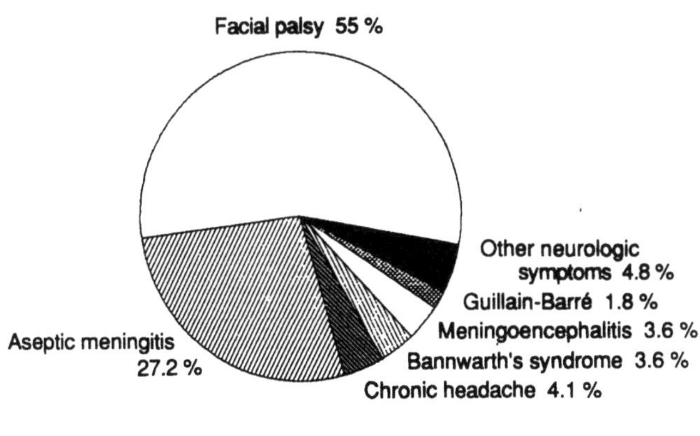

Fig. 1. Clinical spectrum of Lyme neuroborreliosis (n =169). Results of a prospective multicenter study on Lyme borreliosis in childhood (Christen and Hanefeld, unpublished)

of unclear etiology, facial palsy due to Lyme borreliosis was only observed between May and November, a tick bite or an erythema migrans in the head and neck region was recalled by 24 and 33 out of 93 children, respectively; with one exception, facial palsy developed within 3 weeks and 3 months on the ipsilateral side of the tick bite; all but three children with Lyme borreliosis revealed an inflammatory CSF syndrome with lymphocytic pleocytosis (median cell count 128 cells/µl, range 1-960 cells/µl; median protein content 49 mg%, range 11-339 mg%). This was true irrespective of whether clinical signs of meningitis were present or not. Furthermore in the group with unclear etiology, only 14 out of 120 children showed a slight CSF pleocytosis (median cell count 1 cell/µl, range 1-77 cells/µl; median protein content 25 mg%, range 10-61 mg%).

17.5.2.2 Aseptic Meningitis

Aseptic meningitis is the second most frequently reported neurologic disorder in Lyme borreliosis in Europe (Sturfelt and Cavell 1985; Schaad et al. 1986; Jörbeck et al. 1987; Christen et al. 1988; Huber and Baumann 1989; Huppertz and Sticht-Groh 1989; Naglo and Wide 1989) and the USA (Meissner et al. 1982; Williams et al. 1990). This is in accordance with our present observations. One-fourth of the children with Lyme neuroborreliosis presented with aseptic meningitis (Fig. 1). Infection with *B. burgdorferi* accounted for 11.7% of 240 patients with aseptic meningitis investigated consecutively over a 3-year period (Christen and Hanefeld, unpublished); thus Lyme borreliosis seems to be the main cause of aseptic meningitis besides mumps and enterovirus infections (Fig.2). As in facial palsy, meningitis due to Lyme borreliosis occurred only between June and November. Neither clinical findings nor the patient's history was helpful for the diagnosis. A preceding tick bite or erythema migrans was only recalled by one-third of the patients with meningitis due to *B. burgdorferi*. The examination of the CSF revealed no differences for the cell count or protein content ranges with respect to etiology. Therefore, diagnosis of Lyme borreliosis could only be verified if specific IgM antibodies were detected in the CSF.

17.5.2.3 Meningoradiculoneuritis (Bannwarth's Syndrome)

The leading neurologic manifestation in adults, meningoradiculoneuritis (Bannwarth's syndrome), is rarely observed in children (Bannwarth 1941; Pfister et al. 1986;

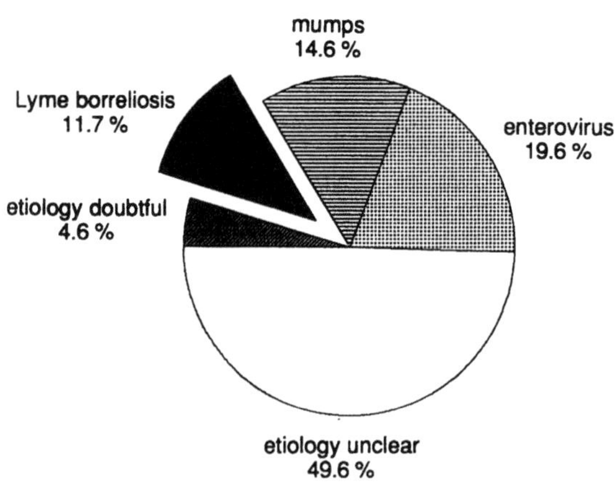

Fig. 2. Etiology of aseptic meningitis in childhood (*n*=240). Results of a prospective multicenter study on aseptic meningitis. Where etiology is doubtful, cross-reactions or double infections are possible (Christen and Hanefeld, unpublished)

Stiernstedt et al. 1986). However, three well-documented cases of lymphocytic meningoradiculoneuritis in children have been reported by Caflisch et al. (1984) and Schmedding et al. (1986). In our series, only 6 out of 169 children with Lyme neuroborreliosis presented with peripheral radiculoneuritis in addition to meningitis, which was thus classified as Bannwarth's syndrome (Fig. 1) (Christen and Hanefeld, unpublished).

17.5.2.4 Other Neurologic Disorders

Unusual neurologic syndromes in childhood attributed to Lyme borreliosis include meningoencephalitis (Schaad et al. 1986; Neumärker et al. 1989), focal encephalitis (Feder et al. 1988), and abducens nerve palsy (Huber and Baumann 1989). Pseudotumor cerebri as an associated symptom of Lyme borreliosis was observed in two children by Raucher et al. (1985) in the USA. In Europe, well-documented cases of pseudotumor cerebri, revealing inflammatory CSF changes besides specific antibodies in the CSF, were described by von Mühlendahl (1988) and Naglo and Wide (1989). In addition, individual children presenting with acute ataxia, isolated oculomotor palsy, recurrent headache (without neck stiffness!), and Guillain-Barré syndrome were observed in our series (Fig. 1) (Christen and Hanefeld, unpublished). Apart from those with Guillain-Barré syndrome, examination of the CSF revealed a lymphocytic pleocytosis and specific IgM antibodies in all children.

Chronic neurologic disorders due to Lyme borreliosis may occur in children as well as in adults. Ackermann et al. (1988) mentioned a 7-year-old child with chronic neurologic manifestations. A well-documented case of progressively destructive chronic meningoencephalitis in a child was reported by Bensch et al. (1987). Stahl et al. (1990) observed anorexia nervosa as the main neurologic feature in a girl suffering from chronic Lyme neuroborreliosis.

17.6 Clinical Differences Between Children and Adults

The whole clinical spectrum of Lyme borreliosis described in adults is also found in children, but there are some striking differences in the relative frequency of certain manifestations. Borrelial lymphocytoma is mainly observed in children and young adults. ACA is described only rarely in children, but is a well-known disorder in elderly patients. Peripheral facial palsy and aseptic meningitis are the most common neurologic disorders found as symptoms of Lyme borreliosis in childhood. In contrast, Bannwarth's syndrome, the leading neurologic feature in adults, seems to occur rarely in pediatric patients.

These clinical differences cannot yet be explained sufficiently. One probable explanation may derive from age-specific differences in the mode of infection. As a result of their different stature and behavior, it is to be expected that in children the head-neck region is relatively more often affected by tick bites than in adults. This could be observed in our series and was especially true for children with facial palsy (Christen et al. 1990). In the same way, Jörbeck et al. (1987) described erythema migrans exclusively located in the head-neck region in children presenting with facial palsy. In contrast, adults are reported to have tick bites or erythema migrans mainly on the lower extremities (Asbrink et al. 1986; Weber and Neubert 1986). Because of the ipsilateral localization of the tick bite or the erythema migrans and the resulting facial palsy, as observed in our series, a direct invasion via the affected nerve by *B. burgdorferi* is conceivable. If that is true, an explanation of the different clinical spectrum in children and adults may, to some extent, be derived from differences in the site of infection.

17.7 Diagnosis of Lyme Borreliosis in Childhood

Children with Lyme borreliosis present predominantly in an early stage of the disease. As in adults, typical erythema migrans, borrelial lymphocytoma, and ACA are diagnostic signs of Lyme borreliosis. In contrast, the leading features of Lyme neuroborreliosis in childhood, such as peripheral facial palsy and aseptic meningitis, are nonspecific. In these disorders, diagnosis of Lyme borreliosis has to be based on positive antibody titers in the CSF. The median interval between the onset of the disease and the hospital admission was no more than 4 days in our series of children with Lyme neuroborreliosis (Christen and Hanefeld, unpublished). This is much shorter than the median of 4 weeks which has been reported for adult patients (see Chap.11). Therefore, testing for IgM antibodies may be essential for the early diagnosis of Lyme borreliosis, especially in children (Stanek and Flamm 1986). In contrast to adults, only 20% of the children with Lyme neuroborreliosis revealed elevated IgG antibody titers on admission to the hospital (Christen and Hanefeld, unpublished). Neurologic disorders due to Lyme borreliosis are usually associated with inflammatory findings in the CSF. Additionally, the seasonal occurrence of Lyme neuroborreliosis from May until November has to be taken into consideration for diagnosis.

17.8 Differential Diagnosis

As far as the different manifestations of Lyme borreliosis are concerned, the differential diagnosis is the same in childhood as in adults with one exception: aseptic meningitis is much more frequent in children than in adults. In this condition, mumpsvirus and enteroviruses have to be regarded as the most frequent infectious agents besides *B. burgdorferi.*

17.9 Treatment of Lyme Borreliosis in Childhood

Antibiotic therapy of Lyme borreliosis in children should follow the same rules as in adults (see Chap. 23–27) with two exceptions:

1. **Oral therapy** with a tetracycline is not permitted in children younger than 12 years of age because of the resulting dental problems. Schaad et al. (1986) reported quite a favorable response to penicillin. However, we have observed relapses in children with erythema migrans after oral therapy with penicillin and erythromycin (Christen and Hanefeld, unpublished). Amoxicillin serves as an alternative for younger children.
2. For **parenteral therapy**, we prefer penicillin G (500 000 IU/kg bw per day intravenously for 10 days). In contrast ot other authors (Finkel 1988; Stechenberg 1988), we also recommend parenteral treatment in children with isolated facial palsy because of the inflammatory CSF findings indicating an associated meningitis in all children in our study. Cefotaxime (200 mg/kg bw per day) and ceftriaxone have been described as effective alternatives, especially in cases of refractory neurologic disorders (Dattwyler et al. 1988), but in pediatrics, ceftriaxone is not suitable as a drug of first choice because of its adverse effects resulting in gastrointestinal disorders (Barson et al. 1985; Jacobs 1988; Schaad et al. 1988).

17.10 Lyme Borreliosis and Pregnancy

It is well known that spirochetes cause congenital infections. Transplacental transmission of *Treponema pallidum, Leptospira,* and *Borrelia recurrentis* is associated with a wide spectrum of adverse outcome of pregnancy including abortion and stillbirth. The same risk might possibly be expected for infection with *B. burgdorferi* during pregnancy. There are several reports available describing Lyme borreliosis during pregnancy without adverse outcome (Markowitz et al. 1986; Weber and Neubert 1986; Ciesielski et al. 1987; Mikkelsen and Palle 1987). There were no differences in outcome according to the time of infection or antimicrobial therapy. It appears that congenital infection with *B. burgdorferi* is rather rare. There is only one well-documented case of congenital Lyme borreliosis in Europe described by Weber et al. (1988), in which the mother suffered from erythema migrans in early pregnancy and the child died on the first day of life. *B. burgdorferi* was detected in the brain by means of monoclonal antibodies and liver by silver staining. Another case of this type was reported by Schlesinger et al. (1985) in the USA.

In a serosurvey on pregnant women and their newborns in Switzerland, prevalence of elevated IgG antibody titers against *B. burgdorferi* was 0.85% in mothers at delivery (Nadal et al. 1989). Abnormalities in the affected newborns could not be attributed to intrauterine infection with *B. burgdorferi.* We also collected the serum of women in labor and the cord blood samples over a 12-month period and tested them for specific IgM and IgG antibodies against *B. burgdorferi* (Table 2) (Christen and Kleinschmidt, unpublished). The study included two obstetrics departments, one with a predominantely rural catchment area (Northeim) and another with an exclusively urban clientele (Berlin). The infection rate was very similar in both groups, 0.8% for specific IgM antibodies and about 7% for specific IgG antibodies. There was a marked age-linked increase in the infection rate as estimated by elevated IgG antibody titers (<20 years of age, 0%; 20–24 years, 6.0%, 25–29 years, 6.3%, >29 years, 10.4%). At time of birth, there were no differences between the children of these mothers and the seronegative group in the type and frequency of perinatal disorders and congenital abnormalities. None of the newborns of seropositive mothers had specific IgM antibodies.

Table 2. Prevalence of elevated IgM and IgG antibody titers against *B. burgdorferi* in women in labor from two obstetrics departments (Christen and Kleinschmidt, unpublished) [a]

	Rural area (Northeim)	Urban area (Berlin)
Study period	March 1987 – February 1988	March 1987 – February 1988
Number of births	617	3124
Women tested	517	2131 [b]
(% of all births)	(83.8%)	(68.2%)
Elevated IgM antibody titers	4	17
(% of all samples tested)	(0.8%)	(0.8%)
Elevated IgG antibody titers	36	17
(% of all samples tested)	(7.0%)	(7.3%) [b]

[a] The serological investigations were performed by Dr.med. H. Eiffert, Dr. vet. med. H. Lotter, Dr. med. M. Scriba and Prof.Dr.med. R. Thomssen (Department of Medical Microbiology, University of Göttingen).

[b] A seasonally representative 10% selection of all samples of this group was tested for IgG-antibodies (*n*=233).

These results indicate a high infection rate for *B. burgdorferi* in young women, not only in rural but also in urban areas. Therefore, in contrast to syphilis, a serological screening for Lyme borreliosis does not seem reasonable. It does not seem to be justified to draw any therapeutic consequences on the basis of these serological findings. On the contrary, manifest Lyme borreliosis during pregnancy requires early and consequent antimicrobial therapy. According to the experiences of Weber et al. (1988), orally administered penicillin does not seem to be sufficient to avoid transplacental transmission of *B. burgdorferi*. Pregnant women with Lyme borreliosis should be treated with intravenous penicillin for 10 – 14 days; amoxicillin or azithromycin might be similarly effective (Weber, personal communication).

17.11 Prevalence of Antibodies Against *Borrelia burgdorferi* in Children

Representative studies on the infection rate in childhood are not yet available. In order to study the prevalence of antibodies against *B. burgdorferi* among children, blood samples of 580 healthy children from Lower Saxony/Germany were collected over a 2-year period (Christen and Bartlau, unpublished). The infection rate, measured by the determination of specific IgG antibody titers, was 2.6% and showed a clear increase beyond the age of 5 years (< 1 year of age, 2.9%; 1–2 years, 1.2%;3–4 years, 0.9%; 5–7 years, 4.9%; > 7 years, 3.6%). Since all newborns were IgM-negative, the fact that some babies were found to be IgG-positive in the first year of life reflects maternal antibody transfer. IgM antibodies were found in 1.1% of the study group. Paul et al. (1989) studied the risk of Lyme borreliosis prospectively in children who had been bitten by *B. burgdorferi*-positive ticks. 15 out of 51 children showed a seroconversion, but only one child developed an erythema migrans.

References

Ackermann R, Rehse-Küpper B, Gollmer E, Schmidt R (1988) Chronic neurologic manifestations of erythema migrans borreliosis. Ann NY Acad Sci 539:16-23

Åsbrink E, Olsson I, Hovmark A (1986) Erythema chronicum migrans Afzelius in Sweden. A study of 231 patients. Zentralbl Bakteriol Hyg (A) 263: 229-236

Åsbrink E, Hovmark A, Olssen I, (1989) Lymphadenosis benigna cutis solitaria – Borrelia lymphocytoma in Sweden. Zentralbl Bakteriol (Suppl) 18:156-163

Bäfverstedt B (1943) Über Lymphadenosis benigna cutis. Acta Derm Venereol (Stockh) 23 (Suppl 11):1-202

Bannwarth A (1941) Chronische lymphocytäre Meningitis, entzündliche Polyneuritis und "Rheumatismus". Arch Psychiatr Nervenkr 113:284-376

Barson WJ, Miller MA, Brady MT, Powell DA (1985) Prospective comparative trial of ceftriaxone vs. conventional therapy for treatment of bacterial meningitis in children. Pediatr Infect Dis J 4:362-368

Bensch J, Olcén P, Hagberg L (1987) Destructive chronic Borrelia meningoencephalitis in a child untreated for 15 years. Scand J Infect Dis 19:697-700

Bloedhorn E (1923) Lymphknoten im Ohrläppchen (Lymphadenosis cutis circum scripta?) Hals-, Nasen- und Ohrenheilkd 7:224-227

Caflisch U, Tönz O, Schaad UB, Aeschlimann A, Burgdorfer W (1984) Die Zecken-Meningoradikulitis – eine Spirochätose. Schweiz Med Wochenschr 144:630-634

Christen HJ, Hanefeld F, Bartlau N, Wassmann K, Thomssen R (1988) Lyme borreliosis in children. A prospective clinical-epidemiological study. Ann NY Acad Sci 539:449-451

Christen HJ, Bartlau N, Hanefeld F, Eiffert H, Thomssen R (1990) Peripheral facial palsy in childhood – Lyme borreliosis to be suspected unless proven otherwise. Acta Paediatr Scand 79:1219-1224

Ciesielski CA, Russel H, Johnson S, Markowitz L, Benach J, Broome C (1987) Prospective study of pregnancy outcome in women with Lyme disease. 27th Interscience conference on antimicrobial agents and chemotherapy, New York, Abstract 39
Culp RW, Eichenfield AH, Davidson RS, Drummond DS, Christofersen MR, Goldsmith DP (1987) Lyme arthritis in children. J Bone Joint Surg (Am)69:96-99
Dattwyler RJ, Halperin JJ, Volkman DJ, Luft BJ (1988) Treatment of late Lyme borreliosis – Randomised comparison of ceftriaxone and penicillin. Lancet I:1191-1194
Donnermann C, Heite HJ (1959) Beitrag zur Symptomatologie der Akrodermatitis chronica atrophicans (Pick-Herxheimer). Arch Klin Exp Dermatol 208:516-527
Eiffert H, Lotter H, Thomssen R (1990) Use of peroxidase-labelled antigen for the detection of antibodies to *Borrelia burgdorferi* in human and animal sera. Scand J Infect Dis 23:78-87
Feder HM Jr, Zalneraitis EL, Reik L Jr (1988) Lyme disease: acute focal meningoencephalitis in a child. Pediatrics 82:931-934
Finkel MF (1988) Lyme disease and its neurologic complications. Arch Neurol 45:99-104
Göhr A (1935) Zur Klinik der Acrodermatitis chronica atrophicans. Thesis, Berlin
Görke W, Felix S, Severien C (1989) Scheinbar zentrale Fazialisparese bei Lyme-Borreliose. Klin Padiatr 210:467-470
Gottron H (1939) Acrodermatitis chronica atrophicans. Zentralbl Haut Geschlechtskr 61:7
Hanefeld F, Christen HJ, Bartlau N, Wassmann K, Thomssen R (1989) Lyme borreliosis in children. Zentralbl Bakteriol (Suppl) 18:192-202
Hauser W (1955) Zur Kenntnis der Acrodermatitis chronica atrophicans. Arch Dermatol Syph 199:350-393
Herxheimer K, Hartmann K (1902) Über Acrodermatitis chronica atrophicans. Arch Dermatol Syph 61:57-76,255-300
Höfer W (1956) Lymphadenosis benigna cutis. Arch Klin Exp Derm 203:23-40
Hollström E (1958) Penicillin treatment of erythema chronicum migrans Afzelius. Acta Derm Venereol (Stockh) 38:285-289
Huber A, Baumann W (1989) Klinische Manifestation der Lyme-Borreliose im Kindesalter. Klin Padiatr 201:133-135
Huppertz HI (1990) Childhood Lyme borreliosis in Europe. Eur J Pediatr 149:814-821
Huppertz HI, Sticht-Groh V (1989) Meningitis due to *Borrelia burgdorferi* in the initial stage of Lyme disease. Eur J Pediatr 148:428-430
Jacobs JC, Rosen JM, Szer IS (1984) Lyme myocarditis diagnosed by gallium scan. J Pediatr 105:950-952
Jacobs RF (1988) Ceftriaxone-associated cholecystitis. Pediatr Infect Dis J 7:434-436
Jörbeck HJA, Gustafsson PM, Lind HCF, Stiernstedt GT (1987) Tick-borne Borrelia-meningitis in children. Acta Paediatr Scand 76:228-233
Kafka J (1953) Statistische Erhebungen über die Fälle von Acrodermatits chronica atrophicans (Herxheimer-Hartmann) an der Universitäts-Hautklinik Giessen 1902-1952. Thesis, Giessen
Kerl W (1914) Erythema chronicum migrans. Arch Dermatol Syph 119:301
Kerl W (1920) Verhandlungen der Wiener Dermatologischen Gesellschaft. Arch Dermatol Syph 125:342-343
Keuth U, Mennicken U (1966) Zur Kenntnis der "peripheren" Facialisparese mit subakut-chronischer lymphozytärer Meningitis (Bannwarth). Z Kinderheilkd 97:49-56
Kishaba RG, Weinhouse E, Chusid MJ, Nudel DB (1988) Lyme disease presenting as heart block. Clin Pediatr (Phila) 27:291-293
Lipschütz B (1913) Über eine seltene Erythemform (Erythema chronicum migrans). Arch Dermatol Syph 118:349-356
Lipschütz B (1923) Weiterer Beitrag zur Kenntnis des "Erythema chronicum migrans". Arch Dermatol Syph 143:365-374
Mach KW, Wilgram GF (1966) Characteristic histopathology of cutaneous lymphoplasia (lymphocytoma). Arch Dermatol 94:26-32
Markowitz LE, Steere AC, Benach J, Slade JD, Broome C (1986) Lyme disease during pregnancy. JAMA 255:3394-3396
Meissner HC, Gellis SE, Milliken JF (1982) Lyme disease first observed to be aseptic meningitis. Am J Dis Child 136:465-467

Mikkelsen AL, Palle C (1987) Lyme disease during pregnancy. Acta Obstet Gynecol Scand 66:477-478

Mühlendahl KE von (1988) Borreliose mit Meningitis als Ursache für Pseudotumor cerebri. Pädiatr Prax 37:675-677

Nadal D, Gundelfinger R, Flueler U, Boltshauser E (1988) Acrodermatitis chronica atrophicans. Arch Dis Child 63:72-74

Nadal D, Hunziker UA, Bucher HU, Hitzig WH, Duc G (1989) Infants born to mothers with antibodies against *Borrelia burgdorferi* at delivery. Eur J Pediatr 148:426-427

Naglo AS, Wide K (1989) Borrelia infection in children. Acta Paediatr Scand 78:918-922

Neumärker KJ, Dudeck U, Plaza P (1989) Borrelien-Enzephalitis und Katatonie im Jugendalter. Nervenarzt 60:115-119

Neumann I von (1899) Verhandlungen der Wiener Dermatologischen Gesellschaft. Idiopathische Atrophie der Haut. Arch Dermatol Syph 49:124-126

Nikulin N (1896) Verhandlungen der dermatologischen und Venereologischen Gesellschaft zu Moskau. Dermatol Z 3:374-375

Olsson I, Engervall K, Asbrink E, Carlsson-Nordlander B, Hovmark A (1988) Tick-borne borreliosis and facial palsy. Acta Otolaryngol (Stockh) 105:100-107

Paschoud JM (1957) Die Lymphadenosis benigna cutis als übertragbare Infektionskrankheit. Hautarzt 5:197-211

Paul H, Ackermann R, Gerth HJ (1989) Infection and manifestation rate of European Lyme borreliosis. Zentralbl Bakteriol (Suppl) 18:44-49

Pette H (1942) Die akut entzündlichen Erkrankungen des Nervensystems. Eine besondere Form der abakteriellen Meningitis (allergische Meningitis). Thieme, Leipzig, pp 302-305

Pfister HW, Einhäupl KM, Wilske B, Preac-Mursic V (1986) Bannwarth's syndrome and the enlarged neurological spectrum of arthropod-borne borreliosis. Zentralbl Bakteriol Hyg (A) 263:343-347

Pirilä V (1951) The penicillin treatment of acrodermatitis atrophicans chronica. Acta Derm Venereol (Stockh) 31:576-591

Raucher HS, Kaufmann DM, Goldfarb J, Jacobson RI, Roseman B, Wolff RR (1985) Pseudotumor cerebri and Lyme disease: a new association. J Pediatr 107:931-933

Schaad UB, Flüeler U, Schaub H, Suter H, Vischer D, Caflisch U, Tschumi A, Wick H, Vest M, Durrer D (1986) Durch *Ixodes-ricinus*-Spirochäten (*Borrelia burgdorferi*) verursachte Krankheitsbilder (Lyme-Krankheit) bei pädiatrischen Patienten in der Schweiz. Schweiz Med Wochenschr 116:1426-1430

Schaad UB, Wegdwood-Krucko J, Tschaepeler H (1988) Reversible ceftriaxone-associated biliary pseudolithiasis in children. Lancet II:1411-1413

Schaltenbrand G (1949) Chronisch aseptische Meningitis. Nervenarzt 20:433-442

Schaltenbrand G (1966) Durch Arthropoden übertragene Erkrankungen der Haut und des Nervensystems. Munch Med Wochenschr 108:1557-1562

Schiller E (1921) Atrophia idiopathica cutis. Mitt Ges Inn Med 20:29-31

Schirduan M (1950) Verdacht auf Toxoplasmose bei Meningitis nach Erythema chronicum migrans bullosum. Arch Dermatol Syph 192:256-260

Schlesinger PA, Duray PH, Burke BA, Steere AC, Stillman T (1985) Maternal-fetal transmission of the Lyme disease spirochete, *Borrelia burgdorferi*. Ann Intern Med 103:67-68

Schmedding E, Verboven M, Lauwers S, Ebinger G, Rom N, Loeb H (1986) Lymphocytic meningoradiculitis (Garin-Bujadoux Bannwarth): from syndrome to disease? Eur J Pediatr 144:497-502

Sedlácek V (1960) Erythema chronicum migrans. Cs Dermatol 35:386-399

Stahl M, Fressl R, Wilske B (1990) Lyme-Neuroborreliose unter dem Bild einer Anorexia nervosa. Munch Med Wochenschr 132:615-619

Stanek G, Flamm H (1986) Borrelia-Infektionen bei Kindern in Mitteleuropa. Pädiatr Grenzgeb 25:113-146

Stechenberg BW (1988) Lyme disease: the latest great imitator. Pediatr Infect Dis J 7:402-409

Steere AC, Malawista SE, Snydman DR, Shope RE, Andiman WA, Ross MR, Steele FM (1977) Lyme arthritis. Arthritis Rheum 20:7-17

Steere AC, Schoen RT, Taylor E (1987) The clinical evolution of Lyme arthritis. Ann Intern Med 107:725-731

Stiernstedt G, Sköldenberg B, Garde A, Kolmodin G, Joerbeck H, Svenungsson B, Carlstörm A (1986) Clinical manifestation of Borrelia infections of the nervous system. Zentralbl Bakteriol Hyg (A) 263:289-296

Strandberg J (1920) Regarding an unusual form of migratory erythema caused by tick bites. Acta Derm Venereol (Stockh) 1:422-427

Sturfelt G, Cavell B (1985) Lyme disease in a 12-year-old girl. Acta Paediatr Scand 74:133-136

Wassmann K, Borg-von Zepelin M, Zimmermann O, Stadler M, Eiffert H, Thomssen R (1989) Determination of immunoglobulin M antibody against *Borrelia burgdorferi* to differentiate between acute and past infections. Zentralbl Bakteriol (Suppl) 18:281-289

Weber K, Neubert U (1986) Clinical features of early erythema migrans disease and related disorders. Zentralbl Bakteriol Hyg (A) 263:209-228

Weber K, Puzik A, Becker T (1983) Erythema-migrans-Krankheit. Beitrag zur Klinik und Beziehung zur Lyme-Krankheit. Dtsch Med Wochenschr 108:1182-1190

Weber K, Schierz G, Wilske B, Preac-Mursic V (1985) Das Lymphozytom – eine Borreliose? Z Hautkr 60:1585-1598

Weber K, Schierz G, Wilske B, Neubert U, Krampitz HE, Barbour AG, Burgdorfer W (1986) Reinfection in erythema migrans disease. Infection 14:32-35

Weber K, Bratzke HJ, Neubert U, Wilske B, Duray PH (1988) *Borrelia burgdorferi* in a newborn despite oral penicillin for Lyme borreliosis during pregnancy. Pediatr Infect Dis J 7:286-289

Williams CL, Strobino B, Lee A, Curran AS, Benach JL, Inamdar S, Cristofaro R (1990) Lyme disease in childhood: clinical and epidemiologic features of ninety cases. Pediatr Infect Dis J 9:10-14

Zellweger H (1946) Über die chronische allergische Meningitis. Helv Paediatr Acta 1:417-426

18 Localized Scleroderma and Lichen Sclerosus et Atrophicus: Manifestations of a Borrelia burgdorferi Infection?

U. NEUBERT, E. ABERER, and T. RUFLI

18.1 Introduction

In 1985 and 1986 Aberer et al., Neubert et al., and Rufli et al. published serological findings showing elevated antibody titers against *Borrelia burgdorferi* in 20%-50% of patients with localized scleroderma (LS) (Aberer et al. 1985; Neubert et al. 1986a,b; Rufli et al. 1986). Reports of negative serological results in LS patients from the United States and Northern Europe, however, led to some scepticism about the significance of these findings (Muhlemann et al. 1986; Hansen et al. 1987; Hoesly et al. 1987). In further studies, elevated immunoglobulin G (IgG) antibody titers against *B. burgdorferi* have also been reported in patients with lichen sclerosus et atrophicus (LSA) (Olsson et al. 1988; Aberer et al. 1989; Neubert 1990; Neubert et al. 1990).

In this chapter, we will compare these controversial findings in order to discuss the possible etiologic role of *B. burgdorferi* in LS and LSA.

18.2 Definitions and Relationship of LS and LSA to Erythema Migrans and Acrodermatitis Chronica Atrophicans

18.2.1 Localized Scleroderma

18.2.1.1 Definition

The term localized or circumscribed scleroderma comprises morphologically quite heterogeneous indurations caused by localized sclerosis in the dermis, subcutis and/or fascia, frequently accompanied or followed by pigmentary alterations. Three main types of LS may be distinguished (Scleroderma Working Group 1990): a plaque-type (morphea in a restricted sense), which is observed in about 70% of LS patients, a linear type, and a profound type. For further information see Jablonska (1975), Ackerman (1982), Braun-Falco et al. (1990) and Scleroderma working Group (1990).

18.2.1.2 LS and Erythema Migrans

The enlarging livid-erythematous border of LS of the plaque type, called "lilac ring", may resemble the anular margin of erythema migrans. The histological pattern in the earliest phase of morphea with perivascular infiltrates of mixed inflammatory cells and only slight homogenization of collagen bundles may be hardly discernible from that of erythema migrans. Transitions from erythema migrans to LS have been reported in three patients (Büchner 1989; Aberer 1991; Neubert et al., submitted for publication).

18.2.1.3 LS and Acrodermatitis Chronica Atrophicans

Acrodermatitis chronica atrophicans (ACA) may be associated with indurated scleroderma-like skin changes within the involved areas, particularly on the legs (references see Donnermann and Heite 1959; Jablonska 1975; Neubert 1992). On the other hand, patients with LS lesions developing locally independent of a coexisting ACA have been repeatedly reported (Ramelet 1987; Langer and Diem et al. 1988, Coulson et al. 1989, Söhnchen et al. 1989). The main histological differences between "true" LS and ACA in its pseudosclerodermatic skin lesions are seen in the elastic tissue, which shows rarefication and atrophy, sometimes followed by hyperplasia in ACA (Korting et al. 1969; Aberer et al. 1991b) whereas in LS it is not altered.

LS lesions present on the extremities appear mostly as linear, sclerotic, sharply demarcated lesions with a prominent lilac rim. In the case of "systemized LS" (Jadassohn 1976; Aberer et al. 1987b) with associated fasciitis and myositis, the distinction from ACA seems to be difficult (Aberer et al. 1991b; Aberer et al. 1987b; Neubert 1989). In Shulman's syndrome, a special form of profound LS, a transition to ACA has also been reported (Neubert 1989; Grahmann et al. 1989).

18.2.2 Lichen Sclerosus et Atrophicus

18.2.2.1 Definition

The earliest lesions of LSA are flat, firm, pink-rose or whitish papules and plaques with central dells resulting from adnexal plugging by cornified cells. These papules tend to coalesce to white atrophic patches. Clinically active lesions show erythematous borders. Predilection sites are the neck, the shoulders, and the anogenital region, especially the labia minora and medial parts of labia maiora as well as the praeputium penis (Jablonska 1975; Ackerman 1982; Braun-Falco et al. 1990).

18.2.2.2 LSA and ACA

The coexistence of ACA and LSA lesions has repeatedly been reported in Europe (e.g. Asbrink et al. 1986; Aberer et al. 1987a). LSA-like lesions can be observed in areas of ACA or at other body sites concomitantly with ACA. Duray and Steere also mention two patients with chronic Lyme borreliosis who showed typical LSA lesions (Duray and Steere 1988). In late stages, LSA and ACA may show remarkably similar histological features (Neubert 1992).

18.2.2.3 LS and LSA

Simultaneous occurrence of LS and LSA may occasionally be observed in the same patient. Each of the diseases usually shows a typical histological pattern; sometimes, however, characteristic features of both dermatoses may be detected in the same histological section. (Jablonska 1975, Uitto et al. 1980; Ackerman 1982).

Table 1. Serum antibodies to *B. burgdorferi* in patients with localized scleroderma. Results of eight working groups

Authors (year)	Country	Serological assay	Patients Seropositive/total (n/n)	%	Controls Seropositive/total (n/n)	%
Aberer et al.(1991a)	Austria	ELISA	13/30	43	9/109	8.3
Halkier-Soerensen et al. (1989)	Denmark	ELISA	4/64	6	7/297	2
Hansen et al. (1987)	Denmark	IFA-ABS	1/20	5	3/100	3
Hoesly et al (1987)	USA	IFA	6/32	19	1/50	2
		ELISA-ABS	1/25	4	0/50	
Lecerf et al. (1989)	France	IFA	0/17		?/?	2.5
Muhlemann et al. (1986)	Gr.Britain	IFA-ABS	0/22		3/46	7
Neubert et al. (1990)	Germany	IFA-ABS	66/210	31	4/100	4
Olsson et al. (1988)	Sweden	ELISA-ABS	2/15	13	9/185	5
Rufli et al. (1986)	Switzerland	IFA	14/38	37	6/43	14

IFA, indirect immunofluorescence assay; ELISA, enzyme-linked immunosorbent assay; ABS, preabsorption with *Treponema phagedenis*;

18.3 Serological Results

18.3.1 Localized Scleroderma

Antibodies against *B. burgdorferi* were investigated in LS patients by eight different research groups from seven different countries of Europe and the USA. The serological results are summarized in Table 1.

Aberer et al. (1991a) reinvestigated the sera of 30 patients with different types of LS using an enzyme-linked immunosorbent assay (ELISA). Thirteen (43%) of the 30 patients showed elevated IgG titers to *B. burgdorferi*. The control group consisted of 109 healthy blood donors 8.3% of whom were seropositive. Western blot studies were positive in 40% of patients, showing a reactivity pattern which can be seen in the course of Lyme borreliosis.

Neubert et al. (1989, 1990) and Neubert (1992) reevaluated clinical and serological data in 210 patients suffering from different types of LS. Sera were investigated by an indirect immunofluorescence assay (IFA) after preabsorption with *Treponema phagedenis* and rheumatoid factor absorbent (ABS). Sera from 100 age- and sex-matched healthy subjects and patients with other dermatoses (n=100) were included as controls. Significantly elevated antibody titers were detected in 66 (31%) patients, whereas 4% of controls were seropositive (p<0.001).

Büchner and Rufli (1988) found elevated antibody titers against *B. burgdorferi* in 14 (37%) of 38 patients with LS and in 14% of controls. These results correspond to earlier investigations in which 15 of 35 (43%) LS patients were seropositive (Rufli et al. 1986). In addition a case report by Maleville et al. (1989), which found seroconversion in a LS patient after therapy, and another one by Büchner (1989) favor the borrelial origin of LS.

In contrast, other authors could not confirm these serological results and questioned the etiologic role of *B. burgdorferi* in LS. For example, Muhlemann et al. (1986) could not show any elevated antibody titers in the sera of 22 LS patients, while Hansen et al. (1987) obtained a positive indirect IFA-ABS test in only one (5%) out of 20 LS patients. In the studies of Hoesly et al. (1987), six (19%) of 32 patients with LS were seropositive, as shown by an indirect IFA, and one (4%) out of 25 as shown by an ELISA-ABS. Also using an ELISA-ABS, Olsson et al. (1988) found increased IgG antibody titers against *B. burgdorferi* in two (13%) of 15 patients. Halkier-Soerensen et al. (1989) observed 6% (four of 64) seropositive LS patients, determined by a *B. burgdorferi*-flagellum-ELISA. Lecerf et al. (1989) could not demonstrate increased antibody titers against *B. burgdorferi* using an indirect IFA in 17 LS patients.

In all these reports no significant differences could be detected between the serological results in LS and in control sera.

18.3.2 Lichen Sclerosus et Atrophicus

In a study by Neubert et al. (1990; Neubert 1990) in 30 LSA patients without ACA, ten (33%) were seropositive as shown by an indirect IFA-ABS. Using an ELISA, Aberer et al. (1989) detected IgG antibodies in 14 (45%) of 35 LSA patients with genital and/or extragenital involvement.

Olsson et al. (1988) also used an ELISA to examine sera of 50 LSA patients. Elevated IgM and/or IgG antibody titers to *B. burgdorferi* were detected in the sera of eight (16%) of the patients in contrast to 4.9% of the sera of 185 healthy controls. The titers of seropositive LSA patients, however, did not exceed the titers of healthy individuals, so that the positive titers were considered insufficient for a definite serological diagnosis of Lyme borreliosis.

18.4 Microbiological Findings

18.4.1 Isolation of Spirochetes from Skin Biopsies

In 1987, Aberer et al. (1987b) reported on the isolation of a slowly growing borrelial strain from a skin biopsy of a patient with linear-systemized morphea, which could be stained with polyclonal antibodies against *B. burgdorferi* using an immunfluorescence method. Weber et al. (1988) described the cultivation of spirochetes in two patients with seronegative LS, one of which represented an atrophoderma of Pasini and Pierini. In another patient with atrophoderma, a *B. burgdorferi* strain could be isolated by Hoffmann et al. (1990).

18.4.2 Demonstration of Spirochetes in Histological Sections

18.4.2.1 Silver Impregnation Methods

Spirochete-like organisms could be shown on histological sections using a Dieterle method in one patient with LS following erythema migrans (Büchner 1989). Maleville et al. (1989) demonstrated spirochetes by a Warthin-Starry method in LS lesions of a

7-year-old child. Abele and Anders (1990) visualized *borrelia*-like organisms in a child with facial hemiatrophy.

A modified Steiner method was used to stain spirochetes in skin biopsies from ten of 25 patients with LS and in skin biopsies from ten of 21 patients with LSA (Ross et al. 1990).

Neubert and de Koning (unpublished) using the Bosma-Steiner stain detected spirochetes resembling borreliae lying between collagen fibers of the deeper dermis in an LS lesion of the plaque type in a 43-year-old female patient showing a positive *Borrelia* serology. The sclerotic lesion softened during the first days of a ceftriaxone therapy.

18.4.2.2 Avidin-Immunoperoxidase Method

Borrelia-like organisms could be visualized in biopsies of four out of 18 patients with LS and six out of 13 patients with LSA using serum of a patient with ACA as antibody source (Aberer and Stanek 1987)

18.4.3 Detection of *B. burgdorferi* by Polymerase Chain Reaction (PCR)

Schempp et al. (1992) reported on a 16-year-old female with a tick bite induced morphea lesion lasting for 10 years. Whereas antibodies to *B. burgdorferi* could not be detected in the peripheral blood by an ELISA and by Western blot analysis, investigation of the affected skin by a nested PCR using primer pairs specific for the flagellin gene revealed *B.-burgdorferi*-specific DNA. A strong and reproducible lymphocyte proliferation in response to ultrasonicated *B. burgdorferi* organisms and, even stronger, to *Treponema denticola* was found.

Neubert and co-workers (unpublished) observed the transition of tick bite induced erythemata migrantia into disseminated LS lesions in a 61-year-old female patient. *B. burgdorferi* was detected in a skin biopsy specimen by direct sequencing of 16S rRNA amplified by PCR after reverse transcription (Adam et al. 1992). Again borrelia serology (IFA-ABS, ELISA) proved to be negative. The LS lesions softened after ceftriaxone therapy.

18.5 Discussion

Clinical observations have shown that Lyme borreliosis may be associated with sclerotic skin lesions. Erythema migrans may be succeeded by morphea, and ACA may be accompanied or followed by LS or LSA. Therefore, it seems conceivable that sclerotic skin lesions may occasionally develop without concomitant ACA as sole manifestations of a *B. burgdorferi* infection, provided that such repeatedly reported associations are not merely accidental. Disappearance of the lilac rim of LS and softening of sclerotic skin lesions after antibiotic penicillin treatment, established in 1949 by Miescher, may be taken as a further, indirect hint for such a connection (Aberer et al. 1991a).

LS and LSA are probably dermatoses of heterogeneous origin. Geographic differences might be responsible for variable factors which play a possible role in the initiation of LS or LSA. It might also be of significance that *Borrelia* strains isolated from the skin of European patients are not as uniform in their antigen expressions as the North American isolates. These differences may affect the pathogenic potentialities of the germs according to their geographic origin.

For the present, the obviously diverging results of serological studies in LS and

LSA patients are the main point of controversy. Several reasons are conceivable as being responsible for these contradictory findings:

1. There are no standardized serological methods for determining antibodies against *B. burgdorferi*, so that the results of the respective investigators are difficult to compare with each other. IFA and ELISA were utilized with various antigens, with and without preabsorption with *Treponema phagedenis*, and with different cut-off levels, dependent on the individual control collective.

2. The composition of control collectives differs according to endemic factors, the incidence of asymptomatic infection, and the significance of levels of titers interpreted as positive.

3. It is well known that not all manifestations of Lyme borreliosis are accompanied by a positive borrelia serology (Dattwyler et al. 1988). In neurological and joint manifestations, for example, a vigorous T-cell response predominates. T-cell responses have been shown in LS lesions as well (Büchner 1989; Aberer et al. 1991b).

4. Seroconversion has been reported in patients with LS after antibiotic therapy by Maleville et al. (1989), Aberer et al. (1991a) , and Neubert et al. (1986b). This brings to mind the significant rise of IgG antibody titers against *B. burgdorferi* shortly after initiation of antibiotic therapy in two patients with ACA (Weber et al. 1984; Langer and Diem 1988).

Irrespective of the controversial serological data, the histological demonstration and the identification of borreliae by culture and by PCR-mediated gene amplification indicate that the pathogen may be present in the affected tissue, even in seronegative patients; the latter has been shown by six different research groups (Aberer et al. 1987b; Weber et al. 1988; Maleville et al. 1989; Hoffmann et al. 1990; Schempp et al. 1992; Neubert et al. submitted for publication).

Considering the heterogeneous nature of LS or LSA and their incidence depending on geographic factors, at least some types of LS and LSA may be related to a *B. burgdorferi* infection.

18.6 Conclusions

Clinical and histological similarities and associations between erythema migrans, ACA, LS and LSA have suggested the idea of regarding *B. burgdorferi* as one of the pathogenic factors causing these different skin disorders. The response to penicillin therapy in some patients with LS as well as the positive serological results in up to 43% of patients with LS and 45% of patients with LSA make a borrelial origin of these two heterogeneous dermatoses in some cases very likely. The histological demonstration of spirochetes in tissue and their isolation from skin biopsies, even in seronegative patients, strongly indicate that spirochetes (presumably *B. burgdorferi*) are involved to some extent in the pathogenesis of some types of LS and LSA. Further evaluations are needed to learn about the significance of these interesting findings.

References

Abele DC, Anders KH (1990) The many faces of borreliosis II. J Am Acad Dermatol 3:401-410

Aberer E. (1991) Cutaneous manifestations of Lyme borreliosis Infection (in press)

Aberer E, Stanek G (1987) Histological evidence for spirochetal origin of morphea and lichen sclerosus et atrophicans. Am J Dermatopathol 9:374-379

Aberer E, Neumann R, Stanek G (1985) Is localized scleroderma a *Borrelia infection*? (letter) Lancet II: 273

Aberer E, Neumann R, Lubec G (1987a) Acrodermatitis chronica atrophicans in association with lichen sclerosus et atrophicans: tubulo-interstitial nephritis and urinary excretion of spirochete-like organisms. Acta Derm Venereol (Stockh) 67:62-91

Aberer E, Stanek G, Ertl M, Neumann R (1987b) Evidence for spirochetal origin of circumscribed scleroderma (morphea). Acta Derm Venereol (Stockh) 67:225-231

Aberer E, Neumann R, Klade H, Reiner M, Stanek G (1989) Screening of dermatological patients for antibodies against *Borrelia burgdorferi*. In: G. Stanek et al. (eds) Lyme Borreliosis II, Zentralbl Bakteriol (Suppl) 18:178-182

Aberer E, Klade H, Stanek G, Gebhart W (1991a) *Borrelia burgdorferi* and different types of morphea. Dermatologica 183:145-154

Aberer E, Klade H, Hobisch G (1991b) A clinical, histological and immunohistochemical comparison of acrodermatitis chronica atrophicans and morphea. Am J Dermatopathol 13:334-341

Ackerman AB (1982) Lichen sclerosus et atrophicus vs. scleroderma. In: Ackerman AB et al. (eds) Differential diagnosis in dermatopathology. Leo and Febiger, Philadelphia, pp 62-63

Adam T, Graf B, Neubert U, Göbel UB (1992) Detection and classification of *Borrelia burgdorferi* by direct sequencing of 16S rRNA amplified after reverse transcription. Med Microbiol Lett 1:120–126

Asbrink E, Brehmer-Andersson E, Hovmark A (1986) Acrodermatitis chronica atrophicans – a spirochetosis. Am J Dermatopathol 8:209-219

Braun-Falco O, Plewig G, Wolff HH, Winkelmann RK (1990) Dermatology. Springer, Berlin Heidelberg New York pp 552-556

Büchner SA (1989) Morphea – eine zeckenübertragene Borreliose der Haut? Ein Beitrag zur Pathogenese der zirkumskripten Sklerodermie. Z Hautkr 64:661-669

Büchner SA, Rufli TH (1988) *Borrelia burgdorferi* antibodies in localized scleroderma. Lecture, symposium "Immunologica Twardziny", Warszawa, November 1988

Coulson IH, Smith NP, Holden CA (1989) Acrodermatitis chronica atrophicans with coexisting morphea. Br J Dermatol 121:263-269

Dattwyler RJ, Volkman DJ, Luft BJ, Halperin JJ, Thomas J, Golightly MG (1988) Seronegative Lyme Disease. Dissociation of specific T- and B-Lymphocyte response to *Borrelia burgdorferi*. N Engl J Med 319:1441-1446

Donnermann C, Heite H-J (1959) Beitrag zur Symptomatologie der Akrodermatitis chronica atrophicans (Pick-Herzheimer). Arch Klin Exp Dermatol 208:516-527

Duray PH, Steere AC (1988) Clinical pathologic correlations of Lyme disease by stage. Ann NY Acad Sci 539:65-79

Grahmann F, Schmidli J, Meier C (1989) Shulman Syndrome as precursor of acrodermatitis chronica atrophicans Herxheimer. Another manifestation of Lyme borreliosis? In: G. Stanek et al. (eds) Lyme Borreliosis II, Zentralbl Bakteriol (Suppl) 18:172-177

Halkier-Soerensen L, Kragballe K, Hansen K (1989) Antibodies to the *Borrelia burgdorferi* flagellum in patients with scleroderma, granuloma annulare and porphyria cutanea tarda. Acta Derm Venereol (Stockh) 69:116-199

Hansen K, Serup J, Hoybye S (1987) Antibodies to *Borrelia burgdorferi* and localized scleroderma. Lancet i: 682

Hoesly JM, Lester EM, Winkelmann RK (1987) Localized scleroderma (morphea) and antibody to *Borrelia burgdorferi*. J Am Acad Dermatol 17:455-458

Hoffmann H, Martin C, Preac-Mursic V (1990) Atrophodermia Pasini et Pierini – seronegative *Borrelia burgdorferi* infection. Abstracts B. IVth international conference on Lyme borreliosis, Stockholm, Sweden, p 89

Jablonska S (1975) Scleroderma and pseudoscleroderma. 2nd edn. Polish Medical Publishers, Warszawa

Jadassohn J (1976) Handbuch der Haut- und Geschlechtskrankheiten. Ergänzungswerk III/A. Springer, Berlin Heidelberg New York, p 482

Korting GW, Hoede N, Holzmann H (1969) Zur Frage des Elasticaverhaltens bei einigen sklerosierenden und atrophisierenden Hautkrankheiten. Hautarzt 20:351-361

Langer K, Diem E (1988) Acrodermatitis chronica atrophicans und sklerodermiforme Hautveränderungen bei Borrelieninfektion. Hautarzt 39:647-651

Lecerf V, Bagot M, Revuz J, Tourain R, Dournon E (1989) *Borrelia burgdorferi* and localized scleroderma (letter). Arch Dermatol 125:297

Maleville J, Picrard GE, Baraton G, Klene C, Sanciaume C, Fontan I, Taieb A (1989) Sclérodermi en plaques multiples: mise en evidence de spirochétes dans le derme. Nouv Dermatol 8:120-121

Miescher V (1949) In- und ausländische Erfahrungen mit neueren Behandlungsmethoden in der Dermatovenerologie. Arch Dermatol Syph (Berl) 189:42

Muhlemann MF, Wright DJM, Black C (1986) Serology of Lyme disease (letter). Lancet II:553-554

Neubert U (1989) Klinische Aspekte der *Borrelia-burgdorferi*-Infektion. Z Hautkr 64:649-656

Neubert U (1990) Zirkumskripte Sklerodermie und Lichen Sclerosus et atrophicans: Welche Rolle spielen Borrelien-Infektionen? In: O. Braun Falco, J. Ring (eds) Fortschritte der praktischen Dermatologie und Venerologie XII. Springer, Berlin Heidelberg New York, pp 470-471

Neubert U, Gerstmeier J, Meurer M, Krieg T (1986a) Spirochetal antibodies in morphea. Lecture, Arbeitsgemeinschaft Dermatologische Forschung. Vienna, Nov 1985. J Invest Dermatol 86:336-337 (abstract)

Neubert U, Krampitz HE, Engl H (1986b) Microbiological findings in erythema (chronicum) migrans and related disorders. Zentralbl Bakteriol Hyg (A) 263:237-252

Neubert U, Kauzmann M, Meurer M, Gerstmeier J, Krieg T (1989) Serological evidence for *Borrelia burgdorferi* as a possible etiologic agent in morphea. Lecture, 18th meeting of the European Society for Dermatological Research, Munich, June 1988. J Invest Dermatol 91:404 (abstract)

Neubert U, Kauzmann M, Meurer M, Krieg T (1990) Serum antibodies to *Borrelia burgdorferi* in localized scleroderma and lichen sclerosus et atrophicans. Abstracts B. IVth international conference on Lyme borreliosis, Stockholm, Sweden, p 88

Neubert U (1992) Zirkumskripte Sklerodermien und Lichen sclerosus et atrophicus – mögliche Folgen einer Borrelieninfektion? In: O.P. Hornstein (ed) Beiträge zur Dermatologie, Band 15. Virale und bakterielle Infektionskrankheiten der Haut. Perimed, Erlangen, pp 232–247

Ramelet AA (1987) Association of acrodermatitis chronica atrophicans and morphea. Dermatologica 175:253-256

Ross SA, Sanchez JL, Taboas JO (1990) Spirochetal forms in the dermal lesions of morphea and lichen sclerosus et atrophicus. Am J Dermatopathol 12 (4) : 357-362

Rufli T, Lehner S, Aeschlimann A, Clamot E, Gigon F, Jeanneret JP (1986) Zum erweiterten Spektrum zeckenübertragener Spirochätosen. Hautarzt 37:597-602

Schempp C, Bocklage H, Lange R, Owsianowski M, Gollnik H, Orfanos CE (1992) Evidence for localized spirochaetal infection in morphoea with absence of systemic antibody response. Arch Dermatol Res 284:22 (abstract).

Scleroderma Working Group of the Dermatological Research Study Group (1990) Classification of circumscribed scleroderma. Multicentre study with 286 patients. Hautarzt 41:16-21

Söhnchen R, Neubert U, Korting HC (1989) Acrodermatitis chronica atrophicans. Nekrobiosis-artige und Morphea-artige Hautveränderungen bei Borrelieninfektion. Case presentation (abstract). Hautarzt 40:726

Uitto J, Santa Cruz DJ, Bauer EA, Elsen AZ (1980) Morphea and lichen sclerosus et atrophicus. J Am Acad Dermatol 3: 271-279

Weber K, Schierz G, Wilske B, Preac-Mursic V (1984) Zur Klinik und Ätiologie der Acrodermatitis chronica atrophicans. Hautarzt 35:571-577

Weber K, Preac-Mursic V, Reimers CD (1988) Spirochetes isolated from two patients with morphea. Infection 16:29-30

19 Ocular Manifestations

U. SCHÖNHERR and F. STRLE

19.1 Introduction

In Lyme borreliosis, the eye may be involved primarily by inflammation of ocular tissues or secondarily by extraocular manifestations. While ocular symptoms due to conjunctivitis or paresis of the cranial nerves are frequent, intraocular inflammations caused by *Borrelia burgdorferi* seem to be rare and are difficult to diagnose.

Steere and coauthors (1985) described the first case of intraocular inflammation in the United States. In this case, spirochetes were found in specimens of vitreous debris obtained at surgery. Since then, Lyme borreliosis has played an increasingly important role in ophthalmology (Mac Donald 1987; Aaberg 1989; Spalton 1990; Winterkorn 1990).

The large spectrum of distinguishable inflammatory processes that can occur within the eye is explained by the existence of boundary structures which modify or limit the extent of intraocular inflammation. Biomicroscopy permits exact localization and determination of the etiology (Naumann and Naumann 1986). Similar to syphilis, Lyme borreliosis seems to be a great imitator of ocular inflammations.

19.2 Clinical Features

19.2.1 Ocular Manifestations

19.2.1.1 Conjunctivitis

According to Steere et al. (1983), conjunctivitis is the most common ocular manifestation of early Lyme disease in the United States (11%; 35 of 314 patients), while Weber and Neubert (1986) reported conjunctivitis in only one of 104 patients with erythema migrans from southern Germany. However, at the University Medical Center in Ljubljana, Slovenia, 23 out of 425 adult patients with typical erythema migrans (5.4%) had conjunctivitis. An additional 14 patients (3.3%) had a history of "red eyes" when they were seen because of erythema migrans (Table 1). The duration of conjunctivitis ranged from 2 days to 6 weeks (median 8 days) before treatment, and 1 – 28 days (median 4 days) after the onset of antibiotic treatment with phenoxymethylpenicillin, doxycycline, or azithromycin. It is not clear at present whether the higher number of patients with conjunctivitis seen in Slovenia than in other parts of Europe represents a selection problem or a higher prevalence of more severe cases. It is interesting to note in this context that the initial illness of Weber and Neubert's German patient mentioned above was more severe and that the patients from Ljubljana had more than one erythema migrans lesion significantly more frequently. Conjunctivitis can be classified

Table 1. Prevalence of conjunctivitis in 425 patients with erythema migrans[a]

	Total number of patients	Conjunctivitis patients (*n*)	%	p value
Localization of erythema migrans				
Head	22	3	13.6	n.s.
Other	403	20	4.9	
Number of erythema migrans lesions				
>1	38	6	15.8	*p*>0.01*
1	387	17	4.4	
Systemic symptoms				
Present	195	15	7.7	n.s.
Absent	230	8	3.5	
Total	425	23	5.4	

n.s., not significant
* Fisher's exact test
[a] Fourteen patients with a history of "red eyes" were not included.

as an early manifestation of Lyme borreliosis because of its close time association with the erythema migrans. It is most likely an early second stage sign (Steere 1989).

19.2.1.2 Iridocyclitis

Observations in the United States have shown that granulomatous inflammation of iris and ciliary body (granulomatous iridocyclitis) with posterior synechia and cellular infiltration of the anterior part of the vitreous body can occur in the second stage of Lyme borreliosis (Eichenfield et al. 1986; Jacobs et al. 1986; Winward et al. 1989; Boutros et al. 1990). A good response to antibiotic treatment is usually reported; however, one patient developed a panophthalmitis leading to blindness following therapy with a systemic corticosteroid and antibiotics (Steere et al. 1985; Kauffmann and Wormser 1990). We have recently seen a patient (patient 18) with iridocyclitis whose clinical picture was similar to the one reported from the United States.

19.2.1.3 Retinal Vasculitis

Inflammation of the retinal vessel system leads to a breakdown of the inner blood retina barrier located in the unfenestrated endothelium of the retinal vessels. This breakdown is followed by an outflow of blood proteins and cells and infiltration of the vitreous body (Naumann and Naumann 1986; Schönherr et al. 1990). The vitreous clouding seen in a Dutch patient seemed to be caused by a retinal vasculitis (Kuiper et al. 1989). Larger local ischemic retinal areas due to occlusive retinal vasculitis can lead to a preretinal neovascularization. We treated a patient with vitreous hemorrhage from preretinal neovascularization, who required Argon laser treatment (patient 11) (Lang et al. 1991). In another patient (patient 10), the retinal vasculitis preceded a facial palsy and showed a chronic course for up to 1 year until it was treated with antibiotics. It seems that the onset of the retinal vasculitis is most likely in the second stage of Lyme borreliosis.

19.2.1.4 Choroiditis

The inflammatory process in choroiditis, a second stage manifestation, is located in the choroid. In our own patients (patients 1-6, 12-17) and other European patients, choroiditis has been shown to be diffuse with inflammatory exudative retinal detachment (Bialasiewicz et al. 1988; Wiegand 1989; Wilk et al. 1989), focal (Bialasiewicz and Schönherr 1990), or disseminated (Schönherr et al. 1990; Krause et al. 1991). All our patients showed a prompt response to antibiotic therapy (Fig. 1 see color plate III after page 118). Asbrink et al. (1989) also mentioned a patient with borrelial lymphocytoma (from which *B. burgdorferi* was isolated), facial palsy, and choroiditis which healed with scarring after antibiotic therapy.

19.2.1.5 Optic Neuropathy

The optic nerve can actually be considered to be an anterior protruding white matter tract of the brain, differing in several respects from a typical peripheral nerve. Diseases of the optic nerve have to be differentiated by typical symptoms and findings in ophthalmoscopy (Naumann and Apple 1986). Optic disk edema is a general nonspecific sign in optic neuropathy. In a papilledema or pseudotumor cerebri, the disk edema is caused by increased pressure of the cerebrospinal fluid (Reik et al. 1979, 1986; Raucher et al. 1985; Schmidt and Ackermann 1985; Pfister et al. 1986; Ackermann et al. 1988; Jacobsen and Frens 1989), typically with good visual acuity in the acute phase. In ischemic optic neuropathy due to occlusion of the posterior ciliary arteries, the disk edema can be segmental with a corresponding segmental visual field loss (Schlechter 1986; Pizzarello et al. 1989). Papillitis is a nondemyelinating second-stage inflammatory process of the optic nerve head leading to sudden loss of vision (Farris and Webb 1988; Boutros et al. 1990). In neuroretinitis, papillitis is associated with vasculitis of the peripapillary retinal vessels (Wu et al. 1986; Bialasiewicz et al. 1989; Lesser et al. 1990; Schönherr et al. 1991; our patients 7-9; Fig. 2 see color plate IV after page 118). In demyelinating neuritis, the optic nerve head is not involved, because the myelin sheath begins behind the globe. Ackermann et al. (1988) noted retrobulbar neuritis in 4 of 44 patients with late borrelial encephalomyelitis. Depending on severity and duration of the different optic neuropathies, atrophy of the optic nerve can develop and result in bilateral poor visual acuity (patient 7) or in almost complete blindness (Stiernstedt et al. 1988; Bialasiewicz et al. 1989).

19.2.1.6 Panuveitis and Panophthalmitis

In panuveitis due to *B. burgdorferi*, all parts of the uvea (iris, ciliary body, and choroid) are involved (Wiegand 1989; Zierhut et al. 1989). In panophthalmitis, all tissues of the eye are infiltrated. One North American patient with Lyme borreliosis developed this manifestation after therapy with a systemic corticosteroid and antibiotics (Steere et al. 1985; Kauffmann and Wormser 1990).

19.2.1.7 Keratitis

Photophobia can be the first symptom of a stromal keratitis, which is believed to be a third stage manifestation. The keratitis can occur years after the onset of Lyme borreliosis, with multiple infiltrates at all levels of the corneal stroma with changing localization, episodic recurrences, good response to topical steroids, and a possible progression into neovascularization, and scarring (Baum et al. 1988; Bertuch et al. 1988; Orlin and Lauffer 1989; Baum and Barza 1989; Kornmehl et al. 1989). The observations made in our own patient 19 correspond well to the observations made in the United States.

19.2.2 Ocular Involvement Due to Extraocular Manifestations

19.2.2.1 Cranial Nerve Palsy

Exposure keratitis can occur due to a lack of surface protection in facial palsy, one of the most common signs in Lyme neuroborreliosis (patient 10) (see Chap. X). Diplopia, mostly caused by oculomotor or abducens nerve palsy, has been observed in a variety of European patients (see Chap.11) and more recently by Reik et al. (1979, 1986), Ackermann et al. (1988), Stiernstedt et al. (1988), and Kehl and Aust (1988). Diplopia may also be due to palsy of the trochlear nerve (Winward et al. 1989). Individual patients with reversible Horner's syndrome (Glauser et al. 1989) or Argyll Robertson pupil (Reik et al. 1986) have been described, but the relevance of these reports has yet to be determined.

19.2.2.2 Orbital Myositis

Orbital myositis has been observed in a child with erythema migrans, aseptic meningitis, facial palsy, and arthritis. There was a good response to a combined systemic corticoid and antibiotic therapy (Seidenberg and Leib 1990).

19.2.2.3 Encephalopathy

Homonymous visual field defects may be caused by occipital cortex lesions due to focal encephalitis or cerebral vasculitis (Hänny and Häuselmann 1987; May and Jabbari 1990).

19.3 Personal Series of Patients

Nineteen consecutive patients with inflammation of intraocular structures and evidence of Lyme borreliosis were seen at the Department of Ophthalmology, Universitiy Erlangen-Nürnberg, Germany (11 patients between 1985 and 1990) and the Department of Infectious Diseases, University Medical Center Ljubljana, Slovenia (8 patients between 1986 and 1990), and are presented in Table 2. The median age of our 19 patients (14 female, 5 male) was 38 years. Fourteen remembered a previous tick bite, and six developed an erythema migrans. Twelve had choroiditis, three neuroretinitis, one iridocyclitis, and three retinal vasculitis, one of whome had a recurring keratitis. Lymphocytic meningitis was present in four patients with diffuse choroiditis and three of them had elevated antibody titers against *B. burgdorferi* in the cerebrospinal fluid. Facial palsy or meningoradiculoneuritis was seen in four patients, and arthralgia and oligoarthritis in nine. In one patient, the diagnosis was made 6 months after the onset of neuroretinitis, when a cerebral demyelinization was shown in magnetic resonance imaging. Despite a good response of the neuroretinitis to antibiotic treatment, an optic atrophy resulted (patient 7). All patients had or developed elevated immunoglobulin G (IgG) antibody titers against *B. burgdorferi* in serum, and ten showed increased IgM antibody titers, which in seven patients disappeared after treatment within 4

Table 2. Personal series of 19 patients with intraocular involvement and without conjunctivitis

Patient No.	Age	Sex	Recent tick bite	Erythema migrans	Ocular signs	Extraocular manifestations	First serology IgM	First serology IgG	Follow-up (months)	Last serology IgM	Last serology IgG	Cerebrospinal fluid	Antibiotic therapy	Visual acuity first examination	Visual acuity last examination
1 [a]	40	m	+	–	Bilateral diffuse choroiditis	Lymphocytic meningitis	neg. [b]	460U [b]	31	neg. [c]	neg. [c]	250 cells, ELISA neg.	Intravenous penicillin 10 days	RE:0.1 LE: 0.1	0.8 0.8
2 [a]	32	f	–	–	Bilateral diffuse choroiditis	Lymphocytic meningitis	1:40 [d]	1:640 [d]	7	neg. [d]	1:320 [d]	91cells, IFA pos.	Doxycycline 200 mg 14 days	RE:0.2 LE: 0.3	0.8 0.8
3 [a]	61	f	+	–	Bilateral focal choroiditis	AV block – second degree meningo-radieulitis	neg. [d]	1:320 [d]	29	neg. [d]	1.80 [d]	n.d.	Intravenous penicillin + oxytetracyclin, 10 days each	RE: - LE: -	1.0 0.8
4 [a]	19	m	–	–	Unilateral focal choroiditis	-	1:80 [d]	1:320 [d]	6	neg. [d]	1.320 [d]	n.d.	Oral docycycline, 200 mg/7 days	RE: 0.4	1.0
5 [a]	26	f	–	–	Bilateral disseminated choroiditis	Lymphadenopathy, oligoarthritis	1:64 [c] pos. WB	1:64 [c]	10	neg. [c]	neg. [c]	17 cells IFA neg.	Oral doxycycline 200 mg/6 weeks	RE: 0.9 LE: 0.1	1.0 1.0
6 [a]	46	f	+	+	Unilateral disseminated choroiditis	Arthralgia, episodical neg. P wave on ECG	neg. [c] pos. WB pos. LPA	1:128 [c]	3	neg. [c]	1:32 [c]	n.d.	Oral doxycycline, 200 mg/4 weeks; ceftriaxone, 2 g/14 days	RE: 0.6	1.0
7 [a]	25	f	–	–	Bilateral neuroretinitis	Focal demyelinating encephalitis	1:640 [d]	1.1280 [d]	11	neg. [d]	1:1280 [d]	7 cells IFA neg.	Oral doxycycline 200 mg/14 days	RE: 0.1 LE: 0.07	0.2 0.1
8 [a]	21	f	+	–	Bilateral neuroretinitis	Headache fever	1:80 [d]	1:160 [d]	8	neg. [d]	1:80 [d]	0 cells IFA neg.	Oral doxycycline, 200 mg/14 days	RE: 0.4 LE: 0.4	1.0 1.0

Table 2. (Continued) Personal series of 19 patients with intraocular involvement and without conjunctivitis

Patient No.	Age	Sex	Recent tick bite	Erythema migrans	Ocular signs	Extraocular manifestations	First serology IgM	First serology IgG	Follow-up (months)	Last serology IgM	Last serology IgG	Cerebrospinal fluid	Antibiotic therapy	Visual acuity first examination	Visual acuity last examination
9 [a]	19	f	+	–	Unilateral neuroretinitis	Headache, fever	neg.[d]	neg.[d]	3	neg.[d]	1:320[d]	n.d.	Oral doxycycline, 200 mg/14 days	RE: 0.3	1.0
10 [a]	25	f	+	–	Bilateral retinal vasculitis	Facial, palsy	neg.[d] pos. LPA	1:320[d]	26	neg.[d]	neg.[d]	n.d.	Oral doxycycline, 200 mg/14 days	RE: 0.1 LE: 1.0	1.0 1.0
11 [a]	43	m	+	–	Bilateral retinal vasculitis	Oligo-arthritis	neg.[d]	1:1280[d]	12	neg.[d]	1:320[d]	n.d.	Oral doxycycline 200 mg/14 days	RE: 1.0 LE: 0.3	1.0 0.6
12 [e]	44	f	+	–	Unilateral diffuse coroiditis	Facial palsy, meningitis, fatigue, arthralgia	n.d.	48 U [f]	49	neg.[g]	neg.[g]	372 cells ELISA pos.[h]	Intravenous penicillin, 20 mega/14 days	LE: 0.1	1.0
13 [e]	34	m	+	+	Bilateral focal choroiditis	-	n.d.	55 U [f]	30	neg.[g]	1:128[g]	ELISA neg.	Intravenous penicillin, 20 mega/14 days	RE: 0.3 LE: 0.4	0.8 0.8
14 [e]	52	f	+	–	Bilateral diffuse choroiditis	Influenza like illness, arthralgia	1:256 [g] pos. WB	1:256[g]	15	neg.[g]	neg.[g]	n.d.	Intravenous penicillin, 20 mega/14 days	RE: 0.1 LE: 0.2	1.0 1.0
15 [e]	51	f	+	–	Unilateral diffuse choroiditis	Meningitis	n.d.	55 U [f]	30	neg.[g]	1:256[g]	87 cells, ELISA pos.[h]	Intravenous penicillin, 20 mega/14 days	RE: 0.1	0.9
16 [e]	54	m	+	+	Bilateral focal choroiditis	Arthralgia paresthesia fatigue	1:512[g] pos. WB	1:512[g]	19	neg.[g]	1:128[g]	IFA neg.	Intravenous ceftriaxone, 2x2 g/14 days	RE: 0.4 LE: 0.3	1.0 1.0

Table 2. (Continued) Personal series of 19 patients with intraocular involvement and without conjunctivitis

Patient No.	Age	Sex	Recent tick bite	Erythema migrans	Ocular signs	Extraocular manifestations	First serology IgM	First serology IgG	Follow-up (months)	Last serology IgM	Last serology IgG	Cerebrospinal fluid	Antibiotic therapy	Visual acuity first examination	Visual acuity last examination
17[e]	36	f	+	+	Unilateral diffuse choroiditis	Arthralgia myalgia	1:256[g]	1:512[g]	7	1:128[g]	1:128[g]	IFA neg.	Oral doxycycline, 200 mg/14 days	LE: 0.2	1.0
18[e]	33	f	+	+	Unilateral iridocyclitis	Arthralgia paresthesia	1:512[g]	1:1024[g]	13	1:128[g]	1:256[g]	IFA neg.	Intravenous ceftriaxone, 2x2 g/14 days	RE: 0.4	0.8
19[e]	51	f	–	+	RE: Keratitis LE: Retinal vasculitis	Oligoarthritis, fatigue	neg.[f,g]	neg.[f,g]	12	neg.[g]	neg.[g]	n.d.	Phenoxymethyl penicillin, 4.5 mega/4 weeks	RE: 0.4 LE: 0.3	0.7 0.9
							1:512[g]	1:512[g]					Not yet done	RE: 0.2 LE: 0.9	

+, yes; –, no; neg., negative; pos., positive; ECG, electrocardiogram; ELISA, enzyme-linked immunosorbent assay; IFA, immunofluorescent assay; LE, left eye; LPA, lymphocyte proliferation assay; RE, right eye; WB, Western blot; n.d., not determined.

[a] Seen at University Eye Hospital, Erlangen-Nürnberg, Germany.

[b] ELISA normal value < 100 U:

[c] IFA (abs) normal value, IgG < 1:16; IgM < 1:8.

[d] IFA normal value, IgG < 1:80; IgM < 1:20

[e] Seen at Department of Infectious Diseases, University Medical Center, Ljubljana, Slovenia.

[f] ELISA normal value, < 48 U.

[g] IFA normal value, IgG < 1:256; IgM < 1:128.

[h] Intrathecal antibody production.

[i] In the right eye, keratitis recurred 2 years later

months. Five patients had no or nonspecific extraocular symptoms, and 12 patients first visited an ophthalmologist because of blurred vision. There was a good response to antibiotic treatment, except in the above-mentioned patient when treatment was initiated 6 months after onset of ocular symptoms (patient 7) and in the patient with recurrence of keratitis (patient 19).

19.4 Diagnosis

In experimentally infected animals, *B. burgdorferi* was cultivated from the eye (Johnson et al. 1984) and isolated from the vitreous body (Duray and Johnson 1986) after intraperitoneal inoculation and was shown to produce an intraocular inflammation (Schuman et al. 1988). Because intraocular material is usually not available from humans, serology is the main aid in diagnosis. Asymptomatic seropositivity can lead to substantial overdiagnosing, particularly in highly endemic regions, but on the other hand a negative test does not rule out Lyme borreliosis (see Chap. 21). For that reason, the presence of erythema migrans and/or other typical manifestation(s) of Lyme borreliosis, the degree and development of the antibody titers against *B. burgdorferi*, and the presence of IgM antibodies are important for establishing the diagnosis. Even the typical clinical picture of Lyme borreliosis does not entirely affirm that all clinical signs and/or symptoms are causally related to the *B. burgdorferi* infection. Anyhow, some patients will remain with suspected, but not provable ocular Lyme borreliosis in whom antibiotic treatment should be considered.

19.5 Differential Diagnosis

The described ocular manifestations are not pathognomonic for Lyme borreliosis, rather they represent the typical spectrum seen in any ocular spirochetosis. For example, this spectrum of manifestations is well known in ocular syphilis (Duke-Elder 1976). Granulomatous iridocyclitis or chorioiditis can be due to bacterial illness (syphilis, tuberculosis, leprosy), fungal infections, or immunologic processes of unclear etiology such as sarcoidosis, Vogt-Koyanagi-Harada syndrome, or Still's disease. Retinal vasculitis and optic neuropathies can also be induced by viruses (varicella-zoster virus) and are seen in encephalitis disseminata. To establish the diagnosis of ocular Lyme borreliosis, other causes have to be ruled out (Naumann and Naumann 1986; Naumann and Apple 1986).

19.6 Therapy

In our 19 patients, we saw a good response to systemic antibiotic therapy with oral doxycycline, intravenous penicillin, and intravenous ceftriaxone. Accompanying systemic corticosteroid therapy under strict clinical control is thought to reduce the extent of damage caused by the inflammatory reaction in severe uveitis and orbital myositis. Keratitis responded well to additional local steroids. In uveitis, it is necessary to keep the pupil dilated (Frauenfelder 1990).

19.7 Conclusion

Ocular inflammation can be one of the first or even the only obvious manifestation of Lyme borreliosis. Ocular manifestations can occur at all stages of the disease. The time interval between tick bite or erythema migrans and ocular manifestations can vary from a few weeks in second-stage choroiditis and neuroretinitis to years in keratitis or third-stage optic neuropathy. Opthalmologists should initiate serological work up for Lyme borreliosis and look for extraocular symptoms of this disease in cases of ocular inflammation. Nonophthalmologists should look for ocular involvement in patients with signs and/or symptoms compatible with Lyme borreliosis, because ocular findings can support the general diagnosis. Long-standing intraocular inflammation results in loss of vision. In our patients, prompt systemic antibiotic treatment was quite effective and led to recovery of visual acuity. Therefore, even suspected ocular Lyme borreliosis should be treated antibiotically. If antibiotic therapy is initiated too late it can lead to a severe loss of vision.

Acknowledgments. We thank Prof. Dr. F.H. Meythaler, Prof. Dr. G.O.H. Naumann, Dr. J. Cimperman, and especially Dr. Gabriele E. Lang for reviewing the manuscript and for valuable dicussion.

References

Aaberg TM (1989) The expanding ophthalmologic spectrum of Lyme disease. Am J Ophthalmol 107:77-80

Ackermann R, Rehse-Küper B, Gollmer E, Schmidt R (1988) Chronic neurologic manifestations of erythema migrans borreliosis. Ann N Y Acad Sci 539:16-23

Asbrink E, Hovmark A, Olsson I (1989) Borrelial lymphocytoma in Sweden. Zentral Bakteriol (Suppl) 18:156-163

Baum J, Barzal M (1989) The expanding ophthalmologic spectrum of Lyme disease. Am J Ophthalmol 107:684-685

Baum J, Barza M, Weinstein P, Groden J, Aswad M (1988) Bilateral keratitis as a manifestation of Lyme disease. Am J Ophthal 105:75-77

Bertuch AW, Rocco E, Schwartz EG (1988) Lyme disease: ocular manifestations. Ann Ophthalmol 20:376-378

Bialasiewicz AA, Schönherr U (1990), Choriokapillaritis (sog. Pigmentepitheliitis) bei *Borrelia burgdorferi*-Serokonversion. Klin Monatsbl Augenheilkd 196:481-483

Bialasiewicz AA, Ruprecht KW, Naumann GOH, Blenk H (1988) Bilateral diffuse chorioditis and exsudative retinal detachments with evidence of Lyme disease (letter) Am J Ophthalmol 105:419-420

Bialasiewicz AA, Huk W, Druschky KF, Naumann GOH (1989) *Borrelia burgdorferi*-Infektion mit beidseitiger Neuritis nervi optici und intrazerebralen Demyelinisierungsherden. Klin Monatsbl Augenheilkd 195:91-94

Boutros A, Rahn E, Nauheim R (1990) Iritis and papillitis as a primary presentation of Lyme disease. Ann Ophthalmol 22:24-25

Duke-Elder S (1976) System of ophthalmology, vol XV. Kimpton, London pp 150-155

Duray PH, Johnson RC (1986) The histopathology of experimentally infected hamsters with the Lyme disease spirochete, *Borrelia burgdorferi*. Proc Soc Exp Biol Med 181:263-269

Eichenfield AH, Goldsmith DP, Benach JL et al. (1986) Childhood Lyme arthritis: experience in an endemic area. J Pediatr 109:753-758

Farris BK, Webb RM (1988) Lyme disease and optic neuritis. J Clin Neuro Ophthalmol 8:73-78

Fraunfelder FT (1990) Lyme disease. In: Fraunfelder FT, Roy FH (eds) Current ocular therapy, Vol 3. J Clin Neuro Ophthalmol 9:225-228

Glauser TA, Brennant PJ, Galetta SL (1989) Reversible Horner's syndrome and Lyme disease. J Clin Neuro Ophthalmol 9:225-228

Hänny PE, Häuselmann HJ (1987) Die Lyme-Krankheit aus der Sicht des Neurologen. Schweiz Med Wochenschr 117:901-915

Jacobs JC, Stevens M, Durray PH (1986) Lyme disease simulating septic arthritis. JAMA 256:1138-1139

Jacobsen DM, Frens DB (1989) Pseudotumor cerebri syndrome associated with Lyme disease. Am J Ophthalmol 107:81-82

Johnson RC, Marek N, Kodner C (1984) Infection of Syrian hamsters with Lyme disease spirochetes. J Clin Microbiol 20:1099-1101

Kauffmann DJH, Wormser GP (1990) Ocular Lyme disease: case report and review of the literature. Br J Ophthalmol 74:325-327

Kehl F, Aust W (1988) Augenbefunde bei Zecken-Borreliose. Klin Monatsbl Augenheilkd 193:66-68

Kornmehl EW, Lesser RL, Jaros P, Rocco E, Steere AC (1989) Bilateral keratitis in Lyme disease. Ophthalmology 96:1194-1197

Krause A, Schönherr U, Schörner C, Brade V, Burmester RG (1991) Gonarthritis, Lymphadenopathie und disseminierte Chorioditis als primäre Manifestation einer Lyme-Borreliose. Kasuistik und Diskussion der diagnostischen Möglichkeiten bei der Lyme-Borreliose. Z Rheumatol 50:10-15

Kuiper H, Koelman JHTM, Jager MJ (1989) Vitreous clouding associated with Lyme borreliosis. Am J Ophthalmol 108:453-454

Lang GE, Schönherr U, Naumann GOH (1991) Retinal vasculitis with proliferative retinopathy in a patient with evidence of *Borrelia burgdorferi* infection. Am J Ophthalmol 111:243-244

Lesser RL, Kornmehl EW, Pachner AR (1990) Neuro-ophthalmologic manifestations of Lyme disease. Ophthalmology 97:699-706

MacDonald AB (1987) Lyme disease, a neuro-ophthalmologic view. J Clin Neuro Ophthalmol 7:185-190

May EF, Jabbari B (1990) Stroke in neuroborreliosis. Stroke 21:1232-1235

Naumann GOH, Apple DJ (1986) Optic nerve. In: Naumann GOH, Apple DJ (eds) Pathology of the eye. Springer, Berlin Heidelberg New York, pp 723-770

Naumann GOH, Naumann LR (1986) Intraocular inflammation. In: Naumann GOH, Apple DJ (eds) Pathology of the eye. Springer, Berlin Heidelberg New York pp 99-184

Orlin SE, Lauffer JL (1989) Lyme disease keratitis. Am J Ophthalmol 107:678-680

Pfister HW, Einhaupl KM, Wilske B, Preac-Mursic V (1986) Bannwarth's syndrome and the enlarged neurological spectrum of anthropod borne borreliosis. Zentralbl Bakteriol Hyg (A) 263:343-347

Pizzarello LD, MacDonald AB, Semlear R, DiLeo F, Berger B (1989) Temporal arteritits associated with *Borrelia* infection. A case report. J Clin Neuro Ophthalmol 9:3

Raucher HS, Kaufman DM, Goldfarb J, Jacobson RI, Roseman B, Wolff RR (1985) Pseudotumor cerebri and Lyme disease, a new association. J Pediatr 107:931

Reik L, Steere AC, Bartenhagen NH, Shope RE, Malawista SE (1979) Neurologic abnormalities of Lyme disease. Medicine (Baltimore) 58:281-294

Reik L, Burgdorfer W, Donaldson JO (1986) Neurologic abnormalities of Lyme disease without erythema chronicum migrans. Am J Med 81:73-78

Schlechter SL (1986) Lyme disease associated with optic neuropathy. Am J Med 81:143-145

Schmidt R, Ackermann R (1985) Durch Zecken übertragene Meningo-Polyneuritis (Garin-Bujadoux-Bannwarth) – Erythema chronicum migrans-Krankheit des Nervensystems. Fortschr Neurol Psychiatr 53:145-153

Schönherr U, Wilk CM, Lang GE, Naumann GOH (1990) IgM *Borrelia burgdorferi* antibody titers in patients with ocular manifestations of Lyme borreliosis. Invest Ophthalmol Vis Sci 31:2804

Schönherr U, Lang GE, Meythaler FH (1991) Bilaterale Leber'sche Neuroretinitis stellata bei *Borrelia burgdorferi* Serokonversion. Klin Monatsbl Augenheildk 198:44-47

Schuman JS, Weinberg RS, Ridderhof J, Tucker M (1988) Experimental Lyme uveitis (abstract). Invest Ophthalmol Vis Sci (Suppl) 29:444

Seidenberg KB, Leib ML (1990) Orbital myositis with Lyme disease. Am J Ophthalmol 109:13-16

Spalton DJ (1990) Lyme disease. Br J Ophthalmol 74:321-322

Steere AC (1989) Lyme disease. N Engl J Med 321:586-596
Steere AC, Bartenhagen NH, Craft JE, Hutchinson GJ, Newman JH, Rahn DW, Sigal LH, Spieler PN, Stenn KS, Malawista SE (1983) The early clinical manifestations of Lyme disease. Ann Intern Med 99:76-82
Steere AC, Duray PH, Kauman DJH, Wormser GP (1985) Unilateral blindness caused by infection with the Lyme disease spirochete, *Borrelia burgdorferi*. Ann Intern Med 103:382-384
Stiernstedt G, Gustavsson R, Karlsson M, Svenungsson B, Sköldenberg B (1988) Clinical manifestations and diagnosis of neuroborreliosis. Ann NY Acad Sci 539:46-55
Weber K, Neubert U (1986) Clinical features of early erythema migrans disease and related disorders. Zentralbl Bakteriol Hyg (A) 263:209-228
Wiegand W (1989) Augenbefunde bei Borreliose – bilaterale Panuveitis mit exsudativer Amotio. Fortsch Ophthalmol 86:659-662
Wilk CM, Bialasiewicz AA, Ruprecht KW, Naumann GOH (1989) Bilaterale akute konfluierende Chorioditis disseminata bei *Borrelia burgdorferi*-Infektion. Klin Monatsbl Augenheildk 194:88-96
Winterkorn JMS (1990 Lyme disease: neurologic and ophthalmic manifestations. Surv Ophthalmol 35:191-204
Winward KE, Smith JL, Culbertson WW, Paris-Hamelin AJ (1989) Ocular Lyme borreliosis. Am J Ophthalmol 108:651-657
Wu G, Lincoff H, Ellsworth RM, Haik BG (1986) Optic disc edema and Lyme disease. Ann Ophthalmol 18:252-255
Zierhut M, Kreissig I, Pickert A (1989) Panuveitis with positive serological tests for syphilis and Lyme borreliosis. J Clin Neuro ophthalmol 9:71-75

20 Borrelial Dermatomyositis-like Syndrome

U. DETMAR and W. MACIEJEWSKI

20.1 Introduction

We have recently observed two patients with what we would like to call "borrelial dermatomyositis-like syndrome." In this chapter, we present the two case reports and discuss our findings.

20.2 Definition

The borrelial dermatomyositis-like syndrome as observed in our two patients is characterized by involvement of skin and muscles in association with elevated antibody titers against *Borrelia burgdorferi* and other abnormalities in laboratory findings. The disorder starts rapidly and presents clinically with widespread, bluish-red, spotty or patchy erythemas with a raised temperature. The patients experience muscular pain and weakness in the extremities.

20.3 Case Presentations

20.3.1 Patient 1

A 43-year-old man was admitted to the hospital early in May 1988. About 3 weeks earlier, in April, he had noticed muscular pain and weakness beginning in his thighs and upper arms. Additionally, he experienced subfebrile temperatures, a feeling of exhaustion, and general weakness. One week later, the first erythemas appeared on the skin of the chest and shoulders, and a significant rise in creatine kinase (CK), glutamate oxalacetate transaminase (GOT) and lactate dehydrogenase (LDH) in the serum were noted by the family physician. Because of increasing skin symptoms and deterioration of his general condition, he was referred to the hospital. The patient's history provided no hints for erythema migrans or any tick bite. The patient was employed as a general maintenance worker for the community and worked outdoors frequently.

Physical examination showed bluish-red, patchy, sometimes spotty or diffuse erythemas with a raised temperature distributed over the entire skin, particularly on the upper arms, thighs, shoulders, and neck. The right arm, the left upper arm, and the right thigh showed diffuse reddening and edematous swelling (Fig. 1, see color plate IV between pp 118–119; Fig. 2). Enlargement of the lymph nodes was not observed.

Neurological examination revealed diminished biceps and triceps reflexes on both sides. The patellar reflex was diminished on the right side only. In other respects, the neurological status was normal. Electromyography of the right musculus (m.) tibialis anterior, m. quadriceps femoris, and m. deltoideus, and electroneurography of the right peroneus and sural muscles showed the following results:

Fig. 2. Borrelial dermatomyositis-like syndrome. Patient 1. Diffuse erythemas on the neck and shoulder

Distal motor latency and nerve conduction velocities were normal. Especially in the proximal extremities (m. quadriceps), pathological spontaneous activity and pseudomyotonic "runs" were noted. The potential of the motor units in the quadriceps and deltoid muscles was polyphasic with normal potential duration. The excitatory pattern during maximal innervation showed an interference pattern with early recruitment. These findings were interpreted as representing a myopathy of medium severity.

The following laboratory values were abnormal: erythrocyte sedimentation rate (ESR) 52 mm/h; white cell count 12200/mm^3, with 89% neutrophils, 4% rod neutrophils, 1% eosinophils, 3% lymphocytes, and 3% monocytes; serum GOT 37 U/l; glutamate pyruvate transaminase (GPT) 25 U/l; CK 310 U/l; LDH 482 U/l; protein electrophoresis: albumin 52.0%, α_1 globulin 4.6%, α_2 globulin 11.6%, β globulin 10.7% and γ globulin 20.8%; complement C3 70 mg; immunoglobulin M (IgM) 962 mg%. Circulating immune complexes were not determined.

Further laboratory investigations yielded the following results: The antinuclear antibody titer was 1:50 (slightly elevated value); the anti-DNA-antibody titer was not elevated; C-reactive protein (CRP) was 48 mg/l; rheumatic factor was slightly elevated (latex 160 U/l, Waaler-Rose 64 U/l); the fluorescence treponema antibodies (FTA) absorption test and the cardiolipin-flocculation test for syphilis did not react, whereas the treponema pallidum hemagglutination (TPHA) was unspecifically positive. The IgG antibody titer against *B. burgdorferi* was strongly increased (IgG-immunofluorescence (IgG-IFT) absorption test 1:2048) while the IgM antibody titer was normal (IgM-IFT abs. 1:16; normal values for IgG and IgM antibody titer ≤ 1:16; method according to Wilske et al. 1984). Serological tests for human immunodeficiency virus (HIV), Elpstein-Barr virus (EBV), influenza A and B, mycoplasma, and chlamydia showed no abnormal results.

Histological examination of the skin demonstrated orthokeratosis, flattened epidermis, and slight atrophy in some areas. In the upper and middle dermis, perivascular edema, dilated vessels with swelling of the walls, and sparse to moderately dense perivascular infiltrates composed of lymphocytes, histiocytoid elements, and numerous plasma cells were seen. Occasionally, neutrophils and eosinophils could be found (Fig. 3). The special staining of the elastica showed a localized loss of elastic fibers. Histological examination of the right m. quadriceps revealed only sparse perivascular lymphocytic infiltrates. A few spirochetal structures could be identified in the Bosma-

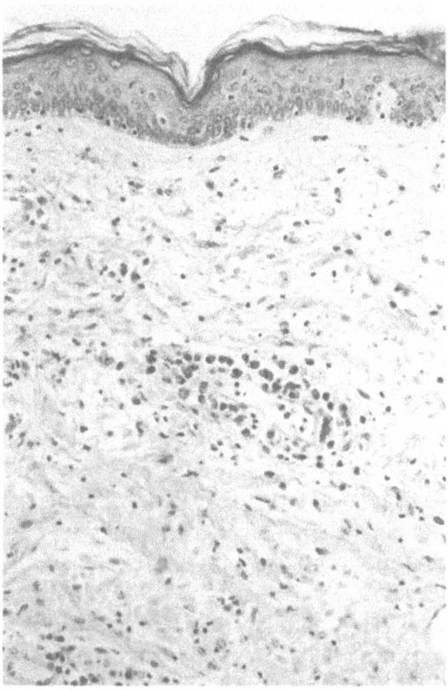

Fig. 3. Flattened epidermis, edematous dermis, dilated vessels with perivascular infiltrates consisting of numerous plasma cells. The biopsy was taken from the bluish-red area of the right upper thigh of patient 1

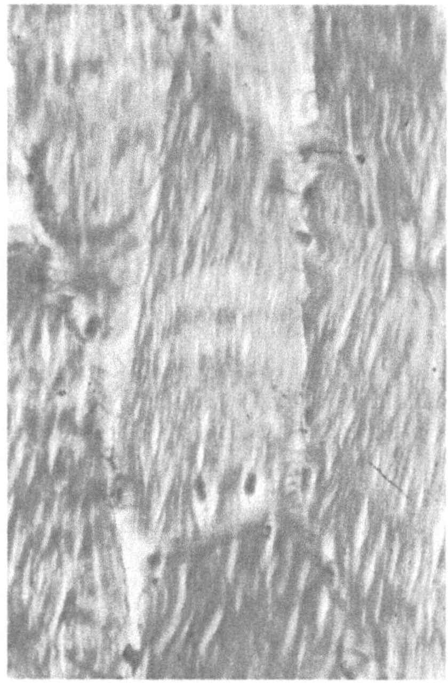

Fig. 4. Presence of spirochetal structures in the muscle biopsy in patient 1 (Bosma-Steiner-staining).

Steiner staining (Fig. 4). We first considered a dermatomyositis, but felt that an erysipelas could not be excluded. The patient was, therefore, treated with prednisolone 125 mg daily for 4 days, with subsequent reduction in the following 29 days. In addition, he received intravenous penicillin 10 ml/U twice daily for the first 7 days. By the end of treatment, the patient was free from signs and symptoms of the illness. No symptoms of any muscular disorder were still found. The patient remained without symptoms afterwards. A serological check-up 1 year later showed that the IgG antibody titer against *B. burgdorferi* had significantly decreased to 1:128.

20.3.2 Patient 2

A 37-year-old electrician had noted skin irritations on the face and chest for approximately 2 weeks when he entered the hospital in April 1988. He also complained about painful redness and swelling of the right wrist and weakness in both legs. Four months earlier, his right knee joint had to be tapped because of a painful swelling; cultures of the effusion revealed no growth. The patient was unable to recall a tick bite, although he frequently walked trough the woods in his leisure time. The physical examination showed patchy, bluish erythemas with raised temperature and diffuse edematous swelling on the face and upper chest. Additionally, bilateral edema of the eyelids were found (Fig. 5). The right wrist was reddened and swollen. A bluish red, infiltrated, dif-

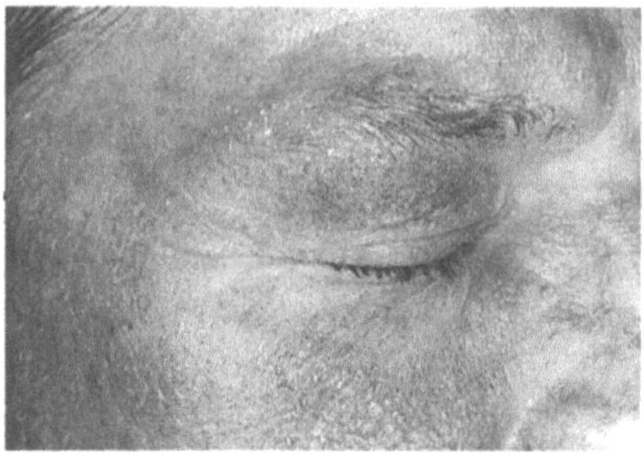

Fig. 5. Borrelial dermatomyositis-like syndrome. Patient 2. Diffuse edematous erythemas on the face with distinct edema of the eyalids

fuse erythema could be found above the right knee. The size of the right lower leg was slightly increased, with the skin showing some atrophy and having a waxy, reflecting surface. Both legs showed numerous disseminated petechial bleedings up to the size of a pinhead. Electromyography of the left m. rectus femoris, the left m. vastus medialis, the right m. rectus femoris, and the right m. tibialis anterior revealed pathological spontaneous activity with positive sharp waves and a high-frequency series of discharges (m. tibialis anterior). Voluntary innervation showed polyphasic, relatively wide-based action potentials of low amplitude. During maximal innervation, the pattern of activity was thinned. No certain demonstrable spontaneous activity in other muscles could be found. During voluntary innervation, the action potentials were regulary formed. Electroneurography of the right m. tibialis anterior showed the motor nerve transmission speed to be 45 m/s with normal action potentials.

The following laboratory values were abnormal: White cell count was $11300/mm^3$ with 78% neutrophils, 2% rod neutrophils, and 20% lymphocytes. In the protein electrophoresis, the albumine was 35.5%, α_1 globulin 3.1%, α_2 globulin 6.7%, β globulin 6.5% and γ globulin 48.2%. The hemoglobin was 11.7 mg%, the erythrocyte count 3.4 million, the platelet count 420000; serum IgG was 4310 mg% and serum IgM 460 mg%; complement C3 was 54 mg% and C4 8 mg%.

Histological examination of the skin showed orthokeratosis, flattening and atrophy of the epidermis, interstitial edema, and dilatation of the blood vessels in the upper and middle dermis. The infiltrates consisting of numerous plasma cells, neutrophils, lymphocytes, and histiocytoid elements were located partly perivascularly, partly in a dis-

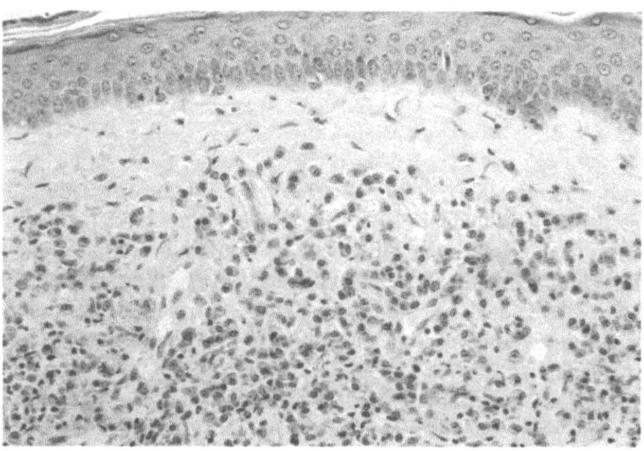

Fig. 6. Flattened epidermis; many dilated vessels and scattered infiltrates consisting mainly of plasma cells in the upper dermis. The biopsy was taken from the forehead of patient 2

seminated pattern. (Fig. 6). Staining of the elastica revealed a significant reduction in the number of elastic fibers or their complete absence in the upper and middle dermis. Histological examination of muscular tissue (m. rectus femoris) showed local inflammatory areas with lymphocytes and monocytes. The immunohistochemical investigation revealed occasional lymphocytes and plasma cells with IgG markers in the interstitial spaces and perivascularly.

B. burgdorferi could be isolated from skin biopsies taken from the chest and the right knee (method according to V. Preac-Mursic et al. 1989). The IgG antibody titer against *B. burgdorferi* was 1:512, and the IgM antibody titer 1:32 (borderline titer, method as stated above). The examination of the cerebrospinal fluid yielded no indication of the production of autochthonous antibodies. Circulating immunocomplexes were detected with a value of 100% (normal level less than 10%; C1q binding) and the Waaler-Rose test was 1:128. The antinuclear antibody titer and the HLA-B 27 were negative.

Antibiotic therapy with intravenous cefotaxime 2 g three times daily for 15 days resulted in a decrease in the edematous erythemas and joint symptoms. The general condition of the patient improved, and he was practically free of all signs and symptoms on discharge from the hospital. A few weeks later, the patient returned for treatment of a newly developed skin eruption. In addition to the persistent diffuse erythema in the facial area, a diffuse erythema with white desquamation now appeared around both hips and on the back. After 14 days therapy with oral tetracycline 1000 mg daily for 14 days, the erythema vanished, leaving behind a cigarette paper-like folded looking area of atrophic skin. On histological examination, this area of skin showed a significant loss of elastic fibers.

20.4 Discussion

The history and findings in our two patients did not coincide completely with any single one of the three definite manifestations of Lyme borreliosis, erythema migrans, borrelial lymphocytoma, or acrodermatitis chronica atrophicans, which have been well known in Europe for a long time (Weber et al. 1984; Asbrink et al. 1985). Because of the general symptoms, consisting of weakness of the muscles, bluish erythemas, and edematous swelling appearing on the skin, a dermatomyositis was first considered. This diagnosis was matched by the pathological findings obtained by electromyography, the elevated levels of the muscle enzymes CK and LDH in the first patient, and the swelling and reddening of the eyelids in the second patient. The asymmetric distribution of the manifestations on the skin, however, with the presence of edematous erythemas not only in the face, but also on various parts of the body, and the identification of plasma cellular infiltrates found on histological investigation did not fit into the picture of dermatomyositis and remained unexplained.

Decreased complement C3 and C4 levels and the presence of circulating immune complexes, as seen in one of our patients, have already been observed in patients with more severe Lyme borreliosis in the United States (Steere et al. 1977; Hardin et al. 1979). These laboratory findings indicate that an immune-mediated disease may play a role in the pathogenesis of the manifestations of Lyme borreliosis presented here. The Lyme serology yielded significantly elevated IgG antibody titers against *B. burgdorferi* in both patients. This, together with the good response to the antibiotic treatment, the isolation of *B. burgdorferi* from the skin of one of our patients, and the decline of the specific antibody titers were strong hints for an infection with *B. burgdorferi*. This dermatomyositis-like syndrome as a manifestation of Lyme borreliosis has not previously been described in Europe. Only in one old report (Beer 1892) is the clinical presentation of a patient with edematous swelling of the face and the extremities mentioned. In that case, the edematous stage of an acrodermatitis chronica atrophicans may have been reported without full documentation. In 1930, Hellerström described a patient with meninigits and a generalized erythema migrans developing within several months, but the skin lesion was different from the ones observed in our two patients.

The general symptoms noted in our patients such as headache, nausea, and painful joints can be observed during the early stages of Lyme borreliosis (Weber und Neubert 1986). These and other symptoms as well as multiple erythema-migrans-like lesions occur much more frequently in the United States than in Europe (Weber 1989). About 50% of 314 patients with erythema migrans observed in the United States by Steere and co-workers (1983) had multiple erythema-migrans-like lesions and some had also evanescent red blotches, urticarial rash, periorbital edema, or diffuse erythema. Sometimes migratory musculoskeletal pain, hepatitis, generalized lymphadenopathy and splenomegaly were also found, lasting for several weeks. The laboratory abnormalities consisted of a high ESR, an elevated serum IgM, or an increased aspartate transaminase level in a portion of the patients. Some of these signs are quite similar to the ones noted in one of our patients. However, the American investigators did not find elevated CK levels or other indications of a true myositis, as proven histologically in our patient 2 (preceding treatment with corticosteroids may have been the reason for the failure to obtain similar histological results in our patient 1). Furthermore, the cutaneous lesions seemed to be somewhat different.

Müller (1969) described a patient with a myositis and some abnormal laboratory values (including a positive antinuclear antibody test) similar to the ones we found in our patients, but the patient suffered from typical acrodermatitis chronica atrophicans, not from skin lesions such as those seen in our patients. All patients with borrelial myositis described so far, presented a somewhat different clinical picture to the one in our patients (Schmutzhard et al. 1986; Atlas et al. 1988) (see Chap. 13). Furthermore, the course in our patients was not intermittent. The clinical manifestations in their entirety thus permit differentation from the one seen in other patients with Lyme borreliosis described so far. It cannot easily be decided with which stage of Lyme borreliosis the borrelial dermatomyositis-like syndrome could be associated. The high IgG antibody titer of patient 1 and the preceding arthritis in patient 2, lack of a recent history of erythema migrans or tick bite, the decreased amount of elastic fibers in the skin, and the onset of the disease of both patients in April point to a stage 3 manifestation of Lyme borreliosis.

20.5 Diagnosis and Differential Diagnosis

20.5.1 Diagnosis

To ascertain the diagnosis of borrelial dermatomyositis-like syndrome, a combination of an elevated IgG antibody titer agains *B. burgdorferi*, the described changes in the skin and muscles, laboratory abnormalities indicating inflammation and an increased serum CK level ought to be found. The cultural isolation of *B. burgdorferi* from the skin and, to a lesser extent, the identification of spirochetes in the tissue help to verify the diagnosis.

20.5.2 Differential Diagnosis

Dermatomyositis, systemic lupus erythematosus, and mixed connective tissue disease must be considered and differentiated by histological and serological examinations.

References

Asbrink E (1985) Erythema chronicum migrans Afzelius and acrodermatitis chronica atrophicans. Acta Derm Venereol (Suppl) Stockh 118:1-63

Atlas E, Nowak SN, Duray PH, Steere AC (1988) Lyme myositis: muscle invasion by *Borrelia burgdorferi*. Ann Intern Med 108:245-246

Beer KD (1892) Verhandlungen der Wiener Dermatologischen Gesellschaft. Arch Dermatol Syph 24:835

Detmar U, Maciejewski W, Link C, Breit R, Sigl H, Robl H, Preac-Mursic V (1989) Ungewöhnliche Erscheinungsformen der Lyme-Borreliose. Hautarzt 40:423-429

Hardin JA, Steere AC, Malawista SE (1979) Immune complexes and the evolution of Lyme arthritis. N Engl J Med 301:1358-1363

Hellerström S (1930) Erythema chronicum migrans Afzelii. Acta Dermatol Venereol (Stockh) 11:315-321

Müller H (1969) Akrodermatitis atrophicans arthropathica mit Pseudosklerodermie, Sklerodermia circumscripta und muskulärer Beteiligung. Haut Geschlechtskr 44:1-12

Preac-Mursic V, Wilske B, Schierz G (1986) European *Borrelia burgdorferi* isolated from humans and ticks. Zentralbl Bakteriol Hyg (A) 263:112-118

Schmutzhardt E, Willeit J, Gerstenbrand F (1986) Meningopolyneuritis Bannwarth with focal nodular myositis. Klin Wochenschr 64:1204-1208

Steere AC, Malawista SE, Hardin JA, Ruddy S, Askenase PW, Andiman WA (1977) Erythema chonicum migrans and Lyme Arthritis. Ann Intern Med 86:685-698

Steere AC, Bartenhagen NH, Craft JE, Hutchinson GJ, Newman JH, Rahn DW, Sigal LH, Spieler PN, Stenn KS, Malawista SF (1983) The early clinical manifestations of Lyme disease. Ann Intern Med 99:76-82

Weber K (1989) Clinical differences between European and North-American Lyme borreliosis – a review. Zentralbl Bacteriol (Suppl) 18:146-155

Weber K, Neubert U (1986) Clinical features of early erythema migrans disease and related disorders. Zentralbl Bakteriol Hyg (A) 263:209-228

Weber K, Schierz G, Wilske B, Preac-Mursic V (1984) European erythema migrans disease and related disorders. Yale J Biol Med 57:463-471

Wilske B, Schierz G, Preac-Mursic V, Weber K, Pfister HW, Einhäupl K (1984) Serological diagnosis of erythema migrans disease and related disorders. Infection 12:331-337

Acknowledgements. We wish to thank Dr. Klaus Weber, Dr. Vera Preac-Mursic, Dr. Bettina Wilske and Dr. Helmut Robl, Munich, for their cooperative support.

V Microbiological Diagnosis of Lyme Borreliosis

21 Microbiological Diagnosis of Lyme Borreliosis

B. WILSKE and V. PREAC-MURSIC

21.1 Bacteriological Diagnosis

21.1.1 Direct Detection of Borrelia burgdorferi

Borrelia burgdorferi may be observed microscopically either as living and unstained organisms or after the application of dyes. Because of the small number of bacteria in specimens from humans, especially in blood and cerebrospinal fluid (CSF), organisms are difficult to find using direct microscopy. Centrifugation of specimens improves detection in many cases. Suitable specimens for detection of *B. burgdorferi* are listed in Table 1. For the direct examination of *B. burgdorferi*, various methods and preparations are currently used.

Table 1. Specimens for the microbiological diagnosis of Lyme borreliosis

Stage	Manifestation	Antibody material	detection follow-up	Detection of *Borrelia burgdorferi* by staining or culture
I	EM	Serum	After 2 and 4–6 weeks	Skin biopsy, blood
II	Lymphocytoma	Serum	After 2 and 4–6 weeks	Skin biopsy, blood
	Carditis	Serum	After 2 and 4–6 weeks	Blood,(heart biopsy)[c]
	LMR	Serum, CSF[a]	After 2 and 4–6 weeks	CSF, blood
III	Arthritis	Serum, synovial fluid	Long-term follow-up[b]	Synovia biopsy, synovial fluid
	CEM	Serum, CSF[a]	Long-term follow-up[b]	CSF, (brain biopsy)[c]
	ACA	Serum	Long-term follow-up[b]	Skin biopsy

EM, erythema migrans; LMR, lymphocytic meningoradiculoneuritis; CSF, cerebrospinal fluid; CEM, chronic encephalomyelitis; ACA, acrodermatitis chronica atrophicans

[a] CSF and serum should be obtained simultaneously for CSF/serum index determination.

[b] Short-term follow-up in patients with short duration of symptoms.

[c] Heart and brain biopsies only in especially indicated cases.

21.1.1.1 Dark-field Microscopy

Dark-field microscopy is useful in specimens where a high concentration of *B. burgdorferi* can be expected, for instance, in blood, suspensions of liver and kidney of infected animals, and cultures containing borreliae (Fig. 1a). Borreliae are difficult to stain and easily lose their characteristic morphology during the fixation and staining procedures. Therefore, dark-field microscopy is the preferred method in bacteriological diagnosis. For dark-field examination, a small drop of blood, CSF, or culture medium is placed on a slide and covered with a cover slip. The slide is examined at a magnification of about 400. The presence of the organisms is frequently detected by observing the erythrocytes or filaments; when these areas are focused upon, the characteristically large spirochetes with broad spirals and very active motility can readily be seen.

21.1.1.2 Staining Methods

Staining techniques often used for the detection of *B. burgdorferi* in CSF, blood, tissue, and in cultures are Giemsa or carbol-fuchsin staining and silver impregnation. Common laboratory Gram-staining is less suitable than Giemsa staining, because bor-

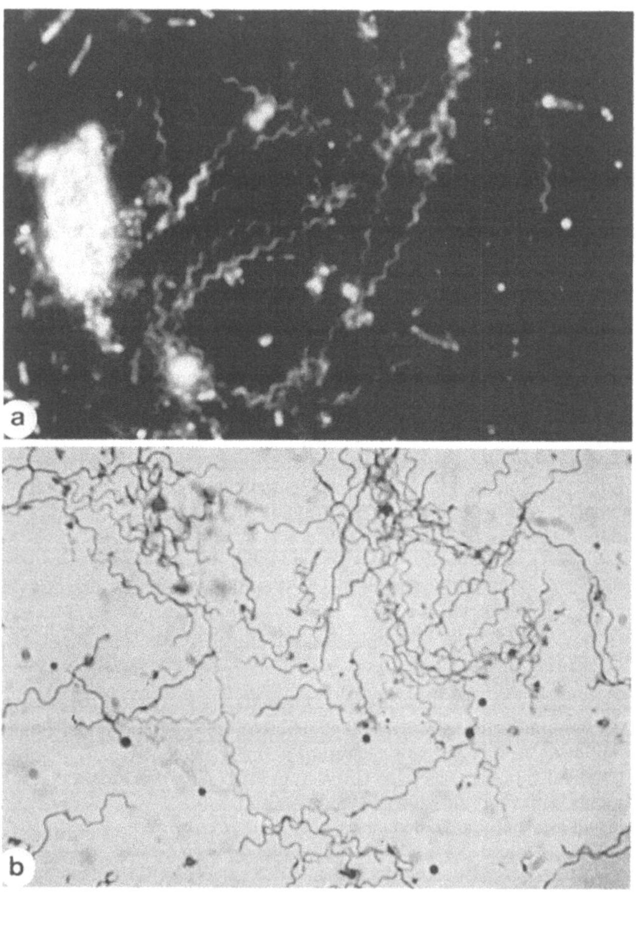

Fig. 1a, b. *Borrelia burgdorferi* in culture. **a** Dark-field microscopy. **b** Carbol-fuchsin stain reliae change their form considerably during fixation and staining procedures. Thus artefacts are often observed. A prolonged Giemsa staining appears to be much better. We prefer staining with carbol-fuchsin for 7 min (Fig.1b) since this retains the morphology and renders the spirochetes clearly visible (Preac-Mursic, unpublished). For demonstrating borreliae in CSF and phagocytic cells, staining with acridine-orange has been used (Bozsik 1985). To detect the organisms in biopsy and autopsy materials, several silver impregnation techniques have been used by various authors (see Chap.6).

Although direct examination may be valuable in establishing a rapid diagnosis, misdiagnosis frequently occurs when artefacts are identified as spirochetes. For instance, forms quite similar to borreliae may result from erythrocyte debris, fibrin filaments, or cellular debris. Such forms can usually be distinguished from borreliae by repeated examination and by culturing. Direct examination is, therefore, not recommended as a single diagnostic procedure. Diagnosis by direct microscopy should be confirmed by cultural, serological, immunological, or other suitable methods.

21.1.1.3 Immunological Detection of B. burgdorferi

Immunological methods are used in several modifications for the detection of either whole borreliae or soluble borrelia antigens and may be applied to different specimens (examples are given in Table 1).

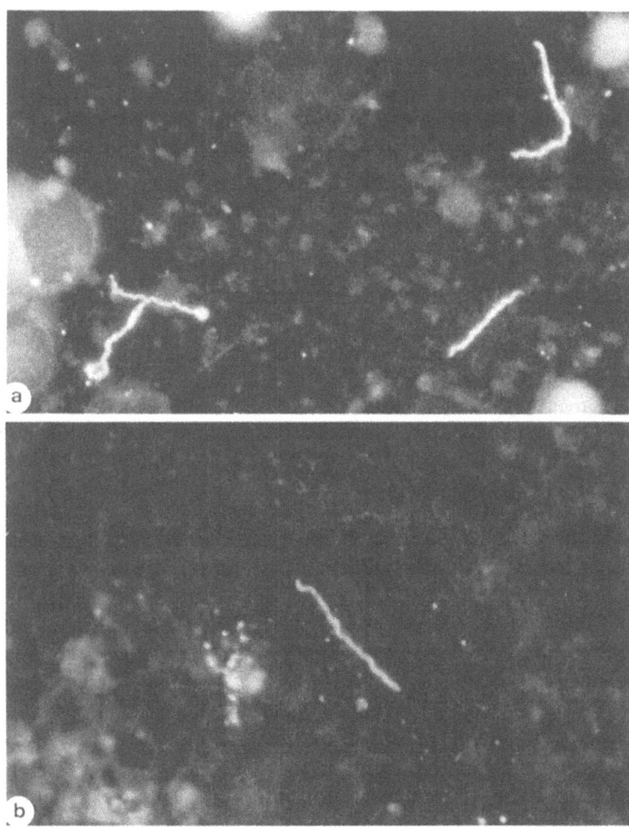

Fig. 2a,b. *Borrelia burgdorferi* in tissue. **a** Borreliae detected in the midgut of *Ixodes ricinus* by direct immunofluorescence. **b** Borreliae detected in the spleen of an infected gerbil (mouse) by indirect immunofluorescence using a rabbit immune serum against *Borrelia burgdorferi*

21.1.1.3.1 Immunofluorescence and Immunoperoxidase Staining

Immunofluorescence tests have been the techniques most widely used to determine the prevalence of borreliae in ticks (Burgdorfer et al. 1982, 1985, 1986, Aeschlimann et al. 1986; Kmety et al. 1986; Stanek et al. 1986; Wilske et al. 1987). Borreliae have also been detected by immunofluorescence in skin biopsies (Park et al. 1986). However, detection of spirochetes by immunohistological methods is time consuming and needs an experienced investigator. This technique can therefore be performed only in special cases and in specialized laboratories.

Borreliae may be detected by direct immunofluorescence (DIF) using fluorescein isiothiocyanate (FITC)-labelled immune sera against *B. burgdorferi* (Fig.2a) or by indirect immunofluorescence (IIF) using first unlabeled *B. burgdorferi*-specific antibodies and second FITC-labeled antibodies (Fig.2b). The DIF has been mainly used for tick surveys, but the IIF may also be performed if borreliae-specific conjugates are not available. For detection of *B. burgdorferi*, polyclonal or monoclonal antibodies (PAB; MAB) have been used. Monoclonal antibodies recognize only part of the borrelial epitopes. Therefore, the risk of false-negative results is higher than with PAB, and a positive result depends on whether the corresponding epitope is expressed by the spirochete or not (Burgdorfer et al. 1985; Park et al. 1986). However, if species-specific epitopes are detected, the immunostain confirms the presence of *B. burgdorferi*. Conjugates with enzymes such as peroxidase instead of fluorescent dyes have been used to detect borreliae in skin biopsies (Aberer et al. 1988). Peroxidase immunostain provides the advantage of permanent slide preparations and the possibility of counter-staining.

21.1.13.2 Antigen Detection in Body Fluids

Antigen detection in body fluids has been reported by Hyde et al. (1989). The authors used Western blots to detect borrelial proteins in the urine of infected hamsters and of patients with Lyme borreliosis. The antigens were identified with monoclonal antibodies against flagella or outer surface proteins of *B. burgdorferi*. However only a few patients were investigated and the diagnostic value of the method remains to be proven.

21.1.2 Cultivation of *B. burgdorferi* In Vitro and In Vivo

B. burgdorferi has been isolated from specimens obtained at all stages of the disease. At advanced stages, the chance of isolating borreliae in specimens (especially from blood and CSF) decreases.

21.1.2.1 Cultivation In Vitro

Specimens. *B. burgdorferi* has been isolated from human blood (Benach et al. 1983) CSF (Steere et al. 1983; Preac-Mursic et al. 1984a; Karlson et al 1990b), and skin biopsies (Steere et al. 1983; Ackermann et al. 1984; Asbrink et al 1984a,b; Berger et al. 1985; Preac-Mursic et al. 1985, 1986; Stanek et al. 1985) synovia biopsies (Schmidli et al. 1988), and autopsy material (MacDonald 1986). The samples should be obtained aseptically before treatment with antibiotics. They must be processed as soon as possible. The biopsy material should be protected from desiccation for instance by adding phosphate buffered saline (PBS)-buffer or 0.85% NaCl solution. After preparation (see below), specimens are immediately placed into the culture medium, and processed as described in Fig. 3. For enrichment, CSF can be centrifuged at 17000 g for 20 min at 22°C and the sediment resuspended in 0.5 ml medium. Blood

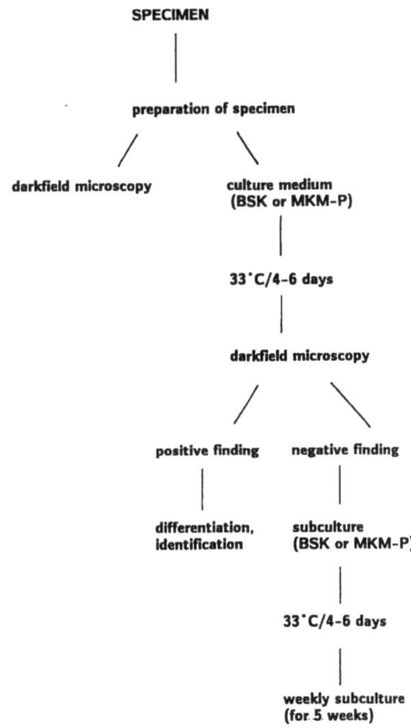

Fig. 3. Procedure to isolate *Borrelia burgdorferi* in culture (BSK, Barbour-Stoenner-Kelly; MKM-P, modified Kelly medium – Preac-Mursic; see Table 2)

(citrated or heparinized) may be centrifuged at 120 g for 10 min to remove cells, the plasma may then be centrifuged at 17000 g for 20 min and the resulting sediment resuspended in 0.5 ml medium (modified from the method of Benach et al. 1983).

To improve the detection and recovery of *B. burgdorferi*, skin and synovial biopsies are dissected with sharp needles and scissors and suspended in 0.5 ml medium. About one-half of each specimen can be examined directly by dark-field microscopy or staining procedures, the remainder is placed into a culture tube. The quality of samples is an important factor for successful isolation of borreliae; also important is the stage of the disease; since it may have an influence on the number of organisms in a specimen. Table 1 depicts specimens suitable for serological and bacteriological diagnosis.

Media and culture conditions. Samples are most frequently cultivated in Barbour-Stoenner-Kelly medium (BSK), (Barbour et al. 1983b) or modified Kelly medium – Preac-Mursic (MKP) (Preac-Mursic et al. 1986) (Table 2) containing CMRL 1066, N-acethylglucosamine, rabbit serum, and bovine serum. These media can also be used with addition of 0.2% agarose, although there does not seem to be a difference in the isolation rate of *B. burgdorferi* between media with and without agarose. It should be kept in mind that dark-field examination may be impaired in media containing agarose. Media may be dispensed in 7-ml test tubes with screw caps and can be stored for 3-4 weeks at 4° C. The culture tubes are usually incubated at 30° – 33° C for 5 weeks. Cultures should be passaged and examined by dark-field microscopy at weekly intervals. They can be discarded if still negative after 5 weeks of incubation. To inhibit bacterial contamination, addition of antibiotics is advisable. We recommend 1 – 5 doses of 30 µg co-trimoxazole for skin and tick cultures (Preac-Mursic et al. 1986). Other investigators use neomycin, kanamycin, and 5-fluocytosin or gentamicin and rifampicin (Barbour et al. 1985b; Johnson et al. 1984). Cultures that still show bacterial contamination are treated with

Table 2. Modified Kelly Medium (MKP) for cultivation of *Borrelia burgdorferi*

Stage I	CMRL 1066 (10x) without glutamine[a]	100 ml
	Aqua bidest.	900 ml
	Neopeptone	3 g
	Hepes	6 g
	Sodium citrate	0.7 g
	Glucose	3 g
	Sodium pyruvate	0.8 g
	N-acetyl glycosamine	0.4 g
	Sodium bicarbonate	2 g
	Adjust pH to 7.6 by adding 1 N NaOH; sterilize by filtration	
Stage II	Gelatine 7% (sterile)	200 ml
	Rabbit serum (heat-inactivated)	70 ml
	Bovine serum albumin	35 ml

[a] Gibco BRL – Life Technologies (Eggenstein, Germany)

antibiotics appropriate for the contaminant. Occasionally it is possible to recover the borreliae from 0.22 µm or 0.45 µm filtrates of contaminated cultures, but only when they contain many organisms (Preac-Mursic et al. 1986). Cultivation of *B. burgdorferi* on solid media is alos possible. This technique is described in Chap. 4.

The rate of recovery of *B. burgdorferi* from samples varies and depends particularly on the stage of the illness. Isolation is most readily accomplished from patients with erythema migrans, especially from the affected skin, rarely from the blood. *B. burgdorferi* has also been isolated frequently from skin biopsies of patients with acrodermatitis chronica atrophicans (ACA) (Asbrink and Hovmark 1985; Preac-Mursic et al. 1985), and less frequently from patients with lymphocytoma (Asbrink et al. 1989; Preac-Mursic, unpublished) and morphea (Weber et al. 1988b). The organism has been isolated only once from synovial fluid (Schmidli et al. 1988). *B. burgdorferi* has been isolated less frequently from CSF than from skin.

The frequency of positive *B. burgdorferi* cultures in our laboratory (Preac-Mursic, unpublished) was:

- From skin biopsy in 70% of patients with erythema migrans and in 50% of those with ACA
- From CSF in 7% – 10% of patients with neurological manifestations.

CSF isolates are more frequently obtained from patients with disease of short duration than from patients with disease of long duration (Karlsson et al. 1990). However, borreliae have been cultivated from the CSF of a patient with neurological symptoms lasting for more than 6 weeks despite high CSF antibody titers (Preac-Mursic et al. 1984 a,b) and from the skin and the CSF of patients after antibiotic treatment (Preac-Mursic et al. 1989). Seropositivity is not a prerequisite for isolation of *B. burgdorferi*; it has been recovered in patients with negative serological findings (Preac-Mursic et al. 1986).

21.1.2.2 Cultivation In Vivo

Inoculation in animals may be helpful in establishing the diagnosis in specimens with few borreliae and when cultures are not available. If cultures are heavily contaminated, purification by filtration or by serial dilutions is not usually possible. In such cases animal passage may be useful. For this purpose, gerbils (*Meriones unguiculatus*) weighing approximately 80 g can be inoculated i.p. with a 0.5 – 1 ml specimen. At daily intervals, the end of the tail is snipped, and blood is examined by dark-field microscopy. When borreliae are demonstrable, usually after 2-4 days, a culture can be established. Isolation of *B. burgdorferi* may also be possible from kidney, liver, and spleen (Preac-Mursic et al. 1990). Because of the complexity of the technique and the long incubation period, this procedure is not recommended for routine use in diagnosis. If specimens can be obtained aseptically, a direct culture is preferable.

21.1.2.3 Differentiation and Identification of B. burgdorferi

For differentiation from other borreliae and identification as *B. burgdorferi*, the isolate must be available in culture. An exception might be immunohistological diagnosis with MAB. The latter procedure, however, is difficult to perform due to the sparseness of borreliae in infected tissue. A culture containing spirochetes with the typical morphology of borreliae (see Chap.4) must be further analyzed for differentiation from other borreliae and for identification as *B. burgdorferi*. DNA hybridization has been performed in a few instances (see Chap.4). The most frequently used methods, however, have been analysis of the protein pattern of a sodium dodecyl sulfate (SDS) whole-cell lysate and the reaction with MAB and monospecific PAB in immunofluorescence or Western blot. SDS polyacrylamide gel electrophoresis (SDS-PAGE) allows the diagnosis regardless of whether the protein pattern is typical for *B. burgdorferi* and whether certain immunodominant proteins are present. Using MAB against the major immunodominant proteins OspA (L32 1F11), pC (L22 1F8), and p100 (L100 1D4)I, a borrelia can be identified as *B. burgdorferi* and differentiated from relapsing fever borreliae (Wilske et al. 1991). Serotyping of strains is mainly based on antigenic heterogeneity of the outer surface protein OspA (see Sect 21.2.2 and Chap 4).

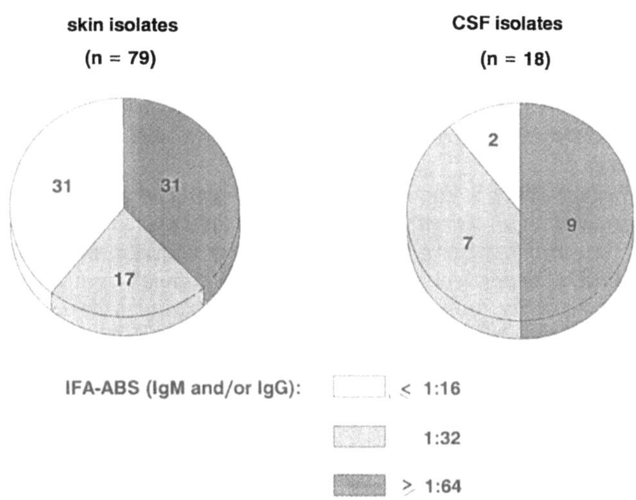

Fig. 4. Serological findings in patients with positive culture (Preac-Mursic and Wilske, unpublished); IFA-ABS, immunofluorescent absorption test

21.1.3 Conclusions

Because of the difficulty in cultivating *B. burgdorferi* from patients and the relatively long incubation time of culture, the diagnosis of the disease is mostly based on serological tests. However, culturing *B. burgdorferi* is for instance indicated in atypical cases of erythema migrans, because serological tests are often negative in the early stage of the disease (Table 5), and culture may be positive in seronegative cases (Fig.4). To confirm the diagnosis in stages II and III of the disease and for therapy, control culture of *B. burgdorferi* is also recommended. In Lyme borreliosis like in other infectious diseases, the diagnosis can be confirmed by detection of the etiologic agent. Attempts to establish the presence of *B. burgdorferi* should be made whenever possible. In the future, DNA detection methods will play a major role in the microbiological diagnosis of Lyme borreliosis. Different DNA probes derived from plasmid or chromosomal DNA have been developed which may be suitable to detect *B. burgdorferi* DNA in specimens from patients (Rosa and Schwan 1989; Schwan et al. 1989; Simpson et al. 1990a; Goodman et al. 1991).

21.2 Serological Diagnosis of Lyme borreliosis

21.2.1 Introduction

The most common diagnostic tool used to determine the etiology of Lyme borreliosis is the detection of specific antibodies against *B. burgdorferi*. Lyme borreliosis serology has many parallels in serological technique and interpretation of results with syphilis serology. Experience with well-established tests for syphilis such as immunofluorescence and passive hemagglutination assays has favored the rapid development of similar serological tests for Lyme borreliosis. In addition, modern techniques such as enzyme-linked immunosorbent assay (ELISA) and Western blot have subsequently been used for the detection of antibodies against *B. burgdorferi*. Commercial tests are now available from American and European companies. At the moment, however, Lyme borreliosis serology appears to be a long way from standardization. Collaborative studies, at least among major laboratories, seem to be urgently needed. This section gives a critical review of our present knowledge of serodiagnosis of Lyme borreliosis and discusses possible approaches for more sensitive and specific serodiagnostic tests.

21.2.2 Antigen Structure of *Borrelia burgdorferi*

Whereas little is known about nonproteinaceous antigens (lipids, carbohydrates), the proteins of *B. burgdorferi* have been investigated in more detail. SDS-PAGE has revealed two major proteins with apparent constant molecular weights of 41 kDa and 60 kDa, respectively (Fig.5). These two proteins have been demonstrable in all the *B. burgdorferi* isolates so far analyzed (Barbour and Hayes 1986; Wilske et al. 1988b). The 41-kDa protein represents the flagella protein which contains genus-wide epitopes (Barbour et al. 1986). Immunological differences have been detected with MAB among the flagella proteins of different strains (Wilske et al. 1990). The 60-kDa protein is the so-called common antigen (CA) which shows broad crossreactivity to many other bacteria. The gene encoding the CA has recently been cloned (Hansen et al. 1988a). This highly conserved protein was believed by Hansen et al. (1988a) to represent a heat

Fig. 5. Protein pattern of 5 *Borrelia burgdorferi* strains (SDS-PAGE, Coomassive blue staining). The reactivity with monoclonal antibodies (H5332 and H3TS) is given at the bottom of the Figure (B31, American-type strain; PBr, PKo, German isolates from CSF and skin; TN, T25, German isolates from ticks)

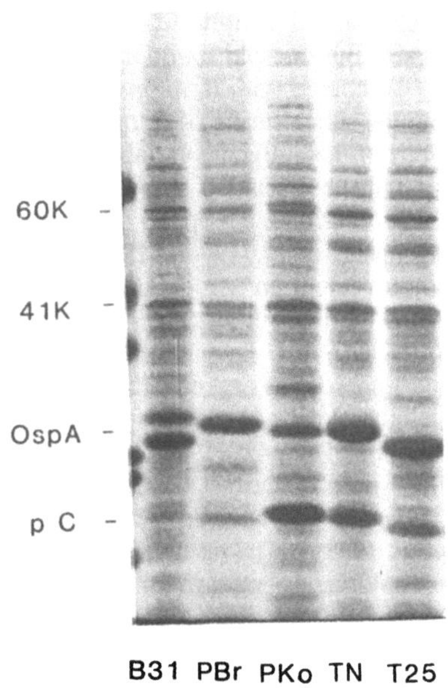

	B31	PBr	PKo	TN	T25
H5332	+	+	(+)	+	-
H3TS	+	-	-	-	-

shock protein of *B. burgdorferi*. However, Cluss and Boothby (1990) have shown that this protein does not belong to the heat shock protein family.

Three additional proteins with lower molecular weights OspA (30 – 32 kDa), OspB (34 – 36 kDa) (Barbour et al. 1983c, 1984) and pC (21 – 22 kDa) (Wilske et al. 1986a) are variable major proteins. OspA and OspB are outer surface proteins. All combinations of OspA, OspB, and pC – except OspB alone – have been observed. These proteins vary in molecular weight and antigenicity from strain to strain. Especially among European isolates, the variability of these proteins is pronounced (Wilske et al. 1988b). However, immunochemical analyses of strains isolated by Anderson et al. (1989) suggest that *B. burgdorferi* strains from the United States might be more heterogeneous than previously described. According to Western blot analyses of OspA and OspB, seven serotypes have been defined among 25 *B. burgdorferi* isolates (Wilske et al. 1988b). A serotyping system has been proposed based solely on analysis of OspA with 8 MAB (Wilske et al. 1991). Skin isolates have turned out to be similar, whereas CSF isolates have been heterogeneous. Besides strain-to-strain variations, variations have also been observed in the expression of OspA, OspB, or pC within the progeny of one and the same isolate during serial in vitro cultures. These variations concern the loss of OspB (Bisset and Hill 1987; Schwan and Burgdorfer 1987; Schwan et al. 1988; Wilske et al. 1986a, 1988b) or quantitative changes between OspA and OspB on one side and pC on the other side (Bisset and Hill 1987; Wilske et al. 1986a, 1988b). Immunological variability of OspB has been reported for the progeny of an uncloned isolate

(Schwan and Burgdorfer 1987) and subsequently also for the progeny of a cloned strain (Bundoc and Barbour 1989).

Compared with *B. hermsii* such phenotypic changes occur with low frequency during the subculturing of *B. burgdorferi* strains (Barbour and Hayes 1986). The implications of the antigenic variability and heterogeneity for serodiagnosis of Lyme borreliosis will be discussed below. Other immunodominant proteins of *B. burgdorferi* have been described, e.g., p100 (Wilske et al. 1988a, 1990), p83 (LeFebre et al. 1990), p39 (Simpson et al. 1990b, 1991) and p17 (Wilske et al. 1988a, 1990). The possible improvement in serodiagnosis using these antigenic components will be discussed below. Brandt et al. (1990) have characterized Triton X-114 fractionated proteins of *B. burgdorferi*. They have shown that proteins that partionate into the detergent phase are lipoproteins and encompass the major *B. burgdorferi* immunogens, except the flagella protein p41. Heat stress proteins of *B. burgdorferi*, with apparent molecular weights of 75, 42, 39 and 27 kDa, respectively, that may play a role for autoimmunity have been described by Cluss and Boothby (1990). More than 100 polypeptides of *B. burgdorferi* have been detected on silver-stained two dimensional (2D) gels by Luft et al. (1989). The surface proteins were studied in more detail. Significant sequence homologies between OspA and a 22-kDa protein may indicate that these proteins are part of a gene family expressed at the surface of the organism. However valid conclusions on the identity of the different polypeptides described by various authors are only obtained if the polypeptides are characterized by MAB or their genes are sequenced.

21.2.3 Crossreactions

B. burgdorferi is antigenically very closely related to other members of the genus *Borrelia*. For instance, immune sera raised against this spirochete react with multiple proteins of relapsing fever borreliae in the Western blot (Wilske et al. 1988b; Luther 1990). Sera from patients with relapsing fever have high titers in the immunofluorescent antibody test (immunofluorescence assay IFA) and in the ELISA for Lyme borreliosis. Vice versa, patients with Lyme borreliosis have comparable titers when tested against relapsing fever borreliae as antigen (Magnarelli et al . 1987). Absorption procedures are in some cases effective, in others not. However, in most instances, relapsing fever can be distinguished on clinical or epidemiological grounds from Lyme borreliosis.

Several authors have described strong crossreactions between *B. burgdorferi* and *Treponema pallidum* (Barbour et al. 1983a; Ackermann et al. 1984, Craft et al. 1984; Russel et al. 1984; Wilske et al 1984; Asbrink et al. 1985; Stiernstedt et al. 1985; Magnarelli et al. 1987; Hansen et al. 1988b; Luther and Moskophidis 1990) in IFA, ELISA, and immunoblot. One major crossreacting protein is the CA p60 (Hansen et al. 1988a). Sera from syphilitic patients are often reactive with this protein, but antibodies against the p60 have also been detected in sera from healthy persons (Wilske et al. 1986a). These observations can be explained by the fact that the p60 of *B. burgdorferi* is a CA that is not only crossreactive with *T. pallidum* and *T. phagedenis*, but also with a wide range of Gram-negative and Gram-positive bacteria including mycobacteria. The gene encoding the common antigen of *B. burgdorferi* shows high homology compared with the genes of a wide range of other bacteria (Hinderson, personal communication). Immunization with ultrasonicate antigens from *T. phagedenis*, *Leptospira grippotyphosa*, *Yersinia enterocolitica* O3, *Escherickia coli*, *Campylobacter jejuni*, *Neisseria meningitidis*, and *Haemophilus influenzae* induced the production of antibodies reactive with the CA (p60) of *B. burgdorferi* (Wilske et al. 1990). Further crossreactive proteins are mainly located in the 60 – 70 kDa region (p60, p75), and there are weak reactions also in the 40, 30, and 20 kDa range. There is evidence that the flagellin p41 has also unspecific epitopes responsible for crossreactions in syphilitic sera and in the sera of healthy controls (Hansen et al. 1988b; Wilske et a. 1986a; Simpson et al. 1990a).

Crossreactions based on nonproteinaceous antigens (carbohydrates and lipids) have not been investigated. However, there are indications that such crossreactions occur. Hechemy et al. (1988) reported that antibodies against *B. burgdorferi* of the immunoglobulin IgG 2 subclass can be removed completely by absorption with *T. phagedenis* (IgG2 antibodies are mainly directed against carbohydrates).

21.2.4 Serodiagnostic Tests

21.2.4.1 Indirect Immunofluorescence Assay

The first serological test developed for the detection of antibodies against the causative agent of Lyme borreliosis, *B. burgdorferi*, was the indirect IFA (Burgdorfer et al. 1982). Subsequently, many American and European authors have reported technical modifications and their experience with this technique as a diagnostic tool in different clinical manifestations (Barbour et al. 1983a; Steere et al. 1983; Craft et al. 1984; Magnarelli et al. 1984; Russel et al. 1984 from the United States and Ackermann et al. 1984; Asbrink et al. 1984a, 1985; Wilske et al. 1984; Stanek et al. 1986; Stiernstedt et al. 1985 from Europe).

In most laboratories, whole washed borreliae (*B. burgdorferi* cells) previously cultivated in modified Kelly medium are used as antigen. The borreliae are either fixed on glass slides by organic solvents such as methanol or acetone (Barbour et al. 1984a; Russel et al. 1984) or by deep-freezing the coated slides at minus 70° C (Steere et al. 1983; Wilske et al. 1984). Important for the quality of the IFA antigen is an optimal cultivation time (exponential growth phase) (Wilske et al. 1984). The borreliae should not aggregate (use of nonclotting strains) or overlap (density of about 50 borreliae per field) to avoid nonspecific staining. Extensive, but careful washings (up to 6 times) of the borreliae prior to coating of slides avoid nonspecific staining without loss of sensitivity (Steere et al. 1983; Wilske et al. 1984). Using polyvalent conjugates, the sera can first be screened for the presence of specific antibodies. Reactive sera should be retested with immunoglobulin class-specific conjugates for the presence of specific IgM or IgG antibodies. Nonspecific results due to crossreactions are often observed in the IFA, and false-positive IgM test results due to rheumatoid factor (RF) may occur (Wilske et al. 1984). Therefore sera should be preabsorbed for crossreactive antibodies with *T. phagedenis* (IFA-ABS). For the IFA detection of IgM, an additional pretreatment of sera with anti-IgG serum (see Sect. 21.2.5.2) is recommended (Wilske et al. 1984).

21.2.4.2 Indirect Hemagglutination Assay

The indirect hemagglutination assay (IHA) for syphilis is a sensitive and specific test for detecting antibodies against *T. pallidum* (*T. pallidum* hemagglutination assay, TPHA). A similar IHA for detecting antibodies against *B. burgdorferi* has been developed (Wilske, unpublished). For this assay, sheep erythrocytes are treated with formalin and tannic acid and are subsequently sensitized with *B. burgdorferi* ultrasonicate antigen. This assay is not easily standardized. Large batches of sensitized cells are prepared after previous antigen titration. The sensitized cells and unsensitized control cells are lyophilized for later use. The IHA is a sensitive screening assay for the detection of both IgM and IgG antibodies against *B. burgdorferi*. Reactive specimens are further investigated by immunoglobulin class-specific assays (IFA or ELISA). The IHA can be recommended for laboratories which have to investigate high numbers of sera. Another advantage of this technique is that sera from animals can also be investigated without the need for species-specific conjugates.

21.2.4.3 Enzyme-linked Immunosorbent Assay

ELISAs have been developed with whole (Magnarelli et al. 1984) or ultrasonicated *B. burgdorferi* cells as antigen (Craft et al. 1984; Russell et al. 1984; Asbrink et al. 1985; Stiernstedt et al. 1985; Stanek et al. 1986; Wilske et al. 1986b). According to Magnarelli (1988), comparable results have been obtained using either whole or ultrasonicated *B. burgdorferi* cells as antigen. The assays are performed using polyvalent conjugates (Magnarelli et al. 1984; Russell et al. 1984) or immunoglobulin class-specific conjugates (Craft et a. 1984; Asbrink et al. 1985; Stiernsted et al. 1985; Stanek et al. 1986; Magnarelli and Anderson 1988; Wilske et al. 1986b, 1989c). Polyvalent assays may be used for screening purposes. Positive test results should be confirmed by immunoglobulin class-specific assays.

Standardization of the ELISA is not easy. The production of large batches of ELISA antigen is recommended. The culture conditions should be standardized with respect to medium and time of culture. SDS-PAGE analysis prior to sonicaton is useful to detect possible antigenic changes. The optimal protein concentration for coating ELISA plates should be determined by titration of the antigen evaluating the P/N ratio using a positve and a negative serum. Coating with high doses of ultrasonicate antigen (10 – 20 μg protein/ml.) results in better discrimination of positive and negative sera than low doses of antigen (Wilske et al. 1989c). In selective ELISAs (see below), lower antigen concentrations are used (Hansen et al. 1988b). Day-to-day variations in test sensitivity can be reduced by using an internal reference serum with arbitrarily defined units as a standard. Actual antibody units of patients sera are calculated from optical density (OD), and serum dilution of reference serum and patients sera, respectively (Craft et al. 1984; Wilske et al. 1986b, 1989c). Similar corrections for standardization of day-to-day variations have been recommended by Rusell et al. (1984), Stanek et al. (1986) and Hansen et al. (1988b). Sera may be tested in a standard dilution or in serial dilutions. Testing a standard dilution provides sufficient information to determine whether a serum is negative, borderline, or positive. A quantitative estimate of the antibody content is also provided in sera with low antibody titers testing the standard dilution. Serial dilutions are used to determine the antibody titer (in units) in sera with high antibody content.

In view of the antigenic variability of *B. burgdorferi*, it is still an open question whether it is sufficient to use only a single strain as antigen or whether several immunologically different strains should be used. Another question is whether a certain strain or serotype has to be recommended. According to Western blots and ELISAs with different strains as antigen, the detection rates for the IgM test have been improved with a skin isolate presenting a major pC component (Wilske et al. 1989b). This protein is an immunodominant protein for the IgM immune response. For the IgG ELISA, no single strain gave significantly better results with a panel of Lyme borreliosis sera. However, individual sera may vary considerably in their reactivity with different strains as antigen either in IgM or IgG ELISA (Wilske et al. 1989b). These findings further underline the need for more standardized ELISAs, possibly by selection or enrichment of certain immunodominant specific antigenic components or use of recombinant antigens for the ELISA (see Sect. 21.2.5.4).

21.2.5 Improvement of Sensitivity and Specifity

Because of *B. burgdorferi* antigens that share common or similar epitopes with spirochetal and even with nonspirochetal microorganisms, researchers have found it necessary to improve the specifity of serodiagnostic tests.

21.2.5.1 Absorption of Crossreactive Antibodies

Stimulated by the benefit of absorbing sera with *T. phagedenis* for the specifity of the *T. pallidum* immunofluorescent antibody test (fluorescent treponemal antibody absorption test (FTA-ABS) for specific diagnosis of syphilis), several authors have tried to improve the specifity of *B. burgdorferi* serology using this absorption procedure (IFA-ABS). The results of the IFA-ABS have been discussed controversially. While Craft et al. (1984) and Russell et al. (1984) have seen no benefit by absorbing sera with *T. phagedenis*, other authors have reported an improvement in both sensitivity and specificity (Asbrink et al. 1984a; Wilske et al. 1984). As a result of absorption, the reactivity of sera from healthy controls is reduced significantly and, therefore, also the cut-off value for the positive test, for example from 1:256 to 1:64 for the IgG IFA (Wilske et al. 1984). However, high concentrations of the sorbent (2 mg lyophilized *T. phagedenis* ultrasonicate/ml serum) are necessary for exhaustive absorption. Using this procedure, crossreacting antibodies are quantitatively absorbed from Lyme borreliosis sera as well as from syphilitic sera. In case of double infection with both *B. burgdorferi* and *T. pallidum*, Lyme borreliosis can be diagnosed by the IFA-ABS. However, the main advantage of the absorption procedure lies in the elimination of crossreacting antibodies with a broad variety of antigens such as the 60-kDa CA, p75, or common epitopes of the flagellin p41. We found for example, that crossreacting immune sera raised against unrelated Gram-negative and Gram-positive bacteria could be rided of the crossreacting antibodies by treatment with *T. phagedenis* ultrasonicate (Wilske, unpublished).

21.2.5.2 Detection of IgM Antibodies

There is often a risk that the sensitivity and specificity of the indirect IFAs or ELISAs for detection of IgM antibodies may be impaired by false-negative results due to competitive inhibition by high titers of IgG antibodies and by false-positive results due to IgM RF activity. If there are false-positive results, IgM RF may bind to IgG antibodies attached to the *B. burgdorferi* antigen on the solid phase, and the μ-chain-specific enzyme conjugate my subsequently be linked to this complex. False-positive IgM tests can, therefore, arise if both IgG antibodies against the test antigen and RF of the IgM class are present. It is important to realize that such false-positive reactions can also be induced by low RF activity (which may not be detected by the RF latex agglutination assay) and also by crossreacting IgG antibodies.

False-positve results due to RF activity and false-negative results due to competitive inhibition by IgG antibodies have been reported for serodiagnostic tests of syphilis (Müller and Moskophidis 1983) and of Lyme borreliosis (Wilske et al. 1984; Hansen and Asbrink 1989). Ziegelmeier et al. (1983) have demonstrated an increase in specificity and sensitivity of IgM antibody detection in viral diseases by previous immunoprecipitation of IgG antibodies with anti-IgG. This method has been used for IgM antibody detection in Lyme borreliosis, too (Wilske et al. 1984). With this procedure, IgG antibodies that can inhibit IgM antibody detection are precipitated by anti-IgG antibodies, and the resulting immunocomplexes bind RF. This technique improves the detection of IgM antibodies in sera from patients with a longer duration of the disease and in CSF, because in both instances coexisting specific IgG antibodies may be present. This technique has recently also been used for the IgM ELISA (Wilske et al. 1989c). Similar good experience with anti-IgG precipitation has also been reported for IgM antibody detection by ELISA in syphilis (Cerny et al. 1985).

Another technique for IgM detection that is widely used in viral serology is the μ-capture assay. In this assay serum IgM, regardless of its specificity, is attached to the ELISA plate coated with μ-chain-specific serum. In a further step, the specific antibodies bound to the solid phase are detected either by enzyme-labeled antigen or unlabeled antigen and corresponding labeled specific antibodies. μ-Capture tests for the detection of antibodies against *B. burgdorferi* have been developed by Wassmann et al. (1989), Berardi et al. (1988) and Hansen et al. (1991). Further investigations are needed to show whether the μ-capture test is superior to the IgM test with previous anti-IgG immunoprecipitation, or vice versa.

21.2.5.3 Antibody Detection in CSF

Detection of a specific intrathecal immune response has proved an important diagnostic tool in infectious diseases involving the nervous system such as neurosyphilis (Müller et al. 1982). However, it is not sufficient to demonstrate the presence of specific antibodies within the CSF because such antibodies may be blood serum-derived, especially when the meninges are inflamed. The antibody level within the CSF, therefore, depends on the following variable parameters: (a) the antibody level within the blood serum, (b) the permeability of the blood – CSF barrier, and (c) on the intrathecal antibody production. Different methods referred to below have been used to demonstrate the intrathecal production of specific antibodies within the CSF.

21.2.5.3.1 CSF/Serum Indices

To estimate intrathecal antibody production, the CSF/serum ratio of *B. burgdorferi*-specific antibodies (RBb) is compared to the CSF/serum ratio of another protein not produced intrathecally. One can use for example (a) the CSF/serum albumin ratio (RAlb) (Stiernstedt et al. 1985), (b) the CSF/serum IgG ratio (RIgG) (Rehse-Küpper et al. 1986; Wilske et al. 1986b), or (c) the CSF/serum ratio for a specific antibody (for instance against tetanus toxoid) which is not intrathecally produced (RTe) (Wilske et al. 1986b). The corresponding CSF/serum indices are designated (a) RBb/RAlb, (b) RBb/RIgG, and (c) RBb/RTe in this section. Specific antibodies should be determined quantitatively, e.g., by ELISA. Antibody titer determinations using the IFA are not suitable.

The different CSF/serum indices mentioned above have certain advantages and disadvantages. A comparison of CSF/serum ratios between the RBb and another exclusively serum-derived protein with the same permeability for the CSF – blood barrier as the antibody should theoretically provide the most sensitive and specific assay. This condition is provided by the RBb/RTe index. Both the albumin and the IgG index determinations (RBb/RAlb and RBb/RIgG) may falsely yield too low values. RAlb values are greater than RIgG values by a factor of about 2.0 due to higher permeability of the albumin compared with IgG in patients without loss of the molecular selectivity of the blood – CSF barrier. In case of impairment of molecular selectivity, however, the factor between RAlb und RIgG may decrease to values of about 1.0. Thus, RBb/RAlb values may be quite similar or up to two times lower than RBb/RIgG values. In the case of RBb/RIgG index calculation, the sensitivity of the index is impaired by the fact that a proportion of the total IgG of the CSF may consist of specific antibody and lowers the ratio in this way. The determination of the RBb/RTe index is, therefore, more sensitive than either of the other indices, but it is restricted to persons with antibodies to tetanus toxoid and, therefore, cannot be used generally. However, other antibodies not intrathecally produced could also be used.

From a practical point of view, we prefer the RBb/RIgG index. If CSF and serum are diluted to the same IgG content, RBb provides a direct measure of the relative concentration of specific antibodies in the CSF because RIgG is 1.0 (Rehse-Küpper et al. 1986). The cut-off values given in the literature for a positive CSF/serum index value range from ≥ 1.0 up to ≥ 4.0 (Hofstadt et al. 1987; Rehse-Küpper et al. 1986; Stiernstedt et al. 1985; Wilske et al. 1986b, 1987). We regard an index of ≥ 4.0 as positive, and indices of ≥ 2.0 and ≤ 4.0 as borderline (Table 3). Patients with neurosyphilis often have high CSF/serum indices for intrathecal *B. burgdorferi* antibody production. this finding is due to crossreacting antibodies. In contrast to patients with Lyme neuroborreliosis, the Borrelia IFA-ABS is negative in the CSF of patients with neurosyphilis (Table 3).

Intrathecal IgM antibody production has been reported by Stiernstedt et al. (1985). However, these authors used an indirect ELISA without considering the possibility of a false-positive reaction due to RF or a false-negative reaction due to competitive inhibition by IgG antibodies. False-positive IgM antibody tests due to RF in the serum of patients with Lyme borreliosis and false-negative IgM antibody tests in the CSF have been reported (Wilske et al. 1984). The use of a μ-capture test or pretreatment of the specimens with anti-IgG serum might overcome these problems.

Table 3. Serodiagnosis in Lyme neuroborreliosis

Patients	IgG antibodies				IgG CSF/serum index			
	Serum		CSF		RBb/RIgG		RBb/RAIb	
	IFA ABS ≥1:64	ELISA ≥15 U/ml	IFA ABS ≥1:4	ELISA ≥0.1 U/ml	2.0-3.9	≥4.0	2.0-3.9	≥4.0
Neuroborreliosis stage II (*n*=76)	66 (87%)	38 (50%)	63 (83%)	70 (92%)	10 (13%)	49[a] (64%)	11 (14%)	47 (62%)
stage III (*n*=5)	5	5	5	5	0	5	1	4
Neurological controls (*n*=55)	0	1 (2%)	0	1 (2%)	1 (2%)	0	1 (2%)	0
Neurosyphilis (*n*=10)	1	10	0	9	2	6	n.d.	n.d.

IFA ABS, immunofluorescence absorption test; CSF, cerebrospinal fluid; RBb, CSF/serum ratio of *B. burgdorferi*-specific antibodies; RIgG, CSF/serum ratio of IgG; RAIb, CSF/serum ratio of albumin

[a] 38 (70%) out of 54 patients with RBb/RIgG ≥4.0 negative for IgM antibodies in the serum or the CSF.

21.2.5.3.2 Isoelectrofocusing and Immunoblot

Intrathecal IgG production can be demonstrated by isoelectrofocusing (IEF) of IgG-matched serum and CSF. Oligoclonal IgG band demonstrable in the CSF, but not in the serum, are indicative for intrathecal IgG production. Specificity of oligoclonal IgG for *B. burgdorferi* antigen has been detected in patients with Bannwarth's syndrome by immunofixation using radiolabeled *B. burgdorferi* antigen (Stiernstedt et al. 1984; Kristoferitsch et al. 1986). Combining IEF of CSF and blotting of the IEF gel to an antigen-loaded nitrocellulose filter (IEF affinity blot), Martin et al. (1987, 1988), Hansen et al. (1990), and Wilske (unpublished) have demonstrated that oligoclonal IgG is reactive with *B. burgdorferi* antigens in Bannwarth's syndrome (Fig.6). Only a

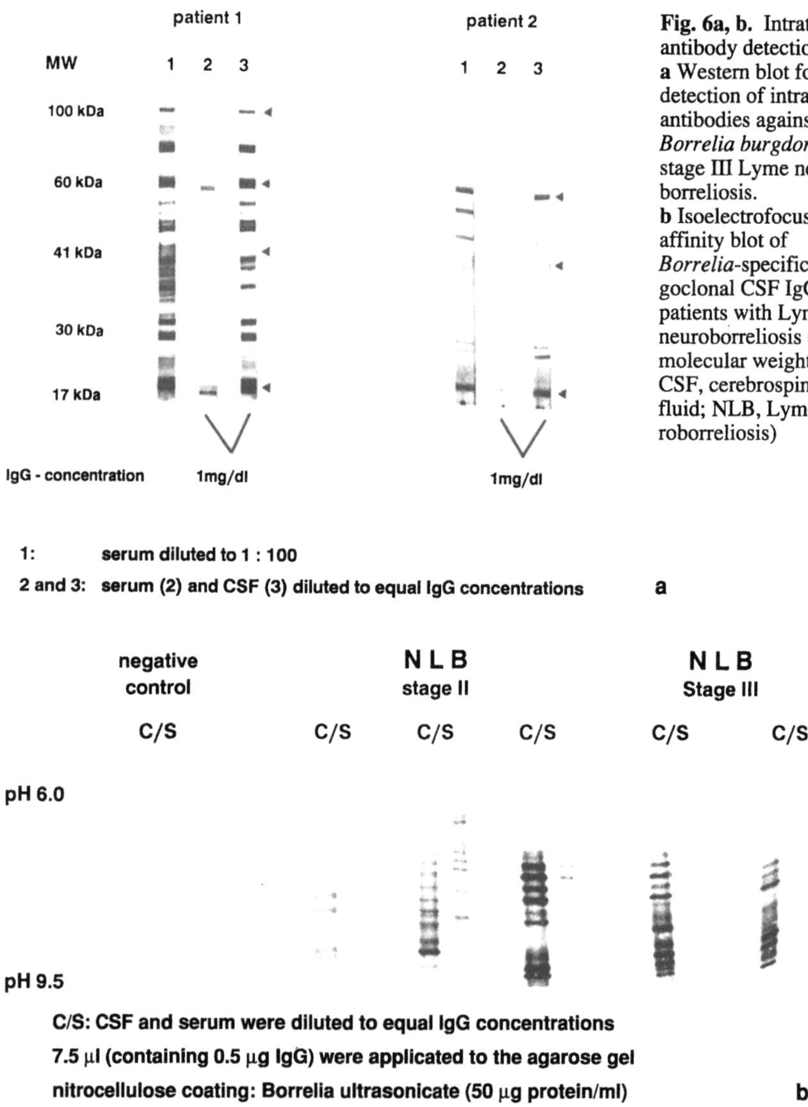

Fig. 6a, b. Intrathecal antibody detection. **a** Western blot for detection of intrathecal antibodies against *Borrelia burgdorferi* in stage III Lyme neuroborreliosis. **b** Isoelectrofocusing affinity blot of *Borrelia*-specific oligoclonal CSF IgG in patients with Lyme neuroborreliosis (MW, molecular weight; CSF, cerebrospinal fluid; NLB, Lyme neuroborreliosis)

proportion of the oligoclonal IgG bands may be reactive with the *B. burgdorferi* antigen, indicating intrathecal immune response to heterologous antigens. However, it cannot be excluded that non-reactive bands have specificity for borrelial components not contained in the test antigens. Oligoclonal *B. burgdorferi* IgG has been also demonstrated in the CSF of patients with chronic progressive encephalomyelitis (CEM) (Wilske, unpublished) (Fig. 6b).

Western blots of CSF and serum adjusted to equal IgG concentrations (Fig.6a) revealed that the intrathecal antibody response is directed against a panel of up to nine different antigenic components of *B. burgdorferi*. Different antibody patterns can be demonstrated in the CSF of patients with Lyme neuroborreliosis. The immunodominant protein for the intrathecal immune response in the second stage of Lyme neuroborreliosis is the flagellum protein p41 (Wilske et al. 1986b). These findings have been confirmed using the IEF affinity blot by Hansen et al. (1990), who demonstrated that the major part of the oligoclonal borrelia-specific bands are reactive with purified flagella. Whether a CSF/serum index determination using isolated p41 as ELISA antigen can provide a more sensitive assay, however, remains to be investigated.

21.2.5.4 Use of Selected Immunodominant Proteins

21.2.5.4.1 Western Blot

The Western blot has been used to analyze the humoral immune response in Lyme borreliosis (Barbour et al. 1983b; Craft et al. 1986; Grodzicki and Steere 1988; Wilske et al. 1986a,b, 1988a, 1989b; Karlsson et al. 1990a). The Western blot allows determination of the antibody pattern to the various polypeptides of *B. burgdorferi*. Borrelial proteins (whole-cell SDS lysates) are electrophoretically separated by SDS-PAGE and blotted to a nitrocellulose filter according to techniques previously described (Barbour et al. 1983a). Using class-specific enzyme-labeled conjugates, the IgM or IgG nature of the borrelial antibodies can be determined (Craft et al. 1986; Wilske et al. 1986a). As with procedures for IgM IFA and IgM ELISA, previous immunoprecipitation of the sera with anti-IgG serum is also recommended for the IgM Western blot (Wilske et al. 1989b).

The IgM and the IgG immune response is almost restricted to p41 (flagellin) in early Lyme borreliosis (Craft et al. 1986). IgM and IgG antibodies recognize both p41 and pC in European patients (Wilske et al., 1986a). The pC is a major 20-kDa protein demonstrable by SDS-PAGE in about 50% of European strains (Wilske et al. 1988b). A major 20-kDa protein is only rarely demonstrable in strains from the United States (Barbour and Hayes 1986; Bisset and Hill 1987). In patients with erythema migrans, IgM antibodies to pC are detected more often than to p41. But p41 was more often recognized by IgM antibodies from patients with meningoradiculitis (Bannwarth's syndrome) (typical patterns are shown in Fig. 7a). This may be partially due to heterogeneity among the pC components of strains causing neurological disease (CSF isolates are antigenically more variable than skin isolates). Using a combination of different strains, the pC-specific immune response has been detected in more cases than using one strain (Wilske et al. 1989b). A special point of interest is the fact that the IgM immune response to pC or OspA in patients with Bannwarth's syndrome has been detected with heterologous strains, but not with the autologous isolate, indicating possible antigenic changes in vivo or in culture (Wilske et al. 1986a, 1988b). The major immunodominant protein for the IgG immune response in patients with both erythema migrans and Bannwarth's syndrome is p41 (Fig.7b); pC is recognized in fewer cases (Wilske et al. 1989b).

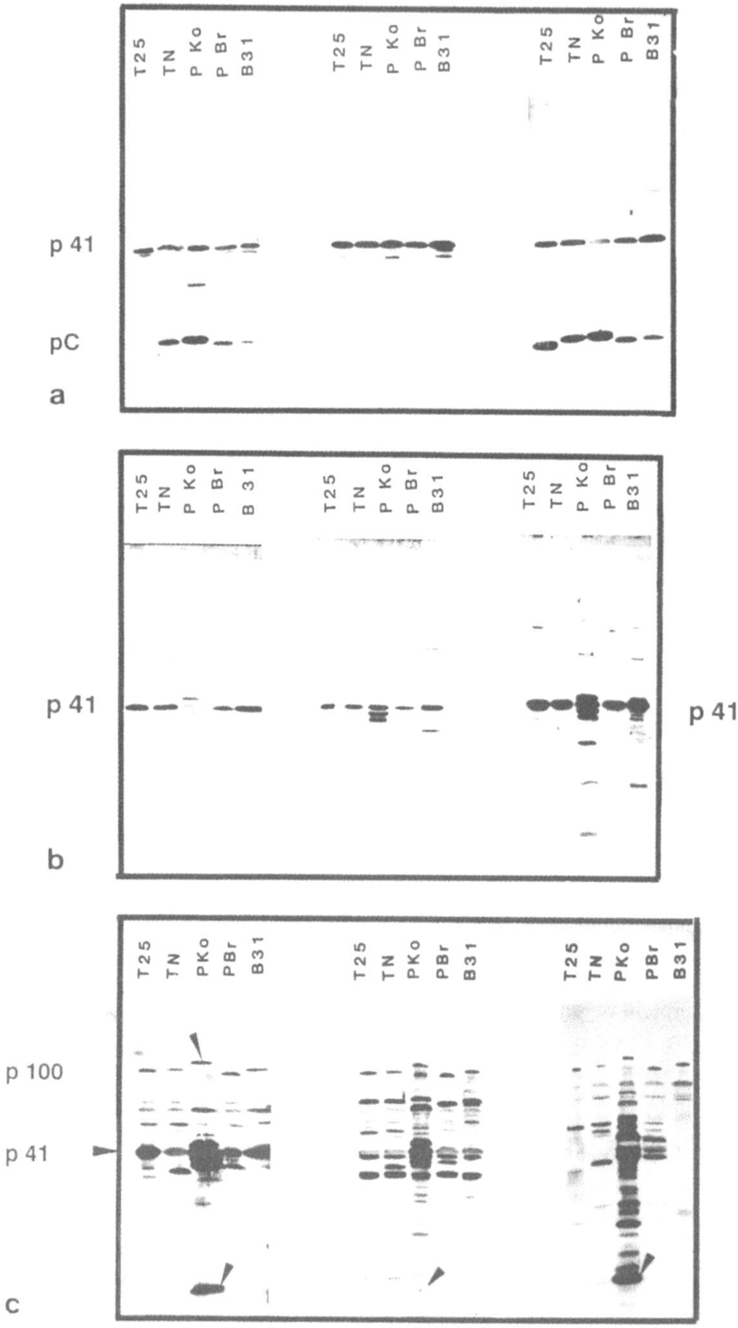

Fig. 7a–d. Humoral immune response to *Borrelia burgdorferi* proteins (Western blot using a 5-strain panel as antigen as explained in Fig.5). **a** IgM blot, three patients with stage II Lyme neuroborreliosis. **b** IgG blot, three patients with stage II Lyme neuroborreliosis. **c** IgG blot, three patients with acrodermatitis chronica atrophicans. **d** IgG blot, three patients with arthritis

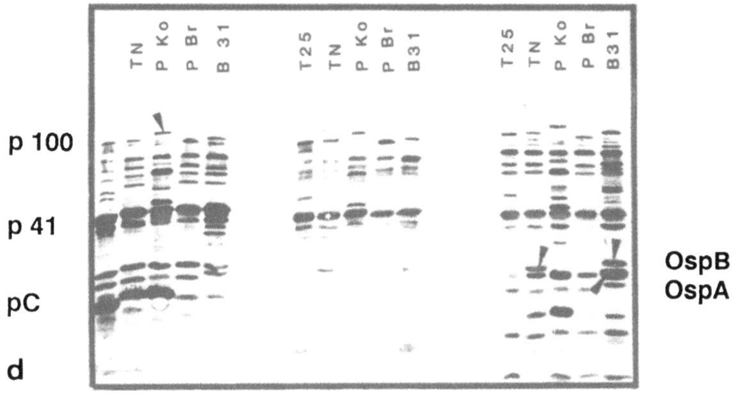

In later stages of Lyme borreliosis, the IgG immune response expands to a broad panel of different polypeptide-specific antibodies. Up to 20 polypeptides are recognized by sera from patients with Lyme arthritis or ACA (Barbour et al. 1984; Wilske et al. 1986a, 1988a). Using a panel of five different strains, considerable differences in the antibody patterns can be demonstrated (Fig. 7c,d). Sera from patients with ACA recognize a broad panel of polypeptides of a skin isolate, but only a smaller number of proteins from strains isolated from ticks and CSF. A comparable predominance of one strain was not found by testing sera from patients with Lyme arthritis. However, significant differences depending on the strain used as antigen suggest antigenic heterogeneity of borreliae causing Lyme arthritis (Wilske et al. 1988a). Significant antigens for the late human immune response are p100 and p17. An immune response to p100 and/or p17 was observed in all 17 cases of late disease (8 cases of ACA and 10 cases of arthritis) examined by Western blot (Wilske et al. 1990). The p100 components of different OspA serotypes are slightly variable in molecular weight size (Fig.7c,d) but, however, have identical epitopes recognized by a MaB L100 1D4 (Wilske et al. 1990).

Usually only IgG antibodies are present in late disease, although, IgM antibodies have also been demonstrated occasionally. According to Craft et al. (1986), the IgM response to p41 may persist until erythema migrans is followed by arthritis. In two patients with arthritis, the authors have even observed a new IgM immune response to p34 (OspB) (Craft et al. 1986). In Europe, a new IgM response to p41, OspA, and OspB after antibiotic treatment has been reported by Röther et al. (1988), and IgM antibodies to pC have been detected in a patient with chronic myositis (Reimers et al. 1989). Bearing in mind the possible antigenic variability of OspA and OspB versus pC, a late IgM immune response to these proteins may be due to antigenic changes of *B. burgdorferi*. On the other hand, antibiotic treatment or change from latent to acute status of the disease possibly make previously hidden antigens accessible to the immune system.

The use of the Western blot in routine serology is limited because the assay is very laborious, not easy to standardize, and often difficult to interpret. Crossreacting and specific proteins of *B. burgdorferi* may have similar molecular weights (Fig. 8, see color plate IV between pp 118–119). Most of the crossreacting proteins have apparent molecular weights in the 60 – 75 kDa range (p60, p66, p75). Another broadly crossreacting antigenic component (p40) is located very closely to the flagella protein. On the other hand the 39-kDa component described by Simpson et al. (1990b) as a highly specific immunodominant protein of *B. burgdorferi* for Lyme disease in patients in the United States could have been mistaken for p41. We have also observed a negative reaction to p41 and a positve reaction to a 39-kDa protein (that may be identical to the p39 described by Simpson) in patients with Lyme borreliosis. However, the diagnostic value of p39 as diagnostic antigen for European patients remains to be investigated. According to Zöller et al. (1991), antibodies to a 39-kDa polypeptide have been found less frequently in sera from European patients with Lyme borreliosis than have antibodies to p41. These authors favor densitometric evaluation of the Western blots for better standardization. However, in view of the fact that specific and crossreacting antigens of *B. burgdorferi* do have very similar molecular weights, the conventional Western blot needs interpretation by an experienced researcher and even then may be difficult to interprete.

21.2.5.4.2 Selective Western Blot

In an attempt to improve the specificity of the Western blot and to improve interpretation, certain immunodominant *B. burgdorferi* proteins (pC, p41, p100) can be selected as antigens. The pC is an immunodominant protein of the early immune response, p41 of both the early and the late immune response, and p100 only of the late immune response. It appears that p100 components of *B. burgdorferi* have species-specific common epitopes (Wilske et al. 1989a) (Table 4). Using a combination of p100, p41, and pC as antigens, crossreacting sera have been negative in the selective blot while sera from patients with Lyme borreliosis (stages II and III) remained positive (Wilske et al. 1989a). American investigators successfully used selective Western blots with recombinant *B. burgdorferi*-specific proteins of apparent molecular weights of 83 kDa (LeFebvre et al. 1990) and 39 kDa (Simpson et al. 1990b) for serodiagnosis of Lyme disease. Simpson et al. (1990) have shown that sera from Lyme disease patients with a significant IFA titer are reactive with p39. They have further shown (Simpson et al. 1991) that in mice inoculated with live borreliae, p39 is more immunogenic than p41. In the future, the availability of a defined panel of recombinant proteins (see Sect.21.2.5.4.4) will hopefully allow development of a standardized Western blot technique.

Table 4. Immunodominant proteins of *Borrelia burgdorferi*

Protein	Crossreactions within the species *Borrelia burgdorferi*	Crossreactions With RFB	With treponemes	Humoral immune response in Lyme borreliosis (stage) I	II	III
P100	+++	(+)	(–)	–	+	+++
P60[a]	+++	+++	++	–	+	+++
P41[b]	+++	++	(+)	+++	+++	+++
OspB[c]	+	(+)	–	–	–	(+)
OspA[c]	++	(+)	–	–	–	(+)
pC	++	(+)	–	+++	++	+

[a] "Common antigen", crossreactive with other bacteria such as *Legionella* and *Pseudomonas* species.
[b] Weak crossreactions with sera from patients with syphilis.
[c] High immunogenicity in artificially infected or immunized animals.
RFB, relapsing fever borreliae

21.2.5.4.3 Selective ELISA

The conventional whole-cell and ultrasonicate ELISAs are hampered by nonspecific reactions and are difficult to standardize. Different attempts have, therefore, been made to improve the ELISA by modifying the antigen. Coleman and Benach (1987) have used a flagella-enriched fraction as antigen. Using this flagella-enriched fraction, detection of the early immune response in patients with erythema migrans is improved. This assay is also superior to ELISAs in which single proteins are used such as p41, p66, OspA, and OspB eluted from SDS-PAGE. Some improvement in sensitivity of the ELISA has also been demonstrated by Grodzicki and Steere (1988) using another flagella-enriched preparation which, however, contained most of the other borrelial polypeptides.

Considerable improvement over the conventional ultrasonicate ELISA has been obtained by Hansen et al. (1988b) using a flagella preparation from a Swedish skin isolate purified by density gradient centrifugation as ELISA antigen. Besides small amounts of OspA and OspB, the preparation showed only the 41-kDa flagella band in a sensitive gold stain. The most important step is density gradient centrifugation leading to complete elimination of the crossreacting CA p60. The flagella have been shown by electron microscopy to be morphologically intact. Using flagella as ELISA antigen, the detection of both IgM and IgG antibodies is improved, especially in cases with short duration of the disease. Diagnostic specificity and sensitivity is increased for IgM and IgG antibody detection. Crossreactivity with syphilitic sera is decreased considerably but could not be completely removed. The increase in specificity is probably due to removing crossreacting antigens, particularly the 60 kDa common antigen (Hansen et al. 1988a).

21.2.5.4.4 Recombinant Antigens

Meanwhile several *B. burgdorferi* genes encoding immunodominant proteins have been cloned in *E. coli*: (a) the OspA of strain B31 (Bergström et al. 1989), a German tick isolate (Wallich et al. 1989), a Swedish skin isolate (Bergström et al. 1990), and a German skin isolate (Zumstein et al. 1990); (b) the p41 genes of B31, a German tick isolate (Gaßmann et al. 1989; Wallich et al. 1990), and a German skin isolate (Wilske et al. 1990); (c) the pC (Wilske et al. 1990) and the p100 genes (Fuchs et al. 1990b) of a German skin isolate; and (d) genes encoding a 83-kDa protein (LeFebvre et al. 1990) and a 39-kDa protein (Simpson et al. 1990b) of North American isolates. The genes encoding the 83 kDa and the 100 kDa protein are variants of the same *B. burgdorferi* protein (Wilske et al. 1992). Recombinant p39 was a sensitive and specific antigen for detection of antibodies to *B. burgdorferi* in Lyme disease patients and infected animals in the Western blot (Simpson et al. 1990, 1991). Recombinant p41 and recombinant pC ELISAs were more sensitive for detection of IgM and IgG antibodies to *B. burgdorferi* in patients with erythema migrans than using the conventional ultrasonicate ELISA (Wilske et al. 1990). In this study, sera from 74 patients with erythema migrans have been compared with sera from 100 blood donors. The p41 and pC ELISAs were positive in 34 (45.9%) for IgG and also in 34 (45.9%) for IgM compared with positive results in only 17 (22.9%) and 20 (27.0%) for IgG and IgM, respectively with conventional ultrasonicate ELISA. The 95% percentile has been used as the diagnostic cut-off, meaning that 5% of the blood donor sera had OD values above the cut-off values of the p41 and/or the pC ELISA and the ultrasonicate ELISA, respectively. Recombinant OspA has not been suitable for detection of antibodies in early manifestations (Wilske et al. 1990). Detection of antibodies against p100 was improved by using recombinant p100 (hyperexpressed in *E. coli*) for the Western blot compared to the conventional Western blot with ultrasonicate antigen (Wilske and Soutschek, unpublished). Extended studies are necessary to examine the value of these and other recombinant antigens for diagnosis.

21.2.6 Interpretation of Serodiagnostic Tests

The optimal serodiagnostic test is 100% sensitive and 100% specific. This is usually not achieved in practice. False-positive results (FP) occur besides true-positive ones (TP), and false-negative results (FN) besides true-negative ones (TN). The specificity of a diagnostic test (TN/(TN+FP)) is tested in a control group not suffering from the disease. In case of Lyme borreliosis, as in other infectious diseases, mostly healthy persons without a history of the disease are used as controls. The specificity of quantitative tests can be influenced by determination of the cut-off for the positive test at different levels (cutt-off level). Determining the cut-off level at the 95% percentile means the test is 95% specific. If the diagnostic value of different assays is compared, the sensitivity should be compared at the same cut-off level. For example, the diagnostic value of the IFA can be improved in patients with Lyme neuroborreliosis by previous absorption of the sera with *T. phagedenis* (IFA-ABS) according to Wilske et al. (1984) (Table 3) or the diagnostic value of the ELISA can be improved using purified flagella as antigen (Hansen et al. 1988b).

The predictive value of a test result depends not only on the sensitivity and specificity of a test result, but also on the prevalence of the disease. For example, the predictive value of a positive *B. burgdorferi* antibody test may be quite low in unselected neurological patients but increases considerably if only patients with lymphocytosis of the CSF and radicular pain are considered. If a screening test like the IHA with high sensitivity but lower specificity is used, the predictive value of the negative test is high, but the predictive value of the positive test is low. On the other hand a test with lower sensitivity but higher specificity, for example, the IFA-ABS or the flagella ELISA, is useful for confirming a positive result. We also recommend that serological results should not simply be reported as negative or positive. More information is provided if results are reported as "within the normal range", "borderline", or "significantly elevated" and if the cut-off percentile is also presented. Determination of the antibody titer is especially useful for serological follow-up studies or for calculation of the CSF/serum index.

21.2.7 Serological Findings in Lyme Borreliosis

Antibody detection in Lyme borreliosis depends mainly on the stage of the disease and the duration of symptoms. Suitable specimens for antibody detection are listed in Table 1. An overview of serological findings in different stages of the disease is given in Table 5. Percentages of seropositivity are derived from data from various European authors (Ackermann et al. 1984; Asbrink et al. 1984 a,b, 1985; Stiernstedt et al. 1985; Weber et al. 1984a, 1988a; Wilske et al. 1984, 1989a; Hansen et al. 1988b; Herzer and Wilske 1986; Herzer et al. 1986, Stanek et al. 1986);

21.2.7.1 Erythema Migrans and Borrelial Lymphocytoma

Significantly elevated antibody titers have been reported to occur in only a minority of European patients (20% – 50%) with erythema migrans (Ackermann et al. 1984; Asbrink et al. 1984a, 1985; Weber et al. 1984b, 1988a; Wilske et al. 1984; Stiernstedt

Table 5. Serological findings in representative manifestations of Lyme borreliosis

	All cases	Seropositive cases		Seropositive cases	
Stage/clinical manifestation	elevated titer (%)[a]	IgM IFA ABS ≥64 (%)[b]	Titer (median)	IgG IFA ABS ≥64 (%)[b]	Titer (median)
I Erythema migrans	20 – 50	50 – 90	128	10 – 50	64
II Meningo-radiculitis	70 – 90	15 – 70	64	50– 100	128–256
III Arthritis, ACA	90 – 100	3 – 7	-	90 – 100	256 – 512

IFA ABS, indirect immunofluorescence absorption test; ACA, acrodermatitis chronica atrophicans
[a] Rates of seropositive patients depend on duration and severeness of symptoms and on previous antibiotic therapy.
[b] Rates of positive IgM or IgG test depend on duration of symptoms.

et al. 1986). American authors have reported higher percentages of significantly increased antibody titers. According to Steere et al. (1983), up to 90% of patients with erythema migrans have shown elevated IgM antibody titers in studies where the sera of convalescent patients were included. IgG antibody titers increase later than IgM antibody titers, as they also do in other infectious diseases. Only about 50% of American patients have elevated IgG titers within 6 weeks of onset of the disease.

Significantly elevated levels of both IgM or IgG antibodies have been reported in a lower percentage of European than American patients. This is probably due to differences in the clinical picture or due to antigenic heterogeneity of European *B. burgdorferi* strains. Patients with symptoms of generalized infection (fever, multiple erythemas) are less frequently seen in Europe. However, 13 of 21 (62%) patients who presented with associated symptoms had significantly elevated IgM or IgG titers, compared with only 7 of 23 (30%) of patients with erythema migrans alone (Wilske et al. 1984). Using the flagella ELISA, a significantly higher rate (68.8% versus 39.9%) of positive IgM test results has been obtained in patients with multiple erythema migrans (n=16) than in patients with solitary erythema migrans (n=91) (Hansen and Lebech 1989).

Probably as a result of patient selection, but also as a result of using different serological techniques and cut-off levels for positive test results, the percentage of seropositive patients reported by different European authors ranges from 12% (Stiernstedt et al. 1985) up to 84% (Neubert et al. 1986). IFA-ABS test results in 87 consecutive unselected patients with erythema migrans, as described by Weber et al. (1988a), using different cut-off levels for the positive test are presented in Table 6. The table also shows that some patients develop significant antibody titers after the beginning of antibiotic therapy. On the other hand, it must be assumed that occurrence of a significant immune response may be inhibited by antibiotic treatment. Patients with a short history of the disease tend to have predominantly an IgM antibody response. IgG antibodies may be detected in some patients as early as 1 week after onset of the erythema migrans, but IgG antibodies usually appear later, if at all (Fig.9). The rate of seropositive findings depends considerably on the duration of symptoms. As shown by Hansen and Asbrink (1989), the detection rate for IgM antibodies has increased from 31.8% to

Table 6. Serological findings in erythema migrans

	IFA ABS ≥ 32				IFA ABA ≥ 64			
	IgM		IgG		IgM		IgG	
	n	%	n	%	n	%	n	%
Erythema migrans (*n*=87) Pretreatment	34	39.1	22	25.3	20	23	12	13.8
Additional positives within 2-6 weeks post treatment	11	12.6	6	6.9	10	11.5	5	5.75
Total	45	51.7	28	32.2	30	34.5	17	19.55
Healthy controls (*n*=273)	2	0.7	5	1.8	1	0.4	3	1.0

IFA ABS, immunofluorescence absorption test

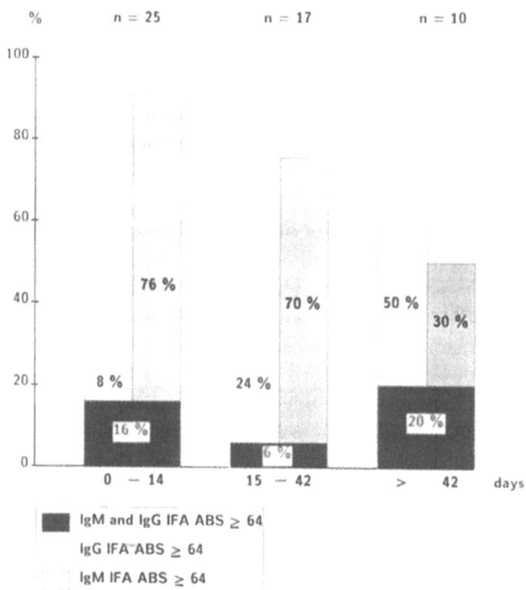

Fig. 9. IgM- and IgG antibody titers in seropositive cases of erythema migrans determined by immunofluorescent absorption test (IFA-ABS)

53.1% and for IgG antibodies from 27.2% to 53.1% in patients with short duration of disease (<1 week) compared with patients with long-lasting symptoms (5-12 weeks). Post-treatment antibody titers decrease below the cut-off level in most patients within 3 – 12 months. However, in some patients, elevated IgM or IgG antibody titers have been observed more than 12 months after treatment (Weber et al. 1988a). Even in erythema migrans in the course of reinfection, a significant antibody response may be missed (Weber et al. 1986). In a case of reinfection during which lymphocytic meningoradiculitis was followed by erythema migrans, a significant increase of IgG antibody titers, but no IgM response was detected (Pfister et al. 1986). Reinfection provides a possible explanation for IgG-positive, but IgM-negative cases of erythema migrans.

Most patients with borrelial lymphocytoma display elevated IgG titers, but IgM antibodies can also be detected (Weber et al. 1984b, 1985; Wilske et al. 1984; Asbrink et al. 1989). However, more patients need to be examined.

21.2.7.2 Acrodermatitis Chronica Atrophicans

Nearly 100% of patients with ACA have significantly elevated IgG antibody titers against *B. burgdorferi* (Ackermann et al. 1984; Weber et al. 1984a; Wilske et al. 1984; Asbrink and Hovmark 1985; Asbrink et al. 1985). In many patients, high IgG antibody levels are observed. Some authors have reported elevated IgM titers in up to 80% of the patients (Asbrink and Hovmark 1985; Neubert et al. 1986). We obtained similar results when native sera were tested for IgM antibodies (Wilske et al. 1984). After previous absorption of the sera with anti-IgG or isolation of the IgM fraction, only 5% of the sera remained positive for IgM antibodies, indicating false-positive IgM titers due to RF activity in the unabsorbed sera (Weber et al. 1984a; Wilske et al. 1984). Antibody titers usually decrease after antibiotic therapy, but the rate of decrease is very slow, and titers often remain significantly elevated for years after treatment (Weber et al. 1984a, 1988a; Wilske et al. 1984).

21.2.7.3 Lyme Neuroborreliosis

In patients with neurological symptoms of the second stage of Lyme disease, elevated serum antibody titers against *B. burgdorferi* are detected in 60% – 90% of patients (Ackermann et al. 1984; Wilske et al. 1984; Stiernstedt et al. 1985; Hansen et al. 1988b). The percentage of seropositive patients is strongly dependent on the duration of the disease. According to Stiernstedt et al. (1985), only 35% of the patients with a short history (<6 weeks) are positive for IgG antibodies in contrast to 65% – 70% of

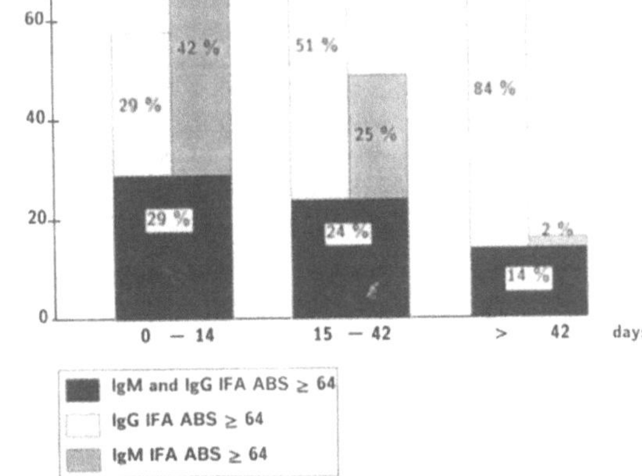

Fig. 10. IgM and IgG antibody titers in seropositive cases of stage II Lyme neuroborreliosis determined by immunofluorescent absorption test (IFA-ABS)

patients with a long history (6–43 weeks) of neurological symptoms. Similar results have been obtained by Hansen et al. (1988b), also using the conventional ultrasonicate ELISA. However, IgG antibody detection has been considerably improved (from 42% sensitivity to 77%) with purified flagella as ELISA antigen (Hansen et al. 1988b). The IFA-ABS has also been shown to be superior than the conventional ultrasonicate ELISA (sensitivity 87% versus 50%) for detection of IgG antibodies (Wilske et al. 1987) (Table 3). In both the flagella ELISA and the IFA-ABS, the nonspecific background due to crossreacting antibodies is reduced. This is especially important if low titers of specific antibodies are to be detected. Serum antibodies of the IgM class are predominantly found in the first 2 weeks when neurological manifestations are present (Wilske et al. 1989c) (Fig.10). In patients with a short history of the disease, only elevated IgM antibody titers may be found; 42% of seropositive patients with neurological symptoms lasting <2 weeks, and 25% of patients with neurological symptoms lasting for 2-6 weeks, are negative for IgG antibodies (Wilske et al. 1989c). More than 6 weeks after onset of neurological symptoms, elevated IgM antibody titers are detected only in low percentage (16% of seropositive patients) and in only 2% without significantly elevated IgG antibodies. On the other hand, a significant IgM response has not been detected in 29%, 51%, and 84% of seropositive patients with neurological symptoms of <2 weeks, 2-6 weeks, and more than 6 weeks duration, respectively (Wilske et al. 1989c). In endemic regions, elevated IgG titers in the serum may be due to subclinical infection and can often be demonstrated in persons at high risk for tick bites (Wilske et al. 1985; Aeschlimann et al. 1988).

Clinically active infection can be confirmed serologically if IgM antibodies are detected. In patients with neurological symptoms who have a negative IgM test, the demonstration of intrathecal specific borrelial antibodies is another important diagnostic tool. As mentioned above, intrathecal specific antibody production may be demonstrated by elevated CSF/serum indices. Table 3 demonstrates the diagnostic value of CSF/serum index determinations using the IgG and the albumin index (RBb/RIgG and RBb/RAlb) in patients with Lyme neuroborreliosis (Wilske, unpublished). Both indices provide camparable results in stage II patients and were significantly elevated in 64% (RBb/RIgG) and 62% (RBb/RAlb) of patients with neurological manifestations, respectively. About 70% of these patients have been found to be negative for IgM antibodies. In cases with borderline values for the RBb/IgG or the RBb/RAlb index, which amounts to an additional 13%-14%, determination of the RBb/RTe index may provide valuable diagnostic aid if the patient has antibodies against tetanus toxoid. All seven patients with borderline RBb/RIgG and antibodies against tetanus toxoid had RBb/RTe values of ≥ 4.0 (data not shown in Table 3). The diagnostic value of CSF/serum index calculations for specific antibodies against *B. burgdorferi* has previously been shown in other studies using the IgG index (Rehse-Küpper et al. 1986) or the albumin index (Stiernstedt et al. 1985). Compared with the data shown in table 3, Stiernstedt et al. (1985) have demonstrated a higher percentage (87%) of patients with elevated IgG CSF/serum indices, probably due to a lower cut-off value for the positive test (≥2.0).

Using the conventional ultrasonicate ELISA, IgG antibodies are detected with higher sensitivity in the CSF than in the serum of affected patients (87% versus 62% according to Stiernstedt et al. 1985, 63% versus 41% according to Hansen et al. (1988b), and 92% versus 50% according to Wilske et al., Table 3). The sensitivity of serum IgG antibody detection is increased using a selective flagella ELISA or using the IFA-ABS test. In some cases, specific antibodies can be detected earlier in the CSF than in the serum. On the other hand, negative CSF findings do not exclude Lyme neuroborreliosis.

The main value of CSF/serum index determinations lies in the possibility of confirming the infection of the nervous system by *B. burgdorferi*. This is especially important in late Lyme neuroborreliosis, which can imitate multiple sclerosis (Ackermann et al. 1985). Demonstration of intrathecal antibody production is mandatory, especially in late Lyme neuroborreliosis, because demonstration of elevated serum antibody titers alone is not sufficient. In the case of borderline index values obtained with one method, determination with a second method is recommended. Furthermore, immunoblotting methods such as Western blot and IEF affinity blot may provide further confirmation of the presence of intrathecal antibody production (Wilske et al. 1986b; Martin et al. 1988; Hansen et al. 1990).

The value of index determination for follow-up studies remains unclear. We have demonstrated high CSF/serum index values of RBb/RIgG of 17.4 and 18.6, respectively, without a significant decrease 26 and 52 weeks, respectively, after onset of symptoms in a patient with a history of Bannwarth's syndrome; the patient had been asymptomatic during this period (Wilske et al. 1986b). The persistence of high index values can be explained differently: (a) triggering of the intrathecal immune response by persistent live borreliae or by borrelial antigen, (b) survival of antibody producing B cells within the CNS, (c) concomitant decline of antibody levels and IgG in serum and CSF, respectively, resulting in a constant CSF/serum index value. Direct detection of specific antibody-producing cells within the CSF may help in diagnosing CNS involvement in *B. burgdorferi* infection. Using a nitrocellulose immunospot technique, Link et al. (1988) have demonstrated B cells producing *B. burgdorferi* antibodies in patients with Lyme neuroborreliosis, but also in neurosyphilis and in control patients with Guillain-Barré and Sjögren syndrome. The results have been obtained in a small number of patients and the diagnostic value of the method remains to be proven.

21.2.7.4 Arthritis

Nearly all patients with Lyme arthritis present with significantly elevated IgG antibody titers (Steere et al. 1983; Herzer and Wilske 1986; Herzer et al. 1989). As in ACA, IgM antibodies are only rarely observed when sera are pretreated with anti-IgG antibodies to prevent nonspecific reactions due to RF activity. IgG antibody titers in the synovial fluid are not significantly different from the corresponding serum titers. IgG matched pairs of joint fluid and blood serum are not significantly different in their IgG antibody titers tested by ELISA (Herzer et al. 1989) or in antibody pattern of the Western blots (Wilske, unpublished). However, these observations do not exclude local antibody production.

Early in the course of the disease, the humoral immune response may be restricted to the p41-flagellar antigen. In late disease, a broad pattern of *B. burgdorferi* peptides is recognized (Craft et al. 1986; Wilske et al. 1988a). Heterogeneous patterns in the IgG Western blot with different strains suggest that antigenically different strains may cause Lyme arthritis (Wilske et al. 1988a).

References

Aberer E, Manitz M, Neumann R, Stanek G (1988) Immunoperoxidase staining of spirochestes in borrelia skin disease. Ann NY Acad Sci 539:362-364

Ackermann R, Kabatzki J, Boisten HP, Steere AC, Grodzicki RL, Hartung S, Runne U (1984) Spirochäten-Ätiologie der Erythema-chronicum-migrans-Krankheit. Dtsch Med Wochenschr 109:92-97

Ackermann R, Gollmer E, Rehse-Küpper B (1985) Progressive Borrelien-Encephalomyelitis. Dtsch Med Wochenschr 26:1039-1042

Aeschlimann A, Chamot E, Gigon F, Jeanneret JP, Kesseler D, Walter C (1986) *B. burgdorferi* in Switzerland. Zentralbl Bacteriol Hyg (A) 263:450-458

Aeschlimann A, Gern L, Zhiova E, Frossard E, Walter A, Fahrer H, Sauvain MJ, van den Linden S, Gerber N (1988) Observations of two high-risk populations from the Swiss Plateau a region heavily infested with Ixodes ricinus/Borrelia burgdorferi complex. Ann NY Acad Sci 539:440-443

Anderson JF, Magnarelli LA, LeFebvre RB, Andreadis TG, McAninch JB, Perng G-C, Johnson RC (1989) Antigenically variable *Borrelia burgdorferi* isolated from cottontail rabbits and Ixodes dentatus in rural and urban areas. J Clin Microbiol 27:13-20

Asbrink E, Hovmark A (1985) Successful cultivation of spirochetes from skin lesion of patients with erythema chronicum migrans Afzelius and acrodermatitis chronica atrophicans. Acta Pathol Microbiol Immunol Scand (B) 93:161-163

Asbrink E, Hederstedt B, Hovmark A (1984a) The spirochetal etiology of erythema chronicum migrans Afzelius. Acta Derm Venereol (Stockh) 64:291-295

Asbrink E, Hovmark A, Hederstedt B (1984b) The spirochetal etiology of acrodermatitis chronica atrophicans Herxheimer. Acta Derm Venereol (Stockh) 64:506-512

Asbrink E, Hovmark A, Hederstedt B (1985) Serological studies of erythema chronicum migrans Afzelius and acrodermatitis chronica atrophicans with indirect immunofluorescence and enzymelinked immunosorbent assays. Acta Derm Venereol (Stockh) 65:509-514

Asbrink E, Hovmark A, Olsson I (1989) Lymphadenosis benigna cutis solitaria-Borrelia lymphocytoma in Sweden. Zentralbl Bacteriol (Suppl) 18:156-163

Barbour AG, Hayes SF (1986) Biology of Borrelia species. Microbiol Rev 50:381-400

Barbour AG, Burgdorfer W, Grunwald E, Steere AC (1983a) Antibodies of patients with Lyme disease to components of the Ixodes dammini spirochete. J Clin Invest 73:504-515

Barbour AG, Burgdorfer W, Hayes SF, Peter O, Aeschlimann A (1983b) Isolation of a cultivable spirochete from Ixodes ricinus ticks of Switzerland. Curr Microbiol 8:123-126

Barbour AG, Tessier SL, Todd WJ (1983c) Lyme disease spirochetes and ixodid tick spirochetes share a common surface antigenic determinant defined by a monoclonal antibody. Infect Immun 41:795-804

Barbour AG, Tessier SL, Hayes SF (1984) Variation in a major surface protein of Lyme disease spirochetes. Infect Immun 45:94-100

Barbour AG, Hayes SF, Heiland RA, Schrumpf ME, Tessier SL (1986) A Borrelia-specific monoclonal antibody binds to a flagellar epitope. Infect Immun 52:549-554

Benach JL, Bosler EM, Hanrahan JP, Coleman JL, Habicht GS, Bast TF, Cameron DJ, Ziegler JL, Barbour AG, Burgdorfer W, Edelman R, Kaslow RA (1983) Spirochetes isolated from the blood of two patients with Lyme disease. N Engl J Med 308:740-742

Berardi VP, Weeks KE, Steere AC (1988) Serodiagnosis of early Lyme disease: analysis of IgM and IgG antibody responses by using an antibody-capture enzyme immunoassay. J Infect Dis 158:754-760

Berger BW, Kaplan MH, Rothenberg IR, Barbour AG (1985) Isolation and characterization of the Lyme disease spirochete from the skin of patients with erythema chronicum migrans. J Am Acad Dermatol 13:444-449

Bergström S, Bundoc VG, Barbour AG (1989) Molecular analysis of linear plasmid-encoded major surface proteins, OspA and OspB, of the Lyme disease spirochaete *Borrelia burgdorferi*. Mol Microbiol 3:479-486

Bissett ML, Hill W (1987) Characterization of *Borrelia burgdorferi* strains isolated from Ixodes pacificus ticks in California. J Clin Microbiol 25:2296-2301

Bozsik BP (1985) *Borrelia burgdorferi* stained with acidic acridine orange. Ann Immunol Hung 25:335-344

Brandt ME, Riley BS, Radolf JD, Norgard MV (1990) Immunogenic integral membrane proteins of *Borrelia burgdorferi* are lipoproteins. Infect Immun 58:989-991

Bundoc VG, Barbour A (1989) Clonal polymorphisms of outer membrane protein OspB of *Borrelia burgdorferi*. Infect Immun 57:2733-2741

Burgdorfer W, Gage KL (1986) Susceptibility of the black-legged tick, Ixodes scapularis, to the Lyme disease spirochete, *Borrelia burgdorferi*. Zentralbl Bakteriol Hyg (A) 263:15-20

Burgdorfer W, Barbour AG, Hayes SF, Benach JL, Grunwald E, Davis JP (1982) Lyme disease – a tickborne spirochetosis ? Science 216:1317-1319

Burgdorfer W, Lane RS, Barbour AG, Gresbrink RA, Anderson JR (1985) The Western black-legged tick, Ixodes pacificus: a vektor of *Borrelia burgdorferi*. Am J Trop Med Hyg 34:925-930

Cerny EH, Farsy CE, Hunter EF, Larsen SA (1985) Rheumatoid factor in syphilis. J Clin Microbiol 22:89-94

Cluss RG, Boothby JT (1990) Thermoregulation of protein synthesis in *Borrelia burgdorferi*. Infect Immun 58:1038-1042

Coleman JL, Benach JL (1987) Isolation of antigenic components from the Lyme disease spirochete: Their role in early diagnosis. J Infect Dis 155:756-765

Craft JE, Grodzicki RL, Steere AC (1984) Antibody response in Lyme disease: evaluation of diagnostic tests. J Infect Dis 149:789-795

Craft JE, Fischer DH, Shumamoto GT, Steere AC (1986) Antigens of *Borrelia burgdorferi* recognized during Lyme disease. J Clin Invest 78:934-939

Fuchs R, Wilske B, Preac-Mursic V, Schierz G, Gueye W, Liegl G (1989) The use of purified immunodominant antigens of Borrelia burgdorferi in IgM- and IgG-ELISA (poster). In: European Society of Clinical Microbiology and Infectious Diseases (ed) 4th European congress of clinical microbiology. Nizza, pp 912/PP43

Fuchs R, Wilske B, Preac-Mursic V, Schierz G (1990a) Purification of *B. burgdorferi* flagellum by use of a monoclonal antibody. Mol Cell Probes 4:451-462

Fuchs R, Soutschek E, Wilske B, Preac-Mursic V (1990b) Identification and molecular characterisation of an immunogenic protein (p100) of *Borrelia burgdorferi* (abstract) in: 4th international conference on Lyme borreliosis, Stockholm

Gaßmann GS, Kramer M, Göbel UB, Wallich R (1989) Nucleotide sequence of a gene encoding the *B. burgdorferi* flagellin. Nucleic Acids Res 17:3590

Goodman JL, Jurkovich P, Kramber JM, Johnson RC (1991) Molecular detection of persistant *Borrelia burgdorferi* in the urine of patients with active Lyme disease. Infect Immun 59:269-278

Grodzicki RL, Steere AC (1988) Comparison of immunoblotting and indirect enzyme-linked immunosorbent assay using different antigen preparations for diagnosing early Lyme disease. J Infect Dis 157:790-797

Hansen K, Asbrink E (1989) Serodiagnosis of erythema migrans and acrodermatitis chronica atrophicans by the *Borrelia burgdorferi* flagellum enzyme-linked immunosorbent assay. J Clin Microbiol 27:545-551

Hansen K, Bangsborg JM, Fjordvang H, Pedersen NS, Hindersson P (1988a) Immunochemical characterization and isolation of the gene for a *Borrelia burgdorferi* immunodominant 60-kilodalton antigen common to a wide range of bacteria. Infect Immun 56:2047-2053

Hansen K, Hindersson P, Pedersen NS (1988b) Measurement of antibodies to the *Borrelia burgdorferi* flagellum improves serodiagnosis in Lyme disease. J Clin Microbiol 26:338-346

Hansen K, Cruz M, Link H (1990) Oligoclonal *Borrelia burgdorferi*-specific IgG antibodies in cerebrospinal fluid in Lyme neuroborreliosis. J Infect Dis 161:1194-1202

Hansen K, Pii K, Lebech A-M (1991) Improved immunoglobulin M serodiagnosis in Lyme borreliosis by using a μ-capture enzyme-linked immunosorbent assay with biotinylated *Borrelia burgdorferi* flagella. J Clin Microbiol 291:166-173

Hechemy KE, Harris HL, Duerr MJ, Benach JL, Reimer CB (1988) Immunoglobulin G subclasses specific to *Borrelia burgdorferi* in patients with Lyme disease. Ann NY Acad Sci 539:162-169

Herzer P, Wilske B (1986) Lyme arthritis in Germany. Zentralbl Bakteriol Hyg (A) 263:268-274

Herzer P, Wilske B, Preac-Mursic V, Schierz G, Schattenkirchner M, Zöllner N (1986) Lyme arthritis: clinical features, serological and radiographic findings of cases in Germany. Klin Wochenschr 64:206-215

Herzer P, Wilske B, Schewe S, Preac-Mursic V, Schierz G (1989) Serologische Diagnostik der Lyme-Arthritis. Aktuel Rheumatol 14:125-129

Hofstad H, Matre R, Nyland H, Ulvestad E (1987) Bannwarth's syndrome: serum and CSF IgG antibodies against *Borrelia burgdorferi* examined by ELISA. Acta Neurol Scand 75:37-45

Hyde FW, Johnson RC, White TJ, Shelburne CE (1989) Detection of antigens in urine of mice and humans infected with *Borrelia burgdorferi*, etiologic agent of Lyme disease. J Clin Microbiol 27:58-61

Johnson SE, Klein GC, Schmid GP, Bowen GS, Feeley JC, Schulze T (1984) Lyme disease: a selective medium for isolation of the suspected etiological agent, a spirochete. J Clin Microbiol 19:81-82

Jonsson M, Bergström S (1990) Molecular characteristics and genetic organization of the osp operon from different European *Borrelia burgdorferi* strains. 4th international conference on Lyme borreliosis, Stockholm

Karlsson M (1990a) Western immunoblot and flagellum enzyme-linked immunosorbent assay for serodiagnosis of Lyme borreliosis. J Clin Microbiol 28:2148-2150

Karlsson M, Hovind-Hougen K, Svenungsson B, Stiernstedt G (1990b) Cultivation and characterization of spirochetes from cerebrospinal fluid of patients with Lyme borreliosis. J Clin Microbiol 28:473-479

Kmety E, Rehacek J, Vyrostekova V (1986) Investigation of ticks for the presence of Borrelia in Czechoslovakia. Zentralbl Bakteriol Hyg (A) 263:468-470

Kristoferitsch W, Sterck AJ, Murray N, Stanek G, Lanschützer H (1986) Oligoclonal antibodies in CSF of patients with meningopolyneuritis Garin-Bujadoux-Bannwarth: Ig class, light chain type and specifity. Zentralbl Bakteriol Hyg (A) 263:307-313

LeFebvre RB, Perng G-C, Johnson RC (1990) The 83-kilodalton antigen of *Borrelia burgdorferi* which stimulates immunoglobulin M (IgM) and IgG responses in infected hosts is expressed by a chromosomal gene. J Clin Microbiol 28:1673-1676

Link H, Baig S, Olsson T, Zachau A, Stiernstedt G (1988) B-cell response at a cellular level in cerebrospinal fluid and blood in Lyme disease and controls. Ann NY Acad Sci 539:389-392

Luft BJ, Jiang W, Munoz P, Dattwyler RJ, Gorevic PD (1989) Biochemical and immunological characterization of the surface proteins of *Borrelia burgdorferi*. Infect Immun 57:3637-3645

Luther B, Moskophidis M (1990) Antigenic cross-reactivity between *Borrelia burgdorferi*, *Borellia recurrentis, Treponema pallidum* and *Treponema phagedenis*. Zentralbl Bakteriol 274:214-226

MacDonald AB (1986) Human fetal borreliosis, toxemia of pregnancy, and fetal death. Zentralbl Bakteriol Hyg (A) 263:189-200

Magnarelli LA (1988) Serological diagnosis of Lyme disease. Ann NY Acad Sci 539:154-161

Magnarelli LA, Anderson JF (1988) Enzyme-linked immunosorbent assays for detection of class-specific immunoglobulins to *Borrelia burgdorferi*. Am J Epidemiol 127:818-825

Magnarelli LA, Meegan JM, Anderson JF, Chappell WA (1984) Comparison of an indirect fluorescent antibody test with an enzyme-linked immunosorbent assay for serological studies of Lyme disease. J Clin Microbiol 20:181-184

Magnarelli LA, Anderson JF, Johnson RC (1987) Cross-reactivity in serological tests for Lyme disease and other spirochetal infections. J Infect Dis 156:183-188

Martin R, Martens U, Krüger H, Sticht-Groh V, Dörries R (1987) Nachweis und Identifizierung *Borrelia-burgdorferi*-spezifischer, oligoklonarer Immunglobuline (IgG) im unkonzentrierten Liquor mit der Immunoblottechnik. In: Peock K, Hacke W, Schneider R, (eds) Neuroimmunologie, spinale Krankheiten, Neuropsychologie, metabolische Enzephalopathien, neurologische Notfälle, interventionelle Neuroradiologie. Springer, Berlin Heidelberg New York, pp440-442 (Verhandlungen der Deutschen Gesellschaft für Neurologie, vol 4)

Martin R, Martens U, Sticht-Groh V, Dörries R, Krüger H (1988) Persistent intrathecal secretion of oligoclonal, *Borrelia burgdorferi*-specific IgG in chronic meningoradiculomyelitis. J Neurol 225:229–239

Müller F (1982) Der 19S-IgM-FTA-Abs-Test in der Serodiagnostik der Syphilis. Technik, Fehlermöglichkeiten und diagnostische Aussagen. Immun Infekt 10:32-34

Müller F, Moskophidis M (1983) Estimation of the local production of antibodies to *Treponema pallidum* in the central nervous system of patients with neurosyphilis. Br J Vener Dis 59:80-84

Neubert U, Krampitz HE, Engl H (1986) Microbiological findings in erythema (chronicum) migrans and related disorders. Zentralbl Bakteriol Hyg (A) 263:237-252

Park HK, Jones BE, Barbour AG (1986) Erythema chronicum migrans of Lyme disease: diagnosis by monoclonal antibodies. J Am Acad Dermatol 15:406-410

Pfister HW, Neubert U, Wilske B, Preac-Mursic V, Einhäupl KM, Borasio GD (1986) Reinfection with *Borrelia burgdorferi*. Lancet II:984-985

Preac-Mursic V, Schierz G, Pfister H-W, Einhäupl K, Wilske B, Weber K (1984a) Isolierung einer Spirochäte aus Liquor cerebrospinalis bei Meningoradiculitis-Bannwarth. Munch Med Wochenschr 126:275-276

Preac-Mursic V, Wilske B, Schierz G, Pfister H-W, Einhäupl K (1984b) Repeated isolation of spirochetes from the cerebrospinal fluid of a patient with meningoradiculitis Bannwarth. Eur J Clin Microbiol 3:564-565

Preac-Mursic V, Wilske B, Herzer P, Schierz G, Bauer H (1985) Acrodermatitis chronica atrophicans: eine Borreliose! Hautarzt 36:691-693

Preac-Mursic V, Wilske B, Schierz G (1986) European *Borrelia burgdorferi* isolated from humans and ticks: Culture conditions and antibiotic susceptibility. Zentralbl Bakteriol Hyg (A) 263:112-118

Preac-Mursic V, Weber K, Pfister HW, Wilske B, Gross B, Baumann A, Prokop J (1989) Survival of *Borrelia burgdorferi* in antibiotically treated patients with Lyme borreliosis. Infection 17:355-359

Preac-Mursic V, Patsouris E, Wilske B, Reinhardt S, Groß B, Mehrain P (1990) Persistence of *Borrelia burgdorferi* and histopathological alterations in experimentally infected animals; comparison with histopathological findings in human Lyme disease. Infection 18:332-341

Rehse-Küpper B, Ackermann R (1986) Demonstration of locally synthesized borrelia antibodies in cerebrospinal fluid. Zentralbl Bakteriol Hyg (A) 263:407-411

Reimers CD, Pongratz DE, Neubert U, Pilz A, Hübner G, Naegele M, Wilske B, Duray PH, de Koning JC (1989) Myositis caused by *B. burgdorferi*: report of four cases. J Neuorol Sci 91:215-226

Rosa PA, Schwan TG (1989) A specific and sensitive assay for the Lyme disease spirochete *Borrelia burgdorferi* using the polymerase chain reaction. J Infect Dis 160:1018-1029

Röther E, Vogt A, Meyer-Düttingdorf E, Peter HH (1988) Wertigkeit der Serodiagnostik für Verlauf und Therapie der Borreliose. Immun Infekt 16:27-28

Russel H, Sampson JS, Schmid GP, Wilkinson HW, Plikaytis B (1984) Enzyme-linked immunosorbent assay and indirect immunofluorescence assay for Lyme disease. J Infect Dis 149:465-470

Schmidli J, Hunziker T, Moesli P, Schaad UB (1988) Cultivation of *Borrelia burgdorferi* from joint fluid three months after treatment of facial palsy due to Lyme borreliosis. J Infect Dis 158:905-906

Schwan TG, Burgdorfer W (1987) Antigenic changes of *Borrelia burgdorferi* as a result of in vitro cultivation. J Infect Dis 156:852-853

Schwan TG, Burgdorfer W, Garon CF (1988) Changes in infectivity and plasmid profile of the Lyme disease spirochete, *Borrelia burgdorferi*, as a result of cultivation. Infect Immun 56:1831-1836

Schwan TG, Simpson WJ, Schrumpf ME, Karstens RH (1989) Identification of *Borrelia burgdorferi* and *B. hermsii* using DNA hybridization probes. J Clin Microbiol 27:1734-1738

Simpson WJ, Garon CF, Schwan TG (1990a) *Borrelia burgdorferi* contains DNA sequences that are species specific and plasmid associated. Infect Immun 58:847-853

Simpson WJ, Schrumpf ME, Schwan TG (1990b) Reactivity of human Lyme borreliosis sera with a 39-kilodalton antigen specific to *Borrelia burgdorferi*. Clin Microbiol 28:1329-1337

Simpson WJ, Burgdorfer W, Schrumpf ME, Karstens RH, Schwan TG (1991) Antibody to a 39-kilodalton *Borrelia burgdorferi* antigen (P39) as a marker for infection in experimentally and naturally inoculated animals. J Clin Microbiol 29:236-243

Stanek G, Wewalka G, Groh V, Neumann R (1985) Isolation of spirochetes from the skin of patients with erythema chronicum migrans in Austria. Zentralbl Bakteriol Hyg (A) 260:88-90

Stanek G, Hirschl A, Kristoferitsch W (1986) IFA and ELISA in serodiagnosis of Lyme borreliosis. Mitt Österr Ges Trop Parasitol 8:1-6

Steere AC, Grodzicki RL, Kornblatt AN, Craft JE, Barbour AG, Burgdorfer W, Schmid GP, Johnson E, Malawista SE (1983) The spirochetal etiology of Lyme disease. N Engl J Med 308:733-740

Stiernstedt G, Sköldenberg B, Vandvik B, Hederstedt B, Gårde A, Kolmodin G, Jörbäck H, Svenungsson B (1984) Chronic meningitis and Lyme disease in Sweden. Yale J Biol Med 57:491-497

Stiernstedt G, Granström M, Hederstedt B, Sköldenberg B (1985) Diagnosis of spirochetal meningitis by enzyme-linked immunosorbent assay and indirect immunofluorescent assay in serum and cerebrospinal fluid. J Clin Microbiol 21:819-825

Stiernstedt G, Eriksson G, Enfors W, Jörbeck H, Svenungson B, Sköldenberg B, Granström M (1986) Erythema chronicum migrans in Sweden: clinical manifestations and antibodies to Ixodes ricinus spirochete measured by indirect immunofluorescence and enzyme-linked immunosorbent assay. Scand J Infect Dis 18:217-224

Wallich R, Schaible UE, Simon MM, Heiberger A, Kramer MD (1989) Cloning and sequencing of the gene encoding the outer surface protein A (OspA) of a European *Borrelia burgdorferi* isolate. Nucleic Acids Res 17:8864

Wallich R, Moter SE, Simon MM, Ebnet K, Heiberger A, Kramer MD (1990) The *Borrelia burgdorferi* flagellum-associated 41-kilodalton antigen (flagellin): Molecular cloning, expression, and amplification of the gene. Infect Immun 58:1711-1719

Wassmann K, Borg-von Zepelin M, Zimmermann O, Stadler M, Eiffert H, Thomssen R (1989) Determination of immunoglobulin M antibody against *Borrelia burgdorferi* to differentiate between acute and past infections. Zentralbl Bacteriol (Suppl) 18:281-289

Weber K, Schierz G, Wilske B, Preac-Mursic V (1984a) Zur Klinik und Ätiologie der Acrodermatitis chronica atrophicans. Hautarzt 35:571-577

Weber K, Schierz G, Wilske B, Preac-Mursic V (1984b) European erythema migrans disease and related disorders. Yale J Biol Med 57:463-471

Weber K, Schierz G, Wilske B, Preac-Mursic V (1985) Das Lymphozytom – eine Borreliose? Z Hautkr 60:1585-1598

Weber K, Schierz G, Wilske B, Neubert U, Krampitz HE, Barbour AG, Burgdorfer W (1986) Reinfection in erythema migrans disease. Infection 14:32-35

Weber K, Preac-Mursic V, Neubert U, Thurmayr R, Herzer P, Wilske B, Schierz G, Marget W (1988a) Antibiotic therapy of early European Lyme borreliosis and acrodermatitis chronica atrophicans. Ann NY Acad Sci 539:324-345

Weber K, Preac-Mursic V, Reimers CD (1988b) Spirochetes isolated from two patients with morphea. Infection 16:25-26

Wilske B, Schierz G, Preac-Mursic V, Weber K, Pfister H-W, Einhäupl K (1984) Serological diagnosis of erythema migrans disease and related disorders. Infection 5:331-337

Wilske B, Münchhoff P, Schierz G, Preac-Mursic V, Roggendorf M, Zoulek G (1985) Zur Epidemiologie der *Borrelia-burgdorferi*-Infektion bei Waldarbeitern aus Oberbayern. Munch Med Wochenschr 127:171-172

Wilske B, Preac-Mursic V, Schierz G, von Busch K (1986a) Immunochemical and immunological analysis of European *Borrelia burgdorferi* strains. Zentralbl Bakteriol Hyg (A) 263:92-102

Wilske B, Schierz G, Preac-Mursic V, von Busch K, Kühbeck R, Pfister HW, Einhäupl K (1986b) Intrathecal production of specific antibodies against *Borrelia burgdorferi* in patients with lymphocytic meningoradiculitis (Bannwarth's syndrome). J Infect Dis 153:304-314

Wilske B, Schierz G, Preac-Mursic V, Bader L, Pfister HW, Einhäupl K, Herzer P (1987) Intrathekale Produktion spezifischer Antikörper gegen *Borrelia burgdorferi* bei Patienten mit Bannwarth-Syndrome. In: Poeck K, Hacke W, Schneider R (eds) Neuroimmunologie, spinale Krankheiten, Neuropsychologie, metabolische Enzephalopathien, neurologische Notfälle, interventionelle Neuroradiologie. Springer, Berlin Heidelberg New York, pp 443-447 (Verhandlungen der Deutschen Gesellschaft für Neurologie, vol 4)

Wilske B, Preac-Mursic V, Schierz G, Gueye W, Herzer P, Weber K (1988a) Immunochemische Analyse der Immunantwort bei Spätmanifestationen der Lyme Borreliose. (Immunochemical analysis of the immune response in late manifestations of Lyme borreliosis) Zentralbl Bakteriol Hyg (A) 267:549-558

Wilske B, Preac-Mursic V, Schierz G, Kühbeck R, Barbour AG, Kramer M (1988b) Antigenic variability of *Borrelia burgdorferi*. Ann NY Acad Sci 539:126-143

Wilske B, Fuchs R, Schierz G, Preac-Mursic V, Liegl G (1989a) Immunodominant proteins for the humoral immune response in Lyme borreliosis (poster). In: European Society of Clinical Microbiology and Infectious Diseases (ed) 4th European congress of clinical microbiology. Nizza, pp 911/PP43

Wilske B, Preac-Mursic V, Schierz G, Liegl G, Gueye W (1989b) Detection of IgM- and IgG antibodies to *Borrelia burgdorferi* using different strains as antigen. Zentralbl Bakteriol (Suppl) 18:299-309

Wilske B, Schierz G, Preac-Mursic V, Pfister HW, Weber K, von Busch K, Baruschke A (1989c) IgM- and IgG immune response to *Borrelia burgdorferi* in erythema migrans and neuroborreliosis. Zentralbl Bakteriol (Suppl) 18:290-298

Wilske B, Preac-Mursic V, Fuchs R, Bruckbauer H, Hofmann A, Zumstein G, Jauris S, Soutschek E, Motz M (1990) Immundominant proteins of *Borrelia burgdorferi*; implications for improving serodiagnosis of Lyme borreliosis. In: Neu HC (ed) New antibacterial strategies. Churchill Livingstone, London

Wilske B, Anderson JF, Baranton G, Barbour AG, Hoving-Hougen K, Johnson RC, Preac-Mursic V (1991) Workshop summary: taxonomy of Borrelia spp. Scand J Infect Dis Suppl 77:108-129

Wilske B, Barbour AG, Bergström S, Burman N, Restrepo BI, Rosa PA, Schwan T, Soutschek E, Wallich R (1992) Antigenic variation and strain heterogeneity in Borrelia spp. Res Microbiol 143:in press

Ziegelmeier R, Bieker R, Behrens F, Vermeer H (1983) Diagnostik akuter Infektionen: IgM-Antikörperbestimmung ohne Rheumafaktor-Interferenz. Laboratoriumsblätter 33:19-25

Zöller L, Burkard S, Schäfer H (1991) Validity of Western immunoblot band patterns in the serodiagnosis of Lyme borreliosis. J Clin Microbiol 29:174-182

Zumstein G, Fuchs R, Preac-Mursic V, Wilske B, Soutschek E (1990) Identification and characterisation by restriction endonucleases of the outer surface protein OspA of *Borrelia burgdorferi* (Poster). In: 4th international conference on Lyme Borreliosis, Stockholm

VI Therapy of Lyme Borreliosis

22 Antibiotic Susceptibility of *Borrelia burgdorferi* In Vitro and In Vivo

V. Preac-Mursic

22.1 Introduction

At present the treatment of Lyme borreliosis is still associated with several unresolved questions. One of the main problems is the establishment of the optimal antibiotic therapy. Recent studies applying in vitro techniques and trials in animals have helped to provide a more solid aproach to a rational antibiotic therapy in patients with Lyme borreliosis. The susceptibility of the causative agent to antimicrobial substances is an important parameter for the evaluation of an antibacterial therapy. Minimal inhibitory concentration (MIC) values which are higher than the tissue level of the antibiotic make the success of the therapy unlikely. To be effective, an antibiotic must achieve concentrations sufficiently high to inhibit or preferably eliminate the bacteria at the site of infection. Knowledge of the susceptibility pattern of *Borrelia burgdorferi* is, therefore, a prerequisite for the selection of an appropriate antibiotic therapy in Lyme borreliosis.

The degree of susceptibility of a microorganism is expressed quantitatively as the lowest concentration of a drug that inhibits the growth of that microorganism. Several strains are usually tested to obtain a correct result. A common measure of the susceptibility is the MIC. The MIC is the lowest concentration of a given antibiotic that results in complete or nearly complete inhibition of growth under specific test conditions (Barry 1976; Walter and Heilmeyer 1975). Accordingly, MIC90 refers to a concentration at which 90% of *B. burgdorferi* cells are inhibited.

22.2 In Vitro Susceptibility of *Borrelia burgdorferi*

We tested the susceptibility to 27 antimicrobials for 20 *B. burgdorferi* strains, isolated from the skin of patients with erythema migrans (*n*=5), acrodermatitis chronica atrophicans (*n*=5), and from cerebrospinal fluid (CSF) of patients with Bannwarth's syndrome (*n*=5). Another five isolates stemmed from Bavarian *Ixodes ricinus* ticks.

22.2.1 Minimal Inhibitory Concentration

The MIC was determined using the broth dilution method. Tubes with modified Kelly medium according to Preac-Mursic (MKP) (Preac-Mursic et al. 1986) (see Chap. 21, Table 2), containing antibiotics in concentrations of 0.015-64 µg/ml were inoculated with actively growing cultures to a final concentration of 10^5 cells/ml. Control tubes with MKP medium with 10^5 cells/ml without antibiotics were included in all tests. The cultures were examined for *B. burgdorferi* by dark-field microscopy for 4 days. Subcultures of inhibited organisms were made and observed for growth for a further 4 days. The MIC was determined as the lowest concentration of antibiotic at which viable borreliae could not be observed in dark-field microscopy or at which 90% of the borreliae were inhibited (Preac-Mursic et al. 1986).

In vitro studies testing the susceptibility of *B. burgdorferi* with antibiotics such as penicillin G, tetracycline and erythromycin, ceftriaxone, and several others have been perfomed by Johnson et al. (1984), Berger et al. (1985), Johnson et al. (1987, 1990), Preac-Mursic et al. (1986, 1987, 1989b), and Luft et al. (1988). The most effective antibiotic in all studies was erythromycin (MIC90 0.03 – 0.06 μg/ml). The MIC range of penicillin G and tetracycline has been found to be between 0.5 and 8 μg/ml and 0.14 and 6 μg/ml, respectively. The minimum bactericidal concentration (MBC) values determined by Johnson et al. (1987) were 6.5 μg/ml for penicillin G, 0.8 μg/ml for tetracycline, and 0.04 μg/ml for ceftriaxone.

As shown in Table 1, the *macrolides* azithromycin, clarithromycin, erythromycin and roxythromycin and the *cephalosporins* ceftriaxone and cefotaxime exert the greatest in vitro effect of the antimicrobials tested with a MIC90 of 0.015 and 0.06 μg/ml and 0.06 and 0.12 μg/ml, respectively. Ceftazidime and the oral cephalosporin cefuroxime also show a good activity. *Imipenem* and the *ureidopenicillins* amoxicillin and mezlocillin are also quite effective against *B. burgdorferi*.

The *tetracyclines* including tetracycline HCL, doxycycline, and minocycline showed a MIC90 of 0.5 and 1 μg/ml. The MIC90 value of 4 μg/ml for penicillin G and

Table 1. Activity in vitro of 27 antibiotics against Borrelia burgdorferi

	MIC (Ug/ml)	
Antibiotic	MIC range	MIC90
Penicillin G	0.50-8.0	4
Oxacillin	0.25-2.0	1
Amoxicillin	0.25-1.0	0.50
Mezlocillin	0.25-1.0	0.50
Cefaclor	0.50-16	6.0
Cefadroxil	1-64	32.0
Cefmenoxine	0.03-0.25	0.12
Cefotaxime	0.06-0.25	0.12
Ceftazidime	0.12-0.25	0.12
Ceftriaxone	0.03-0.25	0.06
Cefuroxime	0.12-0.50	0.25
Azithromycin	0.015-0.03	0.015
Clarithromycin	0.015-0.06	0.015
Erythromycin	0.03-0.12	0.06
Roxythromycin	0.015-0.12	0.03
Tetracycline-HCL	0.25-2.0	1.0
Doxycycline	0.12-1.0	0.50
Minocycline	0.12-1.0	0.50
Ciprofloxacin	1-4	2
Ofloxacin	2-8	4
Lincomycin	0.25-1	0.50
Imipenem	0.06-1	0.25
Chloramphenicol	1-4	2.0
Co-trimoxazole		> 1024
Vancomycin	0.5-8	4
Aminoglycosides		32-128
Antituberculostatic agents		2-32

MIC, minimal inhibitory concentration; from Preac-Mursic et al. (1987 and 1989b)

its MIC range of 0.5 and 8 µg/ml suggest that *B. burgdorferi* is more resistant to penicillin in vitro than was previously believed. *Vancomycin* and the two *quinolones* ciprofloxacin and ofloxacin had a similar activity. The *antituberculostatic* agents streptomycin, rifampicin, isoniazid and ethambutol, the *aminoglycosides*, amikacin, tobramycin, gentamicin, and *co-trimoxazole* were not effective against *B. burgdorferi.*

Experiments evaluating the in vitro susceptibillity of *B. burgdorferi* against various antimicrobials imply that *B. burgdorferi* strains isolated up to now from ticks and patients have very uniform sensitivity patterns against various antibiotics. The results obtained in vitro and also in vivo by Johnson et al. (1984), Berger et al. (1985), Johnson et al. (1987, 1990), Preac-Mursic et al. (1987, 1989b), and Luft et al. (1988) suggest that the sensitivity pattern of *B. burgdorferi* strains is quite stable and need not be checked routinely as is the case with other bacteria.

22.2.2 Killing Kinetics

The killing rate of a given antibiotic for *B. burgdorferi* is dependent on concentration and reaction time. We could show that the kinetics of killing the borreliae differs from antibiotic to antibiotic (see Fig. 1). The time required to kill 50% of *B. burgdorferi* with 1.0 µg of antibiotic was 48 – 72 h for penicillin G and 6 and 24 h for cefotaxime and ceftriaxone. With 1.0 µg cefotaxime and ceftriaxone, a 100% killing effect was achieved after 72 h. The same effect was obtained with 10 µg penicillin G after 96 h.

The outcome of our studies on the killing kinetics and the killing effect correspond to the results reported by Luft et al. (1988). The data of these studies were generally in agreement with previous findings about the sensitivity of *B. burgdorferi* against several antibiotics.

The long generation time of *B. burgdorferi* (7-20 h) and the faculty of antibiotically treated strains to regenerate may be the reason that a prolonged exposure to an antibiotic is necessary for effective killing. It may be of interest that the killing kinetics of *B. burgdorferi* in vitro are remarkably similar to those for *Treponema pallidum* (Eagle 1948; Eagle and Musselman 1948; Nell 1954).

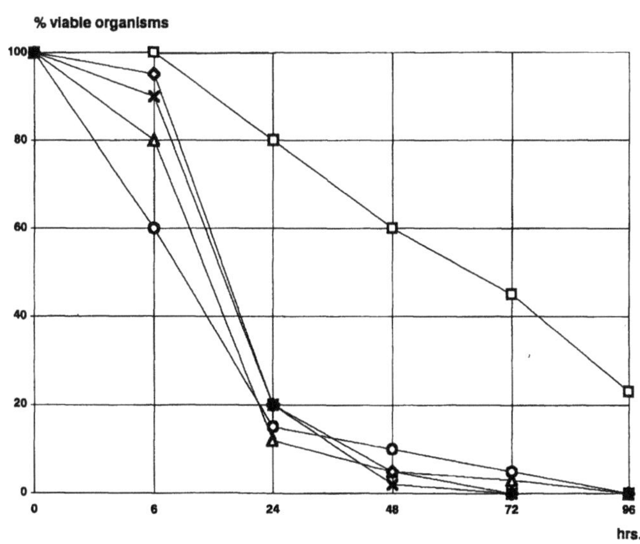

Fig.1. Killing kinetics of *Borrelia burgdorferi* after administration of 1 µg/ml antibiotic at various incubation times in vitro (*square*, penicillin G; *triangle*, erythromycin; *circle* tetracycline; *rhombus*, ceftriaxone; *cross*, cefotaxime)

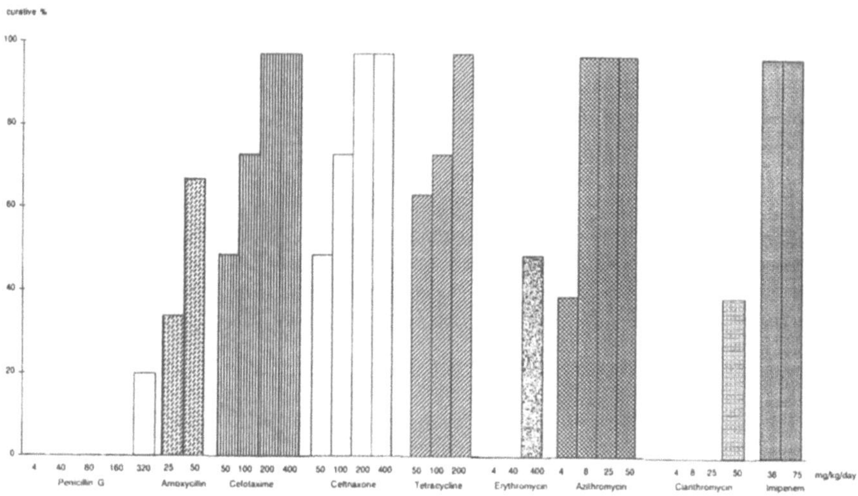

Fig. 2. Therapeutic efficacy of nine antibiotics against *Borrelia burgdorferi* in experimentally infected gerbils

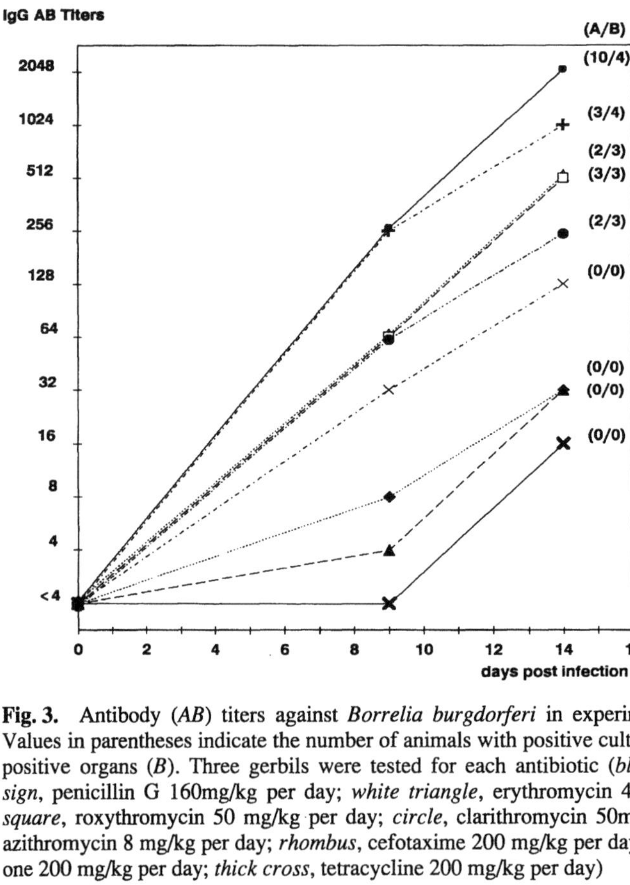

Fig. 3. Antibody (*AB*) titers against *Borrelia burgdorferi* in experimentally infected gerbils. Values in parentheses indicate the number of animals with positive culture (*A*) and the number of positive organs (*B*). Three gerbils were tested for each antibiotic (*black square*, control; *plus sign*, penicillin G 160mg/kg per day; *white triangle*, erythromycin 400 mg/kg per day; *white square*, roxythromycin 50 mg/kg per day; *circle*, clarithromycin 50mg/kg per day; *thin cross*, azithromycin 8 mg/kg per day; *rhombus*, cefotaxime 200 mg/kg per day; *black triangle*, ceftriaxone 200 mg/kg per day; *thick cross*, tetracycline 200 mg/kg per day)

22.3 In Vivo Efficacy – Animal Studies

Results on the in vivo activity of antibiotics against *B. burgdorferi* has so far been published in hamsters by the American group of Johnson et al. (1987) and in gerbil mice by our group (Preac-Mursic et al. 1987, 1989b).

Our experiments were carried out as follows. Gerbils weighing between 70 and 80 g were infected intraperitoneally with 10^7 cells/ml of *B. burgdorferi* isolated from a human skin biopsy. Antimicrobial treatment was initiated 5 days later. Each antibiotic was administered subcutaneously in a single daily dose for 1 week. One week after the final treatment the animals were killed and the liver, spleen, kidney, heart, and blood were cultured as described elsewhere (Preac-Mursic et al. 1987). The results (Fig. 2) show that the in vivo activity corresponded to the activity obtained in vitro, with the exception of the macrolides. However, azithromycin, unlike other macrolides, turned out to be quite effective in vivo. At a dosage of 4 mg/kg per day of azithromycin, 66% of the gerbils were cured and at a dosage of 8 mg/kg per day 100% of the gerbils lacked borreliae. Clarithromycin was markedly less effective, roxythromycin (not shown in Fig. 2) and erythromycin were not effective in curing the experimentally infected gerbils in the tested doses. Tetracycline 50 mg/kg per day cured 66% of the gerbils. The curative rates of cefotaxime and ceftriaxone in the same concentrations were 50%. Using penicillin G, only one gerbil became *B. burgdorferi*-negative in the highest dosage tested. Johnson et al. (1987) tested penicillin G, tetracycline, erythromycin, and ceftriaxone and obtained similar results using hamsters.

Figure 3 depicts the antibody response of the gerbils tested against several antibiotics as partly outlined by Preac-Mursic et al. (1987). The antibody titers show that the gerbils had a good immune response to the infection and their levels seemed to be a reasonable indicator of therapeutic efficacy; high titers correlated mostly with the presence of *B. burgdorferi* in more organs. It should be mentioned that the results obtained

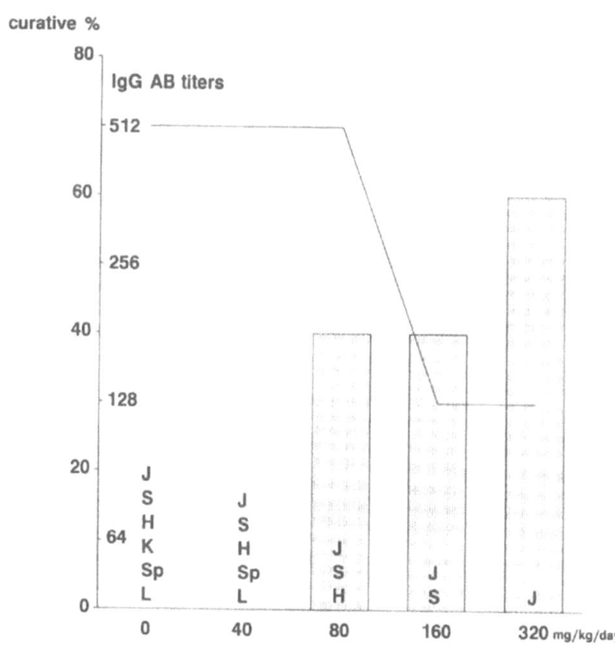

Fig. 4. Therapeutic efficacy of penicillin G against *Borrelia burgdorferi* in experimentally infected gerbils. Letters indicate organs in which *B. burgdorferi* was found (*L*, liver; *K*, kidney; *S*, skin; *Sp*, spleen; *H*, heart; *J*, leg joint). *Shaded areas* indicate curative percentage; *line* indicates level of immunoglobulin G (*IgG*) antibody (*AB*) titers

by Johnson et al. (1987) are based on a single daily subcutaneous injection of the antibiotics, and those obtained by Preac-Mursic et al. (1987) derive from a single subcutaneous injection, with the exception of penicillin G which was administrated once and twice daily. The twice daily dosage showed better results (Fig.4). At a dosage of 80 mg/kg per day, 40% of animals were *B. burgdorferi*-negative and, at 320 mg/kg per day, *B. burgdorferi* was cultivated from 38% of the animals and in fewer organs than with a single daily dose. Antibiotics with a long half-life such as ceftriaxone may be expected to be more favorable in these experiments. However, cefotaxime, which has a shorter half-life than ceftriaxone, behaved similarly in vivo. K. Hansen (personal communication) found that penicillin G 300 mg/kg per day administered three times daily cured almost all experimentally infected gerbils. It must be taken into account that the course of the infection may vary depending on the borrelial strain tested and the susceptibility of the individual animal.

22.4 Clinical Experiences

As with other bacterial infections, several factors influence the outcome of antibiotic therapy in patients with Lyme borreliosis. Some of these factors are the susceptibility of *B. burgdorferi* against a given antibiotic, the penetration of this antibiotic into the infected tissue, and the amount of the antibiotic administered.

B. burgdorferi can persist in humans (Weber 1981; Asbrink 1985; Preac-Mursic et al. 1985; Herzer et al. 1986; Weber et al. 1988) for a long period of time. In human CSF, *B. burgdorferi* may be present for weeks (Preac-Mursic et al. 1984) or months, even after treatment with antibiotics (Preac-Mursic et al. 1989a). Moreover, *B. burgdorferi* has been isolated from animal tissue and organs 1 year and more after infection (Johnson et al. 1988; Preac-Mursic et al. 1990).

A special problem is presented by the blood-brain barrier which is a serious obstacle for the efficacy of many antibiotics. Drug levels in the CSF depend not only on drug dosage, but also on the kind of drug, on the grade of inflammation of the meninges, and on the metabolism and kidney function of the patient (Simon and Stille 1989). It has been shown that levels of antibiotics in CSF vary greatly (Richards et al. 1981; Schoth and Wolters 1987). Finally, in selecting an antibiotic, it is important to consider its toxicity and any clinical side effects.

Penicillins are one of the least toxic group of antibiotics, although a Jarisch-Herxheimer reaction has been observed by Steere et al. (1983a) and Weber (1984) after penicillin therapy in patients with erythema migrans. With regard to their activity against *B. burgdorferi*, there are differences among various penicillin preparations. For oral therapy, the use of amoxicillin is recommended. Its good clinical effect has been reported by Dournon and Assous (1988). Amoxicillin appears to be one of the most effective oral penicillin derivates in vitro (see Table 1) and might, therefore, have advantages over penicillin V. Studies in humans (Berger 1988; Weber et al. 1988) have not, however, been able to prove the superiority of amoxicillin. The often good therapeutic effect of penicillin V and penicillin G, as reported by Sköldenberg et al. (1983), Steere et al. (1980, 1983 a,b), Weber (1984), Asbrink and Olsson (1985), Kristoferitsch et al. (1986) and Weber et al. (1986, 1988), could not be verified in all patients. Oral penicillin V does not result in a sufficiently high CSF concentration (Löwhagen et al. 1983). In a recent study, it has been shown that the CSF concentration of intravenously administered penicillin G 4x3g/day did not reach the MIC90 for *B. burgdorferi* in any of the patients (Pfister et al. 1989). These results suggest that penicillin G therapy may not be able to eliminate *B. burgdorferi* from the CSF.

Analogously, persistence of *Treponema pallidum* after penicillin G therapy has been reported in neurosyphilis (Tramont 1976; Greene et al. 1980). The investigation of the molecular interaction of penicillin G with *B. burgdorferi* and other anaerobic bacteria is still in its early stages (Rosenblatt 1984).

Cephalosporins, like penicillins, inhibit cell wall synthesis and are bactericidal only in the growth phase of the bacteria. Within this group there are differences in tissue penetration, betalactamase stability and affinity to the penicillin-binding protein. The highest antiborrelial activity can be seen in the cefotaxime group. The most effective substances in this group are cefotaxime and ceftriaxone. They show a relatively good tissue penetration (Hell 1989) and penetrate into the central nervous system even through noninflamed meninges (Mullaney and Jahn 1983; Beigue et al. 1984; Trang et al. 1985) The concentration of these two cephalosporins in CSF have been reported to exceed the MIC90 level for *B. burgdorferi* in all patients if administered in the usual doses (Pfister et al. 1991). Dattwyler et al. (1987), Diringer et al. (1987), and Pal et al. (1988) have provided some evidence that ceftriaxone and cefotaxime are effective in patients with meningoencephalitis and late Lyme borreliosis who did not respond to penicillin G therapy. Weber et al. (1990) have not found a significant difference between oral penicillin and ceftriaxone in the treatment of early Lyme borreliosis.

It is also interesting that ceftriaxone maintains a long-term therapeutic level in serum and CSF. This factor can be important in killing the borreliae which have a long generation time of about 7–20 h, because slowly multiplying bacteria may regenerate when the antimicrobial concentration decreases to subinhibitory levels. Cefotaxime behaves similarly to ceftriaxone, but must be administered two, or preferably three times daily compared with the single daily injection needed with ceftriaxone. It is also of importance that these two antibiotics can penetrate into the synovial fluid of an inflamed joint and attain a concentration of 50% – 100% of the serum level (Morgan et al. 1985). Gründer et al. (1980) have found cefotaxime to reach a sufficient concentration in skin to assure a good therapeutic effect in microbial infections of the skin. With their relatively high tissue concentration, good penetration into the CSF, betalactamase stability, a marked antiborrelial efficacy in vitro and in animals, and few side effects, ceftriaxone and cefotaxime have now an important place in the therapy of Lyme borreliosis. For other cephalosporins such as ceftazidime, no clinical data exist. Cefuroxime, a broad-spectrum antibiotic, is not recommended as first-line treatment (Burrascano 1989).

Macrolide antibiotics display a very high acitivity against *B. burgdorferi* in vitro, but the effect of these antimicrobials in vivo vary considerably. Azithromycin, which has a high in vitro activity, is the most effective agent in vivo (Preac-Mursic et al. 1989b). This new macrolide spreads rapidly into human tissues, has a long tissue half-life, and is slowly released after treatment is stopped (Girard et al. 1987). After administration of azithromycin 500 mg per day for 5 days, serum concentrations are consistently below 1 µg/ml, but concentrations in tissues are substantially higher (Shepard et al. 1987). Azithromycin can be given orally once daily, and could turn out to be useful in the treatment of Lyme borreliosis. In a prospective study Strle et al. (1990) compared penicillin V with doxycycline and azithromycin and observed the best therapeutic effect with azithromycin.

The tetracyclines exhibit a considerable efficacy against *B. burgdorferi* in vitro as well as in animal studies. They are apparently more effective than penicillin G according to the results of these studies. Doxycycline and minocycline appear to be slightly more active than tetracycline HCL. The curative dose 50% (CD50) of tetracycline in hamsters is similar to that of ceftriaxone (287 mg/kg and 240 mg/kg) as reported by Johnson et al. (1987). Relatively favorable pharmacokinetics, a good tissue and CSF penetration amounting to 25% – 30%, and a low rate of side effects makes a long-term therapy with doxycycline possible (Datewall et al. 1988). However, failure of tetracycline treatment in several patients with Lyme borreliosis has been reported by Dattwyler et al. (1987).

Thianamycin (imipenem), a new carbopenem antibiotic the pharmacokinetic properties of which very closely resemble those of the penicillins, showed very good antiborrelial activity in vitro and in animals (Preac-Mursic et al. 1987). Several studies have evaluated high plasma and tissue levels of imipenem in healthy volunteers. Limited data have been published on the use of imipenem to treat patients with meningitis or other central nervous system infections. Mondai et al. (1985) reported the penetration of imipenem into the CSF in 12 patients with bacterial meningitis. These results show that the penetrability of imipenem is good when the meninges are inflamed. No clinical data are available concerning the use of thianamycin in the therapy of Lyme borreliosis.

The other antibiotics tested were less suited for the treatment of Lyme borreliosis in accordance with the results of in vitro and in vivo studies, but possibly one of these antibiotics may be effective in combination with others.

22.5 Conclusion

The in vitro and in vivo studies have shown that several groups of antibiotics can be regarded as effective in the therapy of Lyme borreliosis, i.e., cephalosporins, macrolides, penicillins, and tetracyclines. In vitro and in vivo tests gave generally corresponding results, but some discrepancies have been observed, for instance with erythromycin. In strains isolated either from patients or ticks, no significant differences in their response to antibiotics could be ascertained. Furthermore, there are no indications so far that changes in sensitivity will occur under the therapeutic action of antibiotics or any other conditions, but this question will have to be studied in greater detail in the future.

Hence it follows that neither the necessity nor the possibility exists to ensure the optimal therapy in the individual case by aiming for sensitivity testing of the borrelia strain involved. However, the finding that prolonged exposure (72-96 h) is necessary to effectively kill *B. burgdorferi* may have important therapeutic implications. Detailed recommendations for therapeutic measures in Lyme borreliosis are given elsewhere in this book (see Chaps. 23-28).

The drugs in use for treatment of Lyme borreliosis are also effective in sensitivity tests, but there is at least one rather vexing exception: penicillin, which has been shown to be only moderately sensitive according to most in vitro investigations and which does not work at all efficiently in experimentally infected hamsters or gerbils, is considered as clinically quite successful according to a variety of therapeutic studies performed since 1949. The issue is complicated by the fact that the success of an antibiotic therapy will generally be judged solely by the disappearance of clinical symptoms, which is not equivalent to bacteria elimination, since the isolation of borreliae has been achieved after clinical recovery of the patient in several cases (Preac-Mursic et al. 1989a).

Moreover, a variety of factors including stage of disease, virulence and long generation time of *B. burgdorferi*, dormant bacteria, and size of inoculum may influence the efficacy of antibiotic therapy. There may even be technical problems which impairing the results. For instance, the media used for sensitivity testing need the addition of serum and are difficult to standardize. Futhermore, the handling of experimental infections in animals is not without difficulties. The real value of any animal model used depends less on the excellence of the model itself than on the ability of the investigator to interpret the results in full awareness of not only the merits, but also the limitations of the experimental system. Both the animal model and determination of the MICs suffer from a number of shortcomings. However, if investigators are aware of these deficiencies and take them into account when interpreting the results, the predictive value can be significantly enhanced. These and other questions regarding the killing kinetics of various antibiotics are still open, and require further studies in this field.

References

Asbrink E (1985) Erythema chronicum migrans Afzelius and acrodermatitis chronica atrophicans. Early and late manifestions of *Ixodes ricinus*-borne *Borrelia* spirochetosis. Acta Derm Venereol (Stockh) (Suppl) 118:1-63

Asbrink E, Olsson I (1985) Clincal manifestations of erythema chronicum migrans Afzelius in 161 patients. Acta Derm Venereol (Stockh) 65:43-52

Barry AL (1976) The antimicrobic susceptibility test priniciples and practices. Lea and Febiger, Philadelphia

Beigue P, Florett D, Mallet E, Raynand EJ, Safran C, Sarlangues J, Teyssier G (1984) Pharmacokinetics and clinical evaluation of cefotaxime in children suffering with purulent menigitis. J Antimicrob Chemother 14:161-165

Berger BW (1988) Treatment of erythema chronicum migrans of Lyme disease. Ann NY Acad Sci 539: 346-351

Berger BW, Kaplan MH, Rothenberg IR, Barbour AG (1985) Isolation and characterization of the Lyme disease spirochete from the skin of patients with erythema chronicum migrans. J Am Acad Dermatol 13:444-449

Burrascano JJ (1989) Late-stage Lyme disease: treatment options and guidelines. Internal Medicine 10:103-107

Datewall L, Alestig K, Hanner P, Norkrans G, Hagberg L (1988) The use of doxycycline in nervous system *Borrelia burgdorferi* infection. Scand J Jnfect Dis (Suppl) 53:74-79

Dattwyler RJ, Halperin JJ (1987) Failure of tetracycline therapy in early Lyme disease. Arthritis Rheum 30:448-450

Dattwyler RJ, Halperin JJ, Pass H, Luft BJ (1987) Ceftriaxone as effective therapy in refractory Lyme disease. J Infect Dis 155:1322-1325

Diringer MN, Halperin JJ, Dattwyler RJ (1987) Lyme meningoencephalitis – report of a severe, penicillin-resistant case. Arthritis Rheum 30:705-708

Dournon E, Assous M (1988) Amoxicillin for Lyme disease. International congress for infectious disease, Rio de Janeiro, abstract no 645

Eagle H (1948) Speculations as to the therapeutic significance of the penicillin blood level. Ann Intern Med 28:260-278

Eagle H, Musselman AD (1948) The rate of bactericidal action of penicillin in vitro as a function of its concentration, and its paradoxically reduced activity at high concentrations against certain organisms. J Exp Med 88:99-131

Girard AE, Griard D, English AR, Gootz TD, Cimochowski CR, Faiella JA, Haskell SL, Retsema JA (1987) Pharmacokinetic and in vivo studies with azithromycin (CP-62,993), a new macrolide with an extended halflife and excellent tissue distribution. Antimicrob Agents Chemother 31:1948-1954

Greene BM, Miller NR, Bunum TE (1980) Failure of penicillin G in treatment of neurosyphilis. Arch Intern Med 140:1117-1118

Gründer K, Malerczyk V, Petzold D, Seeger K, Seidel G (1980) Concentrations of cefotaxime in skin. Infection 8:280-283

Hell K (1989) Worldwide clinical experience with ceftriaxone. Chemotherapy 35:228-235

Herzer P, Wilske B, Preac-Mursic V, Schierz G, Schattenkirchner M, Zöllner N (1986) Lyme arthritis: clinical features, serological, and radiographic findings of cases in Germany. Klin Wochenschr 64:206-215

Johnson RC, Kodner C, Russel M (1987) In vitro and in vivo susceptibility of the Lyme disease spirochete, *Borrelia burgdorferi* to four antimicrobial agents. Antimicrob Agents Chemother 31:164-167

Johnson RC, Kodner C, Russel M, Duray PH (1988) Experimental infection of the hamster with *Borrelia burgdorferi* in Lyme disease and related disorders. Ann NY Acad Sci 539:258-263

Johnson RC, Kodner C, Jurkovich PJ, Collins JJ (1990) Comparative in vitro and in vivo susceptibilities of the Lyme disease spirochete *Borrelia burgdorferi* to cefuroxime and other antimicrobials. Antimicrob Agents Chemother 34:2133-2136

Johnson SE, Klein GC, Schmidt GP, Feeley JC (1984) Susceptibility of Lyme disease spirochete to seven antimicrobial agents. Yale J Biol Med 57:99-103

Kristoferitsch W, Baumhackl U, Sluga E, Stanek G, Zeiler K (1986) High-dose penicillin in meningopolyneuritis Garin-Bujadoux-Bannwarth. Zentral Bakteriol Hyg (A) 263:357-364

Löwhagen GB, Brarson JE, Kaijser B (1984) Penicillin concentrations in cerebrospinal fluid and serum after intramuscular, intravenous and oral administration to syphilitic patients. Acta Derm Venereol (Stockh) 63:53-57

Luft BJ, Volkmann DJ, Halperin JJ, Dattwyler RJ (1988) New chemotherapeutic approaches in the treatment of Lyme borreliosis. Ann NY Acad Sci 539:352-361

Mondal J, Vittecoq D, Decazes JM, Meulemans A (1985) Penetration of imipenem and cilastatin into cerebrospinal fluid of patients with bacterial meningitis. J Antimicrob Chemother 15:751-755

Morgan JR, Paull A, Sullivan MO, Williams BD (1985) The penetration of ceftriaxone into synovial fluid of the inflamed joint. J Antimicrob Chemother 16:367-371

Mullaney DT, Jahn JE (1983) Cefotaxime therapy. Evaluation of its effect on bacterial meningitis, CSF drug levels and bactericidal activity, Arch Intern Med 143:1705-1708

Nell EE (1954) Comparative sensitivity of treponemes of syphilis, yaws, and bejel to penicillin in vitro, with observations on factors affecting its treponemicidal action. Am J Syph Gonorrhea Vener Dis 38:92-106

Pal GS, Baker JT, Wright DJM (1988) Penicillin-resistant borrelia encephalitis responding to cefotaxime. Lancet I:50-51

Pfister HW, Preac-Mursic V, Wilske B, Einhäupel KM (1989) Cefotaxime versus penicillin G for acute neurological manifestations in Lyme borreliosis: a prospective randomized study. Arch Neurol 46:1190-1194

Pfister HW, Preac-Mursic V, Wilske B, Sörgel F, Schielke E, Einhäupl KM (1991) Randomized comparison of ceftriaxone versus cefotaxime in Lyme neuroborreliosis. J Infect Dis 163:311-318

Preac-Mursic V, Wilske B, Schierz G, Pfister HW, Einhäupl KM (1984) Repeated isolation of spirochetes from the cerebrospinal fluid of a patient with meningoradiculitis Bannwarth. Eur J Clin Microbiol 3:564-565

Preac-Mursic V, Wilske B, Herzer P, Schierz G (1985) Acrodermatitis chronica atrophicans – eine Borreliose. Hautarzt 36:691-693

Preac-Mursic V, Wilske B, Schierz G (1986) European *Borrelia burgdorferi* isolated from humans and ticks. Culture conditions and antibiotic susceptibility. Zentral Bakteriol Hyg (A) 263:112-118

Preac-Mursic V, Wilske B, Schierz G, Holmburger M, Süß E (1987) In vitro and in vivo susceptibility of *Borrelia burgdorferi*. Eur J Clin Microbiol 6:424-426

Preac-Mursic V, Weber K, Pfister HW, Wilske B, Gross B, Baumann A, Prokop J (1989a) Survival of *Borrelia burgdorferi* in antibiotically treated patients with Lyme borreliosis. Infection 17:355-359

Preac-Mursic V, Wilske B, Schierz G, Süß E (1989b) Comparative antimicrobial activities of new macrolides against *Borrelia burgdorferi*. Eur J Clin Microbiol 8:651-653

Preac-Mursic V, Patsouris E, Wilske B, Groß B, Reinhardt S (1990) Persistence of Borrelia burgdorferi and histopathological alterations in experimentally infected gerbils. Infection 18:332-341

Richard ML, Prince RA, Kenaly KA, Johnson JA, Le Frock JL (1981) Antimicrobial penetration into cerebrospinal fluid. Drug Intell Clin Pharm 15:341-368

Rosenblatt JE (1984) Antimicrobic susceptibility testing of anaerobic bacteria. Rev Infect Dis 6 (Suppl) 1:242-248

Schoth PEM,Wolters ECH (1987) Penicillin concentrations in serum and CSF during high-dose intravenous treatment for neurosyphiis. Neurology 37:1214-1216

Shepard RM, Weidler DJ, Garg DC, Madsen PO, Cox CE, Chan KH, Bluestone CD (1987) Human pharmacokinetics of azithromycin (CP-62,993/XZ-450), a new macrolide with an extended half-life and high tissue penetration. Annual meeting of the interscience conference of antimicrobial agents and chemotherapy, New York, abstract 239

Simon C, Stille W (1989) Antibiotika-Therapie in Klinik und Praxis. Schattauer, Stuttgart

Sköldenberg B, Stiernstedt G, Garde A, Kolmodin G, Carlström A, Nord CE (1983) Chronic meningitis caused by a penicillin-sensitive microorganism? Lancet II:75-78

Steere AC, Malawista SE, Newman JH, Spieler PN, Bartenhagen NH (1980) Antibiotic therapy in Lyme disease. Ann Intern Med 93:1-8

Steere AC, Hutchinson GJ, Rahn DW, Sinal LH, Craft JE, De Sanna ET, Malawista SE (1983a) Treatment of the early manifestations of Lyme diesease. Ann Intern Med 99:22-26

Steere AC, Pachner AR, Malawista SE (1983b) Neurologic abnormalities of Lyme disease: successful treatment with high-dose intravenous penicillin. Ann Intern Med 99:767-772

Strle F, Ruzic E, Cimperman J (in press) Erythema migrans: comparison of treatment with azithromycin, doxycycline and phenoxymethylpenicillin. J Antimicrob Chemother

Tramont EC (1976) Persistence of *Treponema pallidum* following penicillin G therapy. JAMA 236:2206-2207

Trang JM, Jacobs RF, Kearns GL, Brown AL, Wells TG, Unterwood FL, Kluza RB (1985) Cefotaxime and desacetylcefotaxime pharmacokinetics in infants and children with meningitis. Antimicrob Agents Chemother 28:791-795

Walter AM, Heilmeyer L (1975) Antibiotika Fibel. In: Otton H, Plempel M, Siegenthaler W (eds) Antibiotika und Chemotherapeutika Therapie mikrobieller Infektionen.

Weber K (1981) Treatment of Lyme disease (letter). Ann Intern Med 94:137

Weber K (1984) Jarisch-Herxheimer-Reaktion bei Erythema-migrans-Krankheit. Hautarzt 35:588-590

Weber K, Neubert U, Thurmayr R (1986) Antibiotic therapy in early erythema migrans disease and related disorders. Zentral Bakteriol Hyg (A) 263:377-388

Weber K, Preac-Mursic V, Neubert U, Thurmayr R, Herzer P, Wilske B, Schierz G, Marget W (1988) Antibiotic therapy of early European Lyme borreliosis and acrodermatitis chronica atrophicans. Ann NY Acad Sci 539:324-345

Weber K, Preac-Mursic V, Wilske B, Thurmayr R, Neubert U, Scherwitz C (1990) A randomized trial of ceftriaxone versus oral penicillin for the treatment of early European Lyme borreliosis. Infection 18:91-96

23 Therapy of Cutaneous Manifestations

K. WEBER

23.1 Introduction

Cutaneous manifestations of Lyme borreliosis have been treated with antibiotics since the late 1940s. Therapy has first been carried out with penicillin and to a lesser extent with tetracycline, erythromycin, and chloramphenicol (see Chap. 1). The initial impression gained from these studies had been that the aforementioned antibiotics are quite effective in ridding patients of their manifestations (Weber 1973). Recent developments have made it necessary to look at the treatment of the cutaneous manifestations from a new standpoint.

23.2 Therapy of Erythema Migrans

23.2.1 Introduction

In treating patients with erythema migrans one should be aware that patients with this manifestation who remain untreated are prone to develop subsequent, later manifestations. Effective therapy should prevent this development and should be carried out even if the erythema migrans has disappeared spontaneously. The latter point reminds us of the situation in syphilis, where disappearance of the primary lesion does not in any way mean cure of the disease.

23.2.2 Pretreatment Evaluation

23.2.2.1 Erythema Migrans Characteristics

Although duration and size of the erythema migrans may have an influence on outcome after therapy in individual patients (Weber et al. 1990), no significant correlation regarding these criteria has been elucidated among groups of patients so far (Steere et al. 1983; Weber et al. 1988a, 1990).

23.2.2.2 Associated Signs and Symptoms

Another important point to consider is that associated symptoms and signs have been found to affect outcome after therapy in the United States (Steere et al. 1983) and in Europe (Weber et al. 1986, 1990). Accordingly, patients with more severe initial disease are at greater risk for developing subsequent manifestations after therapy. Some patients with erythema migrans even have associated complications such as meningitis, meningoradiculoneuritis (Höllström 1951; Weber 1974; Weber and Neubert 1986; Kuiper et al. 1990; see also Chap. 11) or more severe carditis (see Chap. 10). It is important to single out these patients because they should be treated primarily for their complication. For instance, it would not be appropriate to treat an erythema migrans patient with oral penicillin or another antibiotic which does not penetrate sufficiently well into the cerebrospinal fluid when he or she has an established meningitis.

The vast majority of European patients with erythema migrans do not suffer from the aforementioned complications. However, varying numbers of them have associated symtoms, for instance 47% of 73 patients seen in a recent German study (Weber et al. 1990) (see Chap.8). This investigation showed that the subgroups of patients with two and more associated symptoms prior to therapy differed significantly in the development of subsequent manifestations when treated with oral penicillin and ceftriaxone (Weber et al. 1990).

A somewhat different approach to pretreatment evaluation was adopted in a study of 215 patients with erythema migrans in New York carried out by Berger (1984, 1988). Berger distinguishes between a minor and a major form of the illness, but although this qualitative distinction may be appropriate for American patients with erythema migrans, who tend to be sicker than their European counterparts (Weber 1989), this kind of evaluation may be handicapped by subjective interpretation. At present, we prefer the quantitative pretreatment evaluation of associated symptoms.

23.2.2.3 Antibody Titers

Although the determination of immunoglobulin M (IgM) and IgG antibody titers against *Borrelia burgdorferi* is not absolutely necessary in patients with typical erythema migrans, post-treatment evaluation requires examination of the IgM and IgG antibody titer prior to treatment in order to obtain a comparison.

23.2.2.4 Identification of Borrelia burgdorferi

The identification of *B. burgdorferi*, preferably by culture, can only be performed in special laboratories and is not mandatory in typical cases of erythema migrans. However, it may be necessary in atypical cases of erythema migrans to ascertain the diagnosis, and it is desirable in therapeutic trials in which the eradication of *B. burgdorferi* is being investigated.

23.2.3 Post-treatment Criteria for Response

23.2.3.1 Resolution of Erythema Migrans

Studies have shown that the resolution of the erythema migrans is faster after therapy with tetracycline and oral penicillin than with erythromycin (Steere et al. 1983), with tetracycline, erythromycin, and parenteral penicillin than with oral penicillin (Weber et al. 1986), and, in a subgroup of patients with two and more symptoms prior to therapy, with Augmentin (amoxicillin and clavulanic acid) than with oral penicillin ($p<0.05$; Weber and Thurmayr, unpublished). No differences have been found with regard to resolution of the erythema migrans after treatment with oral penicillin, tetracycline, and Augmentin (Weber et al. 1988a; Weber and Thurmayr 1989), with oral penicillin compared with ceftriaxone (Weber et al. 1990), and, in a subgroup analysis, with tetracycline or ceftriaxone compared with oral penicillin (Weber and Thurmayr, unpublished). Obviously, the results of these studies do not follow the same line and are partly contradictory.

The median resolution time of the erythema migrans in our aforementioned German studies was between 1 and 2 weeks, with a range of 0.5 to approximately 20 weeks, and in rare cases even longer. The values found by other European authors have been closer to 1 week after treatment with penicillin (Hollström 1958; Neumann et al. 1986), but in 7.7% of Hollström's 65 patients, resolution was achieved only after 3 weeks.

This criterion of response should thus not be overemphasized. Resolution of the erythema migrans is commonly measured in days, and the patients are asked to make their own observations as to when the erythema migrans has disappeared, because daily follow-ups are usually not feasible. In my experience, patients are often not reliable in making exact observations themselves. Some patients believe that the erythema migrans has disappeared, but examination reveals that their erythema migrans is still present as faint erythema. Conversely, in others, the erythematous ring disappears quickly, but the erythematous center stays on for weeks and even months. In conclusion, resolution of erythema migrans remains an important clinical outcome criterion, but overemphasis is to be avoided.

Erythema migrans may recur after therapy (Hollström 1958; Steere et al. 1980; Hovmark et al. 1990; Weber et al. 1990), a fact which represents failure of treatment.

23.2.3.2 Subsequent Manifestations

Associated signs and symptoms usually resolve within the first 3 weeks, mostly within the first week after initiation of therapy (Weber et al. 1986, 1988, 1990). The symptoms and signs which last longer than 3 weeks after initiation of therapy or occur anew after this time are those which are of major concern. These so-called subsequent or later manifestations (not to be confused with late manifestations such as acrodermatitis chronica atrophicans, ACA) are probably the most important clinical outcome criteria, and their prevention should be attempted. A summary of the findings of two recent studies (Weber et al. 1988a, 1990) is presented in Table 1. Subsequent manifestations have been found to occur in 19% – 27% of 194 German (Weber et al. 1986, 1988a, 1990) and 44% of 184 American patients (Steere et al. 1983). This represents a similar percentage when one considers the number of patients with associated symptoms prior to therapy in both centers. However, only 10% of the 72 American patients have

Table 1. Post-treatment subsequent manifestations in patients with erythema migrans ($n = 194$)[a]

Manifestations	*n*	%
Arthralgia	28	14
Headache	8	4
Sensory disturbances	7	3
Fever	7	3
Fatigue	6	3
Myalgia	6	3
Sleeplessness, irritable	4	2
Palpitations	4	2
Loss of memory	3	2
Difficulty concentrating	3	2
Paresis	3	2
Depression	2	1

[a]From Weber et al. (1988a, 1990)

developed subsequent manifestations in a recent randomized study comparing doxycycline and amoxicillin plus probenecid (Dattwyler et al. 1990).

Subsequent manifestations have been divided into the categories major and minor (Steere et al. 1983). Accordingly, major subsequent manifestations are established meningoencephalitis, more severe myocarditis (second or third degree atrioventricular nodal block or left ventricular dysfunction), or established arthritis lasting more than 2 weeks. Minor subsequent manifestations include facial palsy alone, supraventricular tachycardia, arthritis of less than 2 weeks duration, arthralgia, or headache (Steere et al. 1983). This distinction appears to be somewhat artificial, even though its practical value may be considerable. For instance, established meningoencephalitis includes abnormalities of the cerebrospinal fluid, but this often holds true for facial palsy, too. Facial palsy in patients with erythema migrans should, therefore, be considered as a major subsequent manifestation unless a lumbar puncture has yielded negative results. The distinction between arthritis of long and brief duration may not be clear-cut in all cases since short arthritic attacks can recur even after a considerable period of time due to the intermittent nature of the disease.

Nevertheless, Steere et al. (1983) have found a tendency for tetracycline to prevent "major late manifestations" better than oral penicillin or erythromycin (p=0.07) in a randomized therapeutic trial of their 108 American patients. This finding was based on only a few affected patients in each therapeutic group: three penicillin and four erythromycin recipients versus no tetracycline recipient. This underscores the difficulties that investigators have to deal with, even if they try to perform a study which meets the requirements of modern standards.

In Europe, subsequent manifestations occurring after antibiotic therapy usually belong to the minor type, as defined above (Weber et al. 1986, 1988a, 1990). This has also been the case with the recent American randomized trial (Dattwyler et al. 1990). In the following, if not otherwise specified, subsequent manifestations will be referred to only in general terms.

Table 1 shows that mainly constitutional, musculoskeletal, and neurological subsequent manifestations can be distinguished. Constitutional subsequent manifestations consist of slight to moderate headache, fever, fatigue, and sleeplessness. Arthralgia and myalgia represent musculoskeletal-type manifestations, while all other manifestations are of the neurological type, with the exception of palpitations which may indicate possible cardiac involvement. These subsequent manifestations have been found to last usually between 3 and 12 months, and only rarely for more than 2 years (Weber et al. 1986, 1988a, 1990). They commonly start not later than 3 – 6 months after therapy (Weber et al. 1988a).

Patients with neurological-type subsequent manifestations represent a particularly difficult problem, since it is important to ascertain whether or not their neurological signs and symptoms indicate central nervous system involvement with abnormalities of the cerebrospinal fluid. This problem can often not easily be solved, because symptoms may not appear severe enough to justify lumbar puncture, or patients may simply refuse such a procedure. It is important to be aware of these difficulties when treating patients with erythema migrans. The possibility of an established meningitis arising in a patient who was previously treated with oral penicillin for a typical erythema migrans has been documented in an earlier European report (Weber 1974). In other patients, it might not be possible to establish a suspected meningoradiculoneuritis if the lumbar puncture is performed some months after therapy, since it could then be too late to document elevation of protein content and pleocytosis in the cerebrospinal fluid which may have normalized spontaneously in the meantime. However, more specific methods such as the determination of oligoclonal bands may sometimes be able to indicate a previous infection of the central nervous system. Even so, European patients with established untreated meningoradiculoneuritis have a comparably good prognosis (Krüger et al. 1989), but exceptions do occur (Pfister et al. 1989).

Notwithstanding these problems, a nonrandomized European study has failed to reveal a difference between oral penicillin, tetracycline, and Augmentin in preventing subsequent manifestations (Weber et al. 1986, 1988a). This corresponds to experiences in the United States (Steere et al. 1983; Dattwyler et al. 1990) and to the overall evaluation of a more recent randomized European investigation (Weber et al. 1990). In the latter study, a 12-day course of oral penicillin was compared with a 5-day regimen of parenteral ceftriaxone in 73 patients. Among the subgroups of patients with two and more associated symptoms prior to therapy, ceftriaxone was superior to oral penicillin in preventing subsequent manifestations ($p<0.01$). In addition, a reevaluation of our previous nonrandomized investigations has shown that there was a tendency for minocycline and all tetracyclines, but not for Augmentin to perform better than oral penicillin in preventing subsequent manifestations in the subgroups of erythema migrans patients with two and more symptoms prior to therapy (Table 2).

Subgroup analyses in our erythema migrans patients with one and more or with three and more symptoms prior to therapy revealed a similar tendency; but the results were not so clear-cut in patients with one and more symptoms and too small numbers of patients hampered the analysis of those with three and more symptoms (Weber and Thurmayr, unpublished). Although a subgroup analysis only singles out a certain group of patients, it may provide a way of evaluating the treatment response in patients with erythema migrans without being forced to include huge numbers of patients.

Berger (1984, 1988) has investigated the treatment response by evaluating the prevention of major and minor subsequent manifestations according to Steere et al. (1983) and the necessity for retreatment in the case of recurrence of the illness. Unfortunately, no statistical analysis and no details regarding the interval between onset of therapy with various antibiotics and retreatment have been presented. An interesting observa-

Table 2. Subgroup analyses in antibiotically treated erythema migrans patients with two and more symptoms prior to therapy[a]

Total treatment groups[b]	Subsequent manifestations in subgroup (*n*)	*p* Value
Penicillin (*n*=65) vs. tetracyclines (*n*=36)	19 of 33 4 of 16	0.07
Penicillin (*n*=65) vs. minocycline (*n*=28)	19 of 33 2 of 11	0.06
Penicillin (*n*=65) vs. Augmentin (*n*=20)	19 of 33 4 of 8	n.s.
Penicillin (*n*=33) vs. ceftriaxone (*n*=40)	6 of 7 2 of 11	< 0.01

n, number of patients; n.s., not significant; penicillin, oral penicillin; vs., versus; Augmentin, amoxicillin and clavulanic acid

[a] This evaluation is based on studies of Weber et al. (1988a, 1990) and Weber and Thurmayr (1989) and partly presents hitherto unpublished material (in association with R. Thurmayr).

[b] Total number of patients in parentheses.

tion was that almost exclusively patients with the so-called major form of the illness required retreatment (Berger 1984, 1988). This finding corresponds to the results of others (Steere et al. 1983; Weber et al. 1986, 1990) as regards the significant correlation between the initial severity of the disease and clinical outcome.

In Europe, a few recent noncomparative studies have either failed to detect subsequent manifestations (Neumann et al. 1986) or noted subsequent manifestations in only a few patients with erythema migrans (Asbrink et al. 1986). Asbrink et al. did not connect the symptoms observed with Lyme borreliosis because antibody titers against *B. burgdorferi* were not found to be elevated. However, in my own experience, the fact that antibody titers are not significantly elevated does not rule out subsequent manifestations. This is also supported by observations of Dattwyler et al. (1988). The reasons for the discrepancies in the results among various reports must be determined (for discussion, see Chap. 28).

23.2.4 Antibody Titers After Therapy

Recently, an attempt has been made to evaluate the treatment response by observing the development of antibody titers against *B. burgdorferi* after antibiotic therapy (Weber et al. 1988a, 1990). It was found in these studies that elevated antibody titers normalize in most patients within a few months after initiation of therapy. However, in one study (Weber et al. 1988a), there was a tendency (p=0.06) for elevated IgM antibody titers to persist among oral penicillin recipients (6 out of 21) more often than among tetracycline recipients (0 out of 14). This finding could not be confirmed in our subsequent randomized therapeutic trial of 73 patients, where two ceftriaxone recipients represented an exception (Weber et al. 1990), so that a definite trend cannot be assumed. Nine (6%) of our 151 patients observed in both studies had persistently high IgM antibody titers and five (3%) had persistently elevated IgG antibody titers more than 12 months after therapy. Five of these 14 patients had symptoms such as (long-lasting) arthralgia or headache. We believe that patients with persistently elevated antibody titers are at risk for developing subsequent disease unless their antibody titer declines spontaneously. One of our (asymptomatic) patients with a prolonged IgG antibody response for more than 2 years turned out to have had undiagnosed meningoradiculoneuritis in retrospect when a lumbar puncture was performed 2 years after initial treatment (H.W. Pfister and K. Weber, unpublished). Neumann et al. (1986) have not found any of their 16 oral penicillin recipients to have an elevated IgG antibody titer 4 months after therapy.

23.2.5 Eradication of *Borrelia burgdorferi*

In our most recent therapeutic trial, we attempted to check treatment success by trying to isolate *B. burgdorferi* 3 months after therapy (Weber et al. 1990). The site of the control biopsy was adjacent to the scar of the pretreatment biopsy at the edge of the erythema migrans. One out of ten oral penicillin recipients was still positive, while none of 15 ceftriaxone recipients had a positive post-treatment culture. Prior to therapy, *B. burgdorferi* could be isolated from the skin in 29 of 56 patients (52%). The borreliae were apparently eradicated in most patients.

Obviously, isolation of *B. burgdorferi* after therapy represents a treatment failure and an indication for retreatment. Our oral penicillin recipient had no complaints at the time of post-treatment isolation of *B. burgdorferi*. Preac-Mursic et al. (1989) have recently observed several patients with erythema migrans or meningoradiculoneuritis in whom *B. burgdorferi* could be isolated after antibiotic therapy with penicillin (four patients), doxycyline, and ceftriaxone (one patient each). This is in agreement with investigations in animals in which *B. burgdorferi* could also be isolated after treatment with various antibiotics (Johnson et al. 1987; Preac-Mursic et al. 1987).

23.2.6 Treatment Failures

An attempt has recently been made to define definite and possible treatment failures in patients with erythema migrans (Weber et al. 1990). Accordingly, definite treatment failure can be assumed when *B. burgdorferi* has been isolated, the IgG antibody titers have risen at least fourfold and have persisted at high levels for 6 months or more (a high titer alone without recent fourfold rise may indicate previous infection), if major subsequent manifestations (according to Steere et al. 1983) have developed, or if minor subsequent manifestations have developed which are then promptly cured by retreatment. A possible treatment failure can be assumed when minor subsequent manifestations have developed which are not cured by retreatment or if antibody titers persist at high levels for 6 months or more or at lower but still elevated levels for 12 months or more (Weber et al. 1990). It is of paramount importance to observe patients for a sufficiently long period of time to get an idea about a treatment failure. If retreatment is carried out too early, it might not be justified, and a treatment failure may be assumed erroneously.

23.2.7 Recommendations

23.2.7.1 Antibiotics

23.2.7.1.1 Summary of Observations

Tetracyclines tend to be superior to oral penicillin (Steere et al. 1983; Weber et al. 1988a; Steere 1989; Table 2) and ceftriaxone seems to be superior to oral penicillin in the subgroup of patients with more severe early illness (Weber et al. 1990). No investigation has been performed to compare ceftriaxone with any other oral antibiotic besides oral penicillin. There has been no significant difference between amoxicillin (with clav-

Table 3. Recommended antibiotics for the treatment of erythema migrans

Antibiotic[a]	Daily dosage
Tetracyclines	
Doxycycline/minocycline	2 x 100mg orally
Tetracycline	3 x 500 mg orally
Ceftriaxone	1 x 1 g intramuscularly
Amoxicillin	3 x 500 mg orally
Azithromycin[b]	2 x 250 mg orally
Phenoxymethylpenicillin	3 x 1 million units (0.6 g) orally
Erythromycin	3 x 500 mg orally

[a] The recommended duration of therapy is 14 days, but 5 days for ceftriaxone and 10 days for azithromycin.
[b] Provisional rank, awaits further evaluation.

ulanic acid or probenecid) and oral penicillin or tetracycline (Weber et al. 1988; Dattwyler et al. 1990). Amoxicillin has also been assumed to behave similarly to (oral) penicillin and tetracycline in another study in which no statistical analysis was provided (Berger 1988). Furthermore, amoxicillin was also quite effective in a recent noncomparative study of 67 patients (Morel et al. 1990). Observations with erythromycin are somewhat controversial (Steere et al. 1983; Weber et al. 1986). It is conceivable that the effect of erythromycin strongly depends on the dosage administered and the compliance of the patients, who may experience gastrointestinal discomfort. Presently, several studies are underway to test the efficacy of the newer erythromycin analogues; azithromycin appears to be efficacious (Strle et al. in press; Weber, unpublished), but roxithromycin was less effective than oral penicillin (Hovmark et al. 1990).

23.2.7.1.2 Recommended Antibiotics
Tetracyclines presently seem to be the best choice, but their superiority has still not definitely been proven, not even in tests using minocycline (Weber and Thurmayr 1989) (Table 3). Nevertheless, it seems advisable to use doxycycline or minocycline instead of the usual tetracycline because compliance may be better with these long-acting tetracyclines, although some minocycline (and doxycycline) recipients may suffer from dizziness when taking more than 100 mg daily.

Ceftriaxone appears to be as equally effective as the tetracyclines, even if administered for only 5 days. Ceftriaxone has the disadvantage of having to be administered parenterally, but a single daily intramuscular injection of 1 g may suffice to treat patients with erythema migrans, especially those with more severe initial disease. Ceftriaxone might be particularly well suited for erythema migrans patients with early neurological symptoms who refuse lumbar puncture or in whom central nervous involvement could not be proved. In these patients, treatment should preferably be carried out for 10 days. Patients with established meningitis should be treated as outlined in Chap. 24.

Amoxicillin appears to be a reasonable oral alternative to tetracycline, although it has not been proved to be superior to oral penicillin in therapeutic trials. Amoxicillin performs better in vitro than penicillin G (Johnson et al. 1987; Preac-Mursic et al. 1987). It has not been proved so far that clavulanic acid improves the efficacy of amoxicillin in patients with erythema migrans. Azithromycin might be a promising candidate (Strle et al. in press; Weber, unpublished), but its real value still needs to be determined.

Oral penicillin has a favorable effect in treating erythema migrans (Hollström 1958; Steere et al. 1980) and may, therefore, still be used, but it is not recommended anymore as an antibiotic of first or second choice. The same is true of parenteral penicillin, the superiority of which has not been proved so far either.

In conclusion, tetracycline and ceftriaxone are presently the antibiotics of first and second choice, amoxicillin possibly of third choice, erythromycin and oral penicillin of next choice. Azithromycin awaits further evaluation.

23.2.7.2 Duration of Therapy

Studies have provided some information on the necessary duration of antibiotic therapy. A randomized comparison of a 10-day and a 20-day tetracycline regimen in 49 American patients with erythema migrans failed to reveal a statistically significant difference (Steere et al. 1983). In our randomized study, ceftriaxone was administered for 5 days and oral penicillin for 12 days, and still ceftriaxone performed better in a subgroup of patients (Weber et al. 1990). Proof is lacking at the moment that prolonged treatment time is necessary for patients with erythema migrans. It appears reasonable to favor a course of about 14 days, but 5 days seem to suffice with the ceftriaxone regimen mentioned above.

23.2.7.3 Dosage of Antibiotic

The dosage of the antibiotic is obviously of importance. However, it has not been proved so far that higher doses of penicillin (Weber et al. 1986), penicillin plus probenecid (Berger 1984, 1988) or a different type of tetracycline amounting to a somewhat higher dosage (Weber and Thurmayr 1989) play a decisive role. Comparative statistical analyses have not been performed. Nevertheless, the dosage should be adequately high (Table 3).

23.2.7.4 Retreatment

Retreatment can be carried out similarly to the primary treatment. The same or another antibiotic might be administered for the same or a more prolonged period of time, for instance, for 3 weeks.

23.2.7.5 Treatment in Children

Children should be treated in doses equivalent to those recommended for adults. For children below the age of 12 years, tetracyclines are contraindicated, and here amoxicillin may be the best alternative. One of the newer erythromycin analogues such as azithromycin, once they have become commercially available and found to be effective, may also be used. Erythromycin and oral penicillin might be antibiotics of next choice (see Chap 17).

23.2.7.6 Treatment During Pregnancy

Pregnant women may be treated with ceftriaxone, amoxicillin, erythromycin, or high doses of intravenous penicillin G. In one instance in which oral penicillin was administered during the first trimester of pregnancy, the erythema migrans disappeared, but autopsy of the newborn, who died 1 day after birth, showed an infection of brain and liver with *B. burgdorferi* (Weber et al. 1988b). Thus, oral penicillin is not recommended during pregnancy, although several women on this regimen have delivered normal children (Berger 1984; Weber et al. 1986; Neubert 1987).

23.2.8 Jarisch-Herxheimer Reaction

A Jarisch-Herxheimer reaction may occur shortly after initiation of antibiotic therapy (Steere et al. 1983; Weber 1984; Weber et al. 1986) As with the Jarisch-Herxheimer reaction which occurs in other infectious diseases such as syphilis, fever, chills, fatigue, headache, myalgia and an intensification of the signs and symptoms of the erythema migrans may be observed within a few hours after initiation of therapy. Fever and chills usually subside within hours, but the other symptoms may continue for a few more days or may even start to arise within the first few days after initiation of therapy (Weber 1984; Weber et al. 1986). Some of these signs and symptoms represent intensifications of the signs and symptoms present prior to therapy, while others develop anew. Intensification of the erythema migrans signs and symptoms consists of more intense redness, enlargement, swelling and/or more intensive lesional pain. The Jarisch-Herxheimer reaction has been observed in connection with the use of all antibiotics administered so far. It has been found to occur particularly often during treatment with amoxicillin (Berger 1988; Weber et al. 1988a). A special treatment for the Jarisch-Herxheimer reaction occurring in patients with erythema migrans does not seem to be necessary.

23.3 Therapy of Borrelial Lymphocytoma

Much less is known about the treatment of borrelial lymphocytoma than about treatment of erythema migrans. One has to keep in mind that borrelial lymphocytoma represents a sign which can be present during all stages of Lyme borreliosis. It is particularly common in stage II (see Chap.9), and antibiotic treatment of a stage II sign may require more intensive efforts. However, there is no proof that this is the case with a borrelial lymphocytoma which is not associated with complications such as meningoradiculoneuritis. No comparative study is available.

It has been shown that after treatment with oral penicillin the resolution time in six borrelial lympocytoma patients was a median of 7 (range 4-16) weeks compared with the resolution time of 2 (range 0.5-20) weeks in 65 erythema migrans patients (Weber et al. 1986). This apparently longer resolution time is in agreement with previous reports on the necessity of prolonged penicillin treatment in patients with lymphocytoma or lymphadenosis benigna cutis. However, one should consider that the latter two conditions might have included patients with lymphocytomas of nonborrelial origin. Favorable results with penicillin and tetracycline have been obtained in a Swedish study (Hovmark et al. 1986). Presently, the same recommendations as for patients with erythema migrans are being made (Table 3). Duration of therapy should be about 14 – 20 days.

23.4 Therapy of Acrodermatitis Chronica Atrophicans

23.4.1 Introduction

Although antibiotic therapy of Lyme borreliosis started with a detailed report on the beneficial effect of penicillin on acrodermatitis chronica atrophicans (ACA) (Thyresson 1949), not enough data are available at the moment to make definitive therapeutic recommendations for this late manifestation.

23.4.2 Pretreatment Evaluation

23.4.2.1 Cutaneous Lesions

The extent and duration of ACA are the most important clinical pretreatment variables. In scientific studies, the extent of ACA can be judged by counting one point for the involvement of each of the four extremities and an additional fifth point for involvement elsewhere (Weber et al. 1988a). Other cutaneous signs such as fibrotic nodules or sclerotic changes may be included, too, if present.

23.4.2.2 Extracutaneous Signs and Symptoms

Since lymphadenopathy is common, the lymph nodes should be examined. A difficult problem is concomitant infection of the central nervous system, although this seems to be rather uncommon (Hopf 1966; Kristoferitsch 1989). If central nervous system involvement is suspected, a lumber puncture should be performed because of the therapeutic implications involved. Patients with central nervous system involvement besides their ACA should be treated for this extracutaneous manifestation primarily and followed up accordingly. Other extracutaneous manifestations such as peripheral neuropathy, myositis, or arthritis may also influence the treatment decision.

23.4.2.3 Laboratory Changes

The IgG antibody titer should be determined. An attempt to isolate *B. burgdorferi* might be desirable in atypical cases and in scientific studies. Other laboratory values such as the erythrocyte sedimentation rate and the serum immunoglobulins are of less importance, but may be included (Weber et al. 1986, 1988a).

23.4.3 Post-treatment Criteria for Response

23.4.3.1 Resolution of ACA

The inflammatory changes of the ACA gradually subside to a varying extent after therapy with penicillin, tetracycline, chloramphenicol, azithromycin, and cefotaxime, which are the agents used so far. Atrophy is not influenced by antibiotic therapy. Sclerotic lesions may or may not resolve (Thyresson 1949; Götz and Ludwig 1951; Pirilä 1951; Götz 1952; Ludwig 1955; Weber et al. 1984, 1986, 1988a; Asbrink and Hovmark 1987; Langer and Diem 1988; Hassler et al. 1990; Weber, unpublished). Fibrotic nodules show striking regression (Thyresson 1949; Weber et al. 1984).

Even low doses of penicillin lead to improvement or resolution of inflammatory changes. However, resolution is often not complete or even totally lacking in individual cases (Thyresson 1949; Götz and Ludwig 1951; Pirilä 1951, Ludwig 1955; Betz 1972; Weber et al. 1984, 1988a). The resolution time of the ACA is significantly longer than that of erythema migrans ($p<0.01$; Weber et al. 1986). In a recent investigation, it has been found that the complete resolution of the ACA in 15 of 34 patients (44%) lasted a median of 18 (range 3-44) months after antibiotic therapy, which was carried out usually for 2 – 3 weeks, and sometimes for 7 – 10 days: in particular, the complete resolution time was 18 months in seven oral penicillin recipients, 9 months in six parenteral penicillin recipients, and 14 months in two tetracycline recipients (Weber et al. 1988a). This comparably low success rate seems to have been surpassed by the experiences of Asbrink and Hovmark (1987), who have stated that the inflammatory lesions of their patients treated with 2-3 g phenoxymethylpenicillin daily for 3 weeks usually heal; unfortunately, no details have been presented. Asbrink (1990) later reported that she treated her ACA patients with 200 mg doxycycline daily for 3 weeks.

Incomplete resolution of the ACA can be assumed when inflammatory changes, especially the bluish-red discoloration remain, even if they are rather mild in intensity. These ACA remnants indicate lack of cure. *B. burgdorferi* has been isolated from a lesion of this type 3 years after tetracycline therapy (Weber et al. 1988a). Relapse of ACA at the same site 5 years after therapy with parenteral penicillin was also observed in a patient with seemingly complete resolution of the ACA (Weber et al. 1988a). Patients with ACA need long-term follow-up after therapy to check their clinical response.

23.4.3.2 Subsequent Manifestations

Extracutaneous signs and symptoms may continue or newly develop after antibiotic treatment for ACA. Hopf (1966) and especially Kaiser (1972) were the first to discuss this point in regard to peripheral neuropathy. We have recently extended this observation to arthralgia (arthritis), myalgia (myositis), and lymphadenopathy (Weber et al. 1988a). However, Ludwig (1955) had already described subsequent arthralgia in one patient. In other words, although ACA is a late manifestation of Lyme borreliosis, subsequent manifestations may still occur after therapy.

23.4.3.3 Post-treatment IgG Antibody Titer

The IgG antibody titer against *B. burgdorferi* may or may not decline after antibiotic therapy, but there is usually a tendency for a decrease (Ackermann et al. 1984; Weber et al. 1984, 1988a; Asbrink et al. 1985). The antibiotic treatment used in these studies was somewhat different: Asbrink et al. (1985) observed a significant decline in 15 of 26 patients (58%) 6-13 months after treatment with 2-3 g phenoxymethylpenicillin daily for 3 weeks, while Weber et al. (1988a) found that the decline was not statistically significant within the first 3 years in 29 patients, although 18 of their patients (62%) experienced a significant, at least fourfold, decrease of the IgG antibody titers, which is in agreement with the aforementioned findings of Asbrink et al. (1985). A transient rise of the IgG antibody titer has been noted in individual patients with ACA a few weeks after initiation of therapy (Weber et al. 1984; Langer and Diem 1988). Our patient who suffered a relapse and another patient experienced a fourfold rise several years after the first antibiotic therapy (Weber et al. 1988a), indicating treatment failure. A similar constellation with initial clinical recovery and a significant decline in antibody titer followed by a subsequent increase in IgG antibody titers and clinical relapse was observed in four patients with late Lyme borreliosis (Zöller et al. 1989).

23.4.3.4 Identification of Borrelia burgdorferi

Apart from the single case of post-treatment isolation of *B. burgdorferi* mentioned above, no systematic study has been carried out so far to evaluate the treatment response by identification of *B. burgdorferi*.

23.4.4 Treatment Failures

Incomplete resolution of ACA and possibly of extracutaneous signs and symptoms, the significant rise of antibody titers, and isolation of *B. burgdorferi* represent definite or possible treatment failures. The persistence of the elevated IgG antibody titer cannot be regarded as a sign of treatment failure on the basis of present knowledge.

Retreatment may be necessary in the case of treatment failure. It was already carried out by early investigators because of seeming or apparent incomplete resolution (Thyresson 1949; Götz and Ludwig 1951; Pirilä 1951; Ludwig 1955). In our recent investigation, retreatment has been performed in 13 out of 34 patients (38%) because of ACA remnants, peripheral neuropathy, arthritis, myositis, isolation of *B. burgdorferi*, or relapse (Weber et al. 1988a). Retreatment has been found to be necessary more often in patients with ACA than in patients with erythema migrans (Weber et al. 1986, 1988a; $p<0.01$). However, treatment failures may depend on the kind of antibiotic used and the duration of therapy, so that more data are needed to reevaluate these findings.

23.4.5 Recommendations

Only one preliminary nonrandomized comparative therapeutic trial in 29 patients with ACA has been published so far (Weber et al. 1988a), and no statistically significant difference was noted between oral penicillin, usually rather low doses of parenteral penicillin, and tetracycline. The number of patients was small. In another comparative study on late Lyme borreliosis, a subgroup of eight ACA patients, five of whom had additional arthritis, were either treated with cefotaxime or high doses of penicillin G (Hassler et al. 1990). There was full or partial remission in the four cefotaxime recipients, but no effect in three of four penicillin recipients. These and other findings of noncomparative investigations do not present sufficient data to make final recommendations based on evidence. Preliminary recommendations based on therapeutic experiences in patients with ACA and other manifestations of late Lyme borreliosis are presented in Table 4.

The duration of treatment should be about 2 – 3 weeks. Asbrink and Hovmark (1987) emphasize the importance of a prolonged treatment over 3 weeks on the basis of their therapeutic experiences. Although *B. burgdorferi* cells isolated from patients with ACA tend to grow slowly in vitro (Asbrink and Hovmark 1985; Preac-Mursic et al. 1985), this does not prove the necessitiy of a prolonged antibiotic therapy in ACA as has been suggested (Asbrink and Hovmark 1987). However, a more prolonged treatment may be preferable, even if there are presently not enough data available to prove this point. Retreatment with a different antibiotic should be carried out in the case of treatment failure. Children and pregnant women, who are rarely affected by the ACA, may be treated as outlined above (Sect. 2. 7). Complications in association with the ACA should be treated preferentially.

23.4.6 Jarisch-Herxheimer Reaction

A Jarisch-Herxheimer reaction may also occur in patients with ACA but it is uncommon. A temperature of up to 38.8° C (Götz and Ludwig 1951), exacerbation of a peripheral neuropathy (Weber et al. 1986), and exacerbation of skin and possibly heart manifestations (with subsequent pericarditis) have been observed in relatively rare

Table 4. Preliminary recommendations for the treatment of acrodermatitis chronica atrophicans

Antibiotic[a]	Daily dosage
Tetracyclines	
Doxycycline/minocycline	2 x 100 mg orally
Tetracycline	3 x 500 mg orally
Azithromycin[b]	2 x 250 mg orally
Amoxicillin[c]	3 x 500 mg orally
Ceftriaxone[c]	1 x 2 g intravenously
Cefotaxime[c]	2 (– 3) x 2 g intravenously
Penicillin G	4 x 5 million units intravenously
Phenoxymethylpenicillin	3 x 1 g orally

[a] Recommended duration of therapy is 2 – 3 weeks.
[b] Provisional rank, limited experience in six patients (Weber, umpublished).
[c] Treatment response only reported in a few patients so far (Weber et al. 1986, 1988a; Hassler et al. 1990).

individual patients (Weber et al. 1988a; Herzer 1989). I recently observed the exceptional case of a 40-year old man who developed a temperature of 39.6 °C, anxiety, intensification of the bluish-red erythemas within the preexisting ACA (on the arms), hitherto unrecognized erythema in (seemingly) previously unaffected skin (on the legs), and pronounced enlargement of the axillar lymph nodes a few hours after initiation of therapy with intraveneous penicillin G. These signs and symptoms lasted for several days. After a relapse of the ACA a year later, a similar but weakened reaction (without changes on the legs) lasting 4 days occurred after initiation of doxycycline treatment (unpublished observation).

References

Ackermann R, Boisten HP, Kabatzki J, Runne U, Krüger K, Hermann WP (1984) Serumantikörper gegen *Ixodes-ricinus*-Spirochäte bei Acrodermatitis chronica atrophicans (Herxheimer). Dtsch Med Wochenschr 109:6-10

Asbrink E (1990) Late dermatological manifestations in Lyme borreliosis. In: Einhäupl KM, Felgenhauer K, Herzer P, Stirenstedt G (eds) Lyme borreliosis. Edit Roche, Basel/Grenzach-Wyhlen pp 109-118

Asbrink E, Hovmark A (1985) Successful cultivation of spirochetes from skin lesions of patients with erythema chronicum migrans Afzelius and acrodermatitis chronica atrophicans. Acta Pathol Microbiol Immunol Scand (B) 93:161-163

Asbrink E, Hovmark A (1987) Cutaneous manifestations in *Ixodes*-borne *Borrelia* spirochetosis. Int J Dermatol 26:215-223

Asbrink E, Hovmark A, Hederstedt B (1985) Serologic studies of erythema chronicum migrans Afzelius and acrodermatitis chronica atrophicans with indirect immunofluorescence and enzyme-linked immunosorbent assays. Acta Derm Venereol (Stockh) 65:509-514

Asbrink E, Olsson I, Hovmark A (1986) Erythema chronicum migrans Afzelius in Sweden. A study on 231 patients. Zentralbl Bakt Mikrobiol Hyg (A) 263:229-236

Berger BW (1984) Treating erythema chronicum migrans of Lyme disease. J Am Acad Dermatol 15:459-463

Berger BW (1988) Treatment of erythema chronicum migrans of Lyme disease. Ann NY Acad Sci 539:346-351

Betz E (1972) Nachuntersuchungen über den Behandlungserfolg an den Akrodermatitis chronica atrophicans (Herxheimer) Erkrankten der Universitäts-Hautklinik Köln auf Grund von klinischen und histologischen Befunden. Dissertation, Köln, Germany

Dattwyler RJ, Volkman DL, Luft BJ, Halperin JJ, Thomas J, Golightly MG (1988) Seronegative Lyme disease: dissociation of the specific T-and B-lymphocyte response to *Borrelia burgdorferi*. N Engl J Med 319:1441-1446

Dattwyler RJ, Volkman DJ, Conaty SM, Platkin SP, Luft BJ (1990) Amoxicillin and probenecide versus doxycycline for treatment of erythema migrans borreliosis. Lancet II:1404-1406

Götz H (1952) Die Acrodermatitis chronica atrophicans Herxheimer als neue Indikation für die Behandlung mit Chloromycetin. Hautarzt 3:310-312

Götz H, Ludwig E (1951) Die Behandlung der Akrodermatitis chronica atrophicans Herxheimer mit Penicillin. Hautarzt 2:6-14

Hassler D, Zöller L, Haude M, Hufnagel HD, Heinrich F, Sonntag HG (1990) Cefotaxime versus penicillin in the late stage of Lyme disease – prospective, randomized therapeutic study. Infection 18:16-20

Herzer P (1989) Lyme-Borreliose. Steinkopf, Darmstadt

Hollström E (1951) Successful treatment of erythema migrans Afzelius. Acta Derm Venereol (Stockh) 31:235-243

Hollström E (1958) Penicillin treatment of erythema chronicum migrans Afzelius. Acta Derm Venereol (Stockh) 38:285-289

Hopf HC (1966) Acrodermatitis chronica atrophicans (Herxheimer) und Nervensystem. Springer, Berlin Heidelberg New York (Monographien aus dem Gesamtgebiete der Neurology und Psychiatry, vol 114)

Hovmark A, Asbrink E, Olsson I (1986) The spirochetal etiology of lymphadenosis benigna cutis solitaria. Acta Derm Venereol (Stockh) 66:479-484

Hovmark A, Olsson I, Asbrink E, Olsson E, Halkier-Sörensen L, Hansen K (1990) A comparative study of the efficacy of roxithromycin and phenoxymethylpenicillin in the treatment of erythema migrans (abstract). IVth international conference of Lyme borreliosis, Stockholm

Johnson RC, Kodner C, Russel M (1987) In vitro and in vivo susceptibility of the Lyme disease spirochete, *Borrelia burgdorferi*, to four antimicrobial agents. Antimicrob Agents Chemother 31:164-167

Kaiser M (1972) Neurologische Komplikationen bei Akrodermatitis chronica atrophicans (Herxheimer) und ihre Beeinflussung durch die Penicillintherapie. Dissertation, Göttingen

Kristoferitsch W (1989) Neuropathien bei Lyme-Borreliosis. Springer, Wien New York, pp 114-115

Krüger H, Reuss K, Pulz M, Rohrbach E, Pflughaupt KW, Martin R, Mertens HG (1989) Meningoradiculitis and encephalomyelitis due to *Borrelia burgdorferi* – a follow-up study of 72 patients over 27 years. J Neurol 236:322-328

Kuiper H, de Jongh B, Ramselaar T, Spanjaard L (1990) Evidence of asymptomatic nervous system involvement in localized erythema migrans. IVth internat conference of Lyme Borreliosis, Stockholm, abstract W/Th-P-48

Langer K, Diem E (1988) Acrodermatitis chronica atrophicans und sklerodermiforme Hautveränderungen bei *Borrelia*-Infektion. Hautarzt 39:647-651

Ludwig E (1955) Erfolgreiche Aureomycin- und Terramycinbehandlung als Beitrag zur Klärung der Wirkungsweise von Penicillin bei der Akrodermatitis chronica atrophicans Herxheimer. Dermatol Wochenschr 131:169-178

Morel P, Modai J, Dournon E (1990) Treatment of erythema chronicum migrans (ECM) and prevention of stage 2 Lyme disease (Ld) with oral amoxicillin: a one-year follow-up (abstract). IVth international conference on Lyme Borreliosis, Stockholm

Neubert U (1987) Erythema migrans in der Gravidität. Hautarzt 38:182-183

Neumann R, Aberer E, Stanek G (1986) Treatment and course of erythema chronicum migrans. Zentralbl Bakteriol Hyg (A) 263:372-376

Pfister HW, Wilske B, Preac-Mursic V, Einhäupl KM (1989) Clinical and serological follow up of patients with Bannwarth's syndrome: comparison of patients with and without penicillin treatment. Zentralbl Bakteriol (Suppl) 18:276-279

Pirilä V (1951) The penicillin treatment of acrodermatitis atrophicans chronica. Acta Derm Venereol (Stockh) 31:576-591

Preac-Mursic V, Wilske B, Herzer P, Schierz G, Bauer M (1985) Acrodermatitis chronica atrophicans – eine Borreliose. Hautarzt 36:691-693

Preac-Mursic V, Wilske B, Schierz G, Holmburger M, Süss E (1987) In vitro and in vivo susceptibility of *Borrelia burgdorferi*. Eur J Clin Microbiol 6:424-426

Preac-Mursic V, Weber K, Pfister HW, Wilske B, Gross B, Baumann A, Prokop J (1989) Survival of *Borrelia burgdorferi* in antibiotically treated patients with Lyme borreliosis. Infection 17:355-359

Steere AC (1989) Lyme disease. N Engl J Med 321:586-596

Steere AC, Malawista SE, Newman J, Spieler PN, Bartenhagen HN (1980) Antibiotic therapy in Lyme disease. Ann Intern Med 93:1-8

Steere AC, Hutchinson GJ, Rahn DW, Sigal LH, Craft JE, DeSanna ET, Malawista SE (1983) Treatment of the early manifestations of Lyme disease. Ann Intern Med 99:22-26

Strle F, Ruzic E, Cimperman J (in press) Erythema migrans: comparison of treatment with azithromycin, doxycycline and pheoxymethylpenicillin. J Antimicrob Chemother

Thyresson N (1949) The penicillin treatment of acrodermatitis chronica atrophicans (Herxheimer). Acta Derm Venereol (Stockh) 29:572-621

Weber K (1973) Erkrankungen nach Zeckenbiß und ihre Behandlung. Ther Ggw 112:1402-1409

Weber K (1974) Erythema-chronicum-migrans-Meningitis – eine bakterielle Infektionskrankheit? Munch Med Wochenschr 116:1993-1998

Weber K (1984) Jarisch-Herxheimer-Reaktion bei Erythema-migrans-Krankheit. Hautarzt 35: 588-590

Weber K (1989) Clinical differences between European and North-American Lyme borreliosis – a review. Zentralbl Bakteriol (Suppl) 18:146-155

Weber K, Neubert U (1986) Clincal features of early erythema migrans disease and related disorders. Zentralbl Bakteriol Hyg (A) 263:209-228
Weber K, Thurmayr R (1989) Oral penicillin versus minocycline for the treatment of early Lyme borreliosis. Zentralbl Bakteriol (Suppl) 18:263-268
Weber K, Schierz G, Wilske B, Preac-Mursic V (1984) Zur Klinik und Ätiologie der Acrodermatitis chronica atrophicans. Hautarzt 35:571-577
Weber K, Neubert U, Thurmayr R (1986) Antibiotic therapy in early erythema migrans disease and related disorders. Zentralbl Bakteriol Hyg (A) 263:377-388
Weber K, Preac-Mursic V, Neubert U, Thurmayr R, Herzer P, Wilske B, Schierz G, Marget W (1988a) Antibiotic therapy of early European Lyme borreliosis and acrodermatitis chronica atrophicans. Ann NY Acad Sci 539:324-345
Weber K, Bratzke HJ, Neubert U, Wilske B, Duray PH (1988b) *Borrelia burgdorferi* in a newborn despite oral penicillin for Lyme borreliosis during pregnancy. Pediatr Infect Dis 7:286-289
Weber K, Preac-Mursic V, Wilske B, Thurmayr R, Neubert U, Scherwitz C (1990) A randomized trial of ceftriaxone versus oral penicillin for the treatment of early European Lyme borreliosis. Infection 18:91-96
Zöller L, Haude M, Hassler D, Burkard S, Sonntag HG (1989) Spontaneous and post-treatment antibody kinetics in late Lyme borreliosis. Serodiagnosis Immunther Infect Dis 3:345-353

24 Therapy of Lyme Neuroborreliosis

H.W. Pfister, W. Kristoferitsch, and B. Sköldenberg

24.1 Introduction

Certain antibiotics such as penicillin and tetracycline are effective drugs in the treatment of syphilis and other spirochetal diseases. Penicillin has recently been considered as the drug of choice in Lyme neuroborreliosis. In the following chapter, we will critically review data on the treatment of Lyme neuroborreliosis, with the emphasis on clinical outcome and prognosis after treatment with penicillin or other antibiotics such as cefotaxime, ceftriaxone, or doxycycline. The data will be discussed against the background of the antibiotic susceptibility of *Borrelia burgdorferi* in vitro and the pharmakokinetics of antibiotics concerned.

24.2 Early Neurological Involvement (Stage 2 Lyme Neuroborreliosis)

24.2.1 Penicillin

Even before the borrelial etiology of Lyme borreliosis was clarified by Burgdorfer et al. (1982), several European authors had reported the beneficial effects of penicillin G in the therapy of neurological symptoms associated with erythema migrans. Meningitis was successfully treated with intravenous penicillin G at dosages of 10 million U for 12 days (Hollström 1951), 10.5 million U for 8 days (Bjornstad and Mossige 1955), and 10 million U/day for 8 days (Weber 1974).

In 1983, Sköldenberg et al. published results describing the beneficial effects of intravenous penicillin G on different cerebrospinal fluid (CSF) and clinical parameters in 21 patients with "chronic" meningitis and in 8 patients with various general and neurological signs, including facial nerve palsy. These patients were later demonstrated to have neurological manifestations of Lyme borreliosis (Stiernstedt et al. 1984). In the United States, Steere et al. (1983b) treated 12 patients with neurological manifestations of Lyme borreliosis with penicillin G (20 million U/day for 10 days). The results were compared with those of a historical group of 15 patients treated with prednisone (40 – 60 mg/day). It was shown that the duration of meningitic symptoms such as headache or stiff neck was significantly shorter in patients given penicillin than in those treated with prednisone (mean duration 1 versus 29 weeks). However, there was no difference in the duration for complete recovery of motor deficits (mean duration 8 and 7 weeks, respectively). Steere et al. found that 3 of the 12 patients treated with penicillin continued to have frequent arthralgias, musculoskeletal pain, and fatigue. Nevertheless, high-dose penicillin therapy has been the recommended standard therapy for Lyme neuroborreliosis.

Several European investigators have supported the use of penicillin therapy (Weber et al. 1983; Stiernstedt et al. 1984, 1986; Pfister et al. 1985; Kohlhepp et al. 1986; Kristoferitsch et al. 1986). Sköldenberg et al. (1988) studied 113 patients with Lyme

Table 1. Major prospective trials in Lyme neuroborreliosis

Location Reference	Number of patients	Stage of disease (symptoms)	Study design	Therapeutic regimens	Results
Europe					
Stiernstedt et al.(1986)	35	2, 3(meningitis, radiculoneuropathy, central nervous symptoms)	non randomized	Penicillin G (12 g/day) for 14 days i.v.)	All patients improved or recovered. Decrease of CSF pleocytosis and protein level at follow-up 2 months – 8 years in all patients.
Kristoferitsch et al. (1986)	19	2 (Bannwarth's syndrome)	Historical control group (24 untreated patients)	Penicillin G (20 million U/day for 10 days i.v.)	No clinical differences at follow up 3 weeks and 6 months after onset of disease. Patients treated with penicillin G within 5 weeks after onset of neurological disease had a significantly faster resolution of radicular pain.
Pfister et al. (1988)	21	2 (Bannwarth's syndrome)	Randomized, (double-blind, placebo-controlled)	Penicillin G (20 million U/day for 10 days i.v.) and methylprednisolone (60 mg/day for 7 days p.o.; 11 patients) vs penicillin G (20 million) U/day for 10 days i.v.) and placebo (10 patients)	No clinical differences at the end of therapy. Faster improvement of pain during the first 2 days of therapy in the corticosteroid group
Sköldenberg et al. (1988)	113	2 and 3 (meningitis; central nervous system symptoms, e. g., hemiparesis, ataxia)	Nonrandomized	Penicillin G (9-12 g/day for 10–14 days i.v.; 105 patients); doxycycline (200 mg/day) or cefuroxime (4.5-9 g/day for 10-14 days i.v.; 8 patients)	All patients improved or recovered. Decrease of CSF pleocytosis and protein level at follow up 4 months – 3 years in all patients
Pfister et al. (1989)	21	2 (meningitis, Bannwarth's syndrome)	Randomized	Penicillin G (20 million U/day for 10 days i.v.; 10 patients) vs. cefotaxime (3x2 g/day for 10 days i.v.; 11 patients)	No clinical differences during the treatment period and at follow up 7.7 months (mean duration) after therapy. CSF cefotaxime concentrations reached the MIC90 value in all patients, while none of the penicillin-treated patients had CSF concentrations above the MIC 90 value.
Kristoferitsch et al. (1989)	10	2 (Bannwarth's syndrome)	Historical control group (9 untreated patients)	Ceftriaxone (2 g/day i.v. for 14 days)	Faster improvement of pain in the ceftriaxone group.
Kohlhepp et al. (1989)	75	2,3	Randomized	Penicillin G (20 million U/day for 10 days i.v.) vs doxycycline 100mg/day (first 2 days 200 mg) i.v. for 10 days	All patients improved or recovered; no clinical differences.

Table 1. Continued

Pfister et al. (1991)	33	2,3	Randomized	Ceftriaxone 2g/day i.v. vs cefotaxime 3x2g/day i.v. for 10 days	Similar beneficial effect in both groups
United States					
Steere et al. (1983b)	12	2 (meningitis, cranial neuritis, radiculoneuritis)	Historical control group (15 patients treated with prednisone)	Penicillin G (20 million U/day for 10 days i.v.)	The duration of meningitis symptoms was significantly shorter in the penicillin group (1 vs 29 weeks). No difference in the duration of complete recovery of motor deficits.
Dattwyler et al. (1988)	23	3 (late Lyme borreliosis including neuropathy, encephalopathy)	Randomized	Penicillin G (24 million U/day for 10 days i.v.; 10 patients) vs ceftriaxone (2x2 g/day for 14 days i.v.; 13 patients)	Treatment failures at 3 months after therapy: 5 of 10 penicillin-treated and 1 of 13 ceftriaxone-treated patients.
Logigian et al. (1990)	27	3 (chronic neurological manifestations)	Nonrandomized	Ceftriaxone (2 g/day for 14 days i.v.)	6 months after therapy 17 patients (63%) had improvement, 6 (22%) had improvement, but then relapsed, and 4 (15%) had no change in their condition.

neuroborreliosis (Table 1). In this study, 47 patients were treated with 12 g penicillin G per day, mostly for 14 days; 58 with 9 g penicillin G per day, mostly for 10 days; 5 with 200 mg per day doxycycline; and 3 with 4.5 – 9 g cefuroxime per day. All patients had CSF pleocytosis. Seventy percent of the patients had peripheral nervous system symptoms, predominantly facial nerve palsy and radicular pain. Central nervous system symptoms, including hemiparesis, spastic paraparesis, and ataxia were seen in 13% of the patients. Almost half of the patients were treated more than 4 weeks after the onset of clinical symptoms. Persisting or progressive symptoms over a period of 4 – 11 months occurred in 15% of the patients. All patients improved or recovered. However, the patients with central nervous system symptoms continued to have sequelae for years after treatment. The authors concluded that the antibiotics used were beneficial for the clinical course of the disease and that they were effective in decreasing the CSF cell count and protein levels.

Although the results of these studies indicated that penicillin was an adequate therapy for Lyme neuroborreliosis, reports of treatment failure with penicillin, initially from the United States, followed. Ceftriaxone was effective in treating patients with late Lyme borreliosis who did not respond to penicillin therapy (Dattwyler et al. 1987). In a patient with severe Lyme meningoencephalitis who did not respond to high-dose penicillin G, intravenous administration of chloramphenicol resulted in substantial improvement (Diringer et al. 1987). In addition, Pal et al. reported on a patient with suspected Lyme meningoencephalitis who responded to cefotaxime after failing to respond to penicillin therapy (Pal et al. 1988).

24.2.2 Cefotaxime and Penicillin G

Pfister et al. (1989a) randomly assigned 21 patients with meningoradiculoneuritis (Bannwarth's syndrome) and Lyme meningitis to a 10-day intravenous treatment with either 4x5 million U/day penicillin G (n=10), or 3x2 g/day cefotaxime (n=11) (Table 1). The authors were not able to demonstrate any clinical differences between the cefotaxime and the penicillin recipients, either during the 10-day treatment period or at follow-up examination a mean of 7.7 months after antibiotic therapy. Clinical improvement, which usually occurred on the 3rd and 4th day of therapy, was observed in 8 of 10 patients of the penicillin group and in 9 of 11 patients of the cefotaxime group. Penicillin and cefotaxime were equally effective in treating radicular pain (Pfister et al. 1989a).

At follow-up examinations, none of the 21 patients had evidence of disease activity. Late typical manifestations of Lyme borreliosis were not observed in either group. CSF cell count had returned to normal values in nearly all of the patients. The prominent CSF finding was persistent oligoclonal gammopathy which was evident in the CSF of six patients of each group. Intrathecal production of antibodies against *B. burgdorferi* was still demonstrated in one patient of each group (Pfister et al. 1989a). Persistence of oligoclonal IgG bands as well as specific antibodies within the CSF for months and even years after acute disease is known from other neurological inflammatory diseases (Felgenhauer et al. 1982; Vandvik et al. 1985). However, they do not provide evidence of clinically active disease. The aforementioned study showed that most patients with early neurological manifestations of Lyme borreliosis benefit from a 10-day course of cefotaxime or penicillin G.

In the study of Pfister et al. (1989a), antibiotic concentrations were determined in the CSF and serum on the 8th – 10th day of antibiotic treatment in eight patients of the cefotaxime group and nine patients of the penicillin group. There were no significant differences in the CSF antibiotic concentrations (0.56 ± 0.35 µg/ml for penicillin and 0.65 ± 0.33 µg/ml for cefotaxime). CSF cefotaxime concentrations reached the minimal inhibitory concentration (MIC) 90 level in all patients, while none of the patients treated with penicillin G had CSF concentrations above the MIC 90 value as calculated on the basis of the in vitro studies of Johnson et al. (1987) and Preac-Mursic et al. (1987). CSF antibiotic concentrations above the MIC 90 value observed in all patients treated with cefotaxime offer the most hope for long-term prognosis.

24.2.3 Ceftriaxone

Ten patients with Bannwarth's syndrome were treated with ceftriaxone intravenously (2 g/day for 14 days) in a pilot study by Kristoferitsch et al. in 1989. Therapy was initiated within 4 weeks after onset of neurological symptoms in all patients. The severity of the disease was determined by rating pain, paresis of extremities, and cranial nerve paresis at the onset of therapy, 3 weeks thereafter, and 6 months after the onset of neurological symptoms. The results were compared with those of a historical control group of nine well-matched patients without antibiotic or corticosteroid therapy. Significantly more ceftriaxone recipients ($p<0.05$) improved in their clinical score 3 weeks after the onset of therapy. This was essentially due to a decrease in pain ($p<0.01$). No significant differences could be demonstrated in the course of paresis of extremities or of cranial nerve paresis. No differences were found 6 months after the onset of neurological symptoms. Similar results were found with patients treated with high-dose penicillin G (2 x 10 million U/day for 10 days) when the treatment was initiated within the first 5 weeks after the onset of the neurological symptoms (Kristoferitsch et al. 1986).

24.2.4 Cefotaxime versus Ceftriaxone

In another prospective, randomized, open trial, Pfister et al. (1991) explored the efficacy of ceftriaxone as compared with cefotaxime in early Lyme neuroborreliosis. The authors randomly assigned 33 patients with Bannwarth's syndrome and Lyme meningitis to a 10-day intravenous treatment with either 1 x 2 g/day ceftriaxone (n=17) or 3 x 2 g/day cefotaxime (n=17). Radicular pain, headache, cranial nerve pareses, and pareses of extremities usually began to improve by the 3rd – 5th day of therapy and were often gone at the end of the treatment period. Two patients in each group continued to have radicular pain and cranial nerve pareses during the entire treatment period.

Evaluation of the CSF for the determination of antibiotic concentrations was usually performed on the 3rd day of antibiotic therapy. The lumbar puncture was performed either at the time of expected maximal antibiotic concentrations (0.5 h after infusion for cefotaxime and 3.5 h after infusion for ceftriaxone) or minimal antibiotic concentrations (8 h after infusion for cefotaxime and 24 h after infusion for ceftriaxone). CSF antibiotic concentrations reached the MIC 90 level (0.06 µg/ml for ceftriaxone and 0.12 µg/ml for cefotaxime, according to Preac-Mursic et al. 1987) in nearly all patients examined. The maximum CSF ceftriaxone levels (2.31 ± 1.28 µg/ml, mean ± SD) were significantly higher than the maximum CSF cefotaxime levels (0.73 ± 0.48 µg/ml; n<0.01). However, the minimum CSF antibiotic concentrations did not differ between the groups (1.22 ± 0.52 µg/ml for ceftriaxone and 0.62 ± 0.57 µg/ml for cefotaxime; n>0.05). In conclusion, no clinical differences in treatment effect were noticed in the 33 patients treated with either 2 g ceftriaxone daily or 6 g cefotaxime daily, although discrepant maximal CSF antibiotic concentrations were found (Pfister et al. 1991).

24.2.5 Tetracyclines

Some authors recommend tetracyclines for the treatment of encephalitis caused by *B. burgdorferi* (Bryceson 1988). Anecdotally, therapy with tetracyclines has been reported to be beneficial in Lyme meningitis (Ackley and Lupovici 1986) and in meningoradiculoneuritis (Bannwarth's syndrome) (Dotevall et al. 1988). In a prospective randomized study, intravenous penicillin G (20 million U/day for 10 days) was compared with doxycyline (200 mg on the first 2 days, 100 mg on each of the following 8 days) in 75 patients with Lyme neuroborreliosis (Kohlhepp et al. 1989). In 66 patients, the disease lasted less than 6 months, but 9 patients had chronic illness at the time of therapy. No patient was considered to be a treatment failure. Evaluation 12 months after therapy showed that the treatment regimens were equally effective regarding clinical picture and laboratory findings. Preliminary results of another study performed in Sweden indicated no differences between penicillin G and doxycycline in treating Lyme neuroborreliosis (Karlsson et al. 1990).

Tetracyclines exhibit a slightly higher MIC than cephalosporins (Preac-Mursic et al. 1987), the CSF penetration is slightly less, and they are bacteriostatic antibiotics. Nevertheless, doxycycline may provide an alternative antibiotic in adults in the case of penicillin or cephalosporin allergy.

24.2.6 Corticosteroids

Previous clinical observations in patients with meningoradiculoneuritis (Bannwarth's syndrome) have suggested that corticosteroids provide prompt and sufficient improvement of pain (Schaltenbrand 1966; Pfister et al. 1985). In order to assess the efficacy of corticosteroids in the treatment of radicular pain in Bannwarth's syndrome, Pfister et al. (1988) performed a double-blind, randomized, placebo-controlled trial. The authors randomly assigned 21 patients with Bannwarth's syndrome, who complained of intense radicular pain, to a 7-day treatment with either methylprednisolone (60 mg orally per day) (n=10) or placebo (n=11). In addition, all patients were treated with antibiotics; 19 patients with penicillin G parenterally over 10 days and 2 patients, allergic to penicillin, with doxycycline orally over 4 weeks. In penicillin recipients, radicular pain began to subside 24 – 48 h after the onset of therapy. The pain disappeared completely in nearly all patients after the 4th-5th day of treatment. During the first 2 days of treatment, improvement of radicular pain, especially at night, was significantly faster when corticosteroids were added to the penicillin therapy. At the end of the trial period, radicular pain had almost completely subsided in both groups. Neurological symptoms (facial palsies or pareses of extremities) deteriorated during the treatment period in two patients of each group. The clinical deterioration may indicate that either penicillin is not the antibiotic of choice or the neurological symptoms were due to secondary, immunological, or other mechanisms following the initial borrelial infection. In addition, clinical deterioration may be explained in part by a Jarisch-Herxheimer reaction. In conclusion, Pfister et al. (1988) were not able to demonstrate noticeable differences between the treatment groups in 21 patients.

24.3 Late Neurological Involvement (Stage 3 Lyme Neuroborreliosis)

Controlled clinical treatment trials with stage 3 Lyme neuroborreliosis (chronic progressive encephalitis or encephalomyelitis, chronic peripheral neuropathy, or meningovasculitis) have not been performed. Several European case reports suggest that patients with chronic encephalomyelitis or meningovasculitis associated with Lyme borreliosis gain benefit from high-dose intravenous penicillin or ceftriaxone therapy (Ackermann et al. 1986; Hansen et al 1986; Kohlhepp et al. 1986; Wokke et al. 1987; Kohler et al. 1988; Sköldenberg et al. 1988; May and Jabbari 1990). Progressive central nervous system symptoms improved or were eliminated. However, the exact number of patients with chronic encephalomyelitis who respond to antibiotic therapy is not known.

Chronic peripheral neuropathy usually presents with intermittent limb paresthesias, without radicular pain, and without significant deficits on clinical examination (Halperin et al. 1987). Peripheral nerve dysfunction is based on neurophysiological abnormalities and is reversible with intravenous antibiotic treatment, e.g., penicillin or ceftriaxone. Rapid clinical and neurophysiological improvement following antibiotic treatment was documented in 11 of 12 American patients with chronic peripheral neuropathy (Halperin et al. 1987).

Almost half of the patients with acrodermatitis chronica atrophicans (ACA) suffer from concomitant polyneuropathy (Hopf 1966, 1975; Kristoferitsch et al. 1988). Even before the etiologic agent of Lyme borreliosis was identified, improvement with antibiotic therapy of the polyneuropathy associated with ACA was reported. Hopf reported in 1975 that neurological symptoms due to peripheral neuropathy improved in 64

(69%) of 92 patients after penicillin therapy (4 million U/day over 14 days). Despite antibiotic therapy, a progression or even new development of neurological symptoms has been reported in several cases of ACA (Kaiser 1972; Hopf 1975; Weber et al. 1988). Kaiser described a relapse of the peripheral neuropathy in 5 of 120 patients with ACA (Kaiser 1972).

In a recent study from the United States, 17 patients (63%) with chronic Lyme encephalopathy and/or polyneuropathy (one additional patient had leukencephalitis) showed improvement after a 2 week course of 2 g intravenous ceftriaxone daily, but 10 patients (37%) had relapses or did not improve (Logigian et al. 1990).

24.3.1 Ceftriaxone versus Penicillin

In a prospective, randomized study, Dattwyler et al. (1988) compared the efficacy of penicillin G (24 million/day i.v. over 10 days) with ceftriaxone (2 x 2 g/day intravenously over 14 days) in 23 patients with late manifestations of Lyme borreliosis including chronic peripheral neuropathy and encephalopathy. Ceftriaxone was found to be superior to penicillin G. Five of the ten patients of the penicillin group were considered treatment failures, while only one of 13 patients of the ceftriaxone group did not respond. An additional 31 patients were subsequently treated with ceftriaxone 4 g/day (n=17) or 2 g/day (n=14); there were no differences in these two groups. However, the data of Dattwyler et al. (1988) were difficult to interpret because the patient population included was very heterogeneous, and the treatment periods differed in both groups. In addition, the follow-up period of 3 months may be too short to consider treatment failures in chronic late complications. As is known from neurosyphilis, adequate therapy is defined as a normal CSF cell count and a falling protein content 6 months after treatment (Simon 1985). CSF findings before and after antibiotic therapy were not mentioned in the study of Dattwyler et al. (1988).

24.4 Jarisch-Herxheimer Reaction

Several authors have observed Jarisch-Herxheimer-like reactions during therapy of Lyme neuroborreliosis. Steere et al. (1983b) reported that 3 of the 12 patients treated with penicillin had a Jarisch-Herxheimer like reaction, characterized by intensification of pain and low-grade fever, during the first 18 h of therapy. In a prospective, randomized study comparing penicillin and cefotaxime, clinical deterioration, including development of facial palsy, pareses of extremities, and sensory disturbances, was evident in two patients of each group and occurred mainly on the 3rd day of therapy (Pfister et al. 1989a). It is difficult to decide whether this clinical deterioration during the course of antibiotic therapy was due to drug failure, parainfectious immunological mechanisms, or delayed Jarisch-Herxheimer reaction. As shown in a previous study, corticosteroids did not prevent clinical deterioration in patients with Bannwarth's syndrome (Pfister et al. 1989a).

Weber et al. (1986) noted a delayed Jarisch-Herxheimer-like reaction characterized by fever, chills, fatigue, generalized musculoskeletal pain, localized myalgia, and intensification of symptoms within the erythema migrans (redder lesion, more pain, increase of size, swelling) within a few days after initiation of antibiotic therapy. In the study of Dattwyler et al. (1988) 13 patients noted mild worsening of their joint pains, malaise, headaches, and/or mild fever within the first 3 days of therapy, thus suggesting a mild Jarisch-Herxheimer reaction. Of these patients, one received penicillin, three ceftriaxone (2 g/day) and nine ceftriaxone (4 g/day). Sköldenberg et al. (1988) found a mild Jarisch-Herxheimer reaction with increased fever, malaise, and nausea during the first days of antibiotic treatment in some of their patients.

Typical Jarisch-Herxheimer reaction usually occurs during the first 24 h after onset of therapy (Steere et al. 1983b; Weber 1984), as is also known from syphilis (Bryceson 1976). Whether the Jarisch-Herxheimer reaction may also occur 24 h after antibiotic therapy in Lyme neuroborreliosis is not yet known. The precise role of the Jarisch-Herxheimer reaction in the treatment of Lyme neuroborreliosis remains unclear.

24.5 Patient Selection for Treatment

The decision to initiate antibiotic therapy can be easily made if one of the following clinical settings is present:

1. Typical clinical symptoms of Bannwarth's syndrome; i.e., severe radicular pain which is worst at night, lymphocytic CSF pleocytosis, peripheral pareses of extremities, or cranial nerve palsies, especially facial nerve palsies after a previous arthropod bite and/or erythema migrans.
2. Lymphocytic meningitis with a history of an arthopod bite and/or erythema migrans, elevated antibody titers against *B. burgdorferi* in serum, or specific intrathecal antibody production in the CSF, as shown by the determination of the CSF/serum index, isoelectrofocusing affinity blot, or Western blot analysis (Stiernstedt et al. 1985; Wilske et al. 1986)
3. Other inflammatory infectious conditions (CSF pleocytosis and oligoclonal immunoglobulin G (IgG) bands in the CSF) without a history of arthropod bite or erythema migrans, but with specific intrathecal antibody production.

The decision to initiate antibiotic therapy is obviously more difficult if a neurological disorder of unknown origin is associated with elevated antibody titers against *B. burgdorferi* in the serum, but not in the CSF. Under these conditions, the causal relationship cannot be confirmed. Provided other diseases can be excluded with serological testing, antibiotic therapy seems to be justified if there are elevated serum IgM antibody titers against *B. burgdorferi* or if there is lymphocytic pleocytosis associated with elevated serum IgG antibody titers against *B. burgdorferi*.

24.6 Discussion

As in other spirochetal infections such as neurosyphilis and relapsing fever high-dose intravenous penicillin G has been effective in the treatment of Lyme neuroborreliosis. However, treatment failures with the drug have also been reported. A comparative study of penicillin G and cefotaxime in 21 patients with early Lyme neuroborreliosis did not show any demonstrable clinical superiority of one drug over the other. In other comparative therapeutic trials, ceftriaxone and doxycycline have also been shown to be effective in treating patients with early Lyme neuroborreliosis. In vitro data on the antibiotic susceptibility of B. burgdorferi indicate that ceftriaxone, cefotaxime, and doxycycline are superior to penicillin; however, problems with the in vitro testing of certain antibiotics such as penicillin do exist (see Chap. 22).

In conclusion, regardless of the patients' neurological involvement, we recommend treating Lyme neuroborreliosis with ceftriaxone (2 g/day intravenously for 14 days), cefotaxime (3x2 g/day intravenously for 14 days), or penicillin G (4x5 million U/day intravenously for 14 days). As an alternative in the case of allergy, doxycycline (2 x 100 mg orally for 14 days) may be used. In the case of treatment failure, we prefer to switch to one of the other recommended antibiotics. Presently it is not known whether prolongation of antibiotic therapy would improve the success of treatment.

24.6.1 Problems of Therapeutic Trials

The main limitations of therapeutic trials in early neurological manifestations of Lyme borreliosis include the following:

1. The chances of treatment success are confounded by the problem of spontaneous remissions of acute neurological manifestations, e.g., Bannwarth's syndrome (Pfister et al. 1985, 1989b; Kristoferitsch et al. 1986; Krüger et al. 1989). Kristoferitsch et al. (1986) studied the efficacy of high-dose penicillin therapy in 19 patients with Bannwarth's syndrome compared with a historical untreated control group. They found no clinical differences between the two groups 3 weeks after therapy and 6 months after onset of the neurological disease. Only in a subgroup of patients treated within 5 weeks after the onset of neurological disease it was shown that neuralgic pain improved significantly faster than in the untreated patients. However, there was no difference concerning the improvement of pareses of extremities or cranial nerve pareses.
   Pfister et al. (1987) reexamined 26 patients with Bannwarth's syndrome a mean of 39 months after the onset of the initial disease. The authors could not detect clinical differences in the long-term follow-up between 12 patients treated with analgesics and 14 patients treated with high-dose penicillin. This suggests a favorable spontaneous course of Bannwarth's syndrome. Similar observations were reported by Kristoferitsch (1989) from reexaminations of 8 patients who received high-dose penicillin therapy and 15 patients without antibiotic therapy. The follow-up examinations were carried out at a mean of 29 and 57 months, respectively, after the onset of neurological symptoms.
2. The number of patients with Lyme neuroborreliosis in therapeutical trials has so far been too small to evaluate slight clinical differences between treatment regimens.
3. *B. burgdorferi* can only be cultured in rare cases (Steere et al. 1983a; Preac-Mursic et al. 1986). Therefore, the isolation of borreliae from the CSF is not an adequate method for determining the success of treatment.
4. Longer follow-up periods, probably for many years, are necessary to detect relapses and late complications.
5. The optimal duration of treatment for patients with neurological manifestations of Lyme borreliosis remains unclear. The suggested period of 10 – 14 days has been arbitrarily chosen (Steere et al. 1983b; Stiernstedt et al. 1986; Dattwyler et al. 1988; Sköldenberg et al. 1988; Pfister et al. 1989a). The observation of the persistence of *B. burgdorferi* in the CSF following 10-day intravenous treatment with ceftriaxone may suggest that a treatment period of at least 14 days or more seems to be necessary to treat Lyme neuroborreliosis (Preac-Mursic et al. 1989). The most appropriate duration of therapy will only be revealed by long-term follow-up examinations in comparative studies with different treatment periods.

References

Ackermann R, Rehse-Küpper, Gollmer E (1986) Progressive borrelia encephalomyelitis. Zentralbl Bakteriol Hyg (A) 263:297-300

Ackley A Jr, Lupovici M (1986) Lyme-disease meningitis treated with tetracycline. Ann Intern Med 105:630-631 (letter)

Bjornstad RT, Mossige K (1955) Erythema chronicum migrans med meningo-polyradiculitis. Tidsskr Nor Laegeforen 75:264-265

Bryceson ADM (1976) Clinical pathology of the Jarisch-Herxheimer reaction. J Infect Dis 133:696-704

Bryceson A (1988) Tetracycline for *Borrelia* encephalitis. Lancet I:645 (letter)

Burgdorfer W, Barbour AG, Hayes SF, Benach JL, Grunwaldt E, Davis JP (1982) Lyme disease – a tick-borne spirochetosis? Science 216:1317-1319

Dattwyler RJ, Halperin JJ, Pass H, Luft BJ (1987) Ceftriaxone as effective therapy in refractory Lyme disease. J Infect Dis 155:1322-1325

Dattwyler RJ, Halperin JJ, Volkman DJ, Luft BJ (1988) Treatment of late Lyme borreliosis – randomised comparison of ceftriaxone and penicillin. Lancet II:1191-1194

Diringer MN, Halperin JJ, Dattwyler RJ (1987) Lyme meningoencephalitis: report of a severe, penicillin-resistant case. Arthritis Rheum 30:705-708

Dotevall L, Alestig K, Hanner P, Norkrans G, Hagberg L (1988) The use of doxycycline in nervous system *Borrelia burgdorferi* infections. Scand J Infect Dis (Suppl) 53:74-79

Felgenhauer K, Nekic M, Ackermann R (1982) The demonstration of locally synthetized herpes simplex IgA antibodies in CSF by a sepharose 4B linked enzyme immunoassay. J Neuroimmunol 3:149-152

Halperin JJ, Little BW, Coyle PK, Dattwyler RJ (1987) Lyme disease: cause of a treatable peripheral neuropathy. Neurology 37:1700-1706

Hansen K, Rechnitzer C, Pedersen NS, Arpi M, Jessen O (1986) Borrelia meningitis in Denmark. Zentralbl Bakteriol Hyg (A) 263:348-350

Hollström E (1951) Successful treatment of erythema migrans Afzelius. Acta Derm Venereol (Stockh) 31:235-243

Hopf HC (1966) Acrodermatitis chronica atrophicans (Herxheimer) und Nervensystem. Springer, Berlin Heidelberg New York(Monographien aus dem Gesamtgebiet der Neurologie und Psychiatrie, vol 114)

Hopf HC (1975) Peripheral neuropathy in acrodermatitis chronica atrophicans (Herxheimer). J Neurol Neurosurg Psychiatry 38:452-458

Johnson RC, Kodner C, Russel M (1987) In vitro and in vivo susceptibility of the Lyme disease spirochete, *Borrelia burgdorferi*, to four antimicrobial agents. Antimicrob Agents Chemother 31:164-167

Kaiser M (1972) Neurologische Komplikationen bei Akrodermatitis chronica atrophicans (Herxheimer) und ihre Beeinflussung durch die Penicillintherapie. Dissertation, Göttingen

Karlsson M, Stiernstedt G, Svenungsson B, Lindqvist L, Hammers S (1990) Comparison of intravenous penicillin G and oral doxycycline for treatment of Lyme neuroborreliosis – a preliminary report. Presented at the IV International Conference on Lyme Borreliosis. Stockholm, Sweden, abstract TH-L-9

Kohler J, Kern U, Kasper J, Rehse-Küpper B, Thoden U (1988) Chronic nervous system involvement in Lyme borreliosis. Neurology 38:863-867

Kohlhepp W, Mertens HG, Oschmann P (1986) Acute and chronic illness after tick-bite *Borrelia burgdorferi* infections. Zentralbl Bakteriol Hyg (A) 263:357-364

Kohlhepp W, Oschmann P, Mertens HG (1989) Treatment of Lyme borreliosis. Randomized comparison of doxycycline and penicillin G. J Neurol 236: 464-469

Kristoferitsch W (1989) Neuropathien bei Lyme-Borreliose. Springer, Wien New York

Kristoferitsch W, Baumhackl U, Sluga E, Stanek G, Zeiler K (1986) High dose penicillin in meningopolyneuritis Garin-Bujadoux-Bannwarth. Zentralbl Bakteriol Hyg (A) 263:357-364

Kristoferitsch W, Sluga E, Graf M, Partsch H, Neumann R, Stanek G, Budka H (1988) Neuropathy associated with acrodermatitis chronica atrophicans: clinical and morphological features. Ann NY Acad Sci 539:35-45

Kristoferitsch W, Baumhackl U, Stanek G, Zeiler K (1989) Ceftriaxone therapy in meningopolyneuritis Garin-Bujadoux-Bannwarth. Zentralbl Bakteriol (Suppl) 18:272-275
Krüger H, Reuss K, Pulz M, Rohrbach E, Pflughaupt KW, Martin R, Mertens HG (1989) Meningoradiculitis and encephalomyelitis due to *Borrelia burgdorferi*: a follow-up study of 72 patients over 27 years. J Neurol 236:322-328
Logigian EL, Kaplan RF, Steere AC (1990) Chronic neurologic manifestations of Lyme disease. N Engl J Med 323:1438-1444
May EF, Jabbari B (1990) Stroke in neuroborreliosis. Stroke 21:1232-1235
Pal GS, Baker JT, Wright DJM (1988) Penicillin-resistant *Borrelia* encephalitis responding to cefotaxime. Lancet I:50-51 (letter)
Pfister HW, Einhäupl KM, Garner C, Haberl R (1985) Corticosteriods versus penicillin in the treatment of meningoradiculitis of Bannwarth (Bannwarth's syndrome). J Neurol (Suppl) 231:293 (abstract)
Pfister HW, Einhäupl KM, Wilske B (1987) Langzeitverlauf nach Bannwarth-Syndrom: Vergleich antibiotisch behandelter und symptomatisch behandelter Patienten. Psycho 13: 360-363
Pfister HW, Einhäupl KM, Franz P, Garner C (1988) Corticosteroids for radicular pain in Bannwarth's syndrome. A double-blind, randomized, placebo-controlled trial. Ann NY Acad Sci 539:485-487
Pfister HW, Preac-Mursic V, Einhäupl KM, Wilske B (1989a) Cefotaxime versus penicillin G for acute neurological manifestations of Lyme borreliosis: a prospective randomized study. Arch Neurol 46:1190-1194
Pfister HW, Wilske B, Preac-Mursic V, Einhäupl KM (1989b) Clinical and serological long-term follow up in patients with Bannwarth's syndrome: comparison between untreated and penicillin treated patients. Zentralbl Bakteriol (Suppl) 18:276-279
Pfister HW, Preac-Mursic V, Wilske B, Schielke E, Soergel F, Einhäupl KM (1991) Randomized comparison of ceftriaxone and cefotaxime in Lyme neuroborreliosis. J Infect Dis 163:311-318
Preac-Mursic V, Wilske B, Schierz G (1986) European *Borrelia burgdorferi* isolated from humans and ticks. Culture conditions and antibiotic susceptibility. Zentralbl Bakteriol Hyg (A) 263:112-118
Preac-Mursic V, Wilske B, Schierz G, Holmburger M, Süß E (1987) In vitro and in vivo susceptibility of *Borrelia burgdorferi*. Eur J Clin Microbiol 6:424-426
Preac-Mursic V, Weber K, Pfister HW, Wilske B, Gross B, Baumann A, Prokop J (1989) Survival of *Borrelia burgdorferi* in antibiotically treated patients with Lyme borreliosis. Infection 17:355-359
Schaltenbrand G (1966) Durch Arthropoden übertragene Infektionen der Haut und des Nervensystems. Munch Med Wochenschr 108:1557-1562
Simon RP (1985) Neurosyphilis. Arch Neurol 42:606-613
Sköldenberg B, Stiernstedt G, Garde A, Kolmodin G, Carlström A, Nord CE (1983) Chronic meningitis caused by a penicillin-sensitive microorganism? Lancet II:75-78
Sköldenberg B, Stiernstedt G, Karlsson M, Wretlind B, Svenugsson B (1988) Treatment of Lyme borreliosis with special emphasis on neurological disease. Ann NY Acad Sci 539:317-323
Steere AC, Grodzicki RL, Kornblatt AN, Craft JE, Barbour AG, Burgdorfer W, Schmid GP, Johnson E, Malawista SE (1983a) The spirochetal etiology of Lyme disease. N Engl J Med 308:733-740
Steere AC, Pachner AR, Malawista SE (1983b) Neurologic abnormalities of Lyme disease: successful treatment with high-dose intravenous penicillin. Ann Intern Med 99:767-772
Stiernstedt G, Sköldenberg B, Vandvik B, Hederstedt B, Garde A, Kolmodin G, Jörböck H, Svenungsson B (1984) Chronic meningitis and Lyme disease in Sweden. Yale J Biol Med 57:491-497
Stiernstedt G, Granström M, Hederstedt B, Sköldenberg B (1985) Diagnosis of spirochetal meningitis by enzyme-linked immunosorbent assay and indirect immunofluorescence assay in serum and cerebrospinal fluid. J Clin Microbiol 21:819-825
Stiernstedt G, Sköldenberg B, Garde A, Kolmodin G, Jörbeck H, Svenungsson B, Carlström A (1986) Clinical manifestations of *Borrelia* infections of the nervous system. Zentralbl Bakteriol Hyg (A) 263:289-296

Vandvik B, Sköldenberg B, Forsgren M, Stiernstedt G, Jeansson S, Norrby E (1985) Long-term persistence of intrathecal virus-specific antibody responses after herpes simplex virus encephalitis. J Neurol 231:307-312

Weber K (1974) Erythema-chronicum-migrans-Meningitis – eine bakterielle Infektionskrankheit? Munch Med Wochenschr 116:1993-1998

Weber K (1984) Jarisch-Herxheimer-Reaktion bei Erythema-migrans-Krankheit. Hautarzt 35:588-590

Weber K, Puzik A, Becker T (1983) Erythema-migrans-Krankheit. Dtsch Med Wochenschr 108:1182-1190

Weber K, Neubert U, Thurmayr R (1986) Antibiotic therapy in early erythema migrans disease and related disorders. Zentralbl Bakteriol Hyg (A) 263:377-388

Weber K, Preac-Mursic V, Neubert U, Thurmayr R, Herzer P, Wilske B, Schierz G, Marget W (1988) Antibiotic therapy of early European Lyme borreliosis and acrodermatitis chronica atrophicans. Ann NY Acad Sci 539:324-345

Wilske B, Schierz G, Preac-Mursic V, von Busch K, Kühbeck R, Pfister HW, Einhäupl KM (1986) Intrathecal production of specific antibodies against *Borrelia burgdorferi* in patients with lymphocytic meningoradiculitis (Bannwarth's syndrome). J Infect Dis 153:304-314

Wokke JHJ, van Gijn J, Elderson A, Stanek G (1987) Chronic forms of *Borrelia burgdorferi* infection of the nervous system. Neurology 37:1031-1034

25 Therapy of Joint Manifestations

P. HERZER

25.1 Introduction

Antibiotic therapy is considered the mainstay of treatment for joint manifestations of Lyme borreliosis. However, many questions remain unanswered, and many difficulties hinder the approach to therapy; for example (a) Large-scale placebo-controlled treatment studies are lacking; (b) uncontrolled or comparative studies do not take spontaneous remissions into account; in the report by Steere et al. (1987) of 21 untreated patients with Lyme arthritis, the number of patients with attacks of arthritis decreased by 10% – 20% each year; (c) criteria for judgement of treatment success or failure are defined arbitrarily and heterogenously; and (d) there are uncertainties in the diagnosis of Lyme arthritis and, in particular, as regards arthralgias due to borrelial infection.

25.2 Antibiotics

25.2.1 Treatment of Arthritis

The evidence that antibiotics can cure Lyme arthritis is founded on only one double-blind placebo-controlled study in which 2.4 million U of intramuscular benzathine penicillin was administered weekly for 3 weeks (Steere et al. 1985). Seven of 20 patients treated with penicillin had complete resolution of arthritis within a mean of 4 weeks from the onset of treatment. In contrast, all 20 patients randomized to placebo had persistent disease, which is somewhat surprising with regard to the commonly intermittent course of Lyme arthritis and to the probability that Lyme arthritis also resolves spontaneously. In a subsequent, uncontrolled study reported in the same paper, the administration of intravenous penicillin G, 20 million U/day for 10 days, resulted in immediate improvement and in complete resolution of arthritis within "several weeks" in 11 of 20 patients. In our patients with Lyme arthritis, treatment with 20 million U/day of intravenous penicillin G for 14 days (one patient) or for 21 days (ten patients) led to a similar outcome; four of the 11 patients had persistent arthritis for more than 3 months, and thus were deemed to be treatment failures; the seven patients who had complete resolution of arthritis within 3 months after treatment have remained free of disease for a mean follow-up period of 20 months (Herzer 1988, 1989). Huaux et al. 1988, on the other hand, reported that each of ten patients treated with intravenous penicillin G, 20 million U/day for 10 days, recovered rapidly.

A randomized study comparing intravenous penicillin (24 million U/day for 10 days) and intravenous ceftriaxone (4 g/day for 14 days) in the treatment of late Lyme borreliosis seemed to suggest that ceftriaxone is superior to penicillin (Dattwyler et al. 1988). Three months after treatment, five of ten patients who received penicillin were judged treatment failures, as compared with one of 13 treated with ceftriaxone; in an additional 31 patients treated with ceftriaxone 2 g/day (n=17) or 4 g/day (n=14) the rates of response in both groups resembled that in the patients randomized to ceftriaxone treatment. This study included patients with arthralgias, arthritis, and neurologic symptoms. The apparently high rate of patients with concurrent joint and neurologic manifestations of Lyme borreliosis in this study is uncommon in our experience.

It has been also reported that intravenous cefotaxime (2 x 3 g/day for 8-10 days) is superior to intravenous penicillin G (2 x 10 million U/day for 8-10 days) in the treatment of late Lyme borreliosis (Hassler et al. 1990). The spectrum of the disease in the patients of this study is somewhat unusual compared with previous descriptions of late Lyme borreliosis, in particular with regard to frequent cardiac abnormalities. Because of the many ambiguous symptoms attributed to late Lyme borreliosis in the patients of this study, more information on objective clinical findings would have been of interest. Moreover, when the response rates relating to each individual manifestation in this study are examined, it is revealed that only patients with "peripheral neuropathy" improved significantly more often when using cefotaxime than penicillin. In any case, it seems inappropriate to evaluate treatment outcomes of different manifestations of Lyme borreliosis together, since different pathogenetic factors may be crucial in the particular syndromes.

It has been suggested that oral antibiotics may also be effective in the treatment of Lyme arthritis. Schaad et al. 1986 reported overall beneficial effects of oral antibiotic regimens; seven children with Lyme arthritis responded to oral penicillin given for 2 – 4 weeks, and two children responded to a 3-week regimen of oral doxycycline. In our patients treated with oral penicillin (4 x 1.5 million U/day for 6 – 10 weeks), 4 of 11 had resolution of arthritis within 3 months, as compared with 7 of 11 patients treated with intravenous penicillin G (Herzer 1989). In a randomized trial comparing a 1-month course of doxycycline (2 x 100 mg per day) with one of amoxicillin and probenecid (both 4 x 500 mg per day), 13 of 18 patients (72%) given doxycycline responded, as did 11 of 18 (61%) given amoxicillin and probenecid (Liu et al. 1989). When patients who did not respond to oral antibiotics were treated either with intravenous penicillin (20 million U/day for 14 days) or intravenous ceftriaxone (2 g or 4 g/day for 14 days), 5 of 14 penicillin-treated patients (36%) and 6 of 15 ceftriaxone-treated patients (40%) improved (also reported by Liu et al. 1989).

Regarding the antibiotic treatment of Lyme arthritis, it should be mentioned that mild Jarisch-Herxheimer-like reactions have occasionally been observed (Herzer et al. 1986; Dattwyler et al. 1988).

25.2.1.1 Conclusions

None of the antibiotic regimens used up to now is universally effective in the treatment of Lyme arthritis. There are no convincing or unchallenged data in absolute favor of any one of these regimens. Thus, questions remain as to the optimal antibiotic agent as well as to the optimal duration of therapy. Meanwhile, matters of compliance, cost, and thus the practicability of outpatient treatment may be decisive. Oral antibiotics may be tried first, e.g., 200 mg doxycycline or 4 x 500 mg amoxicillin daily for 1 month. If intravenous therapy is considered necessary, ceftriaxone (2 g/day for 2 weeks) has the advantage that it can be given to outpatients, since its long half-life permits once-a-day administration. Intravenous penicillin should be given in at least four separate doses and cefotaxime three times daily, which usually entails hospitalization.

Apart from the uncertainty about the optimal antibiotic treatment for Lyme arthritis, the question may be raised whether Lyme arthritis is in any case an "infective" arthritis in which live spirochetes are crucial. It is also conceivable that nondegradable spirochetal fragments in the synovium may elicit an immune response which accounts for synovial inflammation – and that the antibiotics may be fighting for a lost cause.

25.2.2 Treatment of Arthralgias

There is no adequate or conclusive experience on the treatment of patients with arthralgias conjecturally due to Lyme borreliosis. In patients who have a history and/or serological findings suggestive of Lyme borreliosis, and who complain of otherwise unexplained musculoskeletal pain, we usually administer 200 mg/day oral doxycyline for 1 month. Only a few of the patients thus treated have shown improvement. All in all, we are still uncertain about the significance of antibiotic therapy for arthralgias – just as we are invariably generally uncertain in diagnosing Lyme borreliosis in patients with these most ambiguous symptoms.

25.3 Antirheumatic Drugs

Treatment with nonsteroidal anti-inflammatory drugs (NSAIDs) may provide relief of symptoms. However, it is unlikely that NSAIDs will modify the course of Lyme arthritis. The use of intra-articular and systemic corticosteroids has been associated with an increased failure of antibiotic therapy (Steere et al. 1985; Dattwyler et al. 1988). There is in any case no rationale for the use of corticosteroids prior to antibiotics. In patients who did not respond to antibiotic treatment, we have occasionally administered intra-articular corticosteroids, but this led only to a temporary improvement.

In one case, the remission of chronic Lyme arthritis was attributed to hydroxychloroquine therapy (Coblyn and Taylor 1981). However, chloroquine proved to be ineffective in both patients with chronic Lyme arthritis treated by us (Herzer et al. 1986; Herzer 1989). We also used sulfasalazine (4 g/day) to treat two patients with chronic Lyme arthritis who did not respond to antibiotic therapy, but neither improved. The use of other second-line antirheumatic drugs or immunosuppressive agents will hardly ever be appropriate in view of the overall good prognosis of Lyme arthritis and in consideration of their potential severe side effects.

25.4 Synovectomy

In patients with chronic Lyme arthritis who are unresponsive to medical treatment, synovectomy may become inevitable. However, synovectomy does not preclude a relapse affecting other joints (Mc Laughlin et al. 1986; Herzer 1989).

References

Coblyn JS, Taylor P (1981) Treatment of chronic Lyme arthritis with hydroxychloroquine. Arthritis Rheum 24:1567-1569

Dattwyler RJ, Volkman DJ, Halperin JJ, Luft BJ (1988) Treatment of late Lyme borreliosis – randomized comparison of ceftriaxone and penicillin. Lancet I:1191-1194

Hassler D, Zöller L, Haude M, Hufnagel H-D, Heinrich F, Sonntag H-G (1990) Cefotaxime versus penicillin in the late stage of Lyme disease – prospective, randomized therapeutic study. Infection 18:16-20

Herzer P (1988) Antibiotische Therapie der Lyme-Arthritis. Klin Wochenschr 66 (Suppl XIII):119

Herzer P (1989) Lyme-Borreliose. Epidemiologie, Ätiologie, Diagnostik, Klinik und Therapie. Steinkopff, Darmstadt, Germany

Herzer P, Wilske B, Preac-Mursic V, Schierz G, Schattenkirchner M, Zöllner N (1986) Lyme arthritis: clinical features, serological, and radiographic findings of cases in Germany. Klin Wochenschr 64:206-215

Huaux JP, Bigaignon G, Stadtsbaeder S, Zangerlé PF, Nagant de Deuxchaisnes C (1988) Pattern of Lyme arthritis in Europe: report of 14 cases. Ann Rheum Dis 47:164-165

Liu NY, Dinerman H, Levin RE, Massarotti E, Molloy PJ, Schoen RT, Taylor E, Steere AC (1989) Randomized trial of doxycycline vs. amoxicillin/probenecid for the treatment of Lyme arthritis: treatment of non-responders with i.v. penicillin or ceftriaxone. Arthritis Rheum (Suppl) 32:S46

Mc Laughlin TP, Zemel L, Fisher RL, Gossling HR (1986) Chronic arthritis of the knee in Lyme disease. Review of the literature and report of two cases treated by synovectomy. J Bone Joint Surg 68A:1057-1060

Schaad UB, Flüeler U, Schaub H, Suter H, Vischer D, Caflisch U, Tschumi A, Wick H, Vest M, Durrer D (1986) Durch *Ixodes-ricinus*-Spirochäten (*Borrelia burgdorferi*) verursachte Krankheitsbilder (Lyme-Krankheit) bei pädiatrischen Patienten in der Schweiz. Schweiz Med Wochenschr 116:1426-1430

Steere AC, Green J, Schoen RT, Taylor E, Hutchinson GJ, Rahn DW, Malawista SE (1985) Successful parenteral penicillin therapy of established Lyme arthritis. N Engl J Med 312:869-874

Steere AC, Schoen RT, Taylor E (1987) The clinical evolution of Lyme arthritis. Ann Intern Med 107:725-731

26 Therapy of Lyme Carditis

W. MAYER-BERGER, M.R. VAN DER LINDE, and D. HASSLER

26.1 Introduction

The therapy of Lyme carditis consists of two components: first, symptomatic treatment of the manifestations and second, causative treatment to eradicate the etiologic agent, *Borrelia burgdorferi*.

Lyme carditis usually takes a favorable course, and it may, therefore, be difficult to demonstrate any positive effect of therapy. In a review of 45 patients with atrioventricular (AV) block, the disorder resolved in all but three of them within 6 weeks (Mc Alister et al. 1989). Incomplete recovery occurred in 5 of 66 European patients (see Chap.10). Third-degree AV block lasting for more than 10 days seems to be extremely rare (Linde et al. 1990). No case of recurrent heart block has been reported to our knowledge. Whether the heart may be involved in the chronic stage of the disease remains controversial, although recent findings seem to support this hypothesis (Hassler et al. 1987; Mc Alister et al. 1989; Linde et al. 1990; Mayer et al. 1990; Stanek et al. 1990).

26.2 Therapy

26.2.1 Symptomatic Therapy

The most frequent abnormality of Lyme carditis is fluctuating AV block (49 out of 66 European patients; see Chap. 10). If Lyme borreliosis is suspected and AV block exists, i.e., first-degree AV block exceeding a PR interval of 0.3, second-degree AV block, or complete heart block, the patient should be hospitalized. Electrocardiographic (ECG) monitoring is obligatory, and insertion of a temporary pacemaker is necessary if there is critical bradycardia with symptoms of hemodynamic instability (Table 1).

Complete heart block is due to a lesion located at or above the level of the AV node in most patients, as is suggested by narrow QRS escape rhythms with rates usually above 40 beats per minute. Electrophysiologic studies usually reveal a supra-hisian block as a prolonged interval between atrium and the bundle of His. In this situation, atropine may be efficacious in bridging an emergency situation until initiation of pacemaker therapy. In two patients, atropine improved the AV block (Steere et al. 1977; Blaauw et al. 1988), while in four others it was not effective (Bedell et al. 1981; Ponsonnaille et al. 1986; Kishaba et al. 1988; Linde et al. 1990). Isoproterenol was only once used with success (Bedell et al. 1981), otherwise it was not helpful (Kishaba et al. 1988; Linde et al. 1990).

An infra-hisian block, seen as a prolonged interval between the bundle of His and ventricle in electrophysiologic studies, consists of a block below the AV node with slow escape rhythm, usually below 40 beats per minute, wide QRS pattern, and tran-

Table 1. Symptomatic therapy of Lyme carditis

Abnormality	Therapeutic measure
AV block (clinically asymptomatic)	Monitoring
AV block (symptomatic and/or hemodynamically compromising high-degree heart block)	Temporary pacemaker
Tachyarrhythmias	Monitoring (caution with negative inotropic and chronotropic drugs)
Perimyocarditis (without rv/lv dysfunction)	Immobilization
Perimyocarditis (with rv/lv dysfunction)	Oxygen, additional drugs (diuretics, vasodilators, caution with digitalis)

AV, atrioventricular; rv/lv, right/left ventricular

sient asystole. This highly dangerous situation necessitates prompt insertion of a temporary pacemaker at an intensive care unit. Temporary pacemakers have been used in 13 of 44 (29%) European patients with high-grade AV block (Chap.10). However, complete heart block rarely persists for more than 1 week. Thus, insertion of a permanent pacemaker should be postponed for at least 10-14 days to avoid an unnecessary procedure, even with unstable escape foci (Linde et al. 1989; Mc Alister et al. 1989). The risk of catheter infection must also be considered. Only one report described an indication for permanent pacemaker therapy because of a symptomatic persistent complete heart block due to Lyme carditis (Linde et al. 1990).

Atrial tachyarrhythmias such as atrial tachycardia, atrial flutter, or atrial fibrillation may also occur (Hassler et al. 1987; Ballmer and Hany 1988; see Chap. 10). Symptomatic treatment of these conditions may lead to complications and should be performed only under ECG monitoring. All of the drugs which are effective in treating these atrial tachyarrhythmias, i.e., digitalis or verapamil, may produce AV conduction disturbances or aggravate an underlying AV block and induce bradyarrhythmia (Table 1).

Patients with perimyocarditis should be immobilized for at least 1 week and observed for the development of frank heart failure; oxygen may be administered. Cardiomegaly has been observed in the United States as well as in Europe at X-ray examination (Steere et al. 1980a; Marcus et al. 1985; Reznick et al. 1986; see Chap.10). In addition, signs of left ventricular dysfunction on radionuclide ventriculography or echocardiography have been reported (Steere et al. 1977, 1980a; Rienzo et al. 1987; Mc Alister et al. 1989; see Chap.10), but since the severity of these findings has subsided within a few days, no specific therapy seemed to be necessary. In the case of congestive heart failure (11 out of 66 European patients; see Chap.10), the general guidelines for therapy should be followed, but digitalis should be used with care, as mentioned above (Table 1).

Pericardial effusion has also been observed (Lavaud et al. 1985; Hansen and Madsen 1986; Lorcerie et al. 1987; Veyssier et al. 1987), but has so far never necessitated pericardiocentesis because of hemodynamic instability.

If patients with complete heart block or congestive heart failure do not respond to antibiotic therapy, corticosteroids have been recommended (Steere 1989). This recommendation is, however, controversial, because undue interference with antibiotic treatment may arise (Dattwyler et al. 1988). Salicylates are no longer in use.

26.2.2 Causative Therapy

In clinically established Lyme carditis, appropriate antibiotic therapy should be initiated. Our recommendation are based on the documented cases of Lyme carditis in the United States and Europe, as well as on experience in the other stage 2 manifestations of Lyme borreliosis. We recognize that our recommendations are preliminary, since the small number of patients has so far made it impossible to carry out systematic trials on the antibiotic therapy of Lyme carditis. Because of ethical problems, double-blind and placebo-controlled studies cannot be carried out, since possible late manifestations may occur with placebo or ineffective therapy (Steere et al. 1980b). The decision for a specific regimen should consider the multisystem pattern of Lyme borreliosis. An intensive search for signs and symptoms other than cardiac ones is obligatory.

In a review on 36 US patients (Mc Alister et al. 1989), four were treated with oral tetracycline, 14 with intravenous penicillin, and one with ceftriaxone. The 66 European patients (Chap. 10) received tetracycline (n=16), penicillin (n=40, with very different regimens) and third-generation cephalosporins (n=11). Nearly all the patients treated showed significant improvement within 6 weeks. Even without antibiotic therapy, the prognosis of Lyme carditis is good (Steere et al. 1980a) (see Chap. 10, Table 2).

In the United States, only one patient showed slight AV conduction delay for 16 months (Mc Alister et al. 1989), and two patients without antibiotic therapy had persistent first-degree AV block (and even episodes of Wenckebach) for up to 5 months (Steere et al. 1977, 1980a). In Europe, one patient with first-degree AV block for 4 – 5 months was reported with tetracycline therapy (Hansen and Madsen 1986), one patient had first-degree AV block for 2 years after treatment with ceftriaxone (Mayer et al. 1990), and a third patient had persistent complete heart block despite therapy with penicillin and a ceftriaxone/corticosteroid combination (Linde et al. 1990).

Studies on the therapy of second-stage neurologic involvement (Steere et al. 1983b; Kohlhepp et al. 1987; Pfister et al. 1989) and on the manifestations of the third stage of Lyme borreliosis (Steere et al. 1985; Dattwyler et al. 1988; Hassler et al. 1990) have shown the efficacy of high doses of penicillin as well as the third-generation cephalosporins ceftriaxone and cefotaxime. Because of the excellent in vitro susceptibility of *B. burgdorferi* to ceftriaxone and cefotaxime (Johnson et al. 1987; Preac-Mursic et al. 1987), these agents have recently been used extensively for the therapy of Lyme borreliosis. Studies by Dattwyler et al. (1988), Pfister et al. (1989), and Hassler et al. (1990) have provided some evidence that these cephalosporins may be more efficacious than penicillin.

According to these experiences and to a recent publication (Steere 1989), the oral regimens used in treating early infection may also be sufficient in first-degree AV block with a PR interval of less than 0.3 s. Outpatient treatment is justified if the patient visits daily for cardiac evaluation, e.g., with ECG, otherwise hospitalization is necessary. However, since AV block tends to fluctuate rapidly and other possible manifestations of Lyme borreliosis may be silent, we prefer intravenous antibiotic therapy, even for first-degree AV block with a PR interval of less than 0.3 s. All other grades of AV block and cardiomegaly should definitely be treated with high doses of penicillin or third-generation cephalosporins (Table 2). Careful follow-up of patients is necessary

Table 2. Antibiotic therapy of Lyme carditis

Clinical set-up	Antibiotic	Dose/day	Duration (days)[a]
AV block or cardiomegaly[b]	Ceftriaxone i.v.	1 x 2 g	10 – 14
	or		
	cefotaxime i.v.	2 x 2 g	10 – 14
	or		
	penicillin G i.v.	20 million U (4 x 5)	14
First-degree AV block < 0.3 s[c]	Tetracycline oral	3 – 4 x 500 mg	10 – 30 (?)
	or		
	doxycycline oral	2 x 100 mg	10 – 30 (?)
	or		
	erythromycin[d] oral	3 – 4 x 500 mg	10 – 30 (?)

i.v., intravenous
[a] Based on clinical response
[b] Usual regimen
[c] Possible alternative
[d] Efficacy controversial

to recognize persistent or late manifestations. ECG recordings and Holter monitoring are useful in documenting intermittent AV block.

To our knowledge, only one patient below the age of 15 years with Lyme carditis has been observed so far. Kishaba et al. (1988) reported on a 12-year-old boy with symptomatic heart block after tick exposure and erythema migrans. The patient needed a temporary pacemaker for 2 weeks. He was treated successfully with penicillin and aspirin. Because of the lack of experience, no recommendation for the antibiotic treatment of children is made in this chapter.

26.2.3 Follow-Up in 66 European Patients

Complete resolution was found in 92% of the 66 European patients; in 82% this occurred shortly after antibiotic therapy, and in 10%, spontaneously (see Chap.10, Table 2). In four patients (6%), minor conduction or rhythm disturbances remained, while in one patient (2%) a persisting complete heart block was found (Linde et al. 1990). This is the only patient in the literature in whom a pacemaker implantation was necessary because of Lyme carditis.

26.2.4 Complications of Antibiotic Therapy

Besides the generally known side effects of antibiotic therapy, one very important consideration is the possibility of a Jarisch-Herxheimer-like reaction. This phenomenon consists of a triad of events: body temperature rises and then falls; existing lesions worsen, especially the dermatological, rheumatological, and neurological signs and symptoms; and vasoconstriction occurs, followed by vasodilatation with falling blood pressure, as has been shown in investigations of patients with relapsing fever, another borrelial disease (Bryceson 1976). Such reactions appear on the first and second day of therapy and are well known in other spirochetal diseases such as syphilis. In louse-borne relapsing fever caused by *B. recurrentis*, particularly severe reactions with fatal outcome have been reported in undernourished Ethiopians (Zein 1987).

The incidence of Jarisch-Herxheimer-like reactions in patients with erythema migrans has been estimated to be between 14% and 23% (Steere et al. 1983a; Weber et al. 1990). Mild reactions were seen in 21% of patients treated with penicillin and in 40% of those treated with cefotaxime for late manifestations of Lyme borreliosis (Hassler et al. 1990). Aggravation of existing cardiac lesions were reported in some patients in Europe (Weber 1984; Hassler et al. 1990).

We recommend careful supervision of patients during the first 3 days of antibiotic treatment in order to recognize a Jarisch-Herxheimer-like reaction. Although prophylaxis of this reaction by a single dose of corticosteroids combined with the first dose of the antibiotic may be controversial (Pfister 1988), as in syphilis, 25-50 mg prednisone may be administered intravenously.

References

Ballmer PE, Hany A (1988) Lyme-Karditis (Lyme carditis). Schweiz Med Wochenschr 118 (10) : 358-362

Bedell SE, Pastor BM, Cohen SI (1981) Symptomatic high grade heart block in Lyme disease. Chest 79:236-237

Blaauw AAM, Braat S, Van Santen-Hoeufft HM, Van der Linden SJ (1988) Cardiologische afwijkingen bij patienten met Lyme-borreliose. Ned Tijdsch Geneeskd 132 (46) : 2111-2114

Bryceson ADM (1976) Clinical pathology of the Jarisch-Herxheimer reaction. J Infect Dis 133 (10) : 696-704

Dattwyler RJ, Halperin JJ, Volkmann DJ, Luft BJ (1988) Treatment of late Lyme borreliosis – randomised comparison of ceftriaxone and penicillin. Lancet II : 1191-1194

Hansen K, Madsen JK (1986) Myocarditis associated with tickborne *Borrelia burgdorferi* infection. Lancet I : 1323-1324

Hassler D, Zipperle G, Ackermann R, Lembke U, Heinrich F (1987) Kardiale Beteiligung auch bei der europäischen Erythema-migrans-Borreliose? Dtsch Med Wochenschr 112:1506-1508

Hassler D, Zöller L, Hufnagel HD, Heinrich F, Sonntag HG (1990) Cefotaxime versus penicillin in the late stage of Lyme disease – prospective, randomized therapeutic study. Infection 18:16-20

Johnson RC, Kodner C, Russel M (1987) In vitro and in vivo susceptibility of Lyme disease spirochete, *Borrelia burgdorferi*, to four antimicrobial agents. Antimicrob Agents Chemother 31:164-167

Kishaba RG, Weinhouse E, Chusid MJ, Nudel DB (1988) Lyme disease presenting as heart block. Clin Pediatr (Phila) 27 (6) : 291-293

Kohlhepp W, Mertens HG, Oschmann P, Rohrbach E (1987) Akute und chonische Erkrankungen bei zeckenvermittelter Borreliose. Nervenarzt 58:557-563

Lavaud P, Etienne J, Chamot E, Atallah G, Moleur P, Brun Y, Touboul P (1985) Bloc auriculo-ventriculaire aige associé une maladie de Lyme. Un cas confirmé par serologie (letter). Press Med 14 (39) : 2020

Linde MR van der, Crijns HJGM, Lie KI (1989) Transient complete AV block in Lyme disease – electrophysiologic observations. Chest 96:219-221

Linde MR van der, Crijns HJGM, de Koning J, Hoogkamp-Korstanje JAA, de Graaf JJ, Piers DA, van der Galien A, Lie KI (1990) Range of atrio-ventricular conduction disturbances in Lyme borreliosis: a report of four cases and review of other published reports. Br Heart J 63:162-168

Lorcerie B, Boutron MC, Portier H, Beuriat P, Ravisy J, Martin F (1987) Manifestations péricardiaques de la maladie de Lyme. Ann Med Interne (Paris) 138 (8) : 601-603

Marcus LC, Steere AC, Duray PH, Anderson AE, Mahoney EB (1985) Fatal pancarditis in a patient with coexistent Lyme disease and babesiosis: demonstration of spirochetes in the myocardium. Ann Intern Med 103:374-376

Mayer W, Kleber FX, Wilske B, Preac-Mursic V, Maciejewski W, Sigl H, Holzer E, Doering W (1990) Persistent atrioventricular block in Lyme Borreliosis, Klin Wochenschr 68:431-435
Mc Alister HF, Klementowicz PT, Andrews C, Fisher JD, Feld M, Furmans S (1989) Lyme carditis: an important cause of reversible heart block. Ann Intern Med 110:339-345
Pfister HW (1988) Wie wird die neurologisch manifeste Lyme-Borreliose behandelt? Nervenarzt 59:687-689
Pfister HW, Preac-Mursic V, Wilske B, Einhäupl KM (1989) Cefotaxime vs penicillin G for acute neurologic manifestations in Lyme borreliosis. Arch Neurol 46:1190-1194
Ponsonaille J, Citron B, Karsenty B, D'Agrosa MC, Peyselon P, Veyre A, Gras H (1986) Myocardite aige au cours d'un syndrome de Lyme. Intérèt de la scintigraphie myocardique au gallium 67. Arch Mal Coeur 79 (13) : 1946-1950
Preac-Mursic V, Wilske B, Schierz G, Holmburger M, Süß E (1987) In vitro and in vivo susceptibility of *Borrelia burgdorferi*. Eur J Clin Microbiol 6 :424-426
Reznick JW, Braunstein DB, Walsh RL, Smith CR, Wolfson PM, Gierke LW, Gorelkin L, Chandler FW (1986) Lyme carditis. Electrophysiologic and histopathologic study. Am J Med 81:923-927
Rienzo RJ, Morel DE, Prager D, Barron L, Post R (1987) Gallium avid Lyme myocarditis. Clin Nucl Med 12:475-476
Stanek G, Klein J, Bittner R, Glogar D (1990) Isolation of *Borrelia burgdorferi* from the myocardium of a patient with longstanding cardiomyopathy. N Engl J Med 322:249-252
Steere AC (1989) Lyme disease. N Engl J Med 321:586-596
Steere AC, Malawista SE, Hardin JA, Ruddy S, Askenase PW, Andiman WA (1977) Erythema chronicum migrans and Lyme arthritis – the enlarging clinical spectrum. Ann Intern Med 86:685-698
Steere AC, Batsford WP, Weinberg M, Alexander J, Berger HJ, Wolfson S, Malawista SE (1980a) Lyme carditis: cardiac abnormalities of Lyme disease. Ann Intern Med 93:8-16
Steere AC, Malawista SE, Newman JH, Spieler PN, Bartenhagen NH (1980b) Antibiotic therapy in Lyme disease. Ann Intern Med 93:1-8
Steere AC, Hutchinson GJ, Rahn DW, Sigal LH, Craft JE, De Sanna ET, Malawista SE (1983a) Treatment of the early manifestations of Lyme disease. Ann Intern Med 99:22-26
Steere AC, Pachner AR, Malawista SE (1983b) Neurologic abnormalities of Lyme disease: successful treatment with high-dose intravenous penicillin. Ann Intern Med 99:767-772
Steere AC, Green J, Schoen RT, Taxlor E, Hutchinson GJ, Rahn DW, Malawista SE (1985) Successful parenteral penicillin therapy of established Lyme arthiritis. N Engl J Med 312:869-874
Veyssier P, Davous N, Kaloustian E, Maitre B, Lallement PY, Serret A (1987) Atteintes cardiaques au cours de la maladie de Lyme. Rev Med Interne 8 (4) : 357-360
Weber K (1984) Jarisch-Herxheimer-Reaktion bei Erythema-migrans-Krankheit. Hautarzt 35:588-590
Weber K, Preac-Mursic V, Wilske B, Thurmayr R, Neubert U, Scherwitz C (1990) A randomized trail of ceftriaxone versus oral penicillin for the treatment of early European Lyme borreliosis. Infection 18:91-96
Zein AZ (1987) Louse-borne relapsing fever (LBRF): mortality and frequencey of Jarisch-Herxheimer-reaction. J Roy Soc Health 4:146-147

27 Therapy of Tick Bite

K. WEBER and W. BURGDORFER

27.1 Introduction

In Europe, various species of ticks can be infected with several different pathogenic microorganisms, e.g., *Borrelia burgdorferi* (see Chap. 5), tick-borne encephalitis virus, and *Rickettsia conorii*. Prerequisite for proper management of a tick bite is the awareness of whether the tick in question stems from an area that is endemic for any or several infectious agents. For instance, in an area endemic for borreliae and tick-borne encephalitis virus, the patient might need passive immunization against the tick-borne encephalitis virus, unless he or she is protected by active immunization; measurements to counteract a possible borrelial infection may also be necessary. The following Chapter will solely be concerned with tick bites in areas which are endemic for the *B. burgdorferi* infection (see Chap. 29).

Transmission of *B. burgdorferi* by its established tick vectors (in Europe, mainly *Ixodes ricinus*) occurs via the saliva or possibly also by regurgitation of infectious gut content. In the majority of infected ticks, spirochetes are only found in the midgut, from where, during the tick's engorgement, they penetrate the gut epithelium and invade via hemolymph the tissues of the salivary glands. Transmission via the saliva occurs during the later stages of the feeding process. This time-consuming development of *B. burgdorferi* has led to speculation that short-term attachment of infected ticks may not be infective. However, immediate transmission of spirochetes via saliva is possible by systemically infected ticks and possibly also by regurgitation of infectious gut material (Burgdorfer et al. 1991). Therefore, attached ticks should be removed immediately upon detection.

The main goal in removing attached ticks is to prevent the borreliae or other infectious material from entering the body through the tick bite wound; one must hope that an infection has not occurred already. Care should be taken to leave the hypostome (commonly referred to as "head") or mouthparts of the tick intact during removal. If these tick remnants are left in the skin, they may become the site of chronic irritation and may lead to nonspecific inflammation. Contrary to a widely held belief, however, this nonspecific reaction is far less important than infection with a pathogenic agent.

27.2 Removal of Ticks

For the proper removal of ticks, the following procedures are recommended:

1. Use blunt forceps or tweezers; pointed forceps are not recommended because they readily puncture partially or fully engorged ticks.
2. Grasp the tick as close to the skin as possible and repeatedly pull upward with increasing pressure. Do not jerk, as this may cause the hypostome to break off. Specially designed forceps allowing the attached tick to be grasped and turned

clockwise and/or counterclockwise to loosen the embedded hypostome have been advocated and are commercially available.

3. Do not squeeze, crush, or puncture the attached tick, as its fluids may be infected with *B. burgdorferi* or other infectious agents, and these may penetrate the bite wound or even the healthy skin.
4. Do not attempt removal of a tick with bare hands, as spirochetes and other agents may enter the body via the mucous membranes or skin. Application of oil, petroleum jelly, fingernail polish, alcohol, or a hot household match are all not recommended because they do not provide proper detachment of ticks and may even induce the tick to salivate or cause injury to it (Needham 1985).

27.3 Considerations After Removal of Tick

The bite wound should be thoroughly disinfected after removal of a tick. If examination of the tick for borreliae is not possible, the affected person may place the tick into a plastic pill box along with a piece of paper recording the date of removal and keep it for at least 3 – 4 weeks as a reminder of a potential source of infection should initial manifestations of Lyme borreliosis such as erythema migrans or flu-like symptoms develop. If such symptoms develop antibiotic therapy should be initiated (see Chap.23).

If examination for borreliae shows the tick to be negative, there is no need for prophylactic antibiotic treatment. If the tick is positive, a patient may want to undergo prophylactic antibiotic therapy, e.g., with doxycycline 200 mg daily or with amoxicillin 500 mg three times daily for 5 – 7 days. The decision to treat should be discussed with the patient because no definite answer can be provided in the individual case as to whether or not infection has already occurred. The time of engorgement of the tick, the type of tick (larva, nymph, or adult), the kind of removal, the presence of borreliae in the salivary glands (which would be an aggravating factor), or possibly the presence of nonpathogenic spirochetes are factors which influence the decision to initiate prophylactic treatment.

References

Burgdorfer W, Anderson JF, Gern L, Lane RS, Piesman J, Spielman A (1991) Relationship of *Borrelia burgdorferi* to its arthropod vectors. Scand J Infect Dis Suppl 77:35-40

Needham GR (1985) Evaluation of five popular methods for tick removal. Pediatrics 75:997-1002

28 Critical Remarks on Antibiotic Therapy

K. WEBER and W. MARGET

28.1 Introduction

In both the early and the late phases of Lyme borreliosis, the aim of antibiotic therapy is to eliminate signs and symptoms of the disease and to eradicate *Borrelia burgdorferi*. Patients with early manifestations of the illness may develop subsequent signs and symptoms and a rise or persistent elevation of their antibody titers against *B. burgdorferi*, despite treatment with antibiotics (Weber 1974; Steere et al. 1983a; Weber et al. 1986, 1990). Patients treated with antibiotics for late manifestations may experience persistence or recurrence of their signs and symptoms, and their IgG antibody titer against *B. burgdorferi* may decline more slowly, if at all, than in patients treated in the early phase of the disease (Weber et al. 1988; Hassler et al. 1990; Logigian et al. 1990). These possible or definite treatment failures might indicate that eradication of *B. burgdorferi* cannot be carried out as easily as the eradication of the causative agents of certain other bacterial infectious diseases.

The following remarks will address some of the problems associated with the treatment of Lyme borreliosis. An evaluation of all conceivable factors would be beyond the scope of this chapter.

28.2 Influence of In Vitro and Animal Studies

Much of our present knowledge on the efficacy of antibiotics in patients with Lyme borreliosis is based on in vitro and animal investigations conducted in the past few years. The results obtained in vitro have provided a considerable body of information about the susceptibility of *B. burgdorferi* to various antibiotics (Johnson et al. 1987; Preac-Mursic et al. 1987) (see Chap.22). Additional in vitro work has shown that it takes longer to kill *B. burgdorferi* cells than to kill other bacteria (Luft et al. 1988) (see Chap. 22). This is due to the fact that *B. burgdorferi*, like *Treponema pallidum*, has a rather long generation time. These observations would suggest that a more prolonged course of antibiotic therapy is necessary in patients with Lyme borreliosis. Furthermore, therapeutic trials have been performed in experimentally infected gerbils and Syrian hamsters which have provided additional information on the most promising antibiotics (Johnson et al. 1987; Preac-Mursic et al. 1987) (see Chap.22).

Discrepancies between the results obtained in vitro and in animals have become apparent and are discussed elsewhere (see Chap.22). In addition, there are discrepancies between some of the in vitro results and the outcome found in human therapeutical trials. For instance, penicillin seems to perform relatively poorly in most in vitro studies, but it seems to work reasonably well in humans with some important exceptions. Erythromycin exerts an excellent effect in vitro, but has been reported to be not so effective in humans (Steere et al. 1983a). Roxithromycin has been found to work quite well in vitro and in gerbils, but failed in humans (Hovmark et al. 1990). These examples emphasize the fact that, as is known from experience with other bacteria, despite the great importance of in vitro and animal experiments, therapeutic trials in humans are decisive in ascertaining the efficacy of a given antibiotic.

28.3 Problems Associated with the Treatment of Lyme Borreliosis

28.3.1 Problems Associated with Certain Parameters of Comparative Studies

Randomized therapeutic trials have only been performed in recent years, and have been either double-blind and placebo-controlled (Steere et al. 1985; Pfister et al. 1988) or open and prospective, e.g., Steere et al. (1983a) and Dattwyler et al. (1988, 1990) in the United States and Kohlhepp et al. (1989), Pfister et al. (1989, 1991), Hassler et al. (1990), Strle et al. (in press) and Weber et al. (1990) in Europe. Most of these studies have drawbacks if evaluated critically.

28.3.1.1 Number of Patients

The number of patients has been small in several studies (Dattwyler et al. 1988; Pfister et al. 1988, 1989; Hovmark et al. 1990). Small numbers of patients may suffice if the difference between groups is large enough (Dattwyler et al. 1988; Pfister et al. 1988; Hovmark et al. 1990). However, such studies should not include other parameters which could lead to depreciation of the results. For instance, the number of the two groups of patients in the randomized part of the study of Dattwyler et al. (1988) was 23, which is too small to permit definite conclusions, especially since others have not obtained such favorable results with ceftriaxone (Logigian et al. 1990) (see Chap. 25).

28.3.1.2 Inclusion Criteria

Difficulties in comparing groups and, therefore, in evaluating outcome correctly may be encountered when patients with different signs and symptoms are included in randomized studies (Dattwyler et al. 1988; Kohlhepp et al. 1989; Hassler et al. 1990). A similar problem arises if patients with erythema migrans have meningitis prior to therapy. Obviously, meningitis is not a correct indication for oral penicillin and certain other oral antibiotics. However, it is often difficult to decide in which patient a lumbar puncture should be performed. Attempts have recently been made in some centers to identify erythema migrans patients with meningitis (Kuiper et al. 1990; Weber et al. 1990), but it has not been reported from others (Steere et al. 1983a).

28.3.1.3 Pretreatment Characteristics

Patient groups have been encountered which differed in an important parameter prior to therapy (Steere et al. 1983b, 1985; Weber and Thurmayr 1989). This may unduly influence outcome.

28.3.1.4 Outcome Criteria

28.3.1.4.1 Clinical Outcome Criteria

All randomized trials for the treatment of Lyme borreliosis have included clinical outcome criteria. Objective clinical outcome criteria have been used in most of these studies, subjective ones in a recent randomized trial (Weber et al. 1990), and both objective and subjective clinical outcome criteria in other investigations (Dattwyler et al. 1988; Hassler et al. 1990). Objective clinical outcome criteria are generally preferred. However, if patients with uncomplicated erythema migrans are treated with appropriate oral antibiotics, they will rarely if ever develop objective subsequent manifestations unless, as shown recently (Hovmark et al. 1990), an antibiotic such as roxithromycin turns out to be ineffective. It may sometimes be preferable to include subjective clinical outcome criteria rather than to abandon the use of any clinical criteria of response, but the evaluation of subjective clinical criteria of response should be handled with great care.

28.3.1.4.2 Laboratory Outcome Criteria

Serological (Weber et al. 1988, 1990; Kohlhepp et al. 1989; Pfister et al. 1989, 1991; Hassler et al. 1990) and bacteriological (Weber et al. 1990) outcome data have been reported in comparative, mainly randomized therapeutic trials only recently. Post-treatment cerebrospinal fluid (CSF) parameters have also been evaluated (Kohlhepp et al. 1989; Pfister et al. 1989, 1991). Although the evaluation of these and other laboratory outcome data have not revealed a statistically significant difference between treatment groups, they are lacking in some of the earlier randomized investigations.

In the following, we shall discuss problems associated with therapeutic trials with reference to both comparative and noncomparative investigations.

28.3.2 Compliance

The frequency with which an oral drug is taken by the patient each day can influence compliance considerably. Administration of the drug four times daily may, theoretically, be most favorable to ensure the desired blood level, but this regimen is quite inconvenient, and a dose may, therefore, be missed by the patient. This may have been a problem in the study of Steere et al. (1983a).

Compliance in studies should preferably be checked by compliance sheets (see, e.g., Weber et al. 1990). Compliance can also be improved by prescribing oral drugs which have to be taken only once or twice daily and which create little discomfort. The problem of compliance has not been addressed appropriately in many therapeutic trials on Lyme borreliosis so far.

28.3.3 Dosage

The dosage of penicillin, tetracycline, and erythromycin has been rather low in all or some of the patients included in trials for the treatment of erythema migrans (Hollström 1958; Steere et al. 1980, 1983a; Weber et al. 1986), acrodermatitis chronica atrophicans (Thyresson 1949; Götz and Ludwig 1951; Pirilä 1951; Ludwig 1955; Weber et al. 1988), late neuropathy (Hopf 1966; Kaiser 1972), and arthritis (Steere et al. 1985) if evaluated from our present point of view. However, the beneficial effect of this low-dose antibiotic therapy could still be demonstrated in many patients.

Higher dosages are expected to yield better results. For instance, Steere et al. (1985) have obtained a somewhat higher success rate in their patients with arthritis who were treated with high doses of intravenous penicillin. Berger (1986) and Weber et al. (1986), treating patients with erythema migrans, and Dattwyler et al. (1988), treating patients with chronic manifestations, could not prove that higher doses of oral penicillin or ceftriaxone, respectively, were better than lower ones, but the groups of patients may have been too small to reveal a significant difference and were not randomly assigned.

Total daily dosages of antibiotics have been divided in different ways. Oral antibiotics for the treatment of erythema migrans were administered four times daily (Steere et al. 1983a), but only three times or less frequently by most other authors. Patients with neurological and/or late manifestations received 20 – 24 million U intravenous penicillin G in two (Hassler et al. 1990), four (Pfister et al. 1989), or six daily doses (Steere et al. 1985; Dattwyler et al. 1988). Discrepancies in the study results may result from the different frequencies with which antibiotics were administered.

28.3.4 Absorption

Some oral antibiotics such as erythromycin and penicillin may be associated with considerable problems of absorption from the gastrointestinal tract. For instance, penicillin V (phenoxymethylpenicillin) may have been absorbed poorly in a patient in a recent investigation who claimed to have taken the prescribed penicillin tablets, but experienced mild to moderate diarrhea. This patient had an antibody titer response similar to patients not taking any antibiotic (Weber et al. 1990, patient a of figures 1 and 2); although she was classified as having had a definite treatment failure, her problem might have been poor absorption rather than poor efficacy of penicillin V.

As mentioned above, erythromycin exerts a high activity in vitro against *B. burgdorferi* (Johnson et al. 1987; Preac-Mursic et al. 1987). Presently, it is not clear whether the seemingly reduced efficacy of erythromycin which has been reported (Steere et al. 1983a) is due to poor absorption or other problems such as poor compliance or too low dosage.

28.3.5 Penetration of Antibiotics into the CSF

Many oral antibiotics such as penicillin V, tetracycline, and erythromycin do not sufficiently penetrate into the CSF. This is of importance in early disease in which a meningitis may already be present. Evidence of meningitis has been obtained in recent investigations in a few patients with erythema migrans (Kuiper et al. 1990; E. Asbrink, personal communication; Weber and Pfister, unpublished observation), but not in a previous study (Putkonen et al. 1962). It is possible that some of the patients with meningitis (Weber 1974; Steere et al. 1983a) or with post-treatment neurological signs and symptoms (Steere et al. 1983a; Berger 1986; Weber et al. 1986, 1988, 1990) might have had silent or subclinical meningitis prior to initial therapy with oral antibiotics for erythema migrans. Thus, it might not have been lack of efficacy, but rather poor penetration into the CSF that was responsible for the treatment failures in these patients.

Even high doses of intravenously administered penicillin G may not reach the desired CSF level (Pfister et al. 1989) as far as can be deduced from the results of recent in vitro studies (Johnson et al. 1987; Preac-Mursic et al. 1987). However, high doses of penicillin G have been used quite successfully in patients with early phase meningitis (Weber 1974; Sköldenberg et al. 1983; Steere et al. 1983b; Kohlhepp et al. 1989; Pfister et al. 1989). On the other hand, ceftriaxone and cefotaxime pentrate sufficiently well into the CSF and have been found to be effective in early phase meningitis (Pfister et al. 1989, 1991). Despite the fact that cefotaxime reaches the desired CSF levels and intravenously administered penicillin G often does not, as anticipated from the aforementioned in vitro studies, Pfister et al. (1989) found no differences between these two antibiotics in their patients with stage 2 meningitis. Thus, other factors besides penetration into the CSF obviously play a role in evaluating the treatment response in such patients (see Chap.24).

28.3.6 Duration of Antibiotic Therapy

It is usually recommended that patients with Lyme borreliosis should be treated for at least 10 days, and in complicated or late disease for about 3 or 4 weeks. However, the necessity of long-term therapy has not been proven in clinical studies so far. Studies in patients with erythema migrans have shown that a 20-day regimen of tetracycline was not significantly better than a 10-day regimen of tetracycline (Steere et al. 1983a) and that a 5-day course of ceftriaxone was better in a subgroup of patients than a 12-day course of penicillin V (Weber et al. 1990). Lack of proof does not mean that a long duration of antibiotic therapy is useless; all it means is that long-term treatment may indeed not be necessary in special situations such as uncomplicated early illness or that the argument lacks power to show the point.

28.4 Conclusion

A critical evaluation of several factors important for the treatment of Lyme borreliosis revealed a variety of problems which have become apparent in most if not all therapeutic trials published so far. However, many of these studies have provided valuable information which may serve as therapeutic guidelines. Additional work needs to be done to determine which is the optimal antibiotic treatment for patients with various stages of Lyme borreliosis.

References

Berger BW (1986) Treating erythema chronicum migrans of Lyme disease. J Am Acad Dermatol 15:459-463

Dattwyler RJ, Volkmann DL, Halperin JJ, Luft BJ (1988) Treatment of late Lyme borreliosis – randomized comparison of ceftriaxone and penicillin. Lancet I :1191-1194

Dattwyler RJ, Volkmann DJ, Conaty SM, Platkin SP, Luft BJ (1990) Amoxicillin and probenecid versus doxycycline for treatment of erythema migrans borreliosis. Lancet II:1404-1406

Götz H, Ludwig E (1952) Die Behandlung der Akrodermatitis chronica atrophicans Herxheimer mit Penicillin. Hautarzt 2:6-14

Hassler D, Zöller L, Haude M, Hufnagel HD, Heinrich F, Sonntag HG (1990) Cefotaxime versus penicillin in the late stage of Lyme disease – prospective randomized therapeutic study. Infection 18:16-20

Hollström E (1958) Penicillin treatment of erythema chronicum migrans Afzelius. Acta Derm Venereol (Stockh) 38:285-289

Hopf HC (1966) Acrodermatitis chronica atrophicans (Herxheimer) und Nervensystem. Springer, Berlin Heidelberg New York (Monographien aus dem Gesamtgebiete der Neurologie und Psychiatrie, vol 114)

Hovmark A, Olsson I, Asbrink E, Olsson E, Halkier-Sörensen L, Hansen K (1990) A comparative study of the efficacy of roxithromycin and phenoxymethylpenicillin in the treatment of erythema migrans (abstract). IVth internat conference on Lyme borreliosis, Stockholm, A p 161

Johnson RC, Kodner C, Russel M (1987) In vitro and in vivo susceptibility of the Lyme disease spirochete, *Borrelia burgdorferi*, to four antimicrobial agents. Antimicrob Agents Chemother 31:164-167

Kaiser M (1972) Neurologische Komplikationen bei Akrodermatitis chronica atrophicans (Herxheimer) und ihre Beeinflussung durch Penicillintherapie. Dissertation, Göttingen

Kohlhepp W, Oschmann P, Mertens H-G (1989) Treatment of Lyme borreliosis. Randomized comparison of doxycycline and penicillin G. J Neurol 236:464-469

Kuiper H, de Jongh B, Ramselaar T, Spanjaard L (1990) Evidence of asymptomatic nervous system involvement in localized erythema migrans. IVth internat conference on Lyme borreliosis, Stockholm, abstract W/Th-P-48, p 84

Logigian EL, Kaplan RF,Steere AC (1990) Chronic neurologic manifestations of Lyme disease. N Engl J Med 323:1438-1444

Ludwig E (1955) Erfolgreiche Aureomycin- und Terramycinbehandlung als Beitrag zur Klärung der Wirkungsweise von Penicillin bei der Akrodermatitis atrophicans Herxheimer. Dermatol Wochenschr 131:169-178

Luft BJ, Volkman DJ, Halperin JJ, Dattwyler RJ (1988) New chemotherapeutic approaches in the treatment of Lyme borreliosis. Ann NY Acad Sci 539:352-361

Pfister HW, Einhäupl KM, Franz P, Garner C (1988) Corticosteroids for radicular pain in Bannwarth's syndrome. Ann NY Acad Sci 539:485-487

Pfister HW, Preac-Mursic V, Wilske B, Einhäupl KM (1989) Cefotaxime vs penicillin G for acute neurologic manifestations in Lyme borreliosis. Arch Neurol 46:1190-1194

Pfister HW, Preac-Mursic V, Wilske B, Schielke E, Sörgel F, Einhäupl KM (1991) Randomized comparison of ceftriaxone and cefotaxime in Lyme neuroborreliosis. J Infect Dis 163:311-318

Pirilä V (1951) The penicillin treatment of acrodermatitis atrophicans chronica. Acta Derm Venereol (Stockh) 31:382-384

Preac-Mursic V, Wilske B, Schierz G, Holmburger M, Süss E (1987) In vitro and in vivo susceptibility of *Borrelia burgdorferi*. Eur J Clin Microbiol 6:424-426

Putkonen T, Mustakallio KK, Salminen A (1962) Erythema chronicum migrans with meningitis: a rare coincidence of two tick-borne diseases? Dermatologica 125:184-188

Sköldenberg B, Garde A, Carlström A, Stiernstedt G, Kolmodin G, Nord CE (1983) Chronic meningitis caused by a penicillin-sensitive microorganism? Lancet II:75-78

Steere AC, Malawista SE, Newman JH, Spieler PN, Bartenhagen NH (1980) Antibiotic therapy in Lyme disease. Ann Intern Med 93:1-8

Steere AC, Hutchinson GJ, Rahn DW, Sigal LH, Craft JE, Desanna ET, Malawista SE (1983a) Treatment of the early manifestations of Lyme disease. Ann Intern Med 99:22-26

Steere AC, Pachner AR, Malawista SE (1983b) Neurologic abnormalities of Lyme disease: successful treatment with high-dose intravenous penicillin. Ann Intern Med 99:767-772

Steere AC, Green J, Schoen RT, Taylor E, Hutchinson GJ, Rahn DW, Malawista SE (1985) Successful parenteral penicillin therapy of established Lyme arthritis. N Engl J Med 312:869-874

Strle F, Ruzic E, Cimperman J (in press), Erythema migrans: comparison of treatment with azithromycin, doxycycline and phenoxymethylpenicillin. J Antimicrob Chemother

Thyresson N (1949) The penicillin treatment of acrodermatitis atrophicans chronica (Herxheimer). Acta Derm Venereol (Stockh) 29:572-621

Weber K (1974) Erythema-chronicum-migrans-Meningitis – eine bakterielle Infektionskrankheit? Munch Med Wochenschr 116:1993-1998

Weber K, Thurmayr T (1989) Oral penicillin versus minocycline for the treatment of early Lyme borreliosis. Zentralbl Bakteriol (Suppl) 18:263-268

Weber K, Neubert U, Thurmayr R (1986) Antibiotic therapy in early erythema migrans disease and related disorders. Zentralbl Bakteriol Hyg (A) 263:377-388

Weber K, Preac-Mursic V, Neubert U, Thurmayr R, Herzer P, Wilske B, Schierz G, Marget W (1988) Antibiotic therapy of early European Lyme borreliosis and acrodermatitis chronica atrophicans. Ann NY Acad Sci 539:324-345

Weber K, Preac-Mursic V, Wilske B, Thurmayr R, Neubert U, Scherwitz C (1990) A randomized trial of ceftriaxone versus oral penicillin for the treatment of early European Lyme borreliosis. Infection 18:91-96

VII Epidemiology of Lyme Borreliosis

29 Epidemiology of Lyme Borreliosis

G. STANEK, N. SATZ, F. STRLE, and B. WILSKE

29.1 Introduction

The aim of an epidemiological consideration of Lyme borreliosis is the description of the frequency of the disease with respect to its causative agent (Flamm 1986). It deals with questions concerning the sources, route, and the target of the infection. All living beings in whom the causative agent exists and multiplies are designated as the infection source. The infection route relates to the possibilities which the agent has of spreading from the source to the next susceptible human host, which represents the target of infection.

29.2 Source of Infection

29.2.1 Agent

The causative agent of Lyme borreliosis, *Borrelia burgdorferi*, was discovered in and isolated from the American deer tick *Ixodes dammini* in the early 1980s (Burgdorfer et al. 1982). Shortly thereafter, it was recovered from patients with Lyme disease. It was recognized as a new species within the genus *Borrelia* and named *B. burgdorferi* in recognition of its discovery by Dr. Willy Burgdorfer (Johnson et al. 1984).

European *Borrelia* isolates show certain differences compared with isolates from the United States regarding molecular weights of major proteins and reactivity of monoclonal antibodies to the outer surface proteins of the *B. burgdorferi*-type strain (Barbour et al. 1984). A certain amount of heterogeneity has been described in European isolates (Stanek et al. 1985b; Wilske et al. 1985b, 1988). A typing system based on the reactivity of monoclonal antibodies (H5332, H3TS) to the outer surface protein B 31 (OspA) has been suggested by Barbour (personal communication), which permits differentiation between three main groups. Interestingly, North American isolates belong almost exclusively to group I, whereas European isolates belong to groups I through III.

Studies from Austria have suggested a relationship of certain types of *B. burgdorferi* to certain geographic regions (Stanek et al. 1990). According to this investigation, serogroup I is prevalent in the western parts of Europe, and groups II and III are prevalent in Central and Eastern Europe (Germany, Austria, and Czechoslovakia). More recently, a serotyping system based on the reactivity of OspA with eight monoclonal antibodies has been proposed, and seven serotypes have been defined (Wilske et al. 1991). Although it is unlikely that every tick or animal isolate is a pathogenic strain, there is presently no technique for differentiating between pathogenic and nonpathogenic strains. However, according to the findings of Wilske et al. (1988, 1991), serotype 2 is prevalent among European isolates from skin, whereas European isolates from ticks and human cerebrospinal fluid are more heterogeneous.

29.2.2 Reservoirs

Feral rodents and other wildlife (Krampitz and Bark 1989) are considered reservoirs. For details the reader is referred to Chap.5.

29.3 Route of Infection

29.3.1 Vectors

Hard ticks, particularly those of the *Ixodes ricinus* complex, and other hematophagous arthropods such as tabanids and fleas are considered to be vectors of *B. burgdorferi*. Involvement of soft (argasid) ticks in maintaining and distributing *B. burgdorferi* is still a matter of conjecture, even though the pigeon tick *Argas persicus* in Italy (Stanek and Simeoni 1989) and *Ornithodorus coriaceus* in California were found infected with this spirochete. Although the latter were identified as vectors in single cases, there is little known about their interactions with *B. burgdorferi*.

I. ricinus is the most common European ixodid tick species. For its geographic distribution and its biological requirements, the reader is referred to Chap.5. The areas in which this tick is found include distinct geographic regions in which ixodid ticks may also be infected with the tick-borne encephalitis virus (Fig.1). Infection with rickettsiae or protozoon agents may also be possible. To the east, another *Ixodes* species, *I. persulcatus*, is abundant und appears to be the prinicipal vector of *B. burgdorferi*.

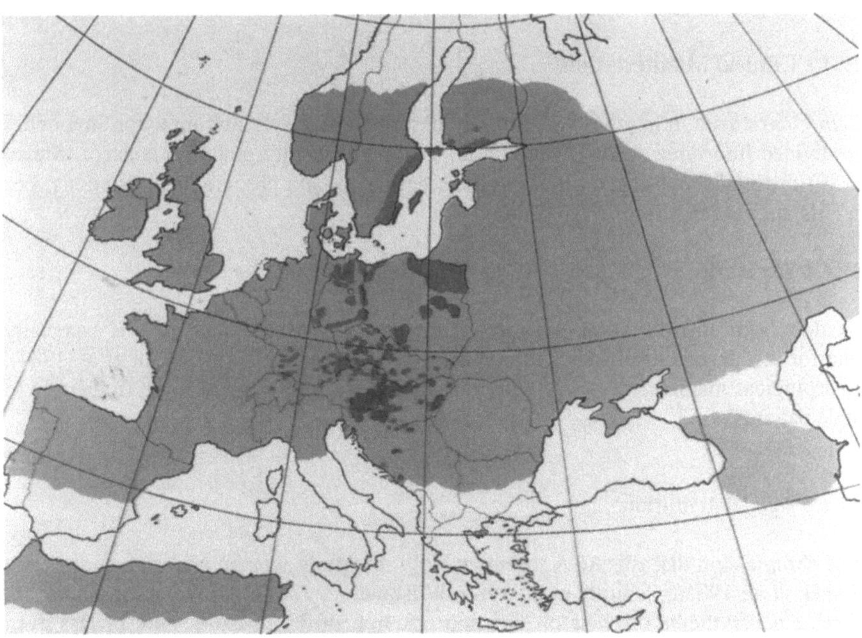

Fig. 1. Geographic distribution of *Ixodes ricinus* corresponding to the distribution of Lyme borreliosis in Europe. Areas where *Ixodes* ticks may also be infected with the tick-borne encephalitis virus are marked in *black*

Compared with the number of ticks infected with the tick-borne encephalitis virus, the prevalence of ticks infected with *B. burgdorferi* is very high. In areas of Europe endemic for tick-borne encephalitis virus every thousandth tick carries the virus, whereas every third tick is on average infected with borreliae. The prevalence of ticks infected with *B. burgdorferi* has been studied in many countries including Austria, Czechoslovakia, Southern Germany, and Switzerland (see Chap. 5).

Transmission of *B. burgdorferi* from ticks to humans is cyclic, whereas tabanids or other hematophagous insects are said to transmit the spirochetes acyclically, i.e., they contaminate their mouth parts through feeding on a parasitimic host and as a result of interrupted feeding may transmit the spirochete to another host.

29.3.2 Contact Infection

Transmission of *Borrelia burgdorferi* by contact infection has been described in the American literature in animals (Burgess and Patrican 1987) and also in Europe, where an erythema migrans was said to have developed from a scratch wound that had been contaminated with manure days before (Rockstroh et al. 1989). Examination of the manure by dark-field microscopy, silver stain, and monoclonal antibodies identified *B. burgdorferi*. These reports suggest that *B. burgdorferi* can be transmitted by contact infection and can live outside an animal host. There is no evidence that *B. burgdorferi* may be transmitted by sexual contact.

29.4 Target of Infection

29.4.1 Clinical Manifestations

Lyme borreliosis in humans was initially described as a systemic infection that could be divided into three stages (Steere et al. 1986). A recently modified concept (Steere 1989) distinguishes between localized, disseminated, and chronic manifestations. For details, the reader is referred to Chap.7.

29.4.2 Sex Ratio

In adults, skin manifestations are more numerous in female patients; this is especially true for acrodermatitis chronica atrophicans (ACA) (Pirilä 1951; Asbrink et al. 1984). Neurological manifestations of Lyme borreliosis are seen slightly more frequently in male patients.

29.4.3 Age Distribution

The chronic skin disease, ACA, occurs mainly in elderly women (Asbrink et al. 1984; Stanek et al. 1985a; Schmidt et al. 1986; Wilske et al. 1987). All the other manifestations, e.g., erythema migrans or meningoradiculoneuritis, reveal a more or less even age distribution. In children, however, meningitis and cranial nerve palsy are more frequently observed than meningoradiculoneuritis (Pfister et al. 1986; Christen et al. 1989; Millner et al. 1989; Pleterski-Rigler 1989).

29.4.4 Seasonal Distribution

A clear seasonal dependence is noticed in patients with erythema migrans in Austria and Slovenia; a peak has constantly been found in July (Fig.2). The same seasonal incidence of erythema migrans was also observed in Germany (Wilske et al. 1987). In Fig. 3, the time interval is shown between an arthropod bite, onset of erythema migrans, and exact clinical diagnosis (the data were obtained in Slovenia). Thus, erythema migrans can be diagnosed throughout the year, but most frequently in early summer and autumn. These data may serve as an example for the situation throughout most of Europe. Some authors, however, have shown that the onset of erythema migrans may differ to some extent. Weber and Neubert (1986) observed a more evenly distributed onset of erythema migrans between May and October (see Fig. 3 in Chap. 7, p 95). In Germany and Austria, neurological disorders were most commonly found in August in 1984 – 1986 (Stanek et al. 1986b). Late manifestations such as ACA and arthritis do not show any seasonal preference.

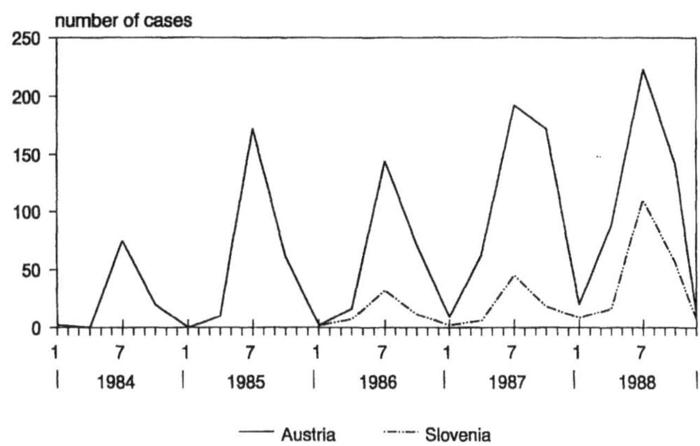

Fig. 2. Seasonal distribution of erythema migrans. Patients were registered at the Hygiene Institute of the University of Vienna, Austria, 1984 – 1988, and at the Clinic for Infectious Diseases in Ljubljana, Slovenia, 1986 – 1988

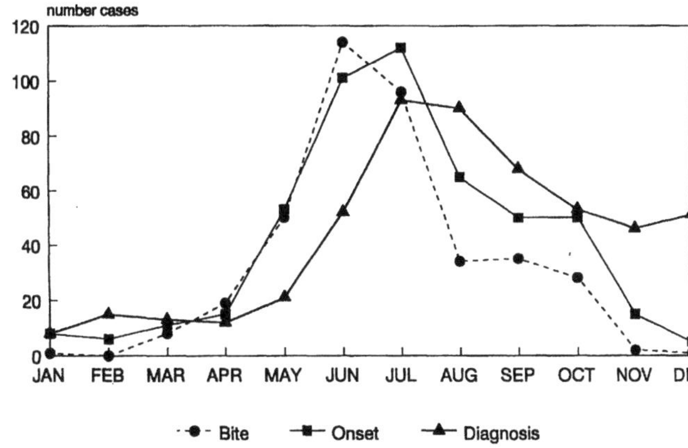

Fig. 3. Time intervals between arthropod bite, onset of erythema migrans, and its exact clinical diagnosis (522 patients). Data were obtained at the Clinic for Infectious Diseases in Ljubljana, Slovenia, in 1988

29.4.5 Incubation Period

The incubation period, i.e., the time period from the transmission of the infectious agent by an arthropod bite to the onset of clinical manifestations, was studied in 169 Lyme borreliosis patients in Bavaria (Wilske et al. 1987). For erythema migrans (*n*=51), neurological disorders (*n*=98), and arthritis (*n*=20), the incubation period was 5-29 days, 20-59 days, and 2-8 months, respectively.

29.4.6 Risk Groups

Several studies (Münchhoff et al. 1986; Pejcoch et al. 1989; Guy et al. 1989) are available on risk groups, e.g., forestry workers, who are regularly exposed to tick bites. These investigations claim that none of the forestry workers had ever been ill with Lyme borreliosis. However, 14%-48% of those studied were found to be seropositive. Whether these elevated antibody titers are nonspecific or represent a response to a latent infection is a matter of speculation.

To establish an index of manifestations by studying persons who are exposed to vectors of *B. burgdorferi* a prospective study was conducted in Austrian army recruits who were on outside duties for 3 months. Recruits were instructed to show each tick bite to a military physician. A first blood sample was taken immediately, and a second blood sample 6 weeks later when the recruit also underwent clinical examination. An erythema migrans was diagnosed in 4% and seroconversion was observed in 22% (Schmutzhard et al. 1988). A similar study was performed in Southwest Germany at a children's holiday camp (Paul et al. 1986). The results were quite similar.

Another risk group is constituted by crosscountry hikers in Switzerland. A study of this subpopulation revealed that 77% had a history of tick bites, 20% had elevated antibody titers to *B. burgdorferi*, and 3.8% showed a seroconversion within 2 years, but only individual persons developed clinical manifestations of Lyme borreliosis (Fahrer et al. 1988).

The following examples show how difficult it is to obtain information on the incidence of Lyme borreliosis by serological methods, even in a relevant study group. Smith et al. (1988) in the United States studied outdoor workers and people who spent leisure time outdoors. The authors found that these persons were 2 and 2.5 times, respectively, more likely to be seropositive than those people who did not engage in outdoor activities. In less than 50% of the seropositives, about 3% of all persons investigated, a history of Lyme borreliosis was recorded. Gern et al. (1989) conducted a study on about 500 persons from the Swiss plateau, an area of Switzerland known to be heavily infested with *I. ricinus* ticks (of which about 20% were found to be infected with *B. burgdorferi*). Screening of sera from individuals aged 14 – 103 years who never had symptoms of Lyme borreliosis revealed seropositivity in 32% by immunofluorescence assay (IFA) and 26.6% by enzyme-linked immunosorbent assay (ELISA).

Risk groups reflect the problems that exist with seroprevalence and serodiagnosis of Lyme borreliosis. Data on seroprevalence rank between 3% and more than 40% in Europe. In Switzerland, seropositivity of the healthy population has been reported to be 10% on average. Satz et al. (1988), however, have shown that about 40% of the population near the city of Berne (Berner Seeland) are seropositive.

In contrast to the high prevalence of antibodies to *B. burgdorferi* in the risk groups, the seroprevalence of viral tick-borne encephalitis infection is comparably low. Of 242 forestry workers in Bavaria, elevated antibody titers to *B. burgdorferi* and to the tick-borne encephalitis virus were found in 16.5% and 2.5%, respectively (Wilske et al.

1985b). However, it should be considered that the differences between seroprevalence values reported by different investigators may also reflect differences in cut-off levels for the seropositive results. This problem is discussed in detail in Chap.21.

29.4.7 Geographical Distribution of Lyme Borreliosis

Lyme borreliosis is endemic in many parts of Europe. Early in the study of Lyme borreliosis, it could be shown that it occurs everywhere where ticks of the genus *Ixodes* are living. From Europe, we have reports about the widespread distribution of now confirmed features of Lyme borreliosis that date back to the 1960s and early 1970s. Interestingly, erythema migrans, lymphocytoma, ACA, and meningoradiculoneuritis were considered as arbovirus infections by some authors (Müller and Schaltenbrand 1973). Although the etiology was unknown, researchers were aware of the widespread distribution of some of the disorders, e.g., ACA and related neuropathies (Hauser 1955; Danda 1963; Hopf 1966). Danda (1963) obtained information about the frequency of ACA from all parts of the world by means of a questionnaire. He could show that Central and Eastern Europe were the locations where ACA was most frequently diagnosed and, interestingly, that this frequency increased during the years 1945 – 1958. Figure 4 displays Danda's data with respect to the total population of each country during this time period. The highest prevalence of ACA was reported from Czechoslovakia and Finland. These data support in part what was later on described by Hopf and Stroux (1968). These authors elaborated that there is a clustering of patients with ACA in Central Europe. This chronic skin disorder was frequently seen in Germany, Czechoslovakia, Austria, and Hungary. Less frequently, patients with ACA were observed in Scandinavia, Poland, Switzerland, Yugoslavia, and in certain areas of European Russia. These findings coincide with what we now know. ACA and associated peripheral neuropathies (Kristoferitsch 1989) are known in Europe and in Eurasia, but information about ACA from North America is rare. There is just one report of a patient with ACA from California (Lavoie et al. 1986).

By 1985 (Stanek et al. 1986a) and 1987 (Stanek et al. 1989a), it had become evident that Lyme borreliosis was a health problem affecting the whole northern hemisphere. Reports from Austria (Stanek et al. 1986, 1988), Belgium (Bigaignon et al. 1989), Czechoslovakia (Kmety et al. 1986; Strnad et al. 1989), Denmark (Hansen et al. 1986), France (Dournon and Assous 1986), Germany (Schmidt et al. 1986; Wilske et al. 1987), Hungary (Lakos 1986), Italy (Trevisan et al. 1986; Cimmino et al. 1989),

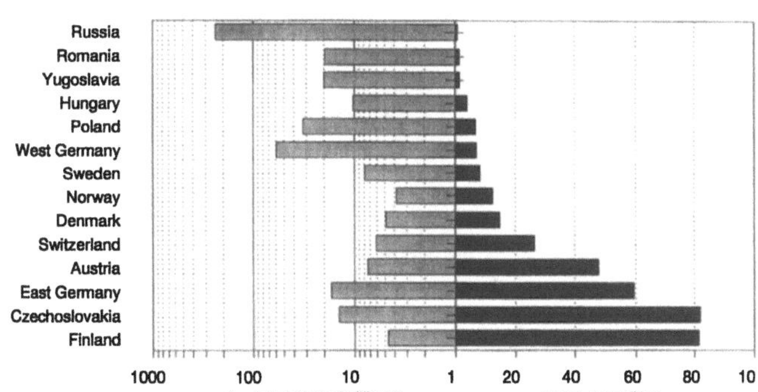

Fig. 4. The relative frequency of acrodermatitis chronica atrophicans in Europe: the *right side* shows the number of patients per one million inhabitants of the respective European country. Calculation based on data published by Danda (1963)

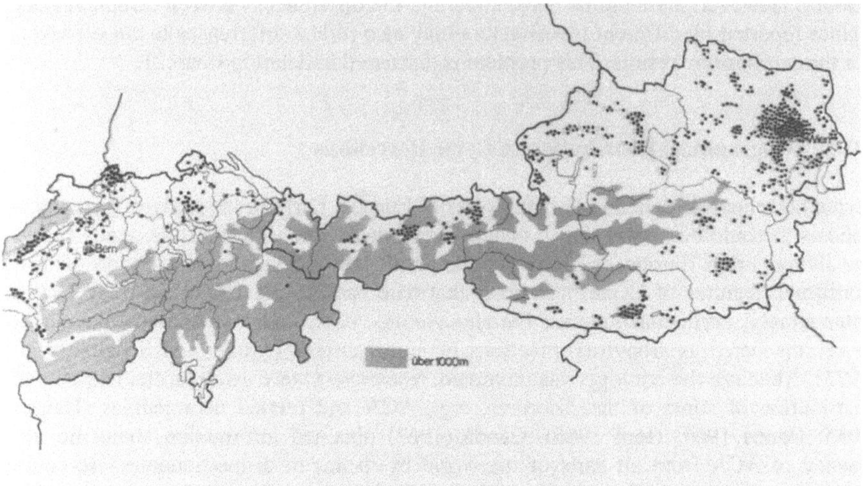

Fig. 5. Patients with erythema migrans in Switzerland and Austria in 1985. *Spots* mark the infection sites (tick or insect bite) as reported by the patients. *Hatched areas* are locations 1000 m above sea level. Note that *hatched areas* are almost free of *spots*. Patients were registered at the Institute of Zoology of the Universitiy of Neuchâtel, Switzerland (Aeschlimann et al. 1986) and at the Hygiene Institiute of the University of Vienna, Austria (Stanek et al. 1986b)

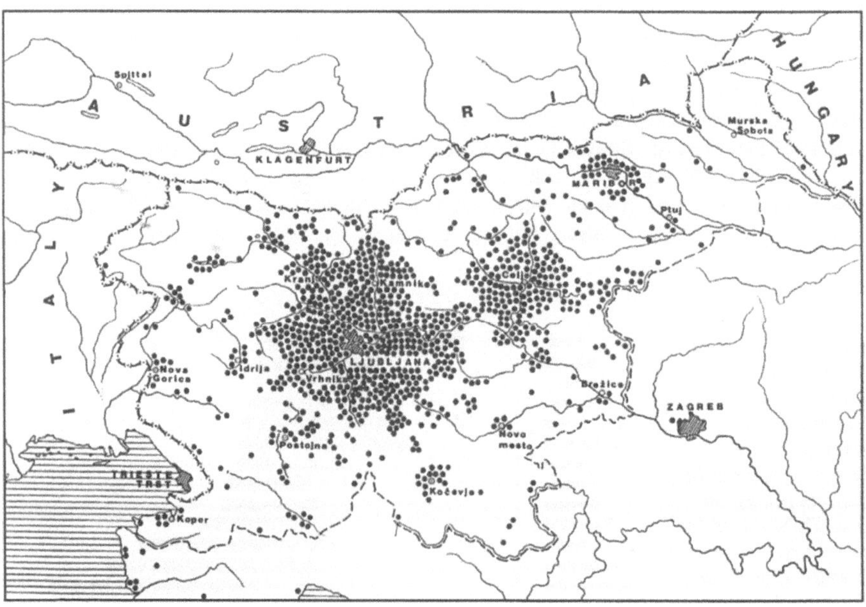

Fig. 6. Patients with erythema migrans in Slovenia in 1988. *Spots* mark the sites of tick bites. Patients were registered at the Clinic for Infectious Disease, Ljubljana, Slovenia

Fig. 7. Lyme Borreliosis all over the northern hemisphere. The distribution of Lyme borreliosis corresponds to the biotopes of *Ixodes* ticks

Luxembourg (Bigaignon et al. 1989), the Netherlands (Houwerzijl et al. 1984; van der Linde et al. 1990), Sweden (Lanner et al. 1989), Switzerland (Aeschlimann et al. 1986), and Slovenia (Strle et al. 1989a,b) revealed that Lyme borreliosis occurs in all parts of these countries. Although there are no current reports on erythema migrans from Finland, the disorder has been known there for decades (Pirilä 1951; Putkonen et al. 1962).

A Swiss study showed that the risk of becoming infested by ticks is limited by an altitude of approximately 1000 – 1200 m above sea level where the primary tick vector, *I. ricinus*, no longer exists (Aeschlimann et al. 1986; Fig. 5). Figure 6 shows the rather even distribution of infection sites for erythema migrans over the entire state of Slovenia in 1988.

Single reports of Lyme borreliosis patients have been published from Great Britain, Norway, and Russia. The first British patients with erythema migrans were reported from the southern part of England (Muhlemann and Wright 1987). A recent study from Russia describes the enormous distribution of infected ticks from the European parts of Russia over the Asian parts to the Pacific coast (Dekonenko et al. 1988). The disease has also been observed in Japan (Kawabata et al. 1987), and well-documented cases were also reported from Hailin County in the area of Beijing, China (Chengxu et al. 1988). Seroprevalence studies undertaken in Africa suggest that Lyme borreliosis may also be present in regions of Central and Southern Africa (Stanek et al. 1989b).

In the United States, Lyme borreliosis had been recognized in 41 states by 1987 (MMWR 1989). Reports on erythema migrans and neurological manifestations of Lyme borreliosis also exist from Canada (Lycka 1986; Doby et al. 1986). Thus, Lyme borreliosis is distributed over the entire northern hemisphere (Fig.7).

29.4.8 Prevalence of Lyme Borreliosis

Since the introduction of a variety of non-standardized commercial tests for the serodiagnosis of Lyme borreliosis, it has become very difficult to obtain relevant data on this disease in a given country. Moreover, there are no reporting systems, because widely accepted case definitions are still not available. The number of cases reported by serodiagnostic laboratories has increased constantly. Interestingly, the numbers of the better known clinical manifestations erythema migrans, neuroborreliosis, ACA, and arthritis have varied. In 1984, for example, 341 patients were identified in Vienna, Austria (Fig.8). Erythema migrans was seen most frequently, meningoradiculoneuritis and arthritis patients ranked second and third, and there were just a few patients with ACA. In 1985, 1303 patients were registered, representing a fourfold increase. This time *B. burgdorferi* infections of the nervous system were the most frequently registered condition, and erythema migrans ranked second (Fig.8). There are similar observations from other laboratories in Europe, e.g., Munich, Budapest, Brussels, and Stockholm, as shown in Fig. 8. Thus, the incidence recorded by different serodiagnostic laboratories does not reflect the true distribution of the manifestations of Lyme borreliosis. For instance, it is evident from clinical observations that erythema migrans is the most frequent manifestation of Lyme borreliosis. However, 40% to more than 60% of patients

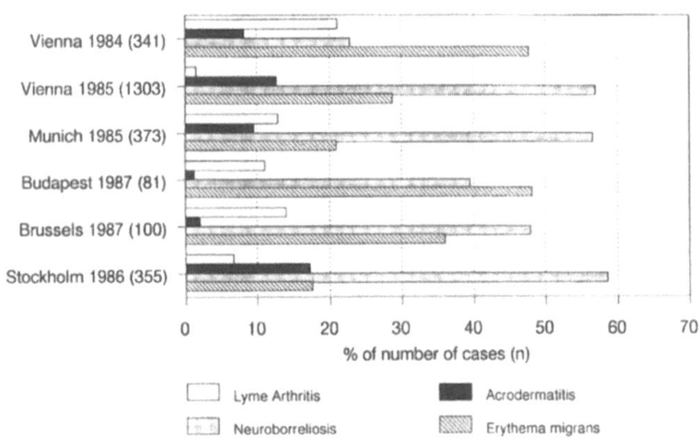

Fig. 8. Differences in the relative frequency of certain clinical manifestations of Lyme borreliosis in different European laboratories and (just shown for Vienna) in different years

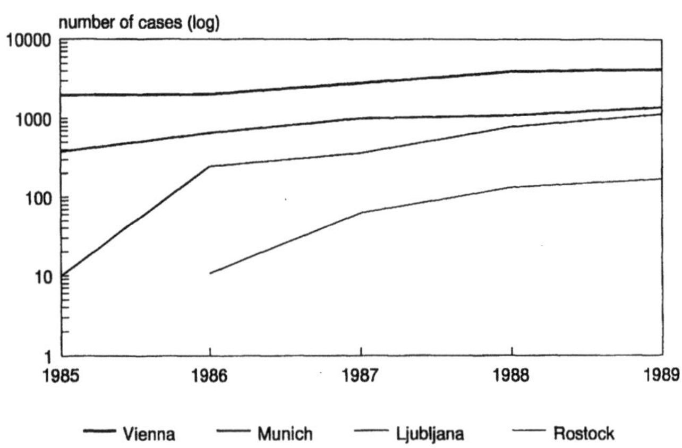

Fig. 9. Lyme borreliosis patients registered during a 5-year period in four different areas of Europe. Although the number of patients varies, the tendency of increase from 1985 to 1989 is obvious. Patients from Ljubljana, Munich, Rostock, and Vienna were registered by F Strle, B Wilske, G Naumann, and G Stanek, respectively

with erythema migrans are seronegative. Erythema migrans patients rank highest when a laboratory has recently started to test samples and get a response from the physicians who were informed about the typical skin lesion. Later on, the clinically clearly identifiable lesion, erythema migrans, will no longer be controlled serologically.

Figure 9 shows the increasing incidence of Lyme borreliosis over a period of 5 years in Vienna, Austria, Ljubljana, Slovenia, and Rostock and Munich, Germany. The number of registered patients increased from two-fold (Vienna) to about 100-fold (Ljubljana) during these years. The exponential increase displays an artefact due to the increased awareness of doctors and patients.

At the Hygiene Institute of the University of Vienna, Austria, 4000 – 5000 patients with Lyme borreliosis have been registered every year since 1988. Based on these data, the incidence rate per 100000 inhabitants per year from three different parts of Europe averages 0.05%. This does not, however, reflect the true incidence, because a large number of Lyme borreliosis cases go unrecognized. It is safe to speculate, therefore, that the incidence of Lyme borreliosis for the total population of Central Europe is about 0.1% – 0.2% per year.

29.5 Conclusion

In Europe and in other parts of the world, Lyme borreliosis is being recognized increasingly by physicians and serodiagnostic laboratories. However, it is currently difficult to present conclusive epidemiological data. There are no widely accepted case definitions for the numerous clinical features due to, or assumed to be linked to, infection with *B. burgdorferi*.

Clinical diagnosis of suspected cases of Lyme borreliosis requires confirmation by the demonstration of the etiologic agent and the recognition of its causative role in the respective disorder. Since serological test results are often not reliable, the true incidence and prevalence of this disease cannot be determined and must await the development of specific and dependable methods to identify actual infection.

References

Aeschlimann A, Chamot E, Gigon F, Jeanneret JP, Kesseler D, Walther C (1986) *B. burgdorferi* in Switzerland. Zentralbl Bakteriol Hyg (A) 263:450-458

Asbrink E, Hovmark A, Hederstedt B (1984) The spirochetal etiology of acrodermatitis chronica atrophicans Herxheimer. Acta Derm Venereol (Stockh) 64:506-512

Barbour AG, Tessier SL, Hayes SF (1984) Variation in a major surface protein of Lyme disease spirochetes. Infect Immun 45:94-100

Bigaignon G, Goubau P, Pierard D, Bukasa K, Sindic CJM (1989) Lyme Borreliosis in Belgium: an epidemiologic study of 102 cases. Zentralbl Bakteriol (Suppl) 18:64-68

Burgdorfer W, Barbour AG, Hayes SF, Benach JL, Grunwaldt E, Davis JP (1982) Lyme disease – a tick-borne spirochetosis? Science 216:1317-1319

Burgess EC, Patrican LA (1987) Oral infection of *Peromyscus maniculatus* with *Borrelia burgdorferi* and subsequent transmission by *Ixodes dammini*. Am J Trop Med Hyg 36:402-407

Chengxu A, Yuxin W, Yongguo Z et al. (1988) Clinical manifestations and epidemiological characteristics of Lyme disease in Hailin county, Heilongjiang province, China. Ann NY Acad Sci 539:302-313

Christen HJ, Bartlau N, Hanefeld F, Thomssen R (1989) Lyme-Borreliose – eine der häufigsten Ursachen der akuten peripheren Fazialisparese im Kindesalter. Monatsschr Kinderheilk 137:151-157

Cimmino MA, Trevisan G, Fumarola D, Crovato F (1989) Lyme arthritis: clinical findings in Italian patients. Zentralbl Bakteriol (Suppl) 18:256-260

Danda J (1963) Die Weltfrequenz der Akrodermatitis chronica atrophicans. Hautarzt 14:337-340

Dekonenko EP, Steere AC, Berardi VP, Kravchuk LN (1988) Lyme borreliosis in the Soviet Union. J Infect Dis 158:748-753

Doby JM, Anderson JF, Couatarmanach A, Magnarelli LA, Martin A (1986) Lyme disease in Canada with possible transmission by an insect. Zentralbl Bakteriol Hyg (A) 263:488-490

Dournon E, Assous M (1986) Lyme disease in France. Zentralbl Bakteriol Hyg (A) 263:464-465

Fahrer H, Sauvain MJ, VdLinden S, Zhioua E, Gern L, Aeschlimann A (1988) Prävalenz der Lyme-Borreliose in einer schweizerischen Risikopopulation. Schweiz Med Wochenschr 118:59-65

Flamm H (ed) (1986) Angewandte Hygiene in Krankenhaus und Arztpraxis. Göschl, Wien

Gern L, Frossard E, Walter A, Aeschlimann A (1989) Prescence of antibodies against *Borrelia burgdorferi* in a population of the Swiss plateau. Zentralbl Bakteriol (Suppl) 18:321-328

Guy EC, Martyn CN, Bateman DE, Heckels JE, Lawton NF (1989) Lyme disease prevalence and clinical importance of *Borrelia burgdorferi* specific IgG in forestry workers. Lancet I:484-485

Hansen K, Rechnitzer C, Pedersen NS, Arpi M, Jessen O (1986) Borrelia meningitis in Denmark. Zentralbl Bakteriol Hyg (A) 263:348-350

Hauser W (1955) Zur Kenntnis der Akrodermatitis chronica atrophicans. Habilitationsschrift Würzburg 1954. Arch Dermatol Syph 199:350-393

Hopf HC (1966) Acrodermatitis chronica atrophicans (Herxheimer) und Nervensystem. Springer, Berlin Heidelberg New York, pp 1-130 (Monographien aus dem Gesamtgebiet der Neurologie und Psychiatrie, vol 114)

Hopf HC, Stroux B (1968) Die geographische Verteilung der Akrodermatitis chronica atrophicans (Herxheimer) in der Umgebung von Würzburg. Z Haut Geschlechtskr 43:41-48

Houwerzijl J, Root JJ, Hoogkamp-Korstanje JAA (1984) A case of Lyme disease with cardiac involvement in the Netherlands. Infection 12:358

Johnson RC, Schmid GP, Hyde FW et al. (1984) *Borrelia burgdorferi* sp.nov.: etiologic agent of Lyme disease. Int J Syst Bacteriol 34:496-497

Kawabata M, Baba S, Iguchi K et al. (1987) Lyme disease in Japan and its possible transmitter the tick *Ixodes persulcatus*. J Infect Dis 156:854

Kmety E, Rehacek J, Vyrostekova V (1986) Investigations of ticks for the presence of borrelia in Czechoslovakia. Zentralbl Bakteriol Hyg (A) 263:468-470

Krampitz HE, Bark S (1989) Patterns of seasonal variation in the infestation of the roe deer, *Capreolus capreolus* L. with borrelia infected *Ixodes ricinus* in Northern Bavaria. Zentralbl Bakteriol (Suppl) 18:21-25

Kristoferitsch W (1989) Neuropathien bei Lyme-Borreliose. Springer, Wien New York

Lakos A (1986) Lyme Borreliosis in Hungary – the first three years. Zentralbl Bakteriol Hyg (A) 263:55-59

Lanner M, Hederstedt B, Asbrink E, Hovmark A, Stiernstedt G (1989) Serodiagnosis of *Borrelia* infections during 1986 in Sweden. Zentralbl Bakteriol (Suppl) 18:69-74

Lavoie PE, Wilson AJ, Tuffanelli DL (1986) Acrodermatitis chronica atrophicans with antecedent Lyme disease in California. Zentralbl Bakteriol Hyg (A) 263:262-265

Linde, MR van der, Crijns HJGM, de Koning J, Hoogkamp-Korstanje JAA, de Graaf JJ, Piers DA, van der Galien A, Lie KI (1990) Range of atrioventricular conduction disturbances in Lyme borreliosis: a report of 4 cases and review of other published reports. Br Heart J 63:162-168

Lycka BAS (1986) Lyme disease in Canada? Can Med Assoc 134:48

Millner M, Schimek MG, Spork D, Schnizer M, Stanek G (1989) Lyme borreliosis in children. A controlled clinical study based on ELISA values. Eur J Pediatr 148:527-530

MMWR (1989) Lyme Disease – United States, 1987 and 1988. MMWR 38 (39):668-672

Muhlemann MF, Wright DJM (1987) Emerging pattern of Lyme disease in the United Kingdom and Irish Republic. Lancet I:260-262

Müller WK, Schaltenbrand G (eds) (1973) Arbovirusinfektionen des Nervensystems in Europa. Thieme, Stuttgart

Münchhoff P, Wilske B, Preac-Mursic V, Schierz G (1986) Antibodies against *Borrelia burgdorferi* in Bavarian forest workers. Zentralbl Bakteriol Hyg (A) 263:412-419

Paul H, Ackermann R, Gerth HJ (1989) Infection and manifestation rate of European Lyme borreliosis in humans. Zentralbl Bakteriol (Suppl) 18:44-49

Pejcoch M, Kralikova Z, Strnad P, Stanek G (1989) Prevalence of antibodies against *Borrelia burgdorferi* in forestry workers of south Moravia. Zentralbl Bakteriol (Suppl) 18:317-320

Pfister HW, Einhäupl KM, Wilske B, Preac-Mursic V (1986) Bannwarth's Syndrome and the enlarged neurological spectrum of arthropod-borne borreliosis. Zentralbl Bakteriol Hyg (A) 263:343-347

Pirilä V (1951) The penicillin treatment of acrodermatitis atrophicans chronica. Acta Derm Venereol (Stockh) 31:576-591

Pleterski-Rigler D (1989) Peripheral facial palsy and *Borrelia burgdorferi* infection in children in Yugoslavia. Zentralbl Bakteriol (Suppl) 18:203-205

Putkonen T, Mustakallio KK, Salminen A (1962) Erythema chronicum migrans with meningitis: a rare coincidence of two tick-borne diseases? Dermatologica 125:184-188

Rockstroh T, Mochmann HP, Stanek G (1989) Lyme borreliosis by contact infection. Zentralbl Bakteriol (Suppl) 18:40-41

Satz N, Ackermann R, Gern L, Aeschlimann A, Ott A, Knoblauch M (1988) Zur Epidemiologie der Infektion mit *Borrelia burgdorferi*. Schweiz Med Wochenschr 118:422-426

Schmidt R, Kabatzki J, Hartung S, Ackermann R (1986) Erythema chronicum migrans disease in the Federal Republic of Germany. Zentralbl Bakteriol Hyg (A) 263:435-441

Schmutzhard E, Stanek G, Pletschette M, Hirschl AM, Pallua A, Schmitzberger R, Schlögl R (1988) Infections after tickbites. Tick-borne encephalitis and Lyme borreliosis – a prospective epidemiological study from Tyrol. Infection 16:269-272

Smith PF, Benach JL, White DJ, Stroup DF, Morse DL (1988) Occupational risk of Lyme disease in endemic areas of New York State. Ann NY Acad Sci 539:289-301

Stanek G, Simeoni J (1989) Are pigeons' ticks transmitters of *Borrelia burgdorferi* to humans? A preliminary report. Zentralbl Bakteriol (Suppl) 18:42-43

Stanek G, Kristoferitsch W, Hirschl A, Wewalka G (1985a) Borrelia-Infektionen in Österreich 1984. Mitt Österr Trop Med Parasitol 7:55-62

Stanek G, Wewalka G, Groh V, Neumann R, Kristoferitsch W (1985b) Differences between Lyme disease and European arthropod-borne infections. Lancet I:905-906

Stanek G, Flamm H, Barbour AG, Burgdorfer W (eds) (1986a) Lyme Borreliosis. Zentralbl Bakteriol Hyg (A) 263:1-501

Stanek G, Flamm H, Groh V, Hirschl A, Kristoferitsch W, Neumann R, Schmutzhard E, Wewalka G (1986b) Epidemiology of *Borrelia* infections in Austria. Zentralbl Bakteriol Hyg (A) 263:442-449

Stanek G, Pletschette M, Flamm H, Hirschl AM, Aberer E, Kristoferitsch W, Schmutzhard E (1988) European Lyme borreliosis. Ann NY Acad Sci 539:274-282

Stanek G, Kristoferitsch W, Pletschette M, Barbour AG, Flamm H (eds) (1989a) Lyme borreliosis II. Zentralbl Bakteriol (Suppl) 18:1-365

Stanek G, Prinz A, Wewalka G, Hirschl AM, Kebela-Ilunga (1989b) Lyme borreliosis in Central Africa. Zentral Bakteriol (Suppl) 18:77-81

Stanek G, Jurkowitsch B, Köchl C, Burger I, Khanakha G (1990) Reactivity of European and American isolates of *Borrelia burgdorferi* with different monoclonal antibodies by means of a microimmunoblot technique. Zentralbl Bakteriol 278:426-436

Steere AC (1989) Lyme disease. N Engl J Med 321:586-596

Steere AC, Bartenhagen NH, Craft JE et al. (1986) Clinical manifestations of Lyme disease. Zentralbl Bakteriol Hyg (A) 263:201-206

Strnad P, Hrazdirova V, Pejoch M, Rehacek J (1989) The incidence of Lyme borreliosis in Southern Moravia, the Brno region, from a neurologist's point of view. Zentralbl Bakteriol (Suppl) 18:60-63

Strle F, Cimperman J, Pejovnik-Pustinek A, Stanek G, Pleterski-Rigler D, Jereb M, Ruzic E (1989a) Lyme borreliosis: epidemiological data for Slovenia. In: Ropac D (ed) Lyme borreliosis in Yugoslavia. Medicinska Akademija Hivatske, Zagreb, pp 35-43 (in Serbocroatian)

Strle F, Pejovnik-Pustinek A, Stanek G, Pleterski D, Rakar R (1989b) Lyme Borreliosis in Slovenia in 1986. Zentralbl Bakteriol (Suppl) 18:50-54

Trevisan G, Crovato F, Marcuccio C, Fumarola D, Scarpa C (1986) Lyme disease in Italy. Zentralbl Bakteriol Hyg (A) 263:459-463

Weber K, Neubert U (1986) Clinical features of early erythema migrans disease and related disorders. Zentralbl Bacteriol Hyg (A) 263:209-228

Wilske B, Münchhoff P, Schierz G, Preac-Mursic V, Roggendorf M, Zoulek G (1985a) Epidemiologie der *Borrelia burgdorferi*-Infektion: Nachweis von Antikörpern gegen Schildzecken-Spirochäten bei Waldarbeitern aus Oberbayern. Munch Med Wochenschr 127:171-172

Wilske B, Preac-Mursic V, Schierz G (1985b) Antigenic heterogeneity of European *Borrelia burgdorferi* strains. Lancet I:1099

Wilske B, Steinhuber R, Bergmeister H, Fingerle V, Schierz G, Preac-Mursic V, Vanek E, Lorbeer B (1987) Lyme-Borreliose in Süddeutschland. Dtsch Med Wochenschr 112:1730-1736

Wilske B, Preac-Mursic V, Schierz G, Kühbeck R, Barbour AG, Kramer M (1988) Antigenic variability of *Borrelia burgdorferi*. Ann NY Acad Sci 539:126-143

Wilske B, Anderson JF, Baranton G, Barbour AG, Hovind-Hougen K, Johnson RC, Preac-Mursic V (1991) Taxonomy of *Borrelia* spp. Scand J Infect Dis (Suppl) 77:108-129

Subject Index

MIX
Papier aus verantwortungsvollen Quellen
Paper from responsible sources
FSC® C105338

If you have any concerns about our products,
you can contact us on
ProductSafety@springernature.com

In case Publisher is established outside the EU,
the EU authorized representative is:
Springer Nature Customer Service Center GmbH
Europaplatz 3, 69115 Heidelberg, Germany

Printed by Libri Plureos GmbH
in Hamburg, Germany